Basic Calculus of Planetary Orbits and Interplanetary Flight

Alexander J. Hahn

Basic Calculus of Planetary Orbits and Interplanetary Flight

The Missions of the Voyagers, Cassini, and Juno

 Springer

Alexander J. Hahn
Department of Mathematics
University of Notre Dame
Notre Dame, IN, USA

ISBN 978-3-030-24870-3   ISBN 978-3-030-24868-0   (eBook)
https://doi.org/10.1007/978-3-030-24868-0

Mathematics Subject Classification (2010): 26Axx 70F15, 81V17, 97Ixx, 97Mxx

This is an artist's depiction of the Cassini spacecraft after it crossed Saturn's ring plane and during the maneuver that
put the craft into orbit around the planet by the firing of its main engine to reduce the spacecraft's velocity. (Image is
free usage courtesy of NASA/JPL.)

This Springer imprint is published by the registered company Springer Nature Switzerland AG
The registered company address is: Gewerbestrasse 11, 6330 Cham, Switzerland

Dedicated in loving memory to our beautiful, accomplished, special daughter Anneliese, who enriched many lives in countless ways. The heavens as God's material creation are the focus of study of this book. But more importantly, Heaven is also a spiritual place and this is where our daughter now peacefully and happily resides.

Preface

Anyone with a sense for mathematics, science, and engineering would surely agree with the assertion that the development of differential and integral calculus is one of humanity's greatest intellectual achievements and that one of the most incredible success stories of modern times has been the astonishing exploration of the inner and outermost reaches of our solar system by sophisticated satellites and spacecraft. It is the primary aim of this text to demonstrate how calculus informs our understanding of the solar system and the navigation of the spacecraft sent to study it.

Basic calculus provides fundamental information about the motion of the planets, their moons, asteroids, comets, and artificial satellites in their orbits, and the way spacecraft travel along their trajectories. The reason that simple, one-variable calculus can be applied to these studies, is the fact that orbiting planets and other celestial bodies move in planes that are essentially fixed (at least during long stretches of time), so that their motion can be analyzed in two dimensions with functions of a single variable. The same is true—in a modified sense—for the trajectories of spacecraft. During the stretches of time when a spacecraft is subject to a single, dominant force, say the gravitational pull of the Sun or a planet, then in the same way, its motion can be modeled with functions of a single variable. When additional forces are involved, then the craft's trajectory is too complex for such an approach. What saves the day, is the way that the missions of spacecraft are traditionally designed. For the most part, the flight of a spacecraft relies on the pull of a single gravitational force and additional forces, such as the thrust of the main rocket engine or a second gravitational force, are operative only intermittently and briefly. A single, dominant gravitational force on a spacecraft acts in the direction of its point of origin (for example, the center of mass of the Sun, a nearby planet, or a moon) and satisfies an inverse square law with regard to the distance involved. It follows from Newton's theory of gravitation that in any such situation, its trajectory is a segment of a conic section. These trajectories can be ellipses, but they can also be parabolas and hyperbolas. As a consequence, the trajectory of a spacecraft is a "patched conic," meaning that it is a conic section as long as a single force acts, followed by a brief, more complicated path when additional forces act, then another conic section when a single force is again operative, and so on. The bottom line is that the trajectory of a spacecraft is a sequence of arcs of conic sections (joined to each other by short, more complicated segments of curves) and that the motion of the craft along each of these arcs can be studied with the methods of basic one-variable calculus. More definitive and far-reaching analyses involving Second Order Differential Equations, Multivariable Vector Calculus, Spherical Harmonic Functions, Probability Theory, Statistical Methods, (and hundreds or thousands of pages of computer code) notwithstanding, much can already be said with single-variable calculus. We'll now turn to a brief outline of the text.

Chapter 1 is a historical essay that provides the context and sets the stage for all that follows. It recalls humanity's efforts to understand the planets, moons, comets, and asteroids and how they move, from the time of the Greeks, to Isaac Newton and the Scientific Revolution, to the discovery of the vast, expanding universe of galaxies in the twentieth century. In describing the decisive contributions of Kepler, Newton, Hubble, and those of other pioneering scientists, the chapter considers some of the elementary mathematical elements involved by introducing the ellipse, parabola, and hyperbola as well as very basic concepts and principles of the physics of motion.

Chapter 2 provides an overview of the information that today's spacecraft have returned from the near and far corners of the solar system. The American and Russian space programs of the last decades of the twentieth century began to explore the planets and their moons. Joined by European, Japanese, Chinese, and Indian contributions in the twenty-first century, this effort has accelerated. The chapter describes the most important of the many unmanned missions that have explored the solar system and provides some of the spectacular images and wealth of information about the planets, their moons, asteroids, and comets that they have captured. Some elements of calculus are introduced toward the end to explain basic aspects of the rocket engines that propel these probes. The study of the flight of the *Juno* spacecraft to Jupiter provides an initial look at the analysis that the last chapter of the text undertakes in detail.

Chapter 3 presents some basic mathematics. The importance of the rectangular or Cartesian coordinate system to the disciplines of geometry, trigonometry, and calculus and their applications cannot be overstated. These disciplines depend on the interplay between geometry and algebra that this coordinate system makes possible. The polar coordinate system makes this connection also. By identifying a location in the plane with respect to a fixed reference point and a ray that emanates from it, it provides a framework that is tailor-made for the analysis of the trajectory of any object that moves in response to a gravitational force. The chapter develops the essentials about the polar coordinate plane, the polar equations for the ellipse, parabola, and hyperbola, the calculus of polar functions, as well as the trigonometry needed along the way.

Chapter 4 applies polar calculus within a comprehensive, self-contained treatment of Newton's theory of gravitation. Newton's treatise Principia Mathematica had been a miracle. It provided a synthesizing and penetrating solution to a question that had occupied many of humanity's best minds for about 3000 years: how do the heavens work? Kepler had discovered the three laws of planetary motion with painstaking observations, but Newton came to recognize the deeper reality. All three of Kepler's laws rest on a combination of mathematical methods, basic laws of motion, and the inverse square law of universal gravitation. The chapter includes a complete analysis of the connection between the magnitude of a centripetal force and the geometry of the trajectory of a point-mass—or a sphere that has its mass radially distributed—on which the force acts.

Chapter 5 discusses the motion of an object that is propelled in an elliptical orbit by a gravitational force. The distance, speed, and direction of the motion of the object (relative to the attracting body) are determined as functions of the elapsed time from periapsis (the point of nearest approach to the attracting body). This study proceeds via the calculus of trigonometric functions and relies on Kepler's equation and its solution. Several concluding sections apply power series in the solution of relevant definite integrals. One of them computes the length of a planet's elliptical orbit and another provides an analysis of the precession of the orbit's perihelion.

Chapter 6 is a discussion of the complex aspects of the design of the trajectory of a spacecraft and the maneuvers that direct it to its target. Applying Newton's analysis and using the *NEAR- Shoemaker*, the *Voyagers*, and *Cassini* missions as contexts, the chapter studies the essentials of gravitational spheres of influence, transfer orbits, orbit insertion, hyperbolic gravity assist flybys, and the ephemerides of trajectories. The study of hyperbolic trajectories and the motion of spacecraft along them is analogous to their motion along elliptical orbits, except that in terms of the calculus involved, hyperbolic functions and the hyperbolic Kepler equation take the place of trigonometric functions and the elliptical Kepler equation.

Even more succinctly put, this book is organized into three components of two chapters each. The first component is primarily historical. It sketches humanity's understanding of our universe from the thoughts of the Greeks to the exploration of the Sun, moon, planets, asteroids, and comets of our solar system by sophisticated modern spacecraft. The second component is mathematical. It presents the calculus of polar functions in detail and applies it to a self-contained development of Newton's theory of gravity. The final component takes on the theory of elliptical orbits and hyperbolic trajectories and applies it to a study of the *NEAR-Shoemaker*, *Voyager*, and *Cassini* missions. The text cites many websites that provide visual details and rich illustrations of the discussions. (If a particular website is no longer active, it should be possible to use relevant key words from the context or from its address to search and find an equivalent or updated alternative.)

As Text for a Course: This text is suitable for a college course for students who have a good understanding of geometry, algebra, trigonometry, precalculus, and the very basics of ordinary one-variable calculus with its limit strategies. Many of today's more advanced high school students will have had course offerings that provide such an understanding. Students who complete this course—including students with majors in engineering disciplines, science, and mathematics—will not only have a sense of the astonishing discoveries that the world's space programs are making, but they will also have a compelling answer to the question "what is this mathematical stuff actually good for." A college course using this text could be a one-semester or two-semester course. A one-semester course could cover the first four chapters and insert reviews of mathematical concepts and details along the way. Such a course would combine the rich history of humanity's understanding of the universe and its modern efforts to explore it, with the basic calculus that is necessary to comprehend both. A course that adds the last two chapters to this agenda by taking on the theory of elliptical orbits and hyperbolic trajectories and its applications to the motion of the bodies of the solar system and the navigation of spacecraft is more than likely a two-semester course. This especially, if it makes extensive use of the sets of problems and topic-expanding discussions—over 70 pages in all—that follow the chapters. These problems and discussions vary in terms of difficulty, so that an instructor needs to assign them with care. Solution sets are posted to Springer's website

https://www.springer.com/us/book/9783030248673.

For readers who wish to test their understanding of the content, whether as students in a formal course or as independent learners, the website hosts solutions to all the odd problems of the text. Full solution sets are available to instructors who adopt this text for a course.

Prerequisites: Coordinate geometry, functions and graphs, basic trigonometry, basic functions, including trigonometric functions, inverses, exponential, and logarithm functions. A working knowledge of the basics of one-variable calculus including derivatives, integrals, and the fundamental theorem.

Conventions and Practices: The physical units used in the text are the meter, kilogram, second of the MKS system and units that are derived from them. The computations are carried out with the calculator https://web2.0calc.com/. Those that are more complex, consider significant figures and round off answers accordingly. In simpler situations, the adherence to such procedures is less strict. It should not come as a surprise that the results of computations that lead to quantitative information about the motion of planets, their moons, asteroids, comets, and spacecraft are only approximations. This begins with the fact that the numerical data on which they rely are themselves approximations. The question of how accurate these approximations are does not get much explicit attention in this text. But there is a general understanding that the following examples convey. Consider the information 1 au $\approx 1.49598 \times 10^{11}$ m (the symbol $\approx$ means "is approximately equal to") and $GM = 1.32712 \times 10^{20}$ m^3/sec^2 used in Chapter

2E. The approximation provides a value of the *astronomical unit* au in meters. The fact that the approximating number is listed with five decimal places tells us that in this case $\approx$ is accurate to within five decimal places (1 au = $1.495978707 \times 10^{11}$ m "on the nose"). The equality $GM = 1.32712 \times 10^{20}$ m^3/sec^2 (where G is Newton's gravitational constant and M the mass of the Sun) is only an approximation. The point is that we'll take the approximation to be an equality within a given computation if it is accurate enough. When greater accuracy is called for in Chapter 5J, $GM = 1.32712440042 \times 10^{20}$ m^3/sec^2 is used. (In case you're wondering, GM has been measured with an accuracy of $GM \approx 1.32712440041939400 \times 10^{20}$ m^3/sec^2.)

The terms velocity and speed are often used interchangeably. The context makes it clear whether a vector or scalar quantity is being discussed. The two aspects of velocity are generally studied separately with the magnitude as speed and the direction in terms of an angle. Angles are understood to be given in radians unless stated otherwise.

Notre Dame, USA   Alexander J. Hahn

Acknowledgements

It is time to thank the many colleagues who have given generously of their expertise and time to assist me with this project. Many thanks to Jeremy Jones of the Jet Propulsion Laboratory, the leader of the navigation team for most of the *Cassini* mission to Saturn, who provided me with much data about the *Cassini*'s trajectory early on. Many thanks also to James McAdams of the Johns Hopkins Applied Physics Laboratory, who led the design team for the *MESSENGER* mission to Mercury. He supplied me with much information about the spacecraft's trajectory correction maneuvers. Thanks also to Ed Bell of the NASA Goddard Space Flight Center for taking the time to clarify matters about data involving the *Voyager* missions. I'd like to express my gratitude to Jon Giorgini, senior engineer with the Solar System Dynamics Group at the Jet Propulsion Laboratory, who went out of his way to explain to me how the HORIZONS ephemerides program determined the data that it provides. Many thanks to Carolyn Porco, the leader of the imaging team for *Cassini*'s entire voyage around Saturn from 2004 to 2017. The passion that she has for "everything *Cassini*" became obvious to the listeners of her magnetic lecture at the University of Notre Dame some years ago. She kept many of us informed about *Cassini* and its many exploits. Carolyn's contributions to the exploration of the outer solar system were recognized when an asteroid was named 7231 Porco. My gratitude goes to Duane Roth (the last leader of the Cassini navigation team) and William Owen, both of the Jet Propulsion Laboratory for their informed answers to my very last *Cassini* trajectory questions.

A warm, very special word of thanks is reserved for Mario Zoccoli, an engineer at NASA Lockheed Martin, who undertook the very time-consuming task of carefully reading the entire manuscript. Within his devoted and valuable effort, he made many valuable suggestions that much improved the final version of this text. My colleagues at Notre Dame, Peter Garnavich and Terry Rettig never failed to answer my questions about the workings of the universe and the solar system. Many thanks to both of them as well.

Last, but most certainly not least, a heartfelt word of appreciation to Elizabeth Loew and Ann Kostant, senior mathematics editors for Springer Publishing, whose continuing encouragement and infinite patience were instrumental in pulling this book project over the finish line. A word of gratitude also to Vishnu Muthuswamy and his Springer production team for the very professional way with which they put this book together.

Finally, a tight hug for my wife Marianne for her unwavering support during the years that I was in orbit myself in concentrated pursuit of planets and spacecraft.

Contents

This chapter is an historical essay that provides the context and sets the stage for all that follows. It recalls the evolution of our understanding of the universe from the Greeks to the Scientific Revolution and beyond. The chapter also turns to some of the basic mathematical elements involved by introducing the ellipse, parabola, and hyperbola and by describing the decisive contributions of Kepler, Newton, Cavendish, and those of other pioneering scientists.

Within the vast expanse of the universe, our solar system of Sun, planets, moons, comets and asteroids is but a tiny collection of whirling specs in the Milky Way galaxy. This galaxy with its many billions of stars is in turn a small cluster in this vast expanse. While it may be tiny in reference to the cosmic scale, it is *our* system and it has been an enduring as well as incredibly challenging focus of study for at least three thousand years. Efforts to understand and organize the night sky go back to the ancient Greeks and the Babylonians before them. The Greeks thought that the stars are fixed on a large celestial sphere that has the Earth at its center. This sphere of stars rotates once a day around the axis that the Earth's center and the northern pole star determine. They grouped the stars into recognizable clusters called constellations. Against the fixed patterns of the constellations they identified a few wandering points of lights that proceeded in one direction and then, for a time, looped back in the other. The Greeks called them planets, and recorded their paths. They regarded the realm of the Moon and above to be perfect and eternal. Spherical objects moved along predictable paths of circles or combinations of circles around the fixed Earth. This stood in sharp contrast to what they observed below the orbiting Moon including the surface of Earth where things were in a constant state of flux and hence beyond organized, predictive understanding. Given the grand design of the Greek cosmos and the later influence of Rome, it is not surprising that many of the names for these celestial objects, Mercury, Venus, Mars, Jupiter, and Saturn for example, come from the Roman names of the gods in the Greek pantheon. The Earth, Sun, and Moon have always been a part of the human experience. Our words for them come from the Old Germanic, Old English, and Old Norse: *erda, eorpe, jord,* and *sonne, sunne, sunna,* and, finally, *mano, mona,* and *mani.*

Figure 1.1 depicts a version of this Greek picture of the universe from the Middle Ages. It adds a place for heaven around the sphere of stars. "Coelum Empireum Habitaculum Dei Et Omnium Electorum" translates to "the empire of heaven, habitation of God and all the elected." The circular orbits of Saturn, Jupiter, Mars, Sun, Venus, Mercury, and Moon follow inside. The Earth is at the center of the scheme. The sphere determined by the orbit of the Moon (labeled Lunae) marks

A. J. Hahn (ed.), *Basic Calculus of Planetary Orbits and Interplanetary Flight*,
https://doi.org/10.1007/978-3-030-24868-0_1

the border between stability and instability. Outside it things are stable and eternal. Inside it all is unstable, always in danger of turning into chaos. The Earth was considered heavy and dirty because of the biblical Fall of Man. If God were to give up control, everything would turn to disorder and chaos. When Greek astronomers observed that the positions of the planets in the heavens differed from those that the simple circular scheme of Figure 1.1 predicted, they eventually replaced this simple orbital scheme with a complex clockwork of circles upon circles. This *Ptolemaic model* of the universe—named after the mathematician and astronomer Claudius Ptolemy—held sway as the accepted explanation of the universe for a millennium and a half. It began to collapse in the 16th

Figure 1.1. Petrus Apianus, a German mathematician, astronomer, and cartographer. The image is taken from his *Cosmographicus Liber*, 1524. This highly respected work on astronomy and navigation was reprinted at least 30 times in 14 languages and remained popular until the end of the 16th century.

and 17th centuries, pushed aside by the insights of the Scientific Revolution. Its overthrow—not
at all easy to achieve—was an effort led by an international cast of mathematicians, astronomers,
and scientists: the Pole Nicolaus Copernicus, the Dane Tycho Brahe, the Italian Galileo Galilei, the
German Johannes Kepler, and the Englishman Isaac Newton. This chapter presents an overview of
the remarkable advances that they and their successors made (but it leaves aside the biographical
details of these brilliant, as well as colorful characters).

1A. Copernicus Moves the Sun to the Center. Copernicus (1473–1543) realized that the
motion of the planets and the Moon in the night sky is much better explained by applying the circular
geometry of the Greeks to planetary orbits that have the Sun—not the Earth—as their center. The
assumption that the Earth—rather than the entire cosmic sphere of the stars—rotates once a day
seemed simpler and more compelling to him. He responded by publishing his own comprehensive
study *De Revolutionibus Orbium Coelestium (On the Revolutions of the Heavenly Spheres)* in 1543.

Let's turn to a description of the basic structure of Copernicus's Sun-centered universe. The Sun
is motionless at the center of an immense, unmoving sphere of fixed stars. The six planets Mercury,
Venus, Earth, Mars, Jupiter, and Saturn (known since their discovery by the ancient astronomers)
orbit the Sun in circles. The radii of these circles increase in the same order. The Moon is in circular
orbit around the Earth. The motion of the Earth has two primary aspects, both illustrated in
Figure 1.2. One is the Earth's daily rotation around an axis through its poles. In the figure, this
axis is represented by the arrow N (defining the direction north). The speed of the rotation is
constant. The other motion is Earth's circular orbit around the Sun. The Earth's axis of rotation

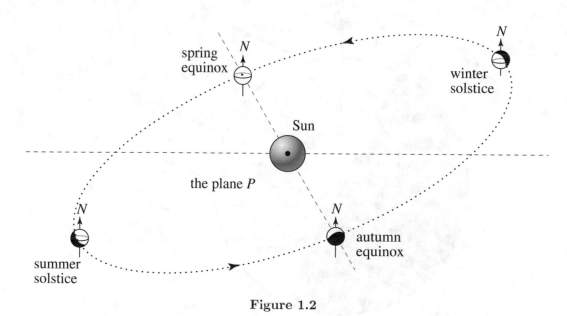

Figure 1.2

is perpendicular to the plane of its equator, but not to the plane of its orbit.

Let P be the plane through the center of the Sun parallel to the plane of Earth's equator, or,
equivalently, perpendicular to the Earth's axis of rotation. The plane P is indicated in Figure 1.2
in perspective by the two dashed lines through the Sun. As the figure shows, the Earth moves
alternately above and below the plane P. There are two occasions when the Earth's center and
hence its equator lie in the plane P. On either of these two days, consider a point on the Earth's
surface at sunrise, follow it around, and notice that exactly one-half of a rotation of the Earth

later, the Sun will set over this point. These are the days of *spring equinox* and *autumn equinox*. On these two days, the time from sunrise to sunset is the same as the time from sunset to sunrise. Since the center of the Earth lies in the plane P at the two equinox positions, the line that joins them—the line of equinoxes—lies in P. This is one of the two dashed lines in the figure. The other is the line in the plane P that is perpendicular to the line of equinoxes. When the Earth is at its lowest point below the plane P, the Sun is highest in the sky at midday in the Northern Hemisphere and shines down on it most directly. This is *summer solstice*. On this day, the Sun is above the horizon for the longest period of time at any location in the Northern Hemisphere. It is the day of longest daylight. From the perspective of Figure 1.2, when the Earth is at its highest point above the plane P, the Sun is lowest in the sky at noon in the Northern Hemisphere, and the smallest portion of this hemisphere is exposed to the Sun. This is *winter solstice*. It is the day on which the Sun is above the horizon for the shortest period of time in the Northern Hemisphere. It is the day of shortest daylight. The figure indicates the dark and sunlit regions of the Earth at each of the four positions we have singled out. These four positions define the seasons. The time the Earth moves from spring equinox to summer solstice is *spring*, the time it moves from summer solstice to autumn equinox is *summer*, the time from autumn equinox to winter solstice is *autumn* or *fall*, and the time from winter solstice to spring equinox is *winter*. Figure 1.3 is a close-up of the summer solstice position of Figure 1.2. The angle between the Earth's equator and the parallel rays of the Sun is approximately $23\frac{1}{2}°$. It is the angle between the plane of the Earth's equator and the plane of its orbit. This angle determines a circle on the surface of the Earth known as the Tropic of Cancer.

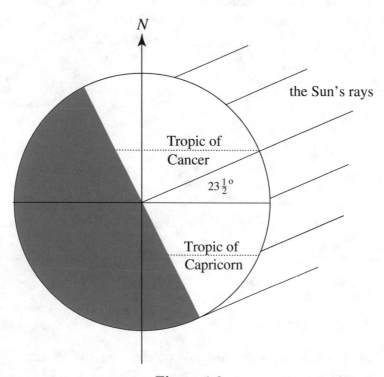

Figure 1.3

The Tropic of Capricorn is the corresponding circle in the Southern Hemisphere. Figures 1.2 and 1.3 tell us that the Earth's axis of rotation is tilted with respect to the plane of its orbit. In Figure 1.4 the Earth's orbital plane is rotated to make this explicit.

Copernicus's geometry of the universe provided a new explanation of the phenomenon of the "precession of the equinoxes" that the ancient astronomers had observed within their Earth-centered perspective as a slow rotation of the line of equinoxes. This explanation is provided by the fact that

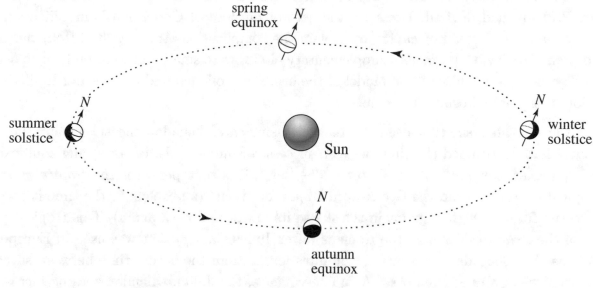

Figure 1.4

the Earth's polar axis of rotation revolves very slowly with respect to the line perpendicular to its orbit. Figure 1.5 captures what happens. This revolution of the axis is extremely slow. By relying on observations of the ancients, Copernicus's *De Revolutionibus* records that one revolution

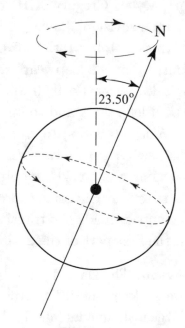

Figure 1.5

of the Earth's axis requires 25,816 years. Today's more accurate value is 25,772 years.

Now comes the all-important question. Was Copernicus's model with the planets in circular orbits around the Sun in sync with what was observed in the heavens? With an accuracy that measured up to the standards of the time? The answer is no! Certainly not if the orbits are taken to be circles with the Sun at their center (or the Earth in the case of the Moon). Copernicus was aware of these inaccuracies, and the *De Revolutionibus* responded by replacing the scheme of simple circular orbits. The off-centered circle—with center an abstract point near, but different from the Sun—became the basis of Copernicus's orbital geometry. But attached to it were smaller circles, so that ultimately Copernicus's scheme of circles was nearly as complicated as the intricate clockwork that the Greek Claudius Ptolemy had devised. However, the central purpose of Copernicus's modifications was completely different. Instead of having to explain the complications that resulted from the Earth-centered point of view, Copernicus's supplementary circles were small and were built in to account for the differences between his basic model of the orbit—his off-centered circle—and Kepler's later description of the actual orbit—the ellipse.

There was another issue that needed attention. Using careful shadow measurements, the Greek astronomers had determined the time between successive summer solstices or spring equinoxes (or autumn equinoxes or winter solstices) to be $365\frac{1}{4}$ days. This time period is the *tropical year*. The word tropical is derived from the Greek word *tropos* for "turn" (and refers to the time it takes the Sun to "turn" from its highest point in the sky to its lowest and back again). This tropical year is the year of the *Julian calendar* of the Romans (after Julius Caesar) that was used in Europe. The problem was that $365\frac{1}{4}$ days was over 11 minutes longer than the actual time between successive summer solstices or spring equinoxes. As a consequence, the Julian calendar became increasingly out of phase with the seasons. By the time Copernicus published his *De Revolutionibus* in 1543, the spring equinox, which was used in determining Easter, the most important feast of the Catholic, Orthodox, and Protestant churches, had moved 10 days from its target date of March 21st. The Council of Trent, convened by the Church in 1545, authorized the pope to take corrective action. Decades later, in 1582, during the papacy of Gregory XIII, the Jesuit astronomer Christopher Clavius (1537–1612) was able to draw up a revision. January 1st was declared to be the beginning of the year and renumbering October 5th as October 15th took care of the 10 day shift. In addition, there was a correction to the "every fourth year is a leap year" strategy of the Julian calendar. Only every fourth centennial year would be a leap year. So 1600 would be a leap year, 1700, 1800, 1900 would not be, but 2000 would again be a leap year, and so on. This leap year convention reduced the calendar year from 365.25 days (365 days 6 hours) to an average of 365.2425 days (365 days 5 hours 49 minutes and 12 seconds). The new *Gregorian calendar* also laid down rules for calculating the date of Easter. This Gregorian calendar was quickly adopted in Catholic countries. Given the divisive and even hostile relationship between the Christian churches, there was a pause of more than a century before the first Protestant countries made the transition to the new calendar. Orthodox Russia and Greece did not do so until the first part of the 20th century.

1B. From Tycho to Kepler to Newton. The Dane Tycho Brahe (1546–1601) provided concrete evidence that the Greek picture was on shaky ground. For over twenty years of the last part of the 16th century he deployed his array of large instruments (and his bare eyes and those of his assistants) to measure everything that happened in the sky with much greater accuracy than ever before. The sudden appearance of a new star—known today to have been a supernova, namely the explosion of an existing star—suggested to him that the heavens are subject to sudden change. A few short years thereafter in 1577 Tycho observed a large comet (see Figure 1.6) and was able to show that it streaked through the skies far beyond the orbiting Moon. Had the comet not been very distant, Tycho would have observed a shift—known as *parallax*—in its position against the fixed stars of

Figure 1.6. The Great Comet of 1577, as seen over Prague. Near the center of the picture, Tycho Brahe wearing a ruffle collar, focuses his attention on a depiction of the comet. Engraving by Jiri Daschitzky. Zentralbibliothek Zürich.

the constellations. Since his instruments detected no shifts, it followed that the Greek concept of an unchanging clockwork of stars and planets could not be correct. Early in the 17th century, Galileo (1564–1642) demonstrated that the phenomenon of projectile motion on Earth's surface can be understood with a combination of experimental and mathematical methods. The fact that at least some of the supposedly chaotic goings-on on Earth could be captured with mathematics contradicted another basic tenet of the Greek understanding of the universe. Galileo fashioned a telescope and pointed it skyward. He saw that our Moon is not perfectly spherical, but that it has mountains and craters. With his discovery of the four large moons of Jupiter (now known as Ganymede, Europa, Io, and Callisto) he found a system of heavenly bodies that did not have the Earth as their center of motion. Galileo also saw that the pattern of moon-shaped phases exhibited by the planet Venus was inconsistent with the picture of the universe that has the Earth at its center. Given the sum of his observations, Galileo became a vocal proponent of the Copernican Sun-centered system. A fixed Sun, however, contradicted a passage in Scripture in which God commanded the moving Sun to stand still. The Catholic Church, already dealing with the threat of the Protestant Reformation, regarded Galileo's views as a challenge to its authority to interpret the Bible. Called to Rome by the Inquisition, Galileo was silenced and placed under house arrest.

Having fallen out of favor with the Danish royals who financed his astronomical observatory, Tycho Brahe, by then a celebrated observer of the heavens, gained employment as astronomer at the court of the Habsburg emperor in Prague. In a move that was to be of major consequence, he called the young astronomer-mathematician Kepler (1571–1630) to assist him. After Tycho's sudden death at the beginning of the 17th century (history tells us that the cause was an exploded bladder), the role of imperial astronomer fell to Kepler. After a few years of painstaking calculations, Kepler hit on the ellipse—a curve studied two thousand years earlier by the Greeks—as the key to the understanding of planetary orbits.

Here is what Kepler discovered. Figure 1.7 shows an ellipse with center O and long diameter AB. The length $a = AO = OB$ is the *semimajor axis* of the ellipse. The point C on the ellipse is chosen so that OC is perpendicular to AB. The points F_1 and F_2 are the two points on the long

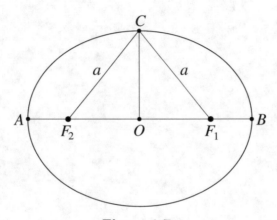

Figure 1.7

diameter with the property that $F_1C = F_2C = a$. These points are the *focal points* of the ellipse. (More detailed information about the ellipse follows in the next section of the chapter.) Focusing on Mars by relying heavily on the massive amounts of observational data that Tycho had collected, and painstakingly checking and rechecking his computations, Kepler showed that the orbit of Mars around the Sun is not a circle, but an ellipse. After a successful five year long "battle with Mars," Kepler published his three laws of planetary motion. They are still the basis of our understanding of the movement of the planets of our solar system.

1. The orbit of any planet P around the Sun S is an ellipse with S at one of the focal points of the ellipse.

2. As the planet P moves in its orbit, the segment SP sweeps out equal areas in equal times.

3. Let a be the semimajor axis of a planet's elliptical orbit and let T be its *period*, namely the time it takes to complete one orbit. Then the ratio $\frac{a^3}{T^2}$ has the same value for all planets.

The elliptical orbits of the planets are all close to being circles. If scaled down to fit on this page they would be indistinguishable from circles. This fact makes it all the more remarkable that Kepler was able to identify them as ellipses. Kepler had observed exceedingly well. But what are the underlying explanations? The road from observation to explanation is long and difficult. It was Isaac Newton who informed us how it is traveled.

In 1687, Newton (1642–1727) collected his deep reflections about the dynamics of the solar system as well as the physical laws and mathematical methods that explain it in his treatise *Philosophiae Naturalis Principia Mathematica* known simply as the *Principia*. This work is among the most celebrated and influential treatises in the history of science. One of Newton's major achievements (although not his achievement alone) is his discovery and formulation of the fundamental concepts and laws that underlie motion:

Law I. The Law of Inertia: An object that is at rest or moving in a straight line with constant speed will continue in this state unless a change is brought about by an external force.

Law II. The action of a force brings about an acceleration of the object, namely a change in the speed and direction of its motion. The directions of the acceleration and the force are the same and the magnitude of the force is proportional to the magnitude of the acceleration, with the mass of the object the constant of proportionality.

Law III. For every action, there is an equal and opposite reaction.

Let's elaborate. Take the meaning of force from everyday experience as any action of pulling or pushing. A force has a numerical *magnitude* that can be measured. For example, an amount of push or pull can be measured by the amount of the displacement it produces in some standardized steel spring. A force also acts in a *direction*. The magnitude and direction together determine the force. Quantities that are determined by a direction and a magnitude are called *vectors* that are represented by arrows. Given a vector quantity, the arrow representing it points in the direction involved and its length is equal to the vector's magnitude. Both aspects of a vector may vary. The motion of a moving point is given by a vector known as the point's *velocity*. Its magnitude is its *speed*.[1] In a similar way, the acceleration of a moving point is also a vector. Newton's second law can be expressed as the vector equation $F = ma$, where F is the force acting on the object, a is the acceleration that the force imparts to it, and m is its mass. Newton's first law is a direct consequence the second: if the magnitude of a force is zero, then the acceleration it produces in the motion of the object is zero, and hence the object's speed and direction remain constant. The third law says that for every force there is always an equal and opposite force. If you push against a wall with your hand, the wall will push back on your hand with an equal and opposite force. If this opposing force were not to exist, your hand would push the wall over (or go through the wall). So forces always occur in pairs of equal magnitude. The two forces do not act on the same object. The push by your hand is a force against the wall. The push by the wall is a force on your hand.

Late in November of 1679, Robert Hooke (1635–1703), a broadly brilliant scientist and Newton's colleague in the prestigious scientific Royal Society of London, began a probing correspondence with Newton about the nature of planetary orbits. What did Newton think about the idea of "compounding the celestiall motions of the planetts [out] of a direct motion by the tangent & an attractive motion towards the central body." Hooke was unable to do anything with this insight.

---

[1] In this text, there are discussions involving the velocity of an object that focus entirely on the magnitude of the velocity (and not on the direction). In such a situation, the term velocity is at times used to refer to the speed of the object. In this regard, the given context will prevent any ambiguities.

He lacked Newton's mathematical genius and (along with most astronomers) did not recognize the significance of Kepler's second law. But Hooke's question seems to be the moment that Newton is first introduced to the thought that the curving motion of a planet could be understood as the simultaneous composite of a tangential motion along a straight line together with one that is the result of an attractive force in the direction of the Sun. In a subsequent letter to Newton of January 1680, Hooke asked again about the nature of the trajectory of an object that is bent away from its inertial linear path by an attractive force that acts in the direction of a fixed point and varies inversely with the square of the distance of the object from this point. Hooke's letter also included the erroneous suggestion that under the action of such an attractive force, a planet's orbital speed would be inversely proportional to its distance from the point of attraction. He concludes "I doubt not but that by your excellent method you will easily find out what that Curve must be, and its propertys, and suggest a physicall Reason of this proportion."

A few months later, the Great Comet of 1680 appeared. Figure 1.8 presents a dramatic rendition of it. Comets, seen as irregular, fleeting, changeable bodies, were thought to follow different laws

Figure 1.8. Lieve Verschuier, *The Great Comet of 1680 over Rotterdam*, oil on panel, 25.5 cm by 32.5 cm, Historisch Museum Rotterdam. Image courtesy of the Hesburgh Library, University of Notre Dame.

of motion than the planets. Kepler was convinced that comets move along straight lines. In fact in the 1670s this was still the prevailing point of view. Newton observed the great comet from December 1680 until March 1681 when it became too faint. He had acquired a telescope for the purpose and kept a careful, almost daily log. If Hooke's question about the motion of the planets had not already done so, it was the puzzle of the comets that moved Newton to think deeply about the comings and goings of the objects in the heavens. What was the explanation of Kepler's elliptical orbits? What forces accounted for the motion of the streaking comets?

These and related questions occupied the minds and conversations of the distinguished men of the Royal Society in London during this time, including Hooke, the astronomers John Flamsteed (1646–1719) and Edmond Halley (1656–1742), and the architect Christopher Wren. After one of these discussions in August of 1684, Halley decided to travel up to Cambridge to consult Newton about these central scientific matters. Newton recalled one of their conversations as follows:

> "the Dr asked him what he thought the Curve would be that would be described by the Planets supposing the force of attraction towards the Sun to be reciprocal to the square of their distance from it. Sr Isaac replied immediately that it would be an Ellipsis. The Doctor struck with joy & amazement asked him how he knew it, why saith he I have calculated it."

Newton did not have a solution to hand to Halley, but his inquiry inspired him to write one up. He buried himself in his quarters at Cambridge from August 1684 until the spring of 1687. In December of 1684, Halley gave an early account of Newton's work to the Royal Society. A year later, Newton had produced a simplified, but still preliminary description of the orbits of planets, moons, comets, and tides for lectures he was to give as professor at Cambridge. By April 1686, Newton had sent Halley the complete text of Book I of the *Principia*. The trajectory of the Great Comet of 1680 continued to present a challenge. As late as June 1686, Newton wrote to Halley that "In Autumn last I spent two months in calculations to no purpose" and that "The third [book of the *Principia*] wants ye Theory of Comets." After another year, Newton succeeded. He fit a parabola to the observed positions of the comet by relying on the assumption—the same assumption that explained the orbits of the planets—that the Sun pulled on the comet along its entire trajectory with a force inversely proportional to the square of the distance between them. He concluded that the comet looped around the Sun and hence that the comet that had approached the Sun and the one that receded from it later were one and the same! The final third book of the *Principia*, *The System of the World* included the study of comets and was completed in the spring of 1687. With financing arranged by Halley, the *Principia* was published in the summer that followed. A second edition of the *Principia* included the study of a comet that Halley had observed in November of 1682. The properties of its orbit were so similar to those of comets that had appeared in 1531 and 1607, that Halley concluded that these comets must be one and the same. Noticing the period of the orbit to be about 76 years, he predicted that the comet would return in the year 1758. This prediction was confirmed (long after Halley's death) and the comet became known as *Halley's comet*.

1C. The Conic Sections. Newton knew that when a comet passes in the vicinity of the Sun or close enough to a planet (especially the massive Jupiter or Saturn), then its path would be deflected by the attractive force of the Sun or the planet. He came to realize that such deflections can follow not only elliptical curves, but parabolic and hyperbolic curves as well. Therefore his mathematical investigations of the dynamics of the solar system needed to consider not only the ellipse, but also the parabola and hyperbola. We'll begin by describing the basic aspects of these curves.

Let's start with the ellipse. Let a and b with $a \geq b$ be two positive constants, and let $c = \sqrt{a^2 - b^2}$. Place the points O, C, and F_1 as indicated in Figure 1.9. If $a > b$, form the triangle $\triangle OF_1C$. Since $a^2 = b^2 + c^2$, this is a right triangle with hypothenuse $CF_1 = a$. Extend the base of this triangle and place the point F_2 as indicated. If $a = b$, then $c = 0$. In this case, take $F_1 = F_2 = O$. The *ellipse* that the constants a and b determine is the set of all points P such that the lengths of the two segments PF_1 and PF_2 add up to $2a$. Notice that if $a = b$, then this is a circle of radius a. The graph of the ellipse is shown in the figure. The length a is the *semimajor axis* and the length b is the *semiminor axis* of the ellipse. The semimajor and semiminor axes are both positive numbers (in spite of the terminology, neither is an axis). The points F_1 and F_2 are the *focal points* of the ellipse and the point O is its *center*. The axis determined by the two focal points is the *focal axis*. In the case of the circle, the *focal axis* can be any axis through $F_1 = F_2 = O$. The

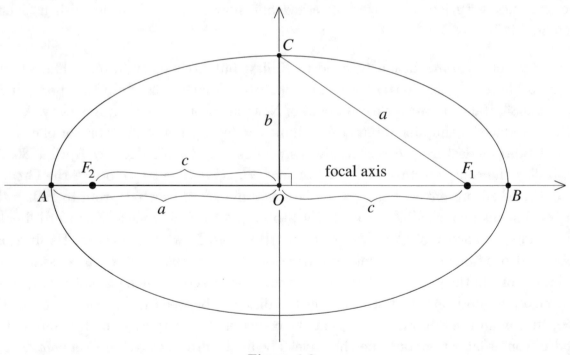

Figure 1.9

eccentricity ε of the ellipse is defined by $\varepsilon = \frac{c}{a}$. Since $c < a$, it follows that $\varepsilon < 1$. If $\varepsilon = 0$, then $c = 0$. So $a = b$ and the ellipse is a circle. The closer ε is to 1, the closer c is to a, the smaller b is relative to a, and the flatter the ellipse is. Consider the focal point F_1. Let A and B be the points of intersection of the ellipse and its focal axis. A look at the figure tells us that B is the point on the ellipse closest to F_1 and that A is the point on the ellipse farthest from F_1. The point B is the *periapsis* and A the *apoapsis* both relative to F_1. Notice that their distances from F_1 are $a - c = a - a\varepsilon = a(1 - \varepsilon)$ and $a + c = a + a\varepsilon = a(1 + \varepsilon)$, respectively. (With respect to the focal point F_2 the roles of B and A are reversed.) In the case of a circle, any point can be the periapsis with the point opposite to it the apoapsis. If the ellipse is an orbit and F_1 the position of the Sun, then B and A are known as the *perihelion* and *aphelion*, respectively. Let the point O be the origin of an xy-coordinate plane as shown in Figure 1.9 and let $P = (x, y)$ be any point in the plane. The formula for the distance between two points in the coordinate plane and a bit of standard algebra show that P is on the ellipse precisely when $\frac{x^2}{a^2} + \frac{y^2}{b^2} = 1$. This is a *standard equation of the ellipse*.

A *parabola* is specified by a line L and a point F not on the line, called *directrix* and *focal point* respectively, as the set of all points that are equidistant from F and L. See Figure 1.10. The line through the focal point perpendicular to the directrix is the *focal axis* of the parabola. The *eccentricity* ε of the parabola is defined to be equal to 1. The point of intersection of the parabola and its focal axis is the point on the parabola closest to F. It is the *periapsis* of the parabola. Take it to be the origin O of an xy-coordinate system as shown in the figure. If c is the distance between F and O, then $F = (0, c)$ and L is the line $y = -c$. Let $P = (x, y)$ be any point in the plane. By the

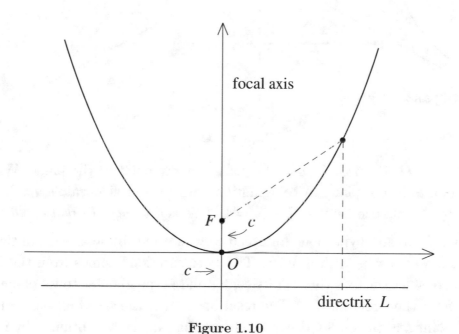

Figure 1.10

distance formula for two points in the plane, $\sqrt{(y - c)^2 + x^2} = y + c$. After simplifying this equation, we see that P is on the parabola precisely when $x^2 = 4cy$. This is a *standard equation of the parabola*.

We'll conclude with the hyperbola. Let a and b be two positive constants and set $c = \sqrt{a^2 + b^2}$. Extend the right triangle with sides a and b and hypotenuse c to the rectangle with center O shown in Figure 1.11. Draw in an axis that bisects the rectangle and place the two points F_1 and F_2 on this axis as shown in the figure. The *hyperbola* that a and b determine is the set of all points P such that the absolute value of the difference in the lengths of the segments PF_1 and PF_2 is equal to $2a$. The length a is the *semimajor axis* of the hyperbola. The two points F_1 and F_2 in the figure are the *focal points* of the hyperbola and the line that they determine is its *focal axis*. The extensions of the two diagonals of the rectangle are both asymptotes of the hyperbola (this means that the hyperbola converges to these two lines as shown in the figure). The *eccentricity* ε of the hyperbola is defined by $\varepsilon = \frac{c}{a}$. Since $c > a$, it follows that $\varepsilon > 1$. Consider the left branch of the hyperbola and its focal point F_1. The point of intersection of the left branch with the focal axis is the *periapsis* of the left branch. The periapsis of the right branch of the hyperbola is defined similarly. Let the midpoint of the segment F_1F_2 be the origin O be of an xy-coordinate system with x-axis the focal axis and

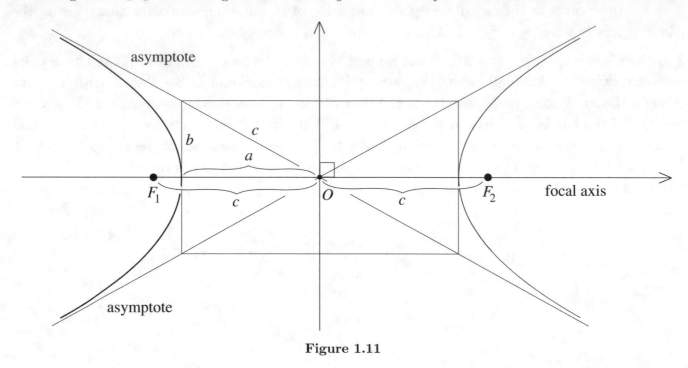

Figure 1.11

y-axis perpendicular to it through O. Let $P = (x, y)$ be any point in the plane. With the distance formula for two points in the plane in combination with several algebraic steps it can be verified that P is on the hyperbola precisely if $\frac{x^2}{a^2} - \frac{y^2}{b^2} = 1$. This is a *standard equation of the hyperbola*.

The ellipse, parabola, and hyperbola are called *conic sections* because each of them arises as the points of intersection of a cone with a plane. The word "section" comes from the Latin for "cut." Take a circle and consider the axis through its center and perpendicular to its plane. Fix a point on the axis distinct from the circle's center. The resulting *cone* is the set of all points that lie on some line through the point and the circle. It can be shown that any curve obtained by intersecting such a cone with a plane is an ellipse, a parabola, a hyperbola (or in some "degenerate" cases, a point or a line) and that any ellipse, parabola, and hyperbola can be obtained in this way. If this plane is taken to be perpendicular to the central axis, then the intersection is a circle (or a point).

1D. Newton's Incisive Insights. We're now ready to describe the essential aspects of what Newton achieved in the *Principia*. We'll describe Newton's methods, but only with a broad brush. Complete proofs of his central assertions from a more modern point of view will follow in a later chapter.

After a presentation of the "method of prime and ultimate ratios," in other words the basics of his differential calculus, Newton turns to the study of the gravitational forces of attraction in the solar system. Initially, he does so in a completely abstract way with a focus on "centripetal force" that is to say any force by which, according to Newton "bodies are drawn, impelled, or any way tend towards a point, as to a centre." Put another way, a force is *centripetal* if it always acts in the direction of a single fixed point, called the *center of force*. The magnitude of a centripetal force is free to vary. Think of a *point-mass* as a particle that is tiny in dimension but has no limit on its mass. Suppose that a point-mass P of mass m is propelled by a centripetal force of variable magnitude F and that this is the only force acting on P. Let S be the fixed center of force and let r_P be the variable distance between P and S. Consider the plane determined by the point S and the direction of the velocity of P at a given time. Since the force vector lies in this plane, it follows that this is the plane in which P moves. We'll refer to the path of the point-mass P as its *orbit* or its *trajectory*, with preference to the former in elliptical situations. The essence of the matter is depicted in Figure 1.12. In this completely abstract setting, Book I of the *Principia* sets

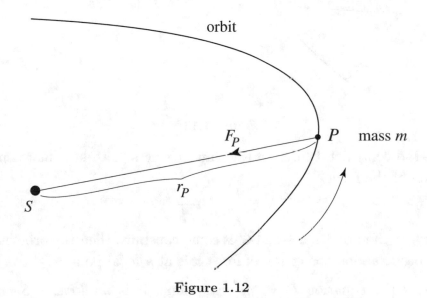

Figure 1.12

out to study the connection between the shape of the orbit of the point-mass P and the magnitude F_P of centripetal force. The key statements of Newton's famous treatise about a point-mass that is propelled by a centripetal force of magnitude F_P with center of force S are these:

Conclusion A. The motion of the point-mass satisfies Kepler's second law: the segment SP sweeps out equal areas in equal times. In particular, if A_t is the area swept out by SP during some time t, then $\frac{A_t}{t}$ is the same constant κ, no matter what t is equal to and no matter where in the orbit this occurs. See Figure 1.13a. Kepler's equal areas in equal times law follows directly from the equality $A_t = \kappa t$. We'll call κ the *Kepler constant* of the orbit.

Conclusion B. If the orbit is an ellipse, a parabola, or a hyperbola and the center of force S is at a focal point, then the magnitude F_P of the force is given by the formula

$$F_P = \frac{8\kappa^2 m}{L} \frac{1}{r_P^2},$$

where m is the mass of the point-mass, L is the *latus rectum* of the orbit (see Figure 1.13b), and r_P is the distance between P and S. Let the orbit be an ellipse with semimajor axis a, semiminor axis b, and period T. Since $ab\pi$ is the area of the ellipse and $L = \frac{2b^2}{a}$, this formula becomes

$$F_P = \frac{4\pi^2 a^3 m}{T^2} \frac{1}{r_P^2}.$$

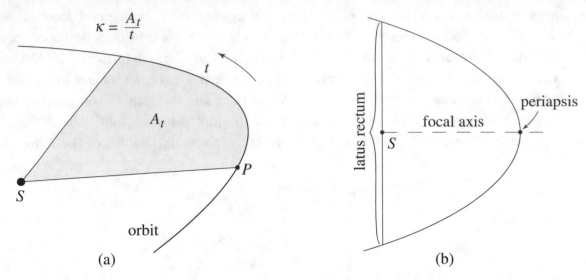

Figure 1.13

Conclusion C. If the centripetal force satisfies an inverse square law, more specifically if F_P is given by an equation of the form

$$F_P = Cm\frac{1}{r_P^2},$$

where m is the mass of the point-mass and C is some constant, then the orbit is either an ellipse, a parabola, or a hyperbola, and the center of force S is at a focal point.

The term $\frac{1}{r^2}$ gives the equation $F = Cm\frac{1}{r^2}$ its name. It is an *Inverse Square Law of Force*. Conclusion C tells us that Kepler's first law of the elliptical orbits of the planets (since planetary orbits are finite in extent) is a mathematical consequence of Newton's inverse square law. In his proofs Newton assumes that the centripetal force acts intermittently (machine gun style) in bursts that are a small fixed time interval apart. In view of the fact that the planets trace out their orbits around the Sun continuously and smoothly, this assumption is counterintuitive. However, it is also

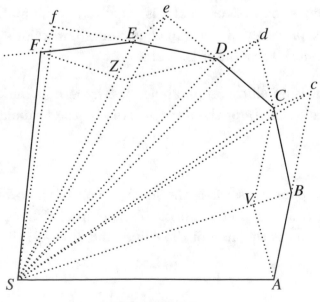

Figure 1.14

ingenious, because it provides an approximation of the orbit of P as a sequence of line segments and (consequently) of the area that P traces out as a sum of triangles. Working with this simplified

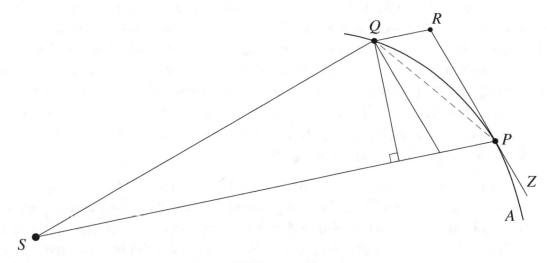

Figure 1.15

triangular geometry, Newton derives approximate versions of the conclusions above. These snap to "on the nose precision" when he lets the time interval between the bursts shrink to zero. This is

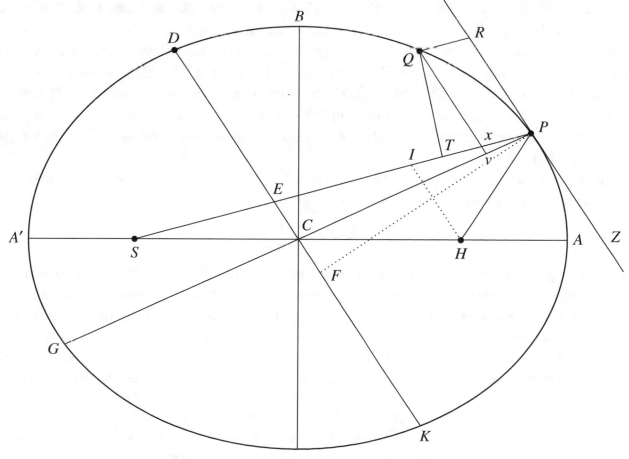

Figure 1.16

the strategy with which Newton derivers his equalities.

The three diagrams above illustrate how Newton proceeds. Figure 1.14 is central to the proof of Kepler's second law as Conclusion A formulates it. It approximates the path of P as the sequence of segments from A to B, B to C, and so on. The key is Newton's observation that the areas of consecutive triangles in the figure are essentially equal. In Figure 1.15, Newton views the motion from P to Q as the composite of the tangential motion from P to R and the accelerated, force-driven motion from R to Q. He then approximates the area of the wedge SPQ as the area of the triangle ΔSPQ and applies Conclusion A. Figure 1.15 is an important component in Figure 1.16 which Newton uses in a lengthy and delicate argument to prove the formula of Conclusion B in the case where the orbit is an ellipse with S at a focal point.

Historical commentary often asserts that even though Newton developed calculus, when it came to the mathematics of his magnum opus, he uses geometry instead. The fact is that Newton derives his equations in the *Principia* by letting things shrink to zero, and in so doing he uses a fundamental strategy of calculus. However, the calculus in the *Principia* is a calculus that lives in geometric constructs and is not today's calculus of functions. We will take up a modern approach to Newton's derivations in Chapter 4 of this text by making use of the calculus of functions in polar coordinates.

While Newton derives his conclusions in the abstract setting of centripetal forces and point-masses, he believes that they apply to the gravitational forces with which more massive bodies in the universe pull on much lighter ones. Suppose that S is a very massive body, that P is a much lighter one, and consider the gravitational force of attraction between them. Since S is massive, the gravitational force of P on S will have only a very small effect, so that S will move only very little. By assuming that the masses of S and P are concentrated at their centers of mass, Newton can take the gravitational force of S on P to be a centripetal force on a point-mass directed to the center of S. He is confident that these considerations apply to the gravitational pull of the Sun on a planet, that of a planet on one of its moons, or that of the Sun or a planet on a comet (at least in a tightly approximate way).

Kepler's observation that the orbits of the planets are ellipses in combination with Conclusion B above provides Newton with evidence that his inverse square law

$$F_P = C_P m \frac{1}{r^2}$$

offers a valid quantitative description of the gravitational force F_P with which a massive body S attracts an object P of smaller mass m at a distance r from S, where C_P some constant depending on P. He becomes convinced that this law is valid not only for the Sun and any planet, for a planet and any of its moons, but indeed, for any two masses anywhere in the universe. Newton now takes a further step. If S exerts a pull on P, then by his third law of motion, P pulls with an equal and opposite force F_S on S. So P pulls on S with a force of magnitude $F_S = F_P$. See Figure 1.17.

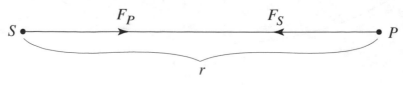

Figure 1.17

The symmetry of the situation requires that F_S should satisfy

$$F_S = C_S M \frac{1}{r^2}$$

where M is the mass of S and C_S is a constant depending on S. Newton puts the matter this way:

> "Since the action of the centripetal force upon bodies attracted is, at equal distances, proportional to the quantities of matter in those bodies, reason requires that it should be also proportional to the quantity of matter in the body attracting."

Since $F_S = F_P$, we see that $C_S M = C_P m$ and hence that $\frac{C_S}{m} = \frac{C_P}{M}$. Now let $G = \frac{C_S}{m} = \frac{C_P}{M}$ and notice that $C_S = Gm$ and $C_P = GM$. Let $F = F_P = F_S$ and substitute to get

$$F = G \frac{mM}{r^2}.$$

This equation is *Newton's Law of Universal Gravitation*. Newton is convinced that the masses m and M and the distance r between them are the essential elements that determine the force, and that the factor G should be a *universal constant*, in other words, a constant that is the same for any two such masses separated by any distance anywhere in the universe.

Newton turns next to any situation in the universe of a body S that is very massive relative to the objects in orbit around it. Let P be an object in an elliptical orbit around S. Let m be the mass of P, and let a and T be the semimajor axis and the period of its orbit. By Newton's Conclusion B, the attractive force of S on P satisfies

$$F = \frac{4\pi^2 a^3}{T^2} m \frac{1}{r^2},$$

where r is the distance between P and the center of S. By his law of universal gravitation,

$$F = G \frac{mM}{r^2},$$

with M the mass of S. After a little algebra, Newton gets

$$\frac{a^3}{T^2} = \frac{GM}{4\pi^2}.$$

Notice that the term $\frac{GM}{4\pi^2}$ on the right has nothing to do with the particulars of the object P and its orbit. In other words, it is the same for any P in orbit around S. It follows that the ratio $\frac{a^3}{T^2}$ of the cube of the semimajor axis a to the square of the period T of the orbit is the same for any body P in orbit around S. This is precisely what Kepler had asserted about the planets orbiting the Sun.

So Newton has shown that Kepler's third law is a consequence of his theory of gravitation! Refer to Conclusions A and C above, and observe that Newton has demonstrated that all three of Kepler's laws are consequences of his theory.

1E. Testing the Moon and Charting the Solar System. To confirm that his conclusions apply in the real world, Newton tests them against available evidence. In particular, he verifies that basic observations about the Moon's orbit around the Earth are consistent with his theory.

Let's begin with some numerical data about the Moon's orbit. By Newton's time, these were much more accurate than the earlier estimates of Copernicus, Kepler, and others. The French had calculated the radius of the Earth at the equator to be $R = 19{,}615{,}800$ Paris feet. (Today's foot, equal to about 94% of a Paris foot, is a little smaller). It was known that the Moon completes an orbit in 27 days, 7 hours, and 43 minutes, or 39,343 minutes. The average distance from the center of the Earth to the center of the Moon was known to be close to $60R$.

For the purpose of corroborating his theory, Newton assumes that the Moon is in a circular orbit of radius $60R$ around the center of the Earth. He takes the Moon in a typical position P and lets it be at Q exactly 1 minute later. In Figure 1.18, the motion of the Moon from

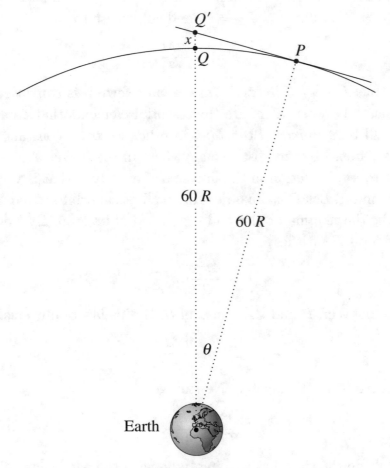

Figure 1.18

P to Q is decomposed into the tangential component PQ' and the component $Q'Q$ in the direction of the Earth. Newton knows that the angle θ is equal to $\frac{360}{39343}$ degrees and is able to compute $1 - \cos\theta = 0.0000000127$ with remarkable accuracy. From the figure, $\cos\theta = \frac{60R}{x+60R}$, so that $1 - \frac{60R}{x+60R} = 1 - \cos\theta = 0.0000000127$. Hence $\frac{x}{x+60R} = 0.0000000127$ and $x = (0.0000000127)(x + 60R)$. Solving for x and taking $R = 19{,}615{,}800$ Paris feet provides the value

$$x = 14.95 \text{ Paris feet.}$$

This estimate for the distance of the "fall" of the Moon toward the Earth in 1 minute is a consequence of observational data alone. Is the value provided by Newton's theory at least approximately the same?

An application of his second law of motion and his law of universal gravitation to an object of mass m on the Earth's surface tells Newton that the gravitational pull of the Earth on the object is $mg = \frac{GmM_E}{R^2}$, where g is the gravitational acceleration near Earth's surface and M_E is the mass of the Earth. So $g = \frac{GM_E}{R^2}$. Applying the same two laws to the gravitational force of the Earth on the Moon informs him that this force is $ma = \frac{GmM_E}{(60R)^2}$, where m is the mass of the Moon and a the acceleration of the Moon's fall toward the Earth. Since $a = \frac{GM_E}{(60R)^2}$, Newton knows that

$$a = \frac{GM_E}{60^2 R^2} = \frac{g}{60^2}.$$

At the latitude of Paris, the gravitational constant g was known to be equal to $g = 30.22$ Paris feet per second^2 (for us, this is equivalent to 32.17 feet/sec^2) and hence to $g = (30.22)(60^2)$ Paris feet per minute^2. Therefore the acceleration of the Moon's fall is $a = 30.22$ Paris feet per minute^2. Since the initial velocity of the Moon's fall toward Earth from Q' to Q is zero, Newton uses elements of his calculus to conclude that the velocity of the Moon along the line from Q' to the center of the Earth is $v = at$, where t is the elapsed time of this fall. This in turn tells him that the distance of this fall is $\frac{1}{2}at^2$. Taking $t = 1$ minute, Newton's theory predicts that the Moon would fall a distance of

$$x = \tfrac{1}{2}a = 15.11 \text{ Paris feet}$$

toward Earth.

Newton's theory has passed the test. The agreement between the observation of $x = 14.95$ Paris feet, and the result $x = 15.11$ Paris feet predicted by Newton's theory is good. The discrepancy can be explained by the fact that simplifying assumptions were made. For example, the Moon's orbit was assumed to be circular and the gravitational effects of the Sun on the Moon were ignored.

Newton was aware that the orbit of the Moon around the Earth is much more complicated than he assumed within his quick calculation above. In fact, Newton needed to fine-tune his earlier thinking. The diagram of Figure 1.17 illustrating that the bodies S and P attract each other mutually with forces of the same magnitude, tells us what is involved. Newton put it this way:

> "I have hitherto explained the motions of bodies attracted towards an immoveable centre, though perhaps no such motions exist in nature. For attractions are made towards bodies; and the actions of bodies attracting and attracted are always mutual and equal, by the third law of motion: so that, if there are two bodies, neither the attracting nor the attracted body can really be at rest; but both as it were by a mutual attraction, revolve about the common center of gravity."

What Newton concludes is that both P and S are in fact in elliptical orbits and that the center of mass of P and S is the relevant focal point for both orbits. This common center of mass is usually referred to as the *barycenter* of the system.

To illustrate more concretely what Newton is saying, let's return to the Earth–Moon system. The center of mass of this system is about 4,900 kilometers (or 3000 miles) from the center of the Earth, or about 1,500 kilometers (or 950 miles) below its surface. What Newton realized is that the centers of mass of both the Moon and the Earth travel along ellipses around the barycenter positioned at a focal point of each ellipse. To get a sense of what is going on, think of the centers of the Earth and Moon as being connected with a horizontally placed lever with fulcrum at the barycenter **B**. See Figure 1.19 and note that the lever is balanced. Now think of the lever as revolving in the horizontal

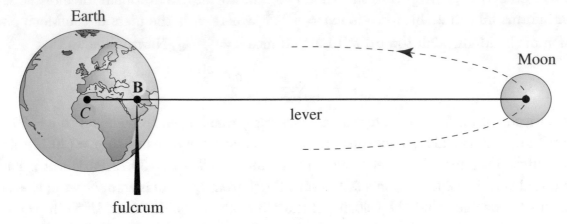

Figure 1.19

plane. This simulates the essential dynamics of the Earth–Moon system. The Moon is in a month-long orbit around **B**. This is a circular orbit in the simulation, but elliptical in fact. The center C of the Earth is also in "orbit" around **B**. In other words, as Figure 1.20 illustrates, the Moon's gravitational pull on Earth causes it to wobble in a monthly elliptical cycle about the barycenter **B**. In terms of Earth's orbit around the Sun, it is the barycenter **B** of the

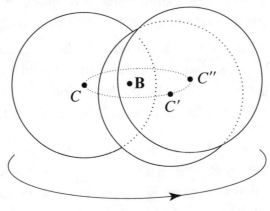

Figure 1.20

Earth–Moon system (rather than the center of the Earth) that describes an elliptical orbit about the Sun. As the barycenter of the Earth–Moon revolves around the Sun and the Moon circles around this barycenter, one would expect for the Moon's path around the Sun to be "loopy." But this is not the case. The reason is that the speed with which the Moon moves around the Earth is much less than the speed of the Earth in its orbit around the Sun. The Moon's orbit around the Sun should be thought of as a flat sine wave that is bent into an elliptical shape around the Sun. (We will take this question up in the Problems and Discussion section of this chapter.)

While continuing to pursue scientific matters after the publication of the *Principia*, Newton also investigated theological questions and experimented with alchemy. Interested in social standing and a secure income, Newton moved to London in 1696 to accept the position of Warden of the Royal Mint, and subsequently that of the more lucrative Master of the Mint. An able administrator, he supervised the recall of England's coinage (that was often forged as the metal in a coin had become

Figure 1.21. William Whitson, A scheme of the solar system with the orbits of the planets and comets belonging thereto. Engraved by John Senex in London, 1720. Size 69 × 60 cm. Division of Maps, Library of Congress.

worth more than its face value) and oversaw the issuance of a more reliable one. His scientific work had largely ended when he resigned from his professorship at Cambridge University in 1702.

William Whitson, a theologian, historian, and mathematician was appointed to succeeded Newton as professor at Cambridge. A leading figure in the popularization of Newton's ideas, Whitson produced a chart of the solar system (see Figure 1.21) that shows the orbits of the six planets known at the time, the trajectories of 24 comets, and summarizes—without the mathematical details—what Newton had discovered and what the astronomer Halley had catalogued. A closer study of the engraving, especially the narrative (zoom in on the electronic version

https://www.loc.gov/resource/g3180.ct003814/

of the Library of Congress) shows that it contains much accurate information. This includes the distances of the planets from the Sun. For example, the semimajor axis of the orbit of the Earth–Moon system around the Sun is listed as 81 million English statute miles. Since the English statute mile of that time (as well as the mile of today) is equivalent to about 1.61 kilometers, this corresponds to 130 million kilometers, or about 87% of the modern value of 150 million kilometers. For some reason, the more accurate result—equivalent to 140 million kilometers—that the astronomers Cassini and Flamsteed achieved in 1672 is bypassed. (See the Problems and Discussion section.) The semimajor axis of the Moon's orbit around the Earth is given as 240,000 miles, the equivalent of 386,400 kilometers, and very close to today's 384,400 kilometers. The periods of the orbits of the planets are also recorded with good accuracy. For Earth's orbit, the chart takes the $365\frac{1}{4}$ days that the Julian calendar assigns to the tropical year. The period of the Moon's orbit around Earth is given as 27 days, 7 hours, and 43 minutes, almost identical to today's value of 27.322 days. Given that their approximate distances from Earth were available, the diameters of the planets could be estimated fairly accurately from the sizes of their telescopic images. For instance, Whitson's chart lists the diameters of Jupiter and Saturn as 83,000 and 68,000 English statute miles respectively. In terms of today's definitive values, this corresponds to 96% for Jupiter and 87% for Saturn. The diameter of the Sun is given as 763,000 English statute miles, about 88% of today's value.

The orbits of the comets on Whitson's chart are numbered. The numbers are placed inside the outer circle, near the next circle (Saturn's orbit). The incoming and outgoing segments of the comets' curving paths are labeled with the same number. The path labeled 3 (look in the 2 o'clock position of the chart) refers to a comet that appeared in 1577, no doubt the comet that Tycho Brahe had observed. One of the curving paths is labeled with the three numbers 14, 16, and 17 (look in the 7 o'clock position). They refer to different passages of Halley's comet. The chart lists its period as $75\frac{1}{2}$ years and predicts its return in 1758. The orbit of the Great Comet of 1680 received the number 1 (just above the 9 o'clock position). It is drawn as a tight parabola. Its period is given as 575 years and the perihelion and aphelion distances are listed as 496,000 miles and 11,000 million miles, respectively. This last distance corresponds to about 18,000 million kilometers. It tells us that Newton and Halley not only had a grasp of the size of the planetary system as then known, but that they seemed to have a remarkable understanding of the vastness of the solar system beyond it!

1F. The Size and Scope of the Solar System. Ancient civilizations observed five planets Mercury, Venus, Mars, Jupiter, and Saturn as points of light wandering against the fixed patterns of the stars of the constellations. They have been a focus of study ever since. And ever since, questions have remained. Had all of these wanderers been detected? Or were some missed? If so, just an isolated few, or possibly swarms of them? And at what distances from the Sun? And about the comets. How many are there? And where do they come from?

For some of the answers to these questions, we'll need a convenient unit for measuring distances in the solar system. The semimajor axis of Earth's orbit has served this purpose. This distance, later known as the *astronomical unit*, was very difficult to measure with accuracy. (See the segment *The Parallax of Mars* of the Problems and Discussion section of this chapter.) In Newton's time, values equivalent to 130 million kilometers and also 140 million kilometers were in use. The value of 150 million kilometers turns out to be very close to today's definitive astronomical unit. The symbol "au" is the abbreviated notation.

In the 1760s and 1770s the two German astronomers Johann Titius and Johann Bode played the following numerical game. Start with 0 and 3, double 3 to get 6, and keep doubling to get the sequence of numbers

$$0 \quad 3 \quad 6 \quad 12 \quad 24 \quad 48 \quad 96 \quad 192 \quad 384 \ldots$$

Add 4 to each of them to get

$$4 \quad 7 \quad 10 \quad 16 \quad 28 \quad 52 \quad 100 \quad 196 \quad 388 \ldots$$

and divide each of these numbers by 10 to arrive at

$$0.4 \quad 0.7 \quad 1 \quad 1.6 \quad 2.8 \quad 5.2 \quad 10 \quad 19.6 \quad 38.8 \ldots$$

Consider a body in orbit around the Sun and let a and T be the semimajor axis and period of its orbit in the units au and year. For Earth, both a and T are equal to 1, so that the ratio $\frac{a^3}{T^2}$ for the Earth is also equal to 1. Kepler's third law tells us that $\frac{a^3}{T^2}$ is equal to 1 for any body in orbit around the Sun. So $a^3 = T^2$ and hence $a = T^{\frac{2}{3}}$ au. Since accurate measurements of the periods of the planets had existed since the time of Tycho Brahe and Kepler, astronomers understood the semimajor axes of the orbits of the planets in terms of the astronomical unit since the 17th century. The specific values for Mercury, Venus, Mars, Jupiter, and Saturn were known to be

Mercury	Venus	Earth	Mars		Jupiter	Saturn
0.39 au	0.72 au	1.00 au	1.52 au	—	5.20 au	9.54 au

Titius and Bode noticed that, except for the gap at 2.8, the distances of the planets from the Sun given by their semimajor axes matched up very well with the numbers of their numerical game.

The English astronomer William Herschel (1738–1822) surveyed the skies in the latter part of the 18th century. The telescopes that he built with the large, precise mirrors that he ground, were superior even to those used at the Royal Observatory. In 1781, he discovered a new planet, the first since ancient times! It was later named Uranus after the Roman god of the sky. The semimajor axis of its elliptical orbit was calculated to be 19.19 au. The fact that this was close to the Titius-Bode number 19.6, seemed convincing evidence that the relationship that Titius and Bode had observed was valid. Not surprisingly, the question "what about the gap at 2.8?" became a pressing concern and the search was on for a planet between Mars and Jupiter with a semimajor axis near 2.8. An association of astronomers was formed to look for such a planet in a systematic way. Remarkably, soon thereafter, in January of 1801, such an object was discovered by an Italian astronomer (who was not a member of this club) peering into the night sky from his observatory in Palermo on the island of Sicily. He named the object Ceres after the ancient Roman goddess and patroness of the island. Unfortunately, the tiny arc that his six weeks of observations drew in the sky was not enough to allow him to pinpoint the orbit. The identification of the orbit was essential. Only by understanding it would other astronomers be able to locate the object and confirm the discovery. Indeed, soon after its discovery, the small, wandering point of light had disappeared. Some astronomers searched for it and others attempted to compute its orbit from the sparse data that was available. All these efforts failed. But a 24 year old German took up the challenge and succeeded! Carl Friedrich Gauss—whose mathematical discoveries later established him as one of history's greatest mathematicians—developed a method for computing an entire orbit from just a few points of observation. Searching along the path that Gauss predicted, Ceres was rediscovered in December of 1801. The orbital elements that Gauss computed placed Ceres in an elliptical orbit between Mars and Jupiter with semimajor axis 2.77 au. A complete triumph of the Titius-Bode scheme? Not quite! With its radius of only 470 kilometers, Ceres was less than 1/3 the size of our Moon and was not regarded to be a legitimate planet. The search for the "real" planet went on. Between 1801 and 1808, astronomers tracked down three more such bodies in this region of the solar system. All were smaller than Ceres. Viewed with the telescopes of the day, they resembled small stars so much that astronomers suggested that they be called *asteroids*, meaning "star-like". The discovery of such asteroids continued and by the year 1900 about 450 had been identified. All were small and it became clear that there was no single large planet between Mars and Jupiter. Instead, there was a swarm of smaller bodies revolving around the Sun in this region. The table

Ceres	Vesta	Pallas	Hygeia
2.77 au	2.36 au	2.77 au	3.14 au

lists the semimajor axes of the elliptical orbits of the four largest of these asteroids.

Astrophysicists have reconstructed the history of the solar system. It began as a cloud of gas and dust in space. Over time, gravity pulled the gas and dust together and the cloud began to spin as it collapsed into a disk. As the disk got hotter and thinner, particles began to stick together to form clumps. In the process some clumps got bigger, eventually forming planets and moons. Near the center of the cloud, rocky material survived the heat and the inner planets formed. In the cooler parts of the disc, farther from the center, massive planets composed mostly of gases with a relatively small rocky core developed. The largest was Jupiter. Jupiter's strong gravitational forces sped up the clumps orbiting in its region of space. Instead of fusing together, they collided and shattered. The debris became the swarm of asteroids, now called the *main asteroid belt* (to distinguish it from smaller swarms of asteroids in the solar system). The asteroids of the main belt form a flat, donut-shaped region between the orbits of Mars and Jupiter. Its cross-section is shown in Figure 1.22. Most asteroids are irregularly shaped and are often pitted or cratered. As they revolve around the Sun they often tumble as they go. The main belt is estimated to contain between one and two million asteroids larger than 1 kilometer in diameter, and many millions of smaller ones. The number of asteroids in the main belt is huge, but they are sparsely distributed over the vast region of space that they occupy. The total mass of the asteroids in the main belt has been estimated to be 4% of the mass of our Moon. The four largest, Ceres, Vesta, Pallas, and Hygeia make up about half of this mass.

By 1846, Uranus had completed nearly one full trip around the Sun since its discovery. Astronomers studying its orbit detected irregularities. The astronomers Urbain Le Verrier (1811–1877) in Paris and John Couch Adams (1819–1892) in Cambridge analyzed these and conjectured them to be the consequence of the gravitational tug of some other planet in orbit beyond Uranus. Sure enough, soon thereafter, a new planet was located near the predicted position. It was later named Neptune after the Roman god of the sea. The semimajor axis of Neptune's orbit was calculated to be 30.07 au, far off the 38.8 au that the Titius-Bode law would have called for. This substantial difference, combined with the lack of a scientific explanation for it, finally put the law to rest.

Continued observations in the late 19th century led some astronomers to speculate that Neptune might not be the only planet that perturbed the orbit of Uranus and that there might well be another

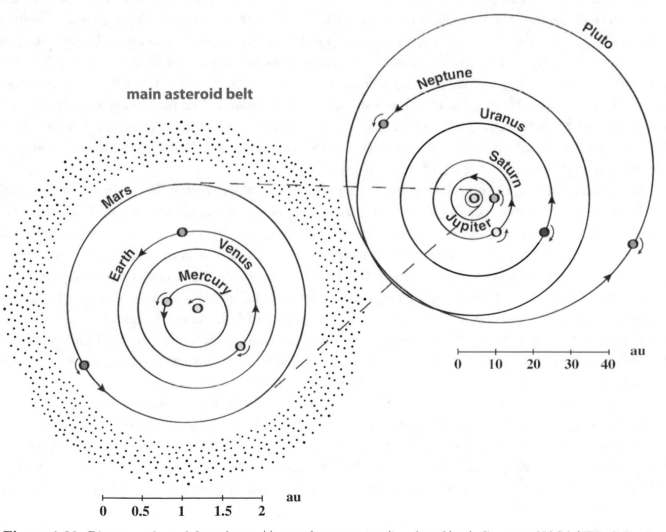

Figure 1.22. Diagram adapted from https://spaceplace.nasa.gov/ice-dwarf/en/. Courtesy NASA/JPL-Caltech

planet that did so. In response, an effort was launched to find one. At an observatory in Arizona, the young astronomer Clyde Tombaugh deployed a sophisticated new telescope to image regions of the night sky on photographic plates. In 1930, after a year of systematically comparing plates, he detected a small point of light that shifted its position slightly from one plate to the next. This point of light became the ninth planet of the solar system. Named Pluto, after the Roman god of the underworld, its orbit had a semimajor axis of 39.53 au and an eccentricity of 0.25, both larger than those of the planets discovered previously. But when its mass turned out to be less that 1/5 of the mass of our Moon, it was realized that it could not have a measurable impact on the orbits of either Uranus or Neptune. With the discovery of Pluto, the picture of the solar system as it was known in the 1930s and 1940s was complete. As seen from a vantage point high above the Earth's north pole, all planets move counterclockwise around the Sun in essentially the same plane. Figure 1.22 provides a diagram of their orbits. Both of its components are to scale.

In the 1950s, the Dutch-American astronomer Gerard Kuiper (1905–1973) proposed the existence of a region of icy, rocky, Pluto-like objects that circle the Sun in elliptical orbits at a distance of 30 to 55 au. And indeed, since the 1990s dozens of such "Kuiper belt objects" have been located, some of them about as large as Pluto. Would they become the tenth, eleventh, ... planets of the solar system? Or should the meaning of the word planet be revisited? In 2006, the International Astronomical Union—the governing body for such issues—met to consider the matter and defined a planet to be an object in orbit around the Sun that is not only massive enough to be rounded into a spherical shape by its own gravitational force, but so massive as to have cleared the vicinity of its orbit of other objects by collision or by attracting them as moons. With this definition, Mercury, Venus, Earth, Mars, Jupiter, Saturn, Uranus, and Neptune qualified as planets, but Pluto did not. Its mass is large enough to give it a spherical shape, but too small to have cleared its orbital path. Pluto was stripped of its status. Along with several similar Kuiper belt objects and the asteroid Ceres, it joined the new category of *dwarf planet*.

Our understanding of the solar system underwent one final expansion. Some theoretical models of the early solar system predicted that the formation of the giant planets would have scattered a vast number of icy objects into the outer solar system and beyond it. Some would have had enough velocity to escape the Sun's gravitational pull, but many others would have been drawn into orbit around it. Accordingly, in 1950, the Dutch astronomer Jan Oort (1900–1992) proposed that in the far reaches of the solar system—at distances from hundreds to hundreds of thousands astronomical units from the Sun—there should be swarms of billions of chunks of icy debris. The most distant part of this region, now known as the *Oort cloud*, is roughly spherical and surrounds the solar system at its gravitational edge. Comets are believed to be icy fragments from both the *Oort cloud* and the *Kuiper belt* that were deflected into the inner solar system by gravitational interactions. Short-period comets, those with periods less than 200 years, are thought to have come from the Kuiper belt. Comets with periods from hundreds to thousands of years must come from much greater distances and give evidence for the existence of the Oort cloud. We saw earlier that both Newton and Halley had some insight into the enormous size of the solar system. The written comment along the trajectory of the Comet of 1680 in Whitson's engraving (this is the parabola labeled 1 in the 9 o'clock position of the outer circle of Figure 1.21) lists the comet's greatest distance from the Sun as 11,200 million miles. This is equivalent to about 120 astronomical units.

1G. The Metric System of Units. In the 18th century, Europe used an array of different units for measuring things. The French foot and the English statute mile that Newton made use of are but two examples. In France alone, hundreds of different units of measurement were in use. The quantity associated with each unit could differ from town to town and from merchant to merchant. These variations gave rise to fraud and hindered commerce and taxation. The metric system was a response to this confusion. In 1790, a year after the start of the French Revolution, a proposal was put to the French National Assembly to create a standard system of units of measurement. On the recommendation of the French Academy of Sciences, the National Assembly accepted the meter as the standard unit of length and the gram as the unit of weight. This was the beginning of a system that the world would come to use—in refined and expanded form—to do its science and commerce.

The *meter* is defined to be equal to one ten-millionth of the distance between the North Pole and the Equator of the Earth. The *kilogram* is declared to be the mass of one thousandth of a cubic meter of water (a few degrees centigrade above freezing). Thereafter, the *second* was added as a unit of time. It is defined in terms of the *solar day*, the time from the instant the Sun is highest in the sky to the instant this occurs a day later. Since the duration of the solar day varies, the average solar day is taken. The second is defined so that the average solar day consists of precisely 86,400 seconds. With this definition, and the understanding that one minute has 60 seconds and one hour has 60 minutes, the average solar day is exactly 24 hours long. In 1832, the famous German mathematician Carl Friedrich Gauss (we met him in the context of the discovery of the asteroid Ceres) promoted this system as the appropriate set of units for the physical sciences. In 1875, the inter-governmental agency General Conference on Weights and Measures (CGPM is its international acronym) was organized. Under its stewardship, the meter–kilogram–second system was extended coherently to the electrical realm with the addition of the *ampere* as the unit of current. In 1960, the CGMP launched the Système International d'Unités, or SI, with its seven coherent base units that include the meter, kilogram, second, and ampere, (as well as units that measure the amount of a substance in terms of elementary particles such as atoms or molecules, thermodynamic temperature, and light intensity). Since the time of their introduction, the definitions of these units have been refined and made more precise. For instance, the second is now defined to be the duration of 9,192,631,770 periods of the radiation frequency at which atoms of the element cesium 133 change from one state to another. This atomic cesium clock is so precise that it looses/gains less than 1 second in a million years. The meter has become the distance traveled by light in a vacuum during a time interval of $\frac{1}{299,792,458}$ of a second. It follows from this definition that light travels exactly 299,792,458 meters in one second.

The system consisting of the meter, kilogram, and second is the international MKS *system of units*. We will usually abbreviate the meter by m, the kilogram by kg, and the second by sec. An overview of the metric units that have been discussed as well as some of the units derived from them follows below. The American equivalents are included for readers whose intuitive sense is more aligned with them. Precise values are expressed in bold type. One big advantage of the MKS system is that its units and derived units (except for units of time) parallel our base 10 decimal number system. This provides this system with a coherence that the American system with its inch, foot, yard, and mile and its blob, slug, and pound mass, etc., does not have.

Length: 1 centimeter = $\frac{1}{100}$ meter, 1 kilometer = **1000** meters, 1 inch = **2.54** centimeters, 1 foot = **0.3048** meters, 1 meter = 3.280840 feet, 1 mile = **1.609344** kilometers, 1 kilometer = 0.621371 miles, and 1 mile = **5280** feet. The centimeter and the kilometer are abbreviated by cm and km, respectively.

Mass: 1 gram = $\frac{1}{1000}$ kilogram, 1 metric ton (or tonne) = **1000** kilograms, 1 slug = 14.593903 kilograms, 1 kilogram = 0.0685218 slugs.

Time: 1 minute = **60** seconds, 1 hour = **60** minutes, 1 day = **24** hours = **86,400** seconds.

Consider an object of 1 kg in mass. Suppose that a constant force imparts an acceleration of $1\,\frac{m}{sec}$ per second to the object. So during each second the object's speed is increased by 1 meter per second. The equation $F = ma$ implies that $1\,kg \cdot 1\,\frac{m}{sec^2}$ is the force on the object. The unit $\frac{kg \cdot m}{sec^2}$ is the basic unit of force in MKS. Appropriately, it is named *newton* and abbreviated by N.

Force: 1 pound force = 4.45359237 newtons, 1 newton = 0.22480894 pounds force, 1 kilonewton = **1000** newtons. The kilonewton is abbreviated by kN.

Recall from section 1F that the astronomical unit is based on the semimajor axis of Earth's elliptical orbit around the Sun. In the 20th century, after the invention of radar and the precise determination of the speed of light, distances in the solar system could be determined accurately. This is done by timing (with atomic clocks) how long it takes a radar beam traveling at the speed of light to travel to an object and bounce back to Earth (or a spacecraft). This approach resulted in the value of 149,597,870.7 kilometers for the semimajor axis of Earth's orbit. This measurement in turn became the basis of the official definition of the *astronomical unit* au as

$$1\,au = 149{,}597{,}870.7 \text{ kilometers.}$$

We turn to the issue of the meaning of the year. The *tropical year* is defined to be the duration of time from one summer solstice to the next (or from one spring equinox to the next). It comprises a complete cycle of the seasons. This is the year the world's calendars are based on. The Gregorian calendar of 365.2425 days had been introduced to correct the 365.25 days that the Julian calendar assigned to the tropical year. The Gregorian calendar is also not completely accurate. The modern value for the average tropical year is approximately 365.242189 days. But from one year to the next, the tropical year can vary by as much as 20 minutes (or 0.012593 days).

With regard to astronomy and Newton's formulas in particular, it is the period of Earth's orbit—the time that the Earth requires to complete exactly one orbit—that is important. Is this equal to one tropical year? The answer is no, because—as was observed in section 1A—the precession of the equinoxes tells us that the position of Earth's summer solstice shifts slightly from one orbit to the next. Take any point along Earth's orbit and measure the time it takes for Earth to return to the same point. The area that the segment from the Earth to the Sun traces out during this time is equal to the area of Earth's ellipse. It follows from Kepler's second law, that the time required for Earth to make one complete revolution around the ellipse does not depend on the starting point. We'll call the time it takes for the Earth to move from one perihelion to the next the *perihelion period*. It too varies slightly. The Earth's *perihelion period* has been calculated to be on average 365.259636 days long. The *perihelion period* is defined in the same way for all the planets and their moons, as well as for comets and asteroids. There is a consideration that complicates things. It has been known for some time that as each planet traces out its elliptical orbit, the ellipse itself moves. Its focal axis—fixed at the Sun—rotates in the same direction as the planet's motion taking the perihelion and aphelion positions with it. See Figure 1.23. This rotation of a planet's ellipse is known as the

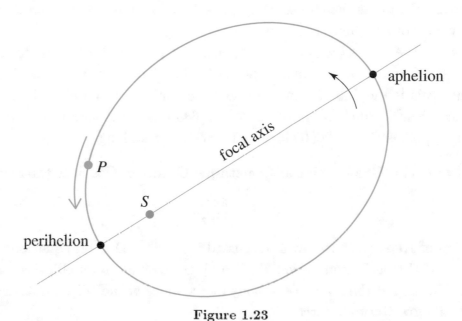

Figure 1.23

precession of perihelion. Careful observations have shown that these rotations are extremely slow. For example, one complete rotation of the Earth's ellipse takes about 112,000 years. During one perihelion period, a planet—given the slow "forward" rotation of its ellipse—moves slightly past

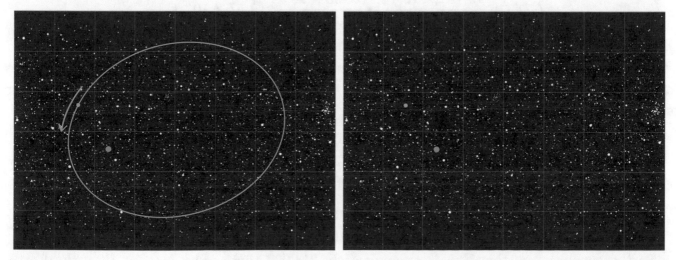

Figure 1.24. On the left the Earth and it's orbit are depicted in blue and the Sun in orange against the night sky. On the right the Earth and Sun are shown in the same position against the same section of sky. Image credit for the starry background: Bruce MacEvoy, Astronomical Files from Black Oak Observatory.

the perihelion position of its previous orbit.

Another measure of a planet's period is the time it takes for it to return from a given position in its orbit—as pinpointed by its location relative to the a frame of reference determined by a set of distant fixed stars—to the same (or most closely the same) position in its next orbit. This duration of time is a planet's *sidereal period* (from the Latin *sidereus* meaning "relating to stars"). For the Earth it is on average equal to 365.256363 days. Figure 1.24 gives a sense of what is involved. For the Earth, the difference between the average perihelion period and the average sidereal period is less than 5 minutes. (Note that the ellipses in Figures 1.23 and 1.24 are exaggerated. The orbits of the planets are much more circular than depicted.)

When it comes to the *year* as a unit of time, this text follows the International Astronomical Union that has defined 1 year to be equal to precisely 365.25 days or $365.25(86,400) = 31,557,600$ seconds. Note that this is the Julian year of ancient Roman times. Simple calculations show that the Earth's tropical year, perihelion period of 365.259636 days, and sidereal period of 365.256363 days are respectively, 0.9999786, 1.0000264, and 1.0000174 years long.

1H. Cavendish and the Gravitational Constants G and g. Consider the rewritten form

$$GM = \frac{4\pi^2 a^3}{T^2}$$

of Newton's version of Kepler's third law as discussed in section 1D. This equation tells us that the semimajor axis a and the perihelion period T of an object moving in an elliptical orbit determines the product GM where M is the mass of the body at the focus of the ellipse that drives the motion of the object with its gravitational force.

Newton drew the following conclusion from this formula. We'll use data from Whitson's chart of Figure 1.21. The values 81×10^6 English statute miles for the semimajor axis of Earth's orbit about the Sun and 365.25 days for its period T provides Newton with the approximation

$$GM_S \approx \frac{4\pi^2 (81 \times 10^6)^3}{365.25^2} \approx \frac{20.98 \times 10^{24}}{13.34 \times 10^4} \approx 1.57 \times 10^{20} \, \tfrac{\text{miles}^3}{\text{day}^2},$$

where M_S is the mass of the Sun.

For the orbit of the Moon around Earth, Whitson's chart provides the value $a \approx 240,000$ miles for the semimajor axis and $T \approx 27.322$ days for the period. This told Newton that with M_E the mass of the Earth,

$$GM_E \approx \frac{4\pi^2 (24 \times 10^4)^3}{27.32^2} \approx \frac{54.57 \times 10^{16}}{7.46 \times 10^2} \approx 7.32 \times 10^{14} \, \tfrac{\text{miles}^3}{\text{day}^2}.$$

It follows directly that

$$\frac{M_S}{M_E} \approx \frac{1.57 \times 10^{20}}{7.32 \times 10^{14}} \approx 214,000.$$

So the conclusion is that Sun is approximately 214,000 times more massive than the Earth. This turns out to be not very accurate. Nor is the assertion that the Sun is "in quantity of Matter 230,000 times as great as the Earth" of Whitson's chart. Newton's data for the orbits of the Earth around the Sun and the Moon around Earth were not precise enough. The principal culprit is the inaccurate value of 81 million miles for the semimajor axis of the Earth's orbit around the Sun. The correct value of 93 million miles would have gotten Newton much closer.

The precise value for the Sun/Earth mass ratio is 332,945. Today's values for GM_E and GM_S also make use of Newton's version of Kepler's third law, but they rely on the precise orbital data that spacecraft and artificial satellites provide. The *MESSENGER* spacecraft (derived from *MErcury Surface, Space ENvironment, GEochemistry,* and *Ranging* that describe its mission) has sent back the precise values of 57,909,050 km for the semimajor axis of Mercury's orbit around the Sun and 87.969 days for the orbital period. This information tells us that $GM_S = 1.32712 \times 10^{20} \frac{\text{m}^3}{\text{sec}^2}$. Bouncing laser beams from Earth off the satellite *LAGEOS* (the name is taken from *LAser GEOdynamic Satellite*) has provided accurate information about its orbit around Earth, and in turn the value $GM_E = 3.98600 \times 10^{14} \frac{\text{m}^3}{\text{sec}^2}$. It follows that $\frac{M_S}{M_E} \approx \frac{1.32712 \times 10^{20}}{3.98600 \times 10^{14}} \approx 332{,}945$.

As an important special case of Newton's second law $F = ma$, consider the acceleration generated by the force of gravity on an object falling near Earth's surface. By the early 17th century Galileo had concluded from his studies in the city of Pisa that this acceleration, nowadays labeled g, is the same no matter what the mass m of the falling object is. The magnitude of the gravitational force $F = mg$ is the *weight* of the object. Since g is now known to be approximately $9.8 \frac{\text{m}}{\text{sec}^2}$ (or about $32 \frac{\text{feet}}{\text{sec}^2}$), the mass m of the body can be determined from its weight.

If the law of universal gravitation holds everywhere in the universe, it should also apply to an object of mass m on or near the surface of Earth. The Earth's matter is distributed in such a way that its density is approximately the same at equal distances from its center. This means that with regard to gravity, all of its mass M_E can be regarded to be concentrated at its center. Inserting the radius r of the Earth into the law of universal gravitation tells us that the Earth's gravitational pull F on the object is $F = G \frac{m M_E}{r^2}$. Since also $F = mg$, it follows that

$$g = G \frac{M_E}{r^2}.$$

The Earth is essentially a sphere, but due to its rotation, it is a sphere that is flattened at the poles and bulging at the equator. See Figure 1.25. In other words, the distance r varies from being smallest at the poles and largest at the equator. So g varies as well, from being largest at the poles

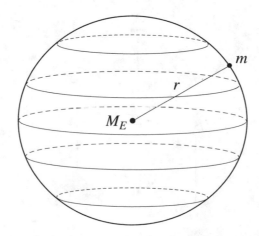

Figure 1.25. The Earth's cross-section through its center and North Pole is not a circle but an ellipse (that is close to a circle). The ellipse is exaggerated and much flatter in the figure than in fact.

and smallest at the equator. So the motion of a thrown object as we observe it every day on the surface of our planet is also impacted by Newton's law of universal gravitation. Using the Earth's average radius $r = 6371$ km and the value $GM_E = 3.98600 \times 10^{14} \frac{m^3}{sec^2}$ provides the approximation

$$g \approx \frac{GM_E}{6{,}371{,}000^2} \approx \frac{3.98600 \times 10^{14}}{(6.371 \times 10^6)^2} \approx 9.82 \frac{m}{sec^2}.$$

The use of precise values for the distance r tells us that g varies from about $9.7639 \frac{m}{sec^2}$ at the top of one the peaks of the Andes in Peru (why would it not instead be at the top of Mt. Everest?) to $9.8337 \frac{m}{sec^2}$ on the surface of the Arctic Ocean. In New York, Washington DC, Chicago, Denver, San Francisco, Madrid, Rome, Tokyo, Sydney, and Buenos Aires, the value of g is close to $9.80 \frac{m}{sec^2}$. In Vancouver, London, Paris, Amsterdam, and Frankfurt it is close to $9.81 \frac{m}{sec^2}$, and a bit farther north, in Oslo, Stockholm, Helsinki, and St. Petersburg for example, it is approximately $9.82 \frac{m}{sec^2}$.

A question at hand is "what are the masses of the Sun and the Earth?" Since, as we have seen, GM_E and GM_S can be tightly computed, this reduces to the determination of the value of the universal constant G. Newton did not think that this was possible:

> "Perhaps it may be objected, according to this philosophy all bodies should mutually attract one another, contrary to the evidence of experiments in terrestrial bodies. But I answer, that the experiments in terrestrial bodies come to no account. For the attraction of homogeneous spheres near their surface are as their diameters. Whence a sphere of one foot in diameter, and of like nature to the Earth, would attract a small body placed near its surface with a force of 20,000,000 less than the Earth would do if placed near its surface. But so small a force could produce no sensible effect."

In other words, the great man regarded it to be an insurmountable task to devise an experiment that would lead to a calculation of G. (There is more about his thinking in this regard in the Problems and Discussions section.)

This time Newton was wrong! About 100 years after Newton completed his *Principia*, in 1798 to be exact, the Englishman Henry Cavendish devised a delicate experiment that provided an estimate for G. As brilliant as Cavendish was as an experimental scientist, as a person he was an even stranger

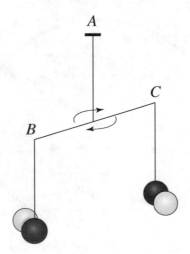

Figure 1.26

bird than the anti-social Newton. He was reclusive, published little, and—even though wealthy—dressed shabbily. He spoke rarely. As one of his contemporaries put it, he "probably uttered fewer words in the course of his life than any man who ever lived fourscore years, not at all excepting the monks of the Trappist Order."

In Cavendish's experimental setup a fine wire is suspended from a fixed point A and a rigid "crossbar" BC is attached to it. Refer to Figure 1.26. From BC in turn, two heavy iron balls are suspended. They are shown in black. Cavendish then moved two more heavy iron balls (shown lightly shaded) into place so that they almost touched the two others. He balanced and controlled his apparatus in the most delicate terms and was able to measure the gravitational force F between the two pairs of balls from the rotation of the axis BC that this force brought about. Since Cavendish knew the masses of the balls and the distances between them, the equation $F = G\frac{mM}{r^2}$ allowed him to derive an estimate for G. Expressed in MKS, Cavendish's experiment provided the value

$$G \approx 6.75 \times 10^{-11} \ \tfrac{\text{m}^3}{\text{kg·sec}^2}$$

The units meter, kilogram, and second can be seen or sensed. They are all of a human size and scale. However, the factor 10^{-11} tells us that the constant G is incredibly small on the human scale, so that gravity is a very weak force. But how is it then that when we jump up, the force of gravity pulls us quickly and powerfully back down to Earth? The answer is that it is the enormously massive Earth that does the pulling!

The results of recent experiments—most of them using refined versions of Cavendish's "torsion balance"—tell us that Newton's assertion that "so small a force could produce no visible effect" has a certain validity. Into the 21st century, the accepted value for G was $6.67259 \times 10^{-11} \ \tfrac{\text{m}^3}{\text{kg·sec}^2}$. A careful experiment carried out in 2010 provided the value $G = 6.67384 \times 10^{-11} \ \tfrac{\text{m}^3}{\text{kg·sec}^2}$. This value is still in common use by astrophysicists. A 10-year experiment produced the value $G = 6.67545 \times 10^{-11} \ \tfrac{\text{m}^3}{\text{kg·sec}^2}$ in 2012. In 2014, the Committee on Data for Science and Technology (CODATA) recommended the value $G = 6.67408 \times 10^{-11} \ \tfrac{\text{m}^3}{\text{kg·sec}^2}$. Also in 2014, a conceptually completely different approach—measuring how several hundred kilograms of tungsten distorts the gravitational effect on super-cooled rubidium atoms—yielded the value $G = 6.67191 \times 10^{-11} \ \tfrac{\text{m}^3}{\text{kg·sec}^2}$. The discrepancies between these results suggest that the measurement of G is a very delicate matter. They also raise obvious questions. Is the correct value of the coefficient of $10^{-11} \ \tfrac{\text{m}^3}{\text{kg·sec}^2}$ when rounded off to the third decimal place, equal to $6.672, 6.673, 6.674,$ or 6.675? Are these differences explained simply as experimental error? Or does the value of G depend on how it is measured or where on Earth the measurement is made? Is the value affected by the changing astronomical environment as Earth moves around the Sun and as the solar system moves within our galaxy? No doubt, further studies and experiments are called for to clarify these questions about the value of G.

The assumption that Newton's law of universal gravitation is accurate has the consequence that GM can be computed with precision for any body of mass M if precise information about an object that is in orbit about the body is available. In such a situation, the uncertainty about G just discussed necessarily implies an uncertainty about M. The websites of the space agency NASA list the Earth's mass as $5.9724 \times 10^{24} \ \text{kg}$ and the Sun's mass as $1,988,500 \times 10^{24} \ \text{kg}$. The definitive values $GM_E = 3.98600 \times 10^{14} \ \tfrac{\text{m}^3}{\text{sec}^2}$ and $GM_S = 1.32712 \times 10^{20} \ \tfrac{\text{m}^3}{\text{sec}^2}$ used above imply that

$G \approx \frac{3.98600 \times 10^{14}}{5.9724 \times 10^{24}} \frac{\text{m}^3}{\text{kg} \cdot \text{sec}^2} \approx 6.67402 \times 10^{-11} \frac{\text{m}^3}{\text{kg} \cdot \text{sec}^2}$ and $G \approx \frac{1.32712 \times 10^{20}}{1.98850 \times 10^{30}} \frac{\text{m}^3}{\text{kg} \cdot \text{sec}^2} \approx 6.67398 \times 10^{-11} \frac{\text{m}^3}{\text{kg} \cdot \text{sec}^2}$.
We can conclude that NASA is working with the approximation $G \approx 6.674 \times 10^{-11} \frac{\text{m}^3}{\text{kg} \cdot \text{sec}^2}$.

1I. The Sun. The Sun is the massive body that holds the solar system together. The entire solar system—the planets with their moons, the asteroid belts, the comets, as well as the objects of the Kuiper belt and the Oort cloud—is driven by the gravitational pull that the Sun exerts on

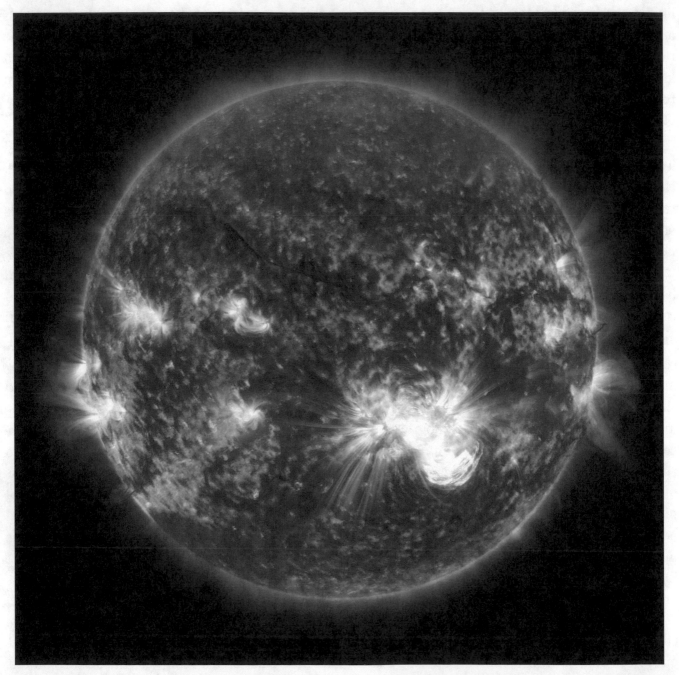

Figure 1.27. NASA's Solar Dynamics Observatory is an Earth satellite launched in 2010 to observe the Sun. Its goal is to understand how the Sun transfers energy into space and to predict the solar activity that influences our technological systems on Earth. Due to its high data transmission rate, the satellite was placed into an orbit that allows it to be in contact with a ground station in Las Cruces, New Mexico. The image was captured on October 2014 in the extreme ultraviolet light-band. Image credit: NASA/SDO and the AIA, EVE, and HMI science teams.

everything. In terms of the Sun's impact, there is the additional fact that without the Sun there would be no life on our planet and, as far as we know, no life anywhere in the solar system.

The Sun has been observed with telescopes since the time of Galileo, but definitive information about it is recent. The Sun has a radius of 695,700 km. The *core* of the Sun is a spherical region at its center of radius about 140,000 km. The extreme pressure and temperature in the core transform hydrogen to helium in a process called nuclear fusion. In the process, relatively small amounts of matter are converted into large amounts of energy. The core is a nuclear fusion reactor that produces almost all of the Sun's energy. (Scientists have been attempting to construct—unsuccessfully so far— a tiny version of such a fusion reactor here on Earth.) The energy produced by the core is transferred through the Sun's successive layers until it radiates into space in the form of heat, light, and charged atomic particles. The core's temperature is estimated to be about 16 million degrees Celsius (also known as centigrade) and that of its surface around 6,000 °C. The image of Figure 1.27, taken in October 2014, shows solar flares and ejections of solar mass. A very intense flare is seen to erupt from the lower half of the Sun. Others arc above the Sun's edge along looping magnetic fields. Such intense bursts of radiation are our solar system's largest explosive events. The dark curving filament in the picture is a huge arc of electrified gas in the Sun's atmosphere. It too is hot, but it looks dark

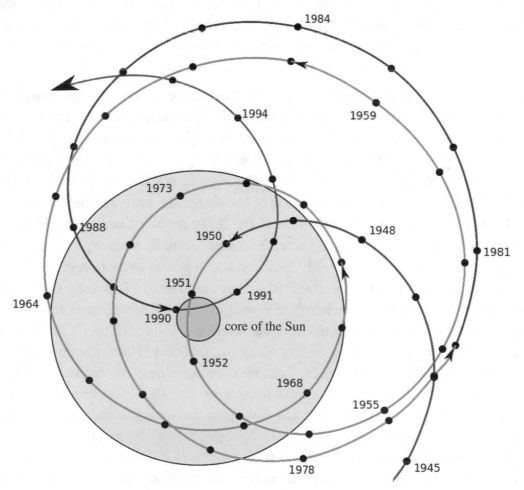

Figure 1.28. Motion of the solar system's barycenter relative to the Sun 1945–1995. Image credit: Carl Smith. See https://commons.wikimedia.org/wiki/File:Solar_system_barycenter.svg

because it is not as hot as the Sun's surface under it.

The center of the gravitational force that holds the solar system together is in fact not Sun but the center of mass, or barycenter, of the entire solar system. Every single object in the solar system, from the enormous Sun to the tiniest speck, exerts a gravitational pull on everything else. The solar system is basically a massive game of tug of war, and all of the pulling balances out at the barycenter of the system. This is the point that is the focal point of the ellipses of the orbits of the Earth–Moon system and all other planet-moon systems. Since the Sun comprises 99.86% of the mass of the solar system, in the larger scheme of things the center of the Sun and the barycenter of the solar system are not far apart. Figure 1.28 shows how the barycenter of the solar system moved relative to the Sun in the years from 1945 to 1994. A better way of thinking about this motion (in analogy with the Earth's movement around the Earth–Moon barycenter) is to regard the Sun to be in an elliptical orbit about, or more descriptively, in an elliptical wobble around, the barycenter of the solar system.

1J. Galaxies and the Expanding Universe. There is one major question that the brilliant scientists of the Scientific Revolution were not able to respond to. The Greek model of the universe places all the stars on a large sphere with center the Earth. Does the consequence that all stars are at the same distance from us correspond—at least more or less—to the facts? What about the constellations that the Greeks identified and measured the motion of the planets against? Are the stars of a particular constellation equally far away?

In the latter part of the 18th century, the French astronomer Charles Messier was focused on his passion of tracking comets with the simple telescopes that existed at the time. As he peered into the night sky he took note of other interesting formations of light. His *Catalogue of Nebulae and Star Clusters* recorded over one hundred, mostly small and fuzzy clusters of stars and clouds of stellar dust and gas that caught his eye. To this day, these objects are still referred to by their *Messier numbers.* In Messier's time, no one had any sense of the size of the universe and the distance of these objects from Earth. The parallax measurements of the 1830s provided some early insights. Friedrich Bessel aimed his split image telescope at a rather faint star (actually a binary system of two stars) on a cloudless night and six months later, with Earth on the opposite side of its orbit, he did so again. When he compared the two images, he observed a very small shift—less than 0.00002 degrees—in the position of the star against the backdrop of a very distant cluster of stars that remained fixed. This parallax measurement not only confirmed the Earth's orbital motion, it also told us that the universe is huge. While it takes light, speeding along at 300,000 km/sec, about 8 minutes to travel the 150 million km from the Sun to Earth, light requires an almost incomprehensible 11 years to reach our planet from the stars that Bessel observed. (See the Problems and Discussions section.) The distance that light travels in one year, the *light-year*, became the unit in which the distances to the stars were measured. Bessel's stars were 11 light-years away from us! The fact that there are close to 63,200 astronomical units in 1 light-year is numerical indication that the size of our solar system is minuscule compared to the size of the universe of stars. In the 1850s, scientists pointed the newly invented spectroscope toward the Sun. By examining the chemical signature its light, they discovered that the Sun was composed of elements that also existed on Earth. This established a physical link between Earth and the heavens. The chemical signatures of the light of some stars were later found to be virtually identical to those of the Sun. From this observation sprang the conclusion that except for the distances involved, the other stars of our universe were just like our Sun.

The discovery that the universe is in fact much much bigger still than the early parallax measurements had shown was made in the 1920s. The American astronomer Edwin Hubble (1889–1953) was working at the Mount Wilson Observatory in Los Angeles with the largest telescope in existence at the time when he began to suspect that the smudge that Messier had recorded as M31 was a separate universe of stars far outside our own. See Figure 1.29. The existence of such galaxies of

Figure 1.29. This spectacular image of galaxy M31 is a digital mosaic of 20 frames taken with a small telescope. Some of the stars in the image are actually stars in our Milky Way galaxy that are well in front of M31. Image credit & Copyright: Robert Gendler. Many thanks to astrophotographer Dr. Gendler for permitting its use.

stars very far from ours was soon confirmed. A headline of the New York Times of January 21, 1925, proclaimed

> "Another Universe Seen by Astronomer. Dr. Hubble Describes Mass of Celestial Bodies 700,000 light-years away."

Measurements involving "standard candles" had provided the confirmation. A standard candle is a class of stars that are known to have the same inherent brightness no matter were they are. A sailor of long ago might have made use of the principle involved. Think of a standard candle as a standard lighthouse. A sailor with a sense of the brightness of the light that it emits, would have been able to approximate his ship's distance from it on a clear night by assessing the brightness of the light that he observed. A much more rigorous and involved use of this principle told Hubble that M31 was about 700,000 light-years away. Since this was far greater than the estimated diameter of our own Milky Way galaxy, M31 had to lie far beyond it. Hubble's discovery had completely changed our understanding of the universe. It has since been shown that M31 is in fact 2.5 million light-years away. Even at this distance, M31 is one of the galaxies nearest our own. Like ours, it has hundreds of billions of stars and is more than 100,000 light-years across.

The image of a comet in Figure 2.23 (in Chapter 2E) shows a small, cigar-shaped sliver of light in the comet's tail. This faint cluster of stars was discovered in 1784 but was not included in "Messier's 100." However, it was included in *The New General Catalogue of Nebulae and Clusters of Stars* compiled about 100 years later. The 7,840 objects that it identifies are known as NGC objects. The faint sliver in the comet's tail is listed as NGC 891. Figure 1.30 shows it "up close." Today we know that NGC 891 is also a large spiraling galaxy similar to M31.

Figure 1.30. This remote-controlled image of NGC 891 was taken in November 2008 from Munich/Germany by Volker Wendel, Stefan Binnewies, and Josef Pöpsel using the telescope at the Skinakas Observatory in Crete, Greece. Many thanks to the photographers for allowing the use of the image. See their very informative website http://www. capella-observatory.com/GalaxiesIndex.htm for many more compelling images.

From Earth, we see M31 at an angle, but from our perspective NGC 891 is exactly edge-on. Also known as "Silver Sliver," NGC 891 spans about 100 thousand light-years across and contains about

500 billion stars. At a distance of about 30 million light-years, it is much farther away than M31. Its flat, thin, galactic disk of stars is cut through the middle by regions of dark, obscuring dust. Apparent in this edge-on view of NGC 891's are filaments of dust that extend hundreds of light-years above and below the dark central line.

The 51st entry in the *Catalogue of Nebulae and Star Clusters* first attracted Messier's interest in 1773 when he located it in the sky near the handle of the Big Dipper constellation. The fuzzy object's structure was not revealed until later. When an astronomer trained a large reflecting telescope on M51 in the middle of the 19th century, its graceful, winding arms were first observed. It was the first cluster of stars identified to have such a spiral structure. It did not become clear that M51 is a huge independent complex of stars far from our own Milky Way until Edwin Hubble first established that distant galaxies existed. We now know that M51 is a *spiral galaxy,* also known as the "Whirlpool" galaxy. It has some 100 billion stars, is about 60,000 light-years across, and is about 30 million light-years away. Spiral galaxies have a spherical structure, called a *bulge*, at their center. The bulge contains mostly older stars. The bulge of a spiral galaxy is surrounded by its *disk*, The spiraling arms within the disk are lanes consisting of dust and gas. In these lanes, hydrogen gas is compressed and clusters of new stars are formed. In Figure 1.31, the red represents infrared

Figure 1.31. This image of M51 captured by the *Hubble Space Telescope*—named in honor of Edwin Hubble—was processed to sharpen details and to bring out the dust lanes and streams that cross in front of its small companion galaxy. Image credits: NASA, ESA, S. Beckwith (STScI) and the *Hubble* Heritage Team (STScI/AURA).

light as well as hydrogen within giant star-forming regions. In blue are regions of light from hot, young stars. The yellow light, visible primarily in the bulge, comes from older stars. Figure 1.31 tells us that from Earth the view of M51 is "face-on." An "edge-on" image of M51 would look very much like the depiction of NGC 891 of Figure 1.30. (Incidentally, M51 and its companion are also catalogued as NGC 5194 and NGC 5195, respectively.) It seems clear from Figure 1.31 that one of the arms of M51 (and possibly both) has been distorted by the gravitational forces unleashed by the companion galaxy. *Hubble*'s image shows NGC 5195 passing behind M51's spiral arm and dust lanes. This passing maneuver occurs in very slow motion. The small galaxy has been gliding past the Whirlpool for hundreds of millions of years. Today's powerful telescopes and the galaxy's face-on position have given astronomers a front seat view of its spiral structure and star-forming processes.

In addition to spiral galaxies, there are also *elliptical galaxies* and *irregular galaxies* in the universe. As the name suggests, irregular galaxies are irregular in shape. They are among the smallest galaxies. Full of gas and dust, lots of star formation goes on within them. This can make

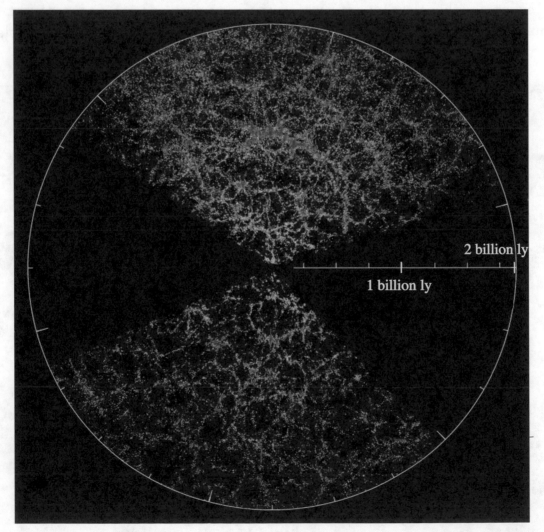

Figure 1.32. This is a survey of galaxies seen from Earth's northern hemisphere by the telescope of the Sloan Digital Sky Survey. Each galaxy is represented by a dot projected onto a plane containing the Earth. The diagram extends from Earth at the center to 2 billion light years at its circumference. The orange areas have higher densities of galaxy clusters. The survey did not extend to the two black regions. Image credit: Sloan Digital Sky Survey.

irregular galaxies very bright. Elliptical galaxies are like a spiral galaxies without the spiraling arms. They are all bulge and consists of older stars. Little or no star formation takes place within them. The light they emit is uniform. Elliptical galaxies range in shape from nearly spherical to elongated spheres with elliptical cross-section. They range in size from tens of thousands light-years in diameter to over a million light-years in diameter. An elliptical galaxy is brightest at the center and its surface brightness decreases in the direction of its boundary. Of the galaxies that astronomers have observed so far about 70 percent have been spiral galaxies. However, given that they consist of older, dimmer stars, elliptical galaxies are generally less bright and more challenging to spot. In large, in-depth surveys of patches of the sky, elliptical galaxies have predominated. The map of galaxies of Figure 1.32 provides a sense of the fabric of the universe. The Sloan Digital Sky Survey (SDSS) that produced it used a 2.5 meter wide-angle optical telescope that gathered not only visible light but also electromagnetic radiation of other wavelengths.

It is important to note that the universe is not static. The light from the standard candles of the galaxies that Hubble observed provided more than distance estimates. They also revealed that the universe is expanding! The fact is that a rapidly moving light is perceived to change its frequency when observed from a fixed location in the same way that a rapidly moving sound is perceived to change its pitch. When a "red shift" was detected in the frequency of the infrared radiation that the standard candles emitted, Hubble knew that the galaxies he studied were receding from our Milky Way and that the velocities with which they did so could be estimated. He discovered that the greater the distance of a galaxy from us, the greater the velocity with which it recedes. In 1929, he realized for any of the galaxies he observed, that if v is its velocity and d its distance, then the ratio $\frac{v}{d}$ is always the same constant. This fact is now expressed as the equality $\frac{v}{d} = H_0$ with the constant H_0 known as *Hubble's constant*. (Notice that in this formula, velocity is a scalar quantity.) Imagine a movie of the expanding universe and think of it as being played backward in time. It will show the galaxies approaching each other. Play it back long enough to picture a moment when all the galaxies are massed together. Reversing the movie again and playing it forward, you will see matter exploding outward in all directions. This is the moment of the creation of the universe that is today referred to as the *Big Bang*. Calculations using the refined estimates of Hubble's constant that the *Hubble Space Telescope* (named after the great astronomer) provided, tell us that the Big Bang occurred from around 13 to 14 billion years ago. In the meantime—in 1915 to be exact—Einstein had combined space and time into a new theoretical framework for the universe. His *Theory of Relativity* gives the universe a four dimensional geometry which is curved by all of the masses that float within it. Instead of being regarded as the consequence of the pull of gravity, the trajectory of a body in space is interpreted to be the result of its response to the curves in this geometry. Think of a rolling golfball responding to the curving surface of a green.

1K. Problems and Discussions. This problem set takes up a number of matters that, while central to this chapter, are only taken up briefly or only in passing. They will now be considered in detail.

1. Copernicus's Measurement of Planetary Distances. Copernicus considered the orbit of the planet Venus. In his study he regarded it to be a circle with the Sun at the center. In

Figure 1.33, S represents the Sun, E designates the Earth in its orbit, and V, V', and V'' are positions of Venus. Observing Venus over time, and measuring the angle $\alpha = \angle VES$ again and again, Copernicus found that the maximum value reached by the angle α is close to $46°$. A look at the figure tells us that α is greatest when the line of sight to Venus is tangent to Venus's circular orbit around S. Let V be the position of Venus when α is a maximum. Since EV is tangent to Venus's orbit, the radius SV is perpendicular to EV. It follows that $\frac{VS}{ES} = \sin \alpha \approx \sin 46°$. Since

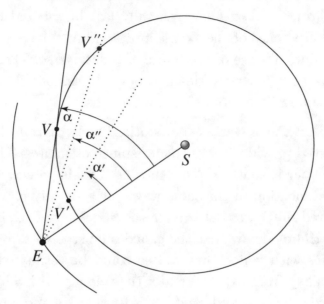

Figure 1.33

Copernicus knows that $\sin 46° \approx 0.72$, he has determined that

$$VS \approx 0.72\, ES.$$

Problem 1.1. For the inner planet Mercury, Copernicus used the same argument to show that the ratio of the distance from Mercury to the Sun to the distance of Earth to the Sun is close to 0.38. Describe how he came to this conclusion.

Copernicus also had accurate estimates of the distances of the outer planets Mars, Jupiter, and Saturn from the Sun in terms of the distance of the Earth from the Sun. Copernicus's calculations

Table 1.1

Planet	Copernicus	Kepler	Modern
Mercury	0.38	0.389	0.387
Venus	0.72	0.724	0.723
Earth	1.00	1.000	1.000
Mars	1.52	1.523	1.524
Jupiter	5.22	5.200	5.202
Saturn	9.17	9.510	9.539

for the outer planets were more complicated, but here too he relied on elementary trigonometry (the law of sines in particular). Within his Sun centered model, Copernicus knew how to express the distance of any planet from the Sun in terms of the distance of the Earth from the Sun. See Table 1.1.

As to the determination of the size of the solar system, the important remaining issue was the difficult question of the distance of Earth from the Sun. Ptolemy's value of around 8 million kilometers for the distance from Earth to the Sun was accepted by Copernicus, and neither Kepler nor Galileo had a much better grasp of it. In the latter part of the 17th century, two of the star astronomers of the time set about the task of calculating the distance between Earth and the Sun. The Italian Giovanni Cassini (1625–1712) was already famous for his observations of Jupiter and Saturn when he was called by the French king to direct the new astronomical observatory in Paris. John Flamsteed (1646–1719), a young astronomer working in central England, would soon become the first Astronomer Royal of the observatory in Greenwich that the king of England was establishing.

2. Earth, Sun, and Mars. The approaches of both Flamsteed and Cassini relied on parallax measurements of the distance between Earth and Mars. They knew that in October of the year 1672 they would be provided with a great opportunity for such a calculation. For a week or two, the Sun, Earth, and Mars would fall in a straight line as depicted in Figure 1.34 and Mars would be near its perihelion position. Both Cassini and Flamsteed knew the eccentricity $\varepsilon_E = 0.017$ of Earth's orbit and with a_E its semimajor axis, that the Earth's distance from the Sun varies from $a_E(1 - 0.017) = 0.983 a_E$ (at perihelion) to $a_E(1 + 0.017) = 1.017 a_E$ (at aphelion) or from about 98% of a_E to about 102% of a_E. This variation fell within their tolerance for error, so they assumed

Figure 1.34

that the distance from the Earth to the Sun is equal to a_E. With Mars they had to be more careful. With a_M the semimajor axis of its orbit and Kepler's value $\varepsilon_M = 0.0926$ for its eccentricity, they knew that the distance from Mars to the Sun varies from $a_M(1 - \varepsilon_M) = a_M(0.9074)$ at perihelion to $a_M(1 + \varepsilon_M) = a_M(1.0926)$ at aphelion, or from about 90% of a_M to about 110% of a_M. However, since Mars would be near its perihelion, they could assume that its distance from the Sun was close to $a_M(1 - \varepsilon_M)$. Therefore they could let d be the distance between Earth and Mars, get the approximation

$$a_E + d \approx a_M(1 - \varepsilon_M)$$

from the diagram in Figure 1.34, and conclude that $a_M \approx \dfrac{a_E + d}{1 - \varepsilon_M}$. Letting T_E and T_M be the periods of the orbits of Earth and Mars, Flamsteed and Cassini could apply Kepler's third law to get

$$\frac{a_E^3}{T_E^2} = \frac{a_M^3}{T_M^2} \approx \frac{(a_E + d)^3}{(1 - \varepsilon_M)^3 T_M^2},$$

and hence

$$\frac{a_E}{T_E^{2/3}} \approx \frac{a_E + d}{(1 - \varepsilon_M)T_M^{2/3}}.$$

Since $a_E \cdot (1 - \varepsilon_M)\left(\frac{T_M}{T_E}\right)^{2/3} \approx a_E + d$, they therefore got $d \approx \left[(1 - \varepsilon_M)\left(\frac{T_M}{T_E}\right)^{2/3} - 1\right]a_E$ and

$$a_E \approx \frac{d}{(1 - \varepsilon_M)\left(\frac{T_M}{T_E}\right)^{2/3} - 1}.$$

Knowing that the periods of the orbits of Earth and Mars were $T_E \approx 365.25$ days and $T_M \approx 686.95$ days, Cassini and Flamsteed got $(1 - \varepsilon_M)\left(\frac{T_M}{T_E}\right)^{2/3} - 1 \approx (0.9074)(1.5237) - 1 = 0.3826$, and they could conclude that

$$a_E \approx 2.61d.$$

It therefore remained for Cassini and Flamsteed to estimate the distance d.

Problem 1.2. Check the computations that allowed Flamsteed and Cassini to conclude that $a_E \approx 2.61d$.

Both Flamsteed and Cassini went about using the method of parallax to make reasonably accurate calculations of the distance d. With the distance d in hand, they could estimate a_E, and hence all the distances between the planets and the Sun. This would answer the central question in astronomy of the time: What is the size of the solar system?

3. The Method of Parallax. The strategy of parallax—already seen to have been an important observational tool of Tycho Brahe—can be used to estimate the distances of the bodies in the solar system (and beyond it) from Earth. We'll illustrate it in the current historical context by using

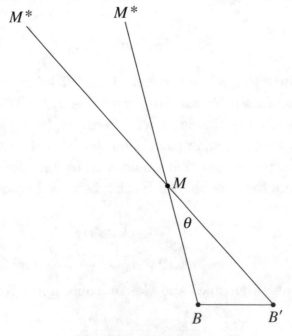

Figure 1.35

it to estimate the distance between Earth and Mars. A lot had changed since Tycho Brahe pointed his astronomical instruments skyward and noted that the new star that had suddenly appeared had no parallax. The telescopes that astronomers of the latter part of the 17th century had available were much more powerful and precise than those that Galileo first used to study the heavens. They were equipped with micrometer eyepieces and telescopic sights that made it possible to measure angular separations to within a small fraction of a degree.

From an observation point B on Earth, an astronomer sights Mars as a point of light in the night sky at a location M^* within a cluster of the fixed stars of a familiar constellation. The sighting of Mars is repeated from a different location B' far from B, with the result that the observed position of M^* within the same star cluster will have shifted slightly. See Figure 1.35. By measuring this shift, an estimate of the angle θ can be obtained. This estimate proceeds as follows. Refer to Figure 1.36a. From the vantage point B, the astronomer fixes a star A in the cluster and carefully measures the

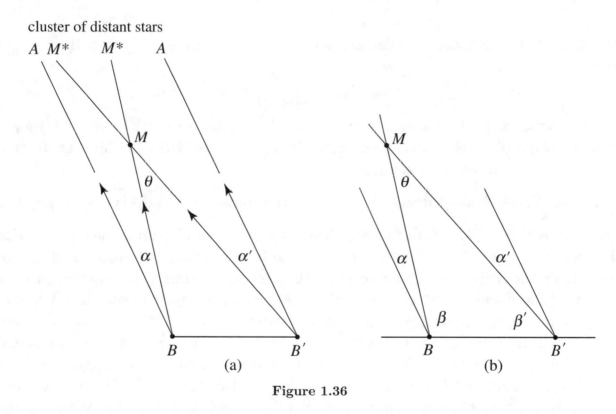

Figure 1.36

angle $\alpha = \angle ABM^*$. After the location of his observation post has changed to B', he locates the star A once more and measures the angle $\alpha' = \angle AB'M^*$. Because A is very far from Earth, the lines of sight BA and $B'A$ are essentially parallel.

Problem 1.3. Use Figure 1.36b to show that $(\alpha + \beta) + (\alpha' + \beta') = \pi$ and hence that $\theta = \alpha + \alpha'$.

The *angle of parallax* $p(M)$ of M relative to the baseline BB' is defined by $p(M) = \frac{1}{2}\theta$. Notice that $p(M) = \frac{1}{2}\theta = \frac{1}{2}(\alpha + \alpha')$ is the average of the measured angles α and α'. Because the angles involved are small, the angle of parallax $p(M)$ and related angles are not measured in degrees, but in seconds. Since 1 degree = 60 minutes and 1 minute = 60 seconds, $1° = 3600''$. Since $180°$ is equal to π radians, the angle of parallax expressed in radians is

$$p(M) \times \frac{1}{3600} \times \frac{\pi}{180} = p(M) \times \frac{\pi}{648,000} \approx p(M) \cdot 4.85 \times 10^{-6} \text{ radians.}$$

Since the distance from Earth to Mars is huge relative to the distance between the observation points B and B', we'll take the distances from B to M and B' to M to be equal and set this distance equal to $d(B,M)$. Note that the triangle $\triangle BMB'$ is approximated by the circular sector BMB' of

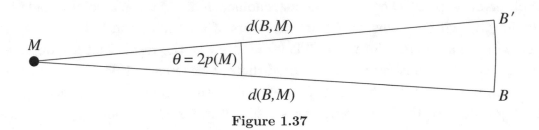

Figure 1.37

Figure 1.37. Using the definition of radian measure of an angle, we get $2p(M)(4.85 \times 10^{-6}) \approx \frac{BB'}{d(B,M)}$. Therefore

$$d = d(B,M) \approx \frac{BB'}{9.7p(M)} \times 10^6,$$

where $p(M)$ is the angle of parallax of Mars in seconds. It follows that the angle $p(M)$ together with the distance BB' provides the approximate distance between Earth and Mars at the time of the measurements of the angles α and α'.

Problem 1.4. Check in all its details the argument that verified $d = d(B,M) \approx \frac{BB'}{9.7p(M)} \times 10^6$.

4. The Parallax of Mars. Cassini and Flamsteed knew that measuring the parallax of Mars would be a delicate task. The angles of parallax would be very small, and Mars would change its observed position not only as a consequence of parallax, but also because of its continuing motion in its orbit. But Cassini and Flamsteed pressed ahead. As was already pointed out, the conditions for detecting and measuring the parallax of the planet in the fall months of 1672 were optimal. Because Mars would be near its perihelion, it would be relatively close to Earth. Since Sun, Earth, and Mars would be aligned, it follows from Figure 1.34 that Mars would be under full sunlight when viewed at night. What Cassini and Flamsteed observed is depicted in Figure 1.38. It shows the dotted sequence of the positions of Mars against the stars of the constellation Aquarius. When a planet is viewed from Earth against a background of stars, it is seen to move in a prevailing direction, but it loops back periodically before proceeding forward again. Figure 1.38 shows a loop in the orbit of Mars in the fall of 1672.

Flamsteed made his measurements during the single night of October 6, 1672 from his observation post in Derby, his hometown in central England. Figure 1.38 tells us that at that time, Mars was at its turnaround from its loop. This meant that a shift in the observed position of Mars between measurements was almost exclusively due to parallax rather than the motion of the planet. Flamsteed first observed Mars late in the evening. The point B of Figure 1.39a marks his location on the surface of our planet. Precisely 6 hours and 10 minutes later, he observed Mars again. The Earth had turned by slightly more than a fourth of a complete rotation, and Flamsteed's observation post

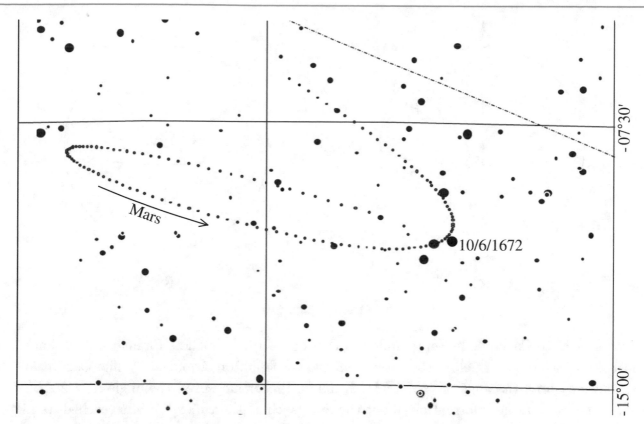

Figure 1.38. Image reproduced with permission from Parker Moreland, who created it from a screenshot made possible by TheSky Astronomy Software © Software Bisque, Inc.

had rotated to the point B' in the figure. Since Flamsteed knew the latitude of his location (the angle between the segment OB from Earth's center O and the equatorial plane) to be about $53°$ he could estimate the length of his baseline BB' to be about 3300 miles, or about 5300 kilometers. As Figure 1.36a illustrates it, Flamsteed peered into the night sky and found M^* near three bright stars in the constellation Aquarius. Letting A be one of them, he measured the angle $\alpha = \angle ABM^*$ and later the angle $\alpha' = \angle AB'M^*$. He found the angle $\theta = \alpha + \alpha'$ to be about 21 seconds and concluded that the angle of parallax $p(M) = \frac{1}{2}\theta$ (with respect to the base line BB') was about 10.5 seconds. This gave Flamsteed the estimate

$$d = d(B,M) \approx \frac{BB'}{9.7p(M)} \times 10^6 \approx \frac{5{,}300}{(9.7)(10.5)} \times 10^6 \approx 52{,}000{,}000 \text{ kilometers.}$$

for the distance d, and in turn the estimate $a_E \approx 2.61d \approx 136{,}000{,}000$ kilometers for the semimajor axis a_E of the Earth's orbit. (Refer to Problem 1.2.)

As the director of the Paris observatory, Domenico Cassini had considerable resources at his disposal. It was his idea to approach the measurement of the parallax of Mars from two different vantage points B and B' on Earth at the same time! This would eliminate the difficulty of having to quantify the motion of the planet during the time between measurements. While Cassini remained in Paris, his colleague Jean Richer was sent on an expedition to Cayenne in French Guiana, a French colony in South America just north of the equator (on the Atlantic coast near the northernmost tip of Brazil). Knowing the latitude and longitude of Cayenne, Cassini could estimate the distance between Paris and Cayenne to be the equivalent of about 6700 kilometers. In Figure 1.39b, B refers

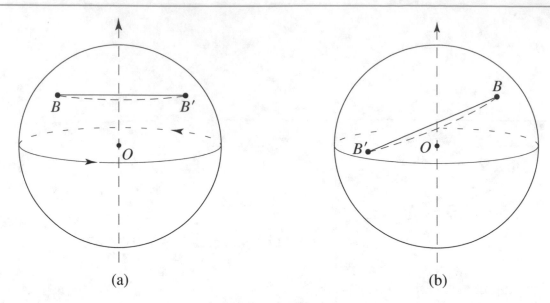

(a)   (b)

Figure 1.39

to Paris and B' to Cayenne. In September and October of 1672, much as Flamsteed had done, both men observed Mars at M^* close to a star A in the constellation Aquarius. While Cassini and his assistants measured the angle $\alpha = \angle ABM^*$ in Paris, Richer measured the angle $\alpha' = \angle AB'M^*$ in Cayenne. By comparing information about the Sun (such as the times the Sun reached its highest position in the sky) at the two locations, they were able to make repeated pairs of measurements at close to the same time. Cassini then waited nearly a year for his colleague to return to Paris with his data! After evaluating their data carefully, they concluded that the angle $\theta = \alpha + \alpha'$ was about 26 seconds. So the corresponding angle of parallax $p(M) = \frac{1}{2}\theta$ with respect to their baseline BB' was approximately 13 seconds. This provided Cassini with the estimate

$$d = d(B, M) \approx \frac{BB'}{9.7\, p(M)} \times 10^6 \approx \frac{6{,}700}{(9.7)(13)} \times 10^6 \approx 53{,}000{,}000 \text{ kilometers},$$

and therefore the value $a_E \approx 2.61d \approx 140{,}000{,}000$ kilometers for the semimajor axis of Earth's orbit. This was slightly better than what Flamsteed had achieved and only about 7% less than today's value of 150,000,000 kilometers.

Problem 1.5. The baseline that Flamsteed used for his parallax measurements involved the two positions B and B' of his observatory in Derby, England with B' obtained from B by the Earth's rotation. Figure 1.40a shows Earth with its center O along with its axis of rotation. The latitude of Derby is given by the angle $\varphi = 52.92°$. At this latitude, Earth's radius is known to be equal to $r_E = 6364.57$ km. Figure 1.40b shows a "top view" of the circle given by the point B and a full rotation of the Earth, along with the angle θ that the two positions B and B' determine.

  i. Show that the radius of the circle of Figure 1.40b is $r_E \cos \varphi$.

  ii. Verify with the law of cosines that the distance from B to B' is equal to $(r_E \cos \varphi)\sqrt{2(1 - \cos \theta)}$.

  iii. Use the fact that the elapsed time between Flamsteed's measurements was 6 hours and 10 minutes to show that $\theta = 92.5°$.

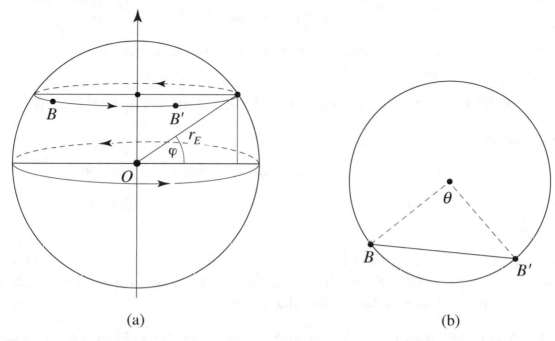

(a)   (b)

Figure 1.40

iv. Conclude that Flamsteed's baseline BB' has a length of 5544 kilometers. (The estimate that was used in the description of Flamsteed's computation was 5300 km.)

5. The Moons of Jupiter and Saturn. In his study of the planet Jupiter in the *Principia*, Newton lists

$$5.578, \quad 8.876, \quad 14.159, \quad \text{and} \quad 24.903$$

for the maximal distances of the four largest moons of Jupiter from Jupiter's center. The unit of distance is the radius of Jupiter. The corresponding periods of the orbits of these moons in days, hours, minutes, and seconds are given as

$$1^d\, 18^h\, 28'\, 36'', \; 3^d\, 13^h\, 17'\, 54'', \; 7^d\, 3^h\, 59'\, 36'', \text{ and } 16^d\, 18^h\, 5'\, 13''.$$

This corresponds to

$$42.48, \quad 85.30, \quad 171.99, \quad \text{and} \; 402.09 \quad \text{hours, respectively.}$$

This information was provided to Newton by the astronomer Flamsteed. These four moons were discovered by Galileo in 1610. Their current names Io, Europa, Ganymede, and Callisto were given to them in the 19th century.

Problem 1.6. Newton verified Kepler's third law for the four moons of Jupiter. Carry out the computations that Newton undertook.

The five largest moons of Saturn were discovered by Huygens and Cassini. The Dutch scientist Huygens discovered Titan in 1655, and Cassini discovered Iapetus in 1671, Rhea in 1672, and Tethys and Dione in 1684. They are discussed by Newton in later editions of the *Principia*. He credits Cassini with the data that he uses. The respective distances from Saturn's center are

$$1\tfrac{19}{20}, \ 2\tfrac{1}{2}, \ 3\tfrac{1}{2}, \ 8, \text{ and } 24,$$

where the unit is the radius of Saturn's outer ring. The periods of the orbits are listed in days, hours, minutes, and seconds as

$$1^{d} \, 21^{h} \, 18' \, 27'', \ 2^{d} \, 17^{h} \, 41' \, 22'', \ 4^{d} \, 12^{h} \, 25' \, 12'', \ 15^{d} \, 12^{h} \, 41' \, 14'', \text{ and } 79^{d} \, 7^{h} \, 48' \, 0''.$$

This corresponds to

$$45.31, \ 65.69, \ 108.42, \ 372.69, \text{ and } 1903.8 \text{ hours, respectively.}$$

Problem 1.7. Newton verified Kepler's third law holds for these five moons of Saturn. Carry out the computations that Newton made use of.

We will now turn to several matters that arise in or are directly related to concerns that Newton considers in his *Principia*. We'll begin by illustrating how Newton would have developed the equation for the maximum speeds of the bodies in the solar system.

6. About Speeds of Objects in the Solar System. Consider a planet, comet, or asteroid P in its elliptical orbit with the Sun S at a focal point of the ellipse. Let a be the semimajor axis, b the semiminor axis, and let $\varepsilon = \frac{c}{a}$, where $c = \sqrt{a^2 - b^2}$, be the eccentricity of the orbit. Let T be the perihelion period of the orbit and κ its Kepler constant. Since the area of an ellipse with semimajor axis a and semiminor axis b is $ab\pi$, it follows that $\kappa = \frac{ab\pi}{T}$.

Figure 1.41 below shows P in five different locations of its orbit. (The ellipse is drawn much flatter than that of any planetary orbit in order to add transparency to our discussion.) The five locations are labeled from 1 to 5 in the figure. The numbers 1 and 5 denote the perihelion and aphelion positions respectively. Let Δt be a short fixed interval of time and consider the five short arcs starting from the five points that P traces out during this time. The five arcs and the thin wedges that the segment SP sweeps out in the process are drawn in as well. Since they are all swept out in the same time Δt, these wedges have the same area by Kepler's second law. Since the wedges get longer as P proceeds from perihelion to aphelion, the arcs get shorter and shorter. Since they are all traced out over the same time, this means that the average speed of P over the arcs decreases from one arc to the next. Pushing Δt to zero shortens the five arcs and pushes the average speed to

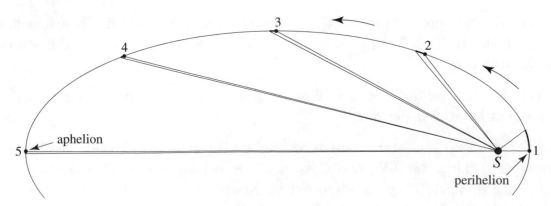

Figure 1.41

the speed at the initial point of each arc. Since the distance from P to S is shortest at perihelion, P achieves its greatest speed v_{max} at perihelion. Similarly, since the distance from P to S is longest at aphelion, P attains its minimum speed v_{min} at aphelion.

Problem 1.8. Figure 1.42 shows a planet (comet, or asteroid) at its perihelion position P, a short stretch of the orbit, and the position Q of the planet a short time Δt later. The segment through P

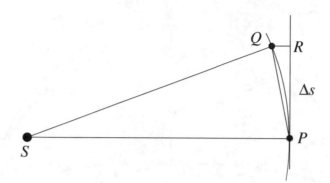

Figure 1.42

is tangent to the orbit at P, and the point R is chosen so that RQ is parallel to SP. We'll let Δs be the length of the segment PR. Since P is the perihelion position, the length of SP is $a - c$.

   i. Provide an expression for the precise value of the *average velocity* v_{av} of the motion of the planet from P to Q. Provide an approximation of v_{av} by using Δs.

   ii. Use both the exact value $\kappa = \frac{ab\pi}{T}$ for Kepler's constant and an approximation for κ that arises from the diagram to verify that the average velocity satisfies $v_{\text{av}} \approx \frac{2ab\pi}{(a-c)T}$.

   iii. What two things happen in (ii) when Δt is pushed to zero that result in the conclusion $v_{\text{max}} = \frac{2ab\pi}{(a-c)T}$?

   iv. Use the equalities $b = \sqrt{a^2 - c^2}$ and $c = \varepsilon a$ to conclude that $v_{\text{max}} = \frac{2\pi a}{T}\sqrt{\frac{1+\varepsilon}{1-\varepsilon}}$.

Problem 1.9. Show that the speed at aphelion is equal to $v_{\text{min}} = \frac{2ab\pi}{(a+c)T} = \frac{2\pi a}{T}\sqrt{\frac{1-\varepsilon}{1+\varepsilon}}$.

The fact that 1 au = 149,597,870.7 kilometers and 1 year as unit of time is equal to 365.25 days, or 365.25(86,400) = 31,557,600 seconds, tells us that

$$1\,\text{au/year} \approx \frac{149{,}597{,}870.7 \text{ km}}{1 \text{ year}} \times \frac{1 \text{ year}}{31{,}557{,}600 \text{ sec}} \approx 4.74\,\text{km/sec}.$$

Problem 1.10. Use the approximations $a = 1$ au for the semimajor axis, $\varepsilon = 0.0167$ for the eccentricity, and $T = 1$ year for the perihelion period of Earth's orbit to show that the maximum and minimum speeds of Earth in its orbit are approximately 30.29 km/sec and 29.29 km/sec, respectively.

Problem 1.11. The elliptical orbit of the comet Halley has semimajor axis $a = 17.83$ au, eccentricity $\varepsilon = 0.967$, and period T = 75.32 years. Use this information to show that its maximum and minimum orbital speeds are approximately 54.43 km/sec and 0.91 km/sec.

Problem 1.12. To gain some familiarity with ellipses, consider the four drawn in Figure 1.43. They are labeled ①, ②, ③, and ④. All have the same semimajor axis a. Their semiminor axes are b_1, b_2, b_3 and a, respectively, where $b_1 < b_2 < b_3 < a$. Their right focal points are color coded to correspond to the color of the ellipse. Explain how these focal points where placed and why the ellipse labeled ④ is a circle. Notice that the closer the focal point is to the periapsis, the flatter

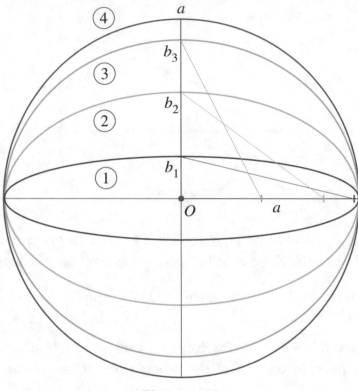

Figure 1.43

the ellipse. Why is this observation consistent with the interpretation of the ellipse as the orbit of a moving object that is subject to the gravitational pull of a massive body at the focal point?

7. About the Earth–Moon System. The shape of the Earth is essentially spherical, but it is a sphere that is flatter at the poles and bulges out at the equator. This shape was brought about by the rotation of the Earth about its axis. Figure 1.44 provides an exaggerated look at this bulging sphere. The Earth's axis is tilted by about 29° relative to the plane of the Moon's orbit about the Earth and about 23.5° relative to Earth's orbital plane. Note that the orbital planes of the Moon and the Earth differed by about 5°. The figure diagrams the Earth as well as the Sun or Moon

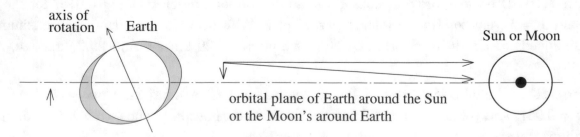

Figure 1.44

(but not to scale) and shows that the gravitational pulls of both the Moon and the Sun on Earth's bulge has a component that pulls Earth's axis of rotation in a direction perpendicular to its orbital plane. This component is substantial enough to produce the very slow gyration of the Earth's axis of rotation that was already discussed in section 1A.

An understanding of the magnitude of the Moon's gravitational force on Earth as well as the location of the barycenter of the Earth–Moon system, requires the determination of the mass of the Moon. This proved to be an elusive task. Since the Moon's density turns out to be much less than that of Earth, this was not merely a matter of estimating the Moon's volume.

8. The Mass of the Moon. The mass of the Moon was a concern that had interested astronomers for a long time. By studying tides—known to be the result of the Moon's gravity—Newton came to the conclusion that the Moon is 20% denser than Earth, and that it's mass is $\frac{1}{40}$ of the Earth's mass. It turned out that he was wrong on both counts. What follows is a study by the astronomer Sir George Airy (1801–1892). Airy was director of the Royal Greenwich Observatory and, like Newton before him, professor of mathematics at Cambridge. Airy set out to quantify what goes on in Figure 1.19. Figure 1.45 depicts the Moon in its orbit around the Moon-Earth barycenter **B**. The mass and the center of mass of Earth are denoted by M_E and C and those of the Moon by M and A, respectively. We'll let d be the distance between C and A and x the

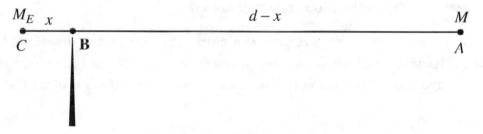

Figure 1.45

corresponding distance between C and **B**. We'll let F be the gravitational force of attraction between Earth and Moon. By Newton's law of universal gravitation, $F = G\frac{MM_E}{d^2}$. Now take the instant when $d - x$ is equal to the semimajor axis a of the Moon's orbit around **B**. By applying Newton's Conclusion B of section 1D with $r_P = a = d - x$ and $m = M$, Airy finds that $F = \frac{4\pi^2 M}{T^2}(d - x)$, where T is the period of the Moon's orbit. By Archimedes's law of the lever, $M_E x = M(d - x)$. So $Md - (M_E + M)x = 0$. Therefore $M_E d + Md - (M_E + M)x = M_E d$, so $(M_E + M)(d - x) = M_E d$, and it follows that $d - x = \frac{M_E d}{M_E + M}$. Airy has now shown that $F = \frac{4\pi^2 M}{T^2}\frac{M_E d}{M_E + M}$.

Problem 1.13. Verify Airy's formula

$$\frac{M}{M_E} = \frac{4\pi^2}{GM_E} \cdot \frac{d^3}{T^2} - 1.$$

The use of Airy's formula requires an accurate value for d at the moment when $d - x$ is equal to the semimajor axis of the Moon's orbit around **B**. When Airy developed this formula in 1849, accurate values for d and $d - x$ were impossible to come by. In fact his formula never provided a precise value for the Moon-mass over Earth-mass ratio.

By the 1900s, a number of astronomers arrived at values for the ratio $\frac{M}{M_E}$ close to $\frac{1}{80}$ (this is $1/2$ the value that Newton had achieved) by using either tidal information or parallax measurements. But the matter was resolved with precision only after the Apollo Moon missions of the late 1960s and early 1970s sent back very accurate data about the orbits of their command-service modules around the Moon. These orbital data allowed for the direct and accurate computation of the value GM for the Moon's mass M. Given that G is known, this meant that M was known, so that the pursuit of the ratio $\frac{M}{M_E}$ became irrelevant. Its precise value turned out to be $\frac{1}{81.3006}$.

Problem 1.14. Use the values $GM_E = 3.98600 \times 10^{14} \frac{m^3}{sec^2}$ and $GM = 4.90279 \times 10^{12} \frac{m^3}{sec^2}$ to compute $\frac{M}{M_E}$. Then take the modern value $M_E = 5.9724 \times 10^{24}$ kg for the Earth's mass to show that the mass of the Moon is $M = 7.3461 \times 10^{22}$ kg.

Problem 1.15. i. Take $GM_E = 3.98600 \times 10^{14} \frac{m^3}{sec^2}$ and $a_M = 3.844 \times 10^8$ m for the semimajor axis of the Moon's orbit to estimate the gravitational force with which the Earth attracts the Moon.

    ii. Let $GM_S = 1.32712 \times 10^{20} \frac{m^3}{sec^2}$ and take the semimajor axis $a_E = 1.49598 \times 10^{11}$ m of Earth's orbit as an approximation for the distance between the Sun and the Moon. Estimate the gravitational force with which the Sun attracts the Moon.

    iii. Confirm that the gravitational force of the Sun on the Moon is greater than that of Earth on the Moon. So why is the Moon orbiting Earth?

    9. On the Orbit of the Moon around the Sun. The simulation described by Figure 1.19 suggests that as the barycenter of the Earth–Moon system revolves around the Sun, the Moon should loop around this barycenter. But this is not the case. The Moon's path around the Sun is in fact

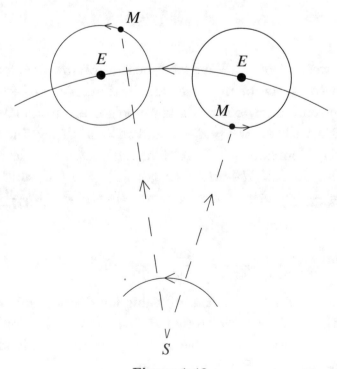

Figure 1.46

not "loopy." The explanation is provided by the comparative speeds of the Earth and the Moon in their respective orbits.

Problem 1.16. Recall the conclusions of Problems 1.8 and 1.9. Use the data $a = 3.84748 \times 10^5$ km for the semimajor axis, $T = 2.36062 \times 10^6$ sec for the period, and $\varepsilon = 0.0549$ for the eccentricity of the Moon's orbit around Earth to show that the maximum and minimum orbital speeds of the Moon in its orbit are 1.082 km/sec and 0.969 km/sec, respectively.

Consider Figure 1.46. Since the positions of the Earth and the Earth–Moon barycenter are very close to each other, they are the same point E in the figure. The Moon is denoted by M. Compare the maximum and minimum speeds of the Moon in its orbit around Earth with that of the Earth around the Sun (refer back to Problem 1.10) and study the figure. Why does the path of the Moon around the Sun not loop around the Earth? Since the Moon is sometimes closer to the Sun than the Earth–Moon barycenter and sometimes farther away, it's path around the Sun can be thought of as a flat sine curve that is bent around the elliptical orbit of the Earth–Moon barycenter.

10. A Speculation of Newton. In the *System of the World* of the *Principia* Newton says

For the attraction of homogeneous spheres near their surfaces are as their diameters. Whence a sphere of one foot in diameter, and of like nature to the Earth, would attract a small body placed near its surface with a force of 20,000,000 less than the Earth would do if placed near its surface. But so small a force could produce no sensible effect. If two such spheres were distant but by $\frac{1}{4}$ inch, they would not even in spaces void of resistance, come together by the force of their mutual attraction in less than a month's time.

Newton speculates about the possibility of estimating the constant G in some sort of experimental setting. He suggests that this would be an impossible task. Is what Newton is saying here correct?

In our discussion, we'll work in the units centimeters-grams-seconds (CGS). The unit of force in this system is the *dyne*, defined as $1\frac{\text{gr}\cdot\text{cm}}{\text{sec}^2}$.

Problem 1.17. Convert the value $G = 6.67384 \times 10^{-11} \frac{\text{m}^3}{\text{kg}\cdot\text{sec}^2}$ as given in MKS to CGS.

In view of the assumption "of like nature to the Earth," we'll start by computing the Earth's density. Recall that the average density of an object is the ratio of the object's mass over its volume.

Problem 1.18. Assume that the Earth is a sphere of radius 6371 km and mass 6×10^{24} kg and estimate its average density in CGS.

Let's turn to have a look at Newton's spheres. Suppose that they are identical and of uniform density equal to the average density of the Earth. This allows the assumption that the entire mass of each sphere is located at its center. Figure 1.47 shows the circular cross sections of the two spheres through their centers in an xy-coordinate plane that has the centimeter as its unit of length. Notice that the centers of the circles are $2c$ cm apart and that the radius of each of the two circles is $(c - d)$ cm.

Problem 1.19. What are c and d equal to for the spheres that Newton describes? What is the mass in grams of each sphere?

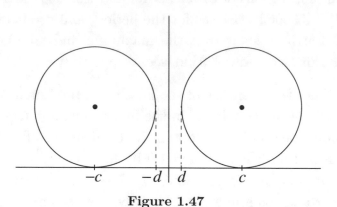

Figure 1.47

Problem 1.20. Apply Newton's law of universal gravitation to derive an expression for the magnitude in dynes of the gravitational force with which the two spheres attract each other. What is the maximum value of the magnitude of this force?

We'll now use a little elementary calculus to examine the acceleration and speed of a moving point. Suppose that a point-mass of mass m is driven along a number line in the positive direction by a constant force F. We'll suppose that the force starts acting at time $t = 0$ and that the point is at rest at the origin at that time. See Figure 1.48. Newton's formula $F = ma$ tells us that the

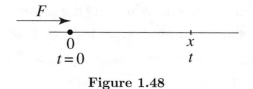

Figure 1.48

acceleration of the point-mass is constant and equal to $a = \frac{F}{m}$. Let $t \geq 0$ be any time into the motion. Let $x(t)$ be the distance of the point-mass from the origin at time t and let $v(t)$ be the velocity of the point-mass at time t. Since the acceleration is equal to the change in the velocity, the derivative $v'(t)$ is equal to a. Since $v(0) = 0$, it follows that $v(t) = a \cdot t$. Since velocity is change in distance, $x'(t) = v(t) = a \cdot t$. Since $x(0) = 0$ it follows that $x(t) = \frac{1}{2}at^2 = \frac{1}{2}\frac{F}{m}t^2$.

Problem 1.21. Suppose that the two spheres of Figure 1.47 start from rest in such a way that the initial distance $2d$ between them is the CGS equivalent of $\frac{1}{4}$ inch that Newton mentions. Suppose that the spheres move frictionlessly as they are pulled toward each other by gravity. How long would it take for the maximum gravitational force of attraction (as described in Problem 1.20) to move the two spheres until they touch? Does your conclusion mesh with Newton's assertions?

11. Parallax and Distances to Stars. Parallax measurements of the sort used by both Flamsteed and Cassini are suitable for estimating the distances of objects within the solar system. But what about the stars? All the stars are far from Earth—much, much farther than any object in our solar system—but they are not all the same distance away from us as the Greek astronomers had thought. Within a given constellation, some are relatively near and others are unimaginably far. Once the distance of Earth from the Sun was known, astronomers could use a baseline determined

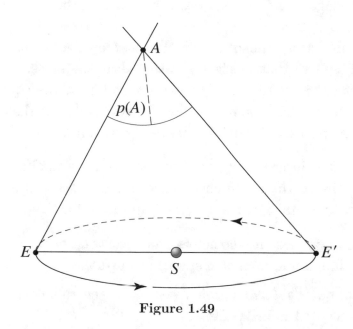

Figure 1.49

by the positions of Earth on opposing sides of its solar orbit. With parallax measurements relative to this baseline, the distances of the nearer stars can be estimated.

Figure 1.49 depicts two such positions E and E' of Earth and the baseline that they determine. A near star A is also shown along with the *angle of stellar parallax* $p(A) = \frac{1}{2}\angle EAE'$. The earlier discussion about parallax extends to the current situation and shows that the distance of A from E is approximated by $d(A,E) \approx \frac{EE'}{9.7p(A)} \times 10^6$, where $p(A)$ is given in seconds and EE' is the distance between E and E'. A standard measure of the distances to the stars is the light-year, abbreviated by ly. This is the distance that light travels in one year. Since light travels about 300,000 km in one second—refer to Chapter 1G—and one year has $365.25(86,400) \approx 31,600,000$ seconds,

$$1 \text{ ly} \approx 300,000 \cdot 31,600,000 \approx 9,480,000,000,000 \text{ km.}$$

In terms of the astronomical unit au $\approx 150,000,000$ km that we use as yardstick to measure distances within the solar system, we get

$$1 \text{ ly} \approx \frac{9,480,000,000,000}{150,000,000} = 63,200 \text{ au.}$$

Problem 1.22. The first successful computations of stellar parallax were carried out for the stars 61 Cygni, Vega, and Alpha Centauri in the late 1830s by three different astronomers working independently. History has credited Friedrich Bessel with his parallax measurement for 61 Cygni as having done it first. The modern values of the angles of stellar parallax for 61 Cygni, Vega, and Alpha Centauri are 0.29, 0.13, and 0.75 seconds, respectively. Take the Sun-Earth distance ES as 1 au and estimate the distances to these stars in light-years.

12. The Start of the Space Race. The launch of the Earth-satellite *Sputnik 1* by the Russians on October 4, 1957 was a dramatic event that inaugurated both the Space Age and the Space Race. Many short wave radio buffs heard its scratchy beeps as it circled the Earth. The response of the United Sates to this challenge to its prestige was the launch of the Earth-satellite *Explorer 1* four months later, on January 31, 1958.

The next two problems study the orbits of *Sputnik 1* and *Explorer 1*. As has already been observed, the Earth is a sphere that is slightly flattened at the poles. So the distance from the Earth's center to its surface varies slightly. The average radius at the equator is 6,378 kilometers and the radius at the poles is 6,357 kilometers. For the two problems that follow, use the average radius of 6,371 kilometers and keep the units consistently within MKS.

Problem 1.23. *Sputnik 1*'s elliptical orbit varied in distance from the Earth's surface from 230 km to 942 km. The satellite circled the Earth once every 96.2 minutes and had a mass of 83.6 kg. It remained in orbit until early in 1958, when it burned up in the Earth's atmosphere.

i. Compute the semimajor axis in kilometers and eccentricity ε of *Sputnik 1*'s orbit as well as the satellite's maximum and minimum speeds in km/sec.

ii. Use information about *Sputnik 1*'s orbit to estimate the value GM for M the mass of the Earth. (Today's accepted value is $GM = 3.986 \times 10^{14}$ m^3/sec^2.)

Problem 1.24. The satellite *Explorer 1*, was launched from Cape Canaveral in Florida, The instruments that it carried comprised about 60% of its total mass of 13.92 kg and included a cosmic-ray detection package, temperature sensors, and micrometeorite erosion gauges. The data they collected were transmitted back to Earth. The orbit of *Explorer 1* took it from a distance of 360 km to a distance of 2,534 km above the Earth's surface. Its orbital period was 114.9 minutes. Solve the previous problem again using the orbital data for *Explorer 1*.

Exploring the Solar System

In the last part of the 20th century, the American and Russian space programs began to concentrate on studies of the planets. Many satellites were placed into Earth orbit and many unmanned missions were sent to explore the inner and outer planets of the solar system. Earth satellites have included spy satellites, satellites that monitor the atmosphere, assess climate and natural resources, and satellites that study the Earth's gravitational and magnetic fields. The exploration of the planets, for example those of Venus by the Russian *Venera* program and Mars by the American *Mariner* program, involved flybys, studies from orbit, and vehicles roaming on their surfaces. The American space effort continues to be organized by the National Aeronautics and Space Administration (NASA) with engineering and scientific expertise provided by the Jet Propulsion Laboratory (JPL) of the California Institute of Technology and the Applied Physics Laboratory (APL) of the Johns Hopkins University. The Russian Space Agency *Roscosmos* has been responsible for the Russian space science and aerospace research programs. A significant development has been the fact that the exploration of space is now no longer driven by the exclusive rivalry between American and Russian space programs, but that it has become an international effort with substantial contributions from the European Space Agency (ESA) and the emerging space programs of Japan, China, and India.

An important example of international cooperation has been the development of the *International Space Station* (ISS). It is a research laboratory in low Earth orbit that has been in operation since its launch in 1998. The ISS is a structure consisting of an assembly of pressurized modular components that has grown to the size of a football field. It serves as a multinational research laboratory in which international crews conduct experiments in biology, physiology, physics, astronomy, meteorology, and a number of other fields in an essentially weightless space environment. Circling the Earth about 400 km above its surface, it is the largest man-made object in space. Five American *Space Shuttles* have been important carriers of scientific equipment into Earth orbit. These reusable, orbit-capable craft with a cargo capacity of around 23,000 kg moved Earth satellites into orbit and lifted the modular components of the ISS into place until the termination of their mission in 2011. They also provided supplies for the ISS and rotated crews. See Figure 2.1. These tasks have since been taken on by Russian rockets and the *Soyuz* spacecraft and the rockets and *Dragon* spacecraft developed by the private American aerospace company *SpaceX*.

The exploration of Mars and Venus has continued with orbiting spacecraft and surface rovers that have gathered substantial information explaining essential aspects of their surfaces, their atmosphere, and their geologic history. Mercury was the last of the inner planets to be studied. In 2011,

© Alexander J. Hahn 2020
A. J. Hahn (ed.), *Basic Calculus of Planetary Orbits and Interplanetary Flight*,
https://doi.org/10.1007/978-3-030-24868-0_2

the spacecraft *MESSENGER* was sent to orbit the planet and to make a careful and extensive photographic record of its terrain. The study of more distant reaches of the solar system began in the early 1970s with the flybys of Jupiter and Saturn by the *Pioneer 10* and *11* missions. This was followed by the journeys of *Voyager 1* and *Voyager 2*. These two spacecraft investigated Jupiter and Saturn in the late 1970s and early 1980s, sending back many incredible images of swirling storms on Jupiter's surface, active volcanoes on one of Jupiter's moons, and intricate and surprising details of Saturn and its rings and moons. *Voyager 2* went on to Uranus and Neptune, and is still the only spacecraft to have visited these planets. Having already traveled farther than any spacecraft—they are now more than 18 billion kilometers from the Sun—the *Voyagers* are on route to escape the solar system. The *Galileo* mission explored Jupiter and its moon system more closely. Launched in 1989, *Galileo* went on a six year interplanetary cruise to Jupiter and studied several asteroids along the way. The craft *Cassini*, named after the 17th century Italian–French astronomer Giovanni Domenico Cassini, was launched in 1997 and reached Saturn in 2004. It spent 13 years in orbit, gathering information about Saturn's complex ring system and its moons. The data that *Galileo* and *Cassini* gathered have provided evidence that some moons of Jupiter and Saturn have large subsurface oceans of water with organic materials, raising the possibility that they might be able to support some form of life. During the last decade, the spacecraft *Dawn* was sent to study the asteroids Ceres and Vesta, *New Horizons* went off to investigate the distant Pluto, the *Rosetta* craft came to a soft landing on a comet, and *Juno* followed *Galileo* to study Jupiter and its environment.

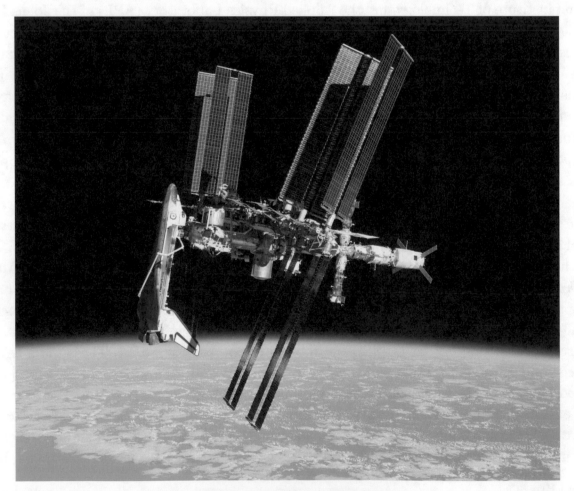

Figure 2.1. The space shuttle *Endeavor* docked at the International Space Station. Photo taken by an astronaut returning to Earth on a Russian *Soyuz* capsule just after it undocked in May of 2011. Image credit: NASA/ESA.

This chapter presents some of the many amazing images and a some of the incredible wealth of data that these and other spacecraft have sent back to Earth. The chapter concludes with a mathematical study of some basic aspects of the engines that drive these craft and considers the basic strategies that underlie the design of their flight paths.

2A. Rockets, Spacecraft, and the _Hubble_. In 1990, a space shuttle carried the _Hubble Space Telescope_ into a nearly circular orbit about 550 km above Earth's surface. The telescope was named after the American astronomer Edwin Hubble who made the dramatic discovery in 1929 that the universe is expanding. This telescope is a 11,000 kilogram silver cylinder that is 13.2 meters long with a radius of 1.2 meters. See Figure 2.2. Its instruments detect not only visible

Figure 2.2. The cylindrical _Hubble_ telescope attached to the cargo bay of a space shuttle. In this first shuttle mission to the _Hubble_ telescope, astronauts installed a set of specialized lenses to correct its initially defective main mirror. Image credit: NASA/JPL-Caltech.

light, but also light in the ultraviolet and infrared range. The *Hubble* has taken thousands of incredible high resolution images, free from the disturbances of Earth's atmosphere and interfering background light. It has provided not only remarkable close up images of objects in the solar system, but has also looked deeply and probingly at distant galaxies. Its observations have resulted in significant advances in astrophysics, including an accurate assessment of the rate of expansion of the universe as well as its age. Between 1993 and 2002, four space shuttle trips repaired, upgraded, and replaced systems of the telescope. A final service mission was completed in 2009 and the telescope will remain in operation until the early 2020s.

The journey of a spacecraft to a planet, asteroid, or comet typically begins on the launch pad of a multi-stage rocket with the craft perched on top within the upper stage. See Figure 2.3. The rocket's

Figure 2.3 Launch of *Dawn* on a *Delta II* rocket. Image credit: NASA/JPL/ESA and the Kennedy Space Center.

engines fire, it lifts off, the lower stages are jettisoned after their fuel is spent, and the spacecraft is placed into a low Earth parking orbit. After a final boost and the separation of the last stage, the craft is sent speeding along on its near-Earth orbit around the Sun. This initial solar orbit is then modified, usually by an intricate sequence of maneuvers that send the spacecraft on its mission. The journey of the craft *Juno* is a recent example. Launched in August 2011 from Cape Canaveral Air Force Station, Florida, *Juno* was sent to explore the planet Jupiter. See Figure 2.4. We will

Figure 2.4. An artist's rendition depicts the spacecraft *Juno* with its main engine firing and the surface of Jupiter in the background. Huge solar panels provide electrical power for the craft. Image credit: NASA/JPL-Caltech.

study some of the essential details of *Juno*'s flight in the last section of this chapter.

We now turn to have a look at some of the treasure trove of information that spacecraft have gathered about the solar system. This includes amazing images and very precise measurements of the sizes and masses of the planets, their moons, and some asteroids and comets, as well as the semimajor axes, eccentricities, and periods of their orbits. The instrumentation and telecommunication systems of spacecraft in orbit around or in flyby near a planet, moon, asteroid, or comet have been the primary source for this information. The radar signals that a spacecraft bounces off an object can be studied. So can the radio signals that a craft sends back to Earth. Since these signals travel with the same 299,792,458 meters per second as the speed of light, they can be timed to provide the corresponding distances with precision. Data about the orbit of a craft around a planet, moon, asteroid, or comet, or about the deflection of a craft's flyby trajectory near them, can be analyzed to give precise information about the masses of these bodies. Accurate measurements can also be made directly from Earth. For instance, a radar pulse directed from Earth to Venus bounces back as a detectable echo and can be used to give a sharp estimate of the distance between the two planets at the time of the measurement. Important additional information is provided by observations made with today's powerful, ground-based telescopes that are equipped with precision optics that dynamically correct for atmospheric conditions and disturbances.

2B. The Inner Planets. This is a discussion about the rocky inner planets of the solar system and their moons. The Earth and our Moon are considered first, then Mars and its two tiny moons, and finally, Venus and Mercury (that have no moons). These planets are small, dense, and metal-rich.

Late in the year 1972, a few hours into the flight of *Apollo 17*, one of the crewmen looked out the window. What he saw inspired him to grab a camera and snap a picture. After the craft

Figure 2.5. "Blue Marble" Earth. Image credit: NASA, NOAA/USAF/DSCOVR.

returned safely ten days later, a technician processed the film and looked at a photograph of the whole, fully illuminated, astonishingly beautiful Earth. This "blue marble" image of Earth was the first photograph taken of the whole, round Earth. It created an immediate sensation and was printed on the front page of nearly every newspaper world-wide. There are now many versions of the blue marble image. The one of Figure 2.5 was taken by a camera on NASA's Deep Space Climate Observatory (*DSCOVR*) in 2015 from about one and a half million kilometers away. North and Central America are visible near the center of the image. The shallow seas around the Caribbean islands have a turquoise hue. The image shows the effects of sunlight scattered by air molecules

Figure 2.6. A supermoon in November of 2016 captured by the *Lunar Reconnaissance Orbiter*. Image credit: NASA's Goddard Space Flight Center/Clare Skelly.

that give it its characteristic bluish tint. *DISCOVR* is a satellite in orbit around a gravitational equilibrium point between Earth and Sun. From its vantage point it has a continuous view of the sunlit half of Earth so that it can monitor the changing makeup of Earth's atmosphere. It also surveys the Sun and sends back information about solar emissions and solar flares.

The leading theory of the origin of our Moon is that a massive body several times its size collided with Earth about 4.5 billion years ago. The resulting debris from both Earth and the impacting body accumulated in a molten state to form the Moon. Within about 100 million years, this mass had crystallized, eventually forming the lunar crust. Since the Moon's atmosphere was too thin to protect it, a steady rain of asteroids, meteoroids, and comets over hundreds of millions of years have ground up its surface into a rubble pile of charcoal-gray dust and rocky debris. The *Apollo* missions provided an extraordinary look at the dusty, rock-strewn, and desolate place our Moon is.

When the Moon is full at a time it when makes its closest pass to Earth, it appears to be up to 14% bigger and 30% brighter and is called a "supermoon." Figure 2.6 is an image of a spectacular

Figure 2.7. The orbiting Moon as captured by a Chinese *Chang'e* spacecraft in October 2014. Compare the Earth's sparkling brilliance with the Moon's dull glow. Image credit: Chinese National Space Administration, Xinhuanet.

supermoon captured in November 2016. The Moon has a radius of 1,740 km. Its solid, iron-rich inner core is 240 km in radius. It makes a complete orbit around Earth in about 27 Earth days and spins around its axis of rotation exactly once during this time. So it is the same hemisphere of the Moon that always faces Earth. The opposite hemisphere—never seen from Earth—is often referred to as the "far side" of the Moon. The far side of the Moon was seen for the first time in 1959 when the Russian *Luna 3* spacecraft returned the first images. The light and dark areas of the Moon represent rocks of different composition and age. Figures 2.6 and 2.7 tell us that the large dark plains of basalt—rock formed by ancient volcanic eruptions—are much more prevalent on the Moon's visible side. Since the Moon has no natural satellite whose orbits could be observed, its mass was difficult to establish. Only after the command-service modules of several of the Apollo missions that orbited the Moon in the late 1960s and early 1970s returned precise data about their orbits was the Moon's mass determined accurately.[1] The Chinese National Space Administration is currently embarking on an ambitious Moon exploration program with end goal to return lunar rock samples to Earth. Named *Chang'e* after a mythological Chinese Moon goddess, the program made history in the early days of January in 2019 with the first ever soft landing of a rover on the far side of the Moon. An earlier craft had orbited the Moon and taken the image of Figure 2.7.

The "red" planet Mars has received a lot of attention. More than half-a-dozen rolling scientific

Figure 2.8. *Curiosity* takes a selfie on Mars. Image credit: NASA/JPL-Caltech/MSSS.

---

[1]See the paragraph *The Mass of the Moon* in the Problems and Discussions section of Chapter 1 for more about the mass of the Moon and paragraph *The Race to the Moon* in the Problems and Discussions sections of the current chapter for more information about the incredible Apollo Program.

laboratories have been sent to study its surface and atmosphere and to look for preconditions and possible signs of life.[2] With regard to life, even primitive life, these explorations have thus far been negative, but they have determined that beneath a region of one of Mars's cracked and pitted plains there is about as much water as in Lake Superior, the largest of the Great Lakes. The rover *Curiosity* is one of the most recent and most scientifically advanced vehicles ever sent to another planet. It has been crawling around Mars for several years now. Figure 2.8 shows a composite of

Figure 2.9. *Hubble* tracks Phobos around Mars in May of 2016. Image credit: NASA's Goddard Space Flight Center.

dozens of images taken by one of *Curiosity*'s cameras in August of 2015. The camera is mounted at the end of the rover's robotic arm. As with any selfie, only a part of the arm holding the camera is included in the photo (the upward pointing segment in the middle) but shadows of the rest of the rover's robotic arm are visible on the ground. The fact that the rover's wheels are about

---

[2]It bears pointing out that life is incredibly adaptable. In the years 2011–2015, scientists made the surprising discovery of never-before-seen multi-cellular life forms, including worms and crustaceans, in pockets of water 5,000 years old in South African goldmines that are 2 miles below Earth's surface. They exist without sunlight in unbearable heat. See https://www.cbsnews.com/news/looking-for-life-on-mars-at-the-bottom-of-a-south-africa-gold-mine/.

50 centimeters in diameter and 40 centimeters wide gives a sense of the size of things. *Curiosity* is looking out from the crest of a rocky 6-meter hill that it climbed. A mountain is visible at some distance behind it. The iron in the dusty soil paints Mars's surface in its reddish color.

The dry lake and river beds that *Curiosity* roamed over give evidence that liquid water once flowed on Mars's surface. In one of the lake beds, the instrumentation that *Curiosity* carries identified organic molecules in rocks billions of years old. They point to the possibility that some life forms exist, or may have existed, on Mars. The thin atmosphere of Mars is composed mostly of carbon dioxide. The recent detection of methane in the atmosphere and the fact that Mars exhibits seasonal cycles indicates the planet is alive at least in a geologic sense. The fact that its thin atmosphere caused the the planet's climate to cool left most of its water locked up in ice. Recently, a radar instrument on board the European Space Agency's (ESA) orbiter *Mars Express* found evidence of an existing body of liquid water. A lake about 20 km across is believed to be embedded under the planet's south polar ice cap.

Mars has two small moons, both discovered in 1877. Their names Phobos and Deimos come from Greek mythology. They are the twins representing fear and terror who accompanied their father Ares, the god of war, into battle. The Greek god Ares is the god Mars of the Roman pantheon. The *Hubble Space Telescope* took images of Mars—see Figure 2.9—that captured Phobos during its orbital trek. Over the course of 22 minutes, *Hubble* took 13 separate exposures, allowing astronomers to create a time-lapse video showing a part of the little moon's orbit. Because the moon is tiny, it appears star-like in *Hubble*'s picture sequence. Figure 2.10 provides images of the

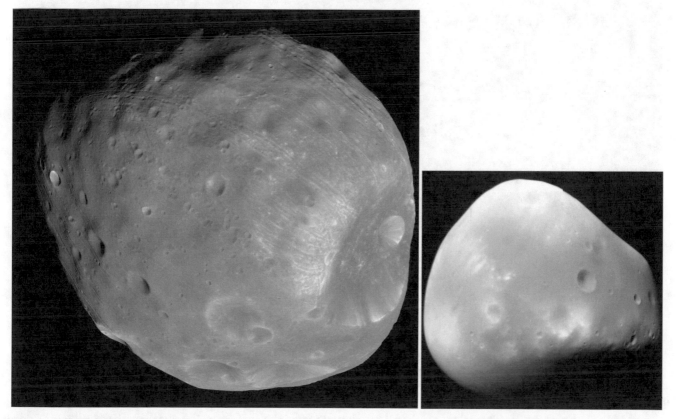

Figure 2.10. The images of Phobos on the left and Deimos below were taken by the high resolution imaging camera of the *Mars Reconnaissance Orbiter* in April 2008 and March 2009. Image credit: HiRISE, MRO, LPL (U. of Arizona), NASA.

two moons in roughly their relative sizes. Since they look like asteroids, Phobos and Deimos, were long thought to have been asteroids that were captured by Mars's gravitational pull. But more recent observations of their compositions and orbits suggest that they may have formed from debris generated by the impact of one or more larger bodies with Mars.

Venus is the second planet from the Sun. It is Earth's closest planetary neighbor and is similar to Earth in size. Unlike Earth, however, it has no moons. Since it is always covered by a thick, unbroken veil of clouds, its surface could not be seen from Earth even with the most powerful

Figure 2.11. This high resolution image of the surface of Venus was computer generated using radar data from the orbiting Magellan spacecraft. The colors are based on pictures from the surface of Venus transmitted by two Russian *Venera* landers. Image credit: NASA/JPL-Caltech, Magellan Project.

telescopes. But in the early 1990s, the imaging radar of the *Magellan* spacecraft and earlier color pictures transmitted by Russian *Venera* landers combined to produce spectacular high resolution computer generated images of the planet and its surface. Figure 2.11 is one of them. Venus's surface is covered with craters, volcanoes, mountains, and big lava plains. The bright area that extends around the middle of the planet represents a large highland region. The atmosphere of Venus consists mainly of carbon dioxide along with clouds of sulfuric acid droplets. It traps the heat of the Sun and the heat that the planet releases and creates a greenhouse-like effect that makes Venus the hottest planet in our solar system. Its surface temperatures—higher than 470° Celsius (880 degrees Fahrenheit)—are hot enough to melt lead. The few spacecraft that have been sent to soft landings on Venus were able to transmit valuable information about its surface. But having to operate in the scorching heat, the instrumentation of these craft quickly failed, so that these transmissions were brief.

The planet Mercury has been known since recorded history began, but until the spacecraft *MESSENGER* (derived from the phrases MErcury Surface, Space ENvironment, GEochemistry and Ranging that describe its mission) went to study it in 2011, it was the least understood of the inner planets. The high resolution image of Figure 2.12 was recorded when *MESSENGER* looked

Figure 2.12. Mercury as captured by *MESSENGER*. Image credit: MESSENGER, NASA, JHU APL, CIW.

down on the planet from an altitude of about 27,000 km on one of its early flybys. In March of 2011, *MESSENGER*'s main engine fired to slow it down and to enter it into orbit around Mercury. Once in orbit, it became *MESSENGER*'s primary mission to gain a broad scientific perspective on the solar system's innermost planet. This included information about its core and its surface. Mercury with its large metallic core—it has a radius of about 2,000 km or about 80% of the planet's radius—is the second densest planet in the solar system after Earth. Data from *MESSENGER*'s instruments provided evidence that this core consists in part of liquid iron. In November 2012, *MESSENGER* discovered both water ice and organic compounds in craters in the shadows near Mercury's north pole. In February 2013, NASA assembled a highly detailed 3D map of Mercury from the thousands of images that the craft had captured. This photographic record revealed with great accuracy how the planet's surface was shaped and scarred by many collisions with meteoroids and comets and by extensive volcanic activity in the past. Its surface has large areas of relatively smooth terrain, but it also features large stretches of cliffs and escarpments, some of them hundreds of miles long and up to a mile high. They were created after Mercury's formation as the planet's interior cooled and contracted. In April of 2015, after having completed more than 4,000 orbits of the planet, *MESSENGER* ran out of fuel and carried out a programmed crash onto Mercury's surface.

There had been concerns about the feasibility of sending a spacecraft to Mercury and inserting it into orbit around it. Since Mercury orbits the Sun closely, a spacecraft sent in the direction of Mercury would move in the general direction of the Sun, and would therefore be accelerated by the Sun's strong gravitational field. The craft's flight path from its initial solar orbit to Mercury would therefore have to be carefully conceived. The thinking was that its execution would involve the extensive use of its thrusters and consume appreciable amounts of propellant. Extra propellant means an increase in the weight of the craft and a more powerful—and more costly—rocket at launch. We will see in the Problems and Discussions section of this chapter how the ingenious design of *MESSENGER*'s trajectory got around these potentially mission-preventing problems.

In addition to the spectacular images and geologic information that spacecraft have provided about the planets and their moons, their telemetric systems have also sent back accurate quantitative data about them and their orbits. Table 2.1 collects basic data about the orbits of the inner planets and Table 2.2 does so for their moons. Recall that neither Venus nor Mercury have moons. Table 2.3 lists the sizes of the planets and their moons along with their masses. All this information is taken from NASA/JPL websites. The values for *GM* are obtained by inserting orbital data of both natural

Table 2.1. Orbital data from NASA/JPL websites, e.g., https://solarsystem.nasa.gov/planet-compare/. The orbit periods are the sidereal periods (defined in Chapter 1G) in Earth years of 365.25 days, and "angle of orbit plane to Earth's" refers to the angle between the orbit plane of the planet and the orbit plane of Earth (defined in Chapter 6N).

Planet	semimajor axis in km	in au	eccentricity	orbit period in years	average speed in km/sec	angle of orbit plane to Earth's
Mercury	57,909,227	0.387	0.20563593	0.2408467	47.362	7.00°
Venus	108,209,475	0.723	0.00677672	0.6151973	35.021	3.39°
Earth	149,598,262	1.000	0.01671123	1.0000174	29.783	0.00°
Mars	227,943,824	1.524	0.0933941	1.8808476	24.077	1.85°

and artificial satellites into Newton's version of Kepler's third law. The estimates for the masses of the planets and moons are obtained by inserting a standard value for G. See Chapter 1H in this regard. As the planetary systems move through the solar system and attract each other, their basic

Table 2.2. Orbital data for the moons from NASA/JPL websites, e.g., https://ssd.jpl.nasa.gov/?sat_elem.

Planet	moon	semimajor axis in km	eccentricity	orbit period in Earth days
Earth	Moon	384,400	0.549	27.322
Mars	Phobos	9,378	0.0151	0.31891
	Deimos	23,459	0.0005	1.26244

quantitative characteristics change slightly—very slightly—over time. So the numerical parameters listed in the tables are continually remeasured. The values that are presented for them often differ

Table 2.3. Size and mass data taken from NASA/JPL websites and W. M. Folkner, J. G. Williams, D. H. Boggs, R. S. Park, and P. Kuchynka, The Planetary and Lunar Ephemerides DE430 and DE431, *The Interplanetary Network Progress Report*, vol 42–196, February 15, 2014, https://ipnpr.jpl.nasa.gov/progress_report/42-196/196C.pdf (see Table 8). For the moons refer to https://ssd.jpl.nasa.gov/?sat_phys_par.

	diameter in km	precise: GM in m^3/sec^2	estimate: mass M in kg
Mercury	4879	$2.20317800 \times 10^{13}$	3.30×10^{23}
Venus	12,104	$3.24858592 \times 10^{14}$	4.87×10^{24}
Earth	12,756	$3.98600435 \times 10^{14}$	5.97×10^{24}
Moon	3475	4.90280×10^{12}	7.35×10^{22}
Mars	3933	$4.28283752 \times 10^{13}$	6.42×10^{23}
Phobos	23	7.127×10^{5}	1.07×10^{16}
Deimos	12	1.01×10^{5}	1.48×10^{15}

a little from one listing to the next.

2C. The Outer Planets. The outer planets Jupiter, Saturn, Uranus, and Neptune in order of their increasing distances from the Sun are much larger and less dense than the inner planets. They are rich in hydrogen and helium and are often referred to as gas giants.

A look at Figure 2.13 tells us that the planet Jupiter is a colorful ball of wind-driven bands of clouds and swirling storms in hues of white, gray, orange, and brown. During an early investigation of Jupiter's upper atmosphere, astronomers studied Jupiter's eclipse of the moon Ganymede depicted in the figure. As Ganymede emerges from behind Jupiter, the sunlight that it reflects flows through Jupiter's atmosphere. The fact that some of this light is blocked provided information about the density of Jupiter's cloud cover. As Ganymede continues its emergence, this reflected light brightens progressively and the study of the changes in the various colors informed scientists about the composition of Jupiter's upper atmosphere. Visible in the center of the figure is Jupiter's great red spot. This is a huge high pressure cyclone that is trapped between two atmospheric jet streams.

It has winds of close to 600 km per hour and it is twice as wide as the entire Earth. Telescopic evidence over the years has confirmed that the great red spot has been churning in Jupiter's skies for hundreds of years. Up to 90% of Jupiter's swirling atmosphere consists of hydrogen gas. Most of the remaining 10% is helium with smaller amounts of ammonia. Hydrogen predominates in the interior of the planet as well. Scientists think that the pressure is so great at depths of about halfway to the planet's center, that hydrogen assumes a metal-like state there. They believe that Jupiter's fast rotation turns this region into a dynamo that generates the electrical currents that drive the planet's powerful magnetic field.

The four moons that Galileo discovered early in the 17th century are Jupiter's four largest. Soon after their discovery, a German astronomer turned to Greek mythology to name them Ganymede,

Figure 2.13. The *Hubble Space Telescope* catches Jupiter as it eclipses its moon Ganymede. Ganymede is the largest moon in the solar system. Image credit: NASA/ESA/LPL(U. of Arizona).

Europa, Io, and Callisto after the friends and lovers of Zeus (the Greek name of the Roman god Jupiter). These names came into general use only much later.

The *Galileo* spacecraft was launched by NASA in 1989 with mission to explore Jupiter and its moon system closely. It took the craft six years to reach the planet. Once in orbit, *Galileo* flew past Jupiter and its moons for eight years before it was sent into a programmed suicidal dive into Jupiter's atmosphere. Over sixty more moons—all tiny—have now been identified in orbit around the planet. The four that Galileo had discovered contain in excess of 99% of the total mass of all of Jupiter's moons. The most exotic of Jupiter's moons is Io. With its hundreds of active volcanoes it is the most geologically alive object in the solar system. The probing cameras and instruments of the *Galileo* spacecraft have provided much information about Io's surface. Most of the surface consists of extensive plains coated with sulfur and sulfur-dioxide frost. Volcanoes have created lava

Figure 2.14. *Galileo*'s high resolution image of Io from 1999. It is a mosaic made with the camera's infrared, green, and violet filters and approximates what the human eye would see. Image credit: NASA/JPL/LPL(U. of Arizona).

flows that are hundreds of kilometers long. Several of the dark, flow-like features correspond to hot spots on Io's surface and may be active lava flows. Many rugged mountains, uplifted by the extensive compression of Io's crust—some of them taller than Earth's tallest peaks—dot its sulfuric plains. There are scores of plateaus of layered materials and many large bowl-shaped volcanic depressions. Some of the volcanoes that rise from these plains spew plumes of sulfuric materials as high as 500 km into Io's thin, patchy atmosphere. There are no visible pitted scars or craters of the sort that meteorites or asteroids form on impact, suggesting that flows of volcanic materials covered them with new deposits as they occurred. The craft's high resolution camera painted an image of Io that shows a sulfuric yellow surface with the splashes of black, brown, green, orange, and red that Io's active volcanoes have colored in. See Figure 2.14. Io looks like an abstract golden spherical Christmas ornament as it draws its nearly circular orbits around Jupiter.

Even more interesting is Jupiter's moon Europa, like Io roughly the size of Earth's moon. Let's start with a very brief description of one of the leading hypotheses for the origin of life as it began on Earth over three billion years ago. In places on the ocean floor the Earth's crust was being pulled apart, causing mountain ridges to rise. On some of these ridges, fields of vents formed from which non-acidic, hot water burst through. Carbonate mineral deposits from this water grew over time into steep, white "chimneys" that rose from the sea floor like organ pipes. The rocks of the vents were porous and packed with tiny holes filled with water. These pockets contained essential chemicals and minerals and were ideal places for metabolic processes to begin. They acted as "cells". The heat, the chemicals, and the water acting and interacting over long time spans could have produced the molecules that are critical to life as we know it. Eventually these pockets created their own membranes, became true cells, and escaped from the porous rock into the open water. In time, these chimneys became home to dense communities of microorganisms that thrived in the water of the vents. Now back to the moon Europa. The *Galileo* mission has uncovered strong evidence that Europa has a deep ocean of liquid water beneath its icy shell. Scientists have long considered it possible that there might also be volcanic activity on Europa and that there could be vents through which mineral-laden hot water could emerge from the sea floor. Could something like the chemical and biological dynamic that—hypothetically—led to life within Earth's oceans be possible on Europa? Given its abundant water, rocky sea floor, and the energy and chemistry of hot vents and volcanic activity, might Europa have what it takes to support simple organisms?

Jupiter and its moons are again the objects of intense exploration. The craft *Juno* was launched in 2011, reached Jupiter in July of 2016 and achieved orbit around the planet. Its camera began to take high resolution images of its surface—see Figure 2.15—and its scientific instruments began to peer below the dense cover of clouds to assess Jupiter's gravitational force, magnetic field, radiation belts, and atmospheric dynamics. In its flights over Jupiter's wind-driven weather systems, *Juno* measured the variations in Jupiter's gravitational pull. These measurements gave the *Juno* science team insight into the movement of the masses in Jupiter's interior and in turn information about its structure. Recent data has revealed that Jupiter's colorful, wind-sculpted bands extend 3,000 km deep into its interior. Even though this is much less than the 70,000 km of the planet's radius, this meant that the weather layer of Jupiter is significantly more massive and deep than had been previously thought. It is estimated that it contains about one percent of Jupiter's mass. By contrast,

Figure 2.15. This image was taken taken in April of 2018 at periapsis of *Juno*'s 12th orbit. The spacecraft was about 16,500 km from Jupiter's surface at the time. The image extends from the great red spot to the dynamic bands in the direction of the south pole. Image credit: NASA/JPL-Caltech/SwRI/MSSS/Kevin M. Gill.

Earth's atmosphere is less than one millionth of Earth's total mass. The question as to whether Jupiter has a rocky core remains open, but indications are that it does. *Juno*'s mission has been extended until July 2021. The space agency NASA is planning to send another craft to Jupiter in the 2020s on a mission to look for answers to the question about the possible existence of some form of life on the moon Europa. During repeated close flybys, the craft's sophisticated instruments are to image Europa's icy shell and investigate the interior of its ocean.

The *Cassini* mission to Saturn was an international collaboration of NASA with the European and Italian Space Agencies. The spacecraft was launched in 1997 and achieved orbit around Saturn in 2004. It remained in orbit for 13 years gathering information about the planet. The flight of *Cassini* has informed all aspects of what we know about Saturn and has completely revised our understanding of its ring system and its moons.

Saturn's basic structural features are similar to those of Jupiter. Its low average density gives evidence that it consists mostly of hydrogen. Its atmosphere consists of about 95% hydrogen gas with the rest mostly helium. Ammonia crystals in its upper atmosphere give it its pale yellow hue. Saturn's atmosphere exhibits a banded pattern similar to Jupiter's, but Saturn's bands are much fainter and calmer and lack the multicolored intensity of Jupiter's bands. Wind speeds on Saturn can reach 1,800 km per hour. Occasional large oval cyclones have been observed on Saturn for some time and *Cassini* was witness to a huge storm that churned around and encircled the planet's northern hemisphere. Saturn's north pole is dominated by a large bluish hexagon. See Figure 2.16. This stable pattern, first observed in 1988 by scientists who analyzed the data provided by the flybys of the *Voyagers* in 1980 and 1981, has a massive hurricane (many times larger than the largest hurricanes on Earth) whirling inside it. The eye of the hurricane is located at the pole, and the boundary of the hexagon is formed by a curving jet stream. The shape is influenced by the turbulence of the flow that swirls between fluid masses rotating at different speeds. Saturn's gravitational field—information about it comes from a study of the way it pulls on its moons and rings as well as the deflections it causes in the path of the orbiting *Cassini*—provided insight about its interior. At a depth of about 1,000 km below the clouds and at a temperature of about 700° Celsius, the planet's hydrogen behaves like a liquid rather than a gas. At a distance about halfway between Saturn's cloud tops and its center the hydrogen is believed to be in a fluid metallic state at a temperature of about 5700° Celsius. Electrical currents within this metallic hydrogen layer are believed to be the source of Saturn's magnetic field. This magnetic field is only about 5% of the strength of Jupiter's. A metal and rock mixture of possibly 10 to 20 Earth masses is thought to form the planet's dense central core.

Saturn's configuration of rings is one of the most stunning astronomical sights in the solar system. This ring system is around 250,000 km wide but only a few tens of meters thick. Consult Figures 2.16 and 2.17. The rings consist of billions of particles, ranging in size from grains of sand to large boulders. They are largely ice-particles, but the rings also draw in rocky fragments from their travels through the solar system. As Figure 2.16 illustrates, the ring system is divided into seven distinct segments. The standard designation for the ring segments from the innermost to the outermost are given by the letters D, C, B, followed by a gap, and then by the letters A, F, G, and E. The fact that the rings have different densities is apparent from both Figure 2.16 (the denser rings are brighter) and Figure 2.17 (the denser rings cast darker shadows). The gap is visible as a black ring known as the *Cassini Division*. It separates the B ring (the brightest and densest of the rings) from the A ring. It is caused by the gravitational pull of Saturn's moon Mimas. The moon Thetys

Figure 2.16. In November of 2012, Saturn's north polar hexagon basked in the Sun's light. Many smaller storms dot the north polar region. Saturn's signature rings—broken by the shadow that Saturn casts—surround the planet. Image credit: NASA/JPL-Caltech/Space Science Institute, *Cassini* imaging team.

orbits at a distance of around 300,000 km from Saturn's center, well outside Saturn's main bright rings near the middle of the very faint outer E ring.

The Dutch scientist Christiaan Huygens discovered the first moon of Saturn in the year 1655. It was later named Titan. Cassini (the astronomer, not the spacecraft) made the next four discoveries: Iapetus in 1671, Rhea in 1672, Dione in 1684, and Tethys in 1684. Mimas and Enceladus were both discovered by William Herschel in 1789. These seven major moons together comprise over 99% of

Figure 2.17. The rings of Saturn are accompanied by the moon Thetys. The shadows they cast give a sense of their stratification. The figure also shows how thin the rings are. APOD: July 22, 2005. Image credit: NASA/JPL-Caltech/Space Science Institute, *Cassini* imaging team.

the total mass of all of Saturn's sixty-some moons. Titan is by far the largest of the moons. Slightly larger than the planet Mercury, it alone contains more than 96% of the total mass of Saturn's moons. Its gravitational force affects the orbits of nearby moons. The *Cassini* craft used Titan's gravitational pull to perform fuel-free corrections of its orbit. The moons of Saturn took their names from the Titans of Greek mythology. With the god Kronos—Saturn to the Romans—in the lead, the Titans had battled the Olympian gods unsuccessfully for control of the Greek pantheon. The moons discovered later were named after other Roman and Greek mythological characters.

Titan and Enceladus are Saturn's most interesting moons and have received much attention. In 2005 *Cassini*, already in orbit around Saturn, released a probe named *Huygens* to study Titan. The probe's camera filmed its descent before it landed successfully and softly on the moon's surface. The descent of *Huygens* in combination with *Cassini*'s many close flybys provided substantial information. It was discovered that Titan has Earth-like landscapes featuring dry river networks, steep canyons, dune lands, and even lakes and seas, and that its hazy, nitrogen-rich atmosphere with traces of methane rains complex organic chemicals to its surface. Images of Titan's surface taken by the *Huygens* probe show chunks of water ice and rounded pebbled shapes scattered over a flat, sandy orange plain. Some of the largest lakes near the north pole consist primarily of liquid ethane and methane, and are of the size of the American Great Lakes. *Cassini* and *Huygens* also found clear evidence that a global ocean of water existed beneath Titan's thick, frozen crust.

The fact that Saturn's moon Enceladus is one of the whitest and brightest objects in the solar system that reflects nearly 100 percent of the sunlight that strikes it, meant that its surface could

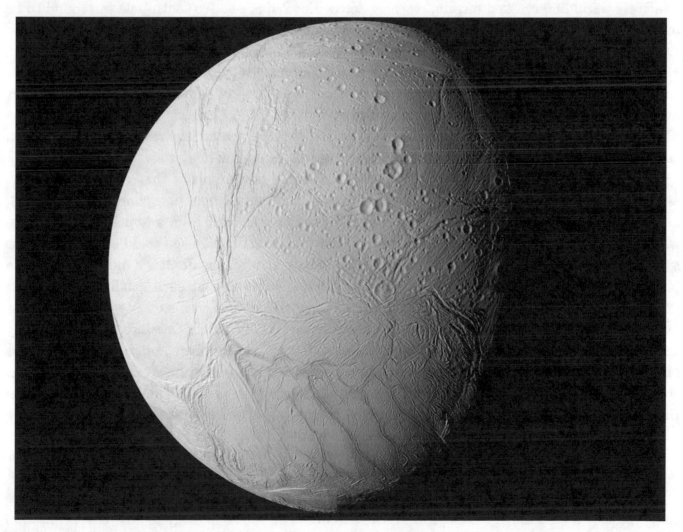

Figure 2.18. Image of Enceladus captured by *Cassini* in October 2008. The "tiger stripes" are color-enhanced in greens and blues. Image credit: NASA/JPL-Caltech/Space Science Institute, *Cassini* imaging team.

only be nearly pure, pristine water ice. See Figure 2.18. It was known from the reports of the *Voyagers* that the surface of Enceladus had relatively few craters, but *Cassini* observed that near the south pole its surface is nearly free of craters altogether. Scientists concluded that Enceladus's southern surface had to be geologically very young and that some ongoing activity was melting or covering its icy outer layer. *Cassini*'s studies of Enceladus have begun to unravel the moon's remarkable mysteries. In its orbital motion around Saturn, Enceladus always faces the same side of the planet. This means that it spins on its axis at the same rate that it orbits Saturn. By inspecting hundreds of photos from more than seven years of *Cassini*'s mission researchers tracked the moon's spin with precision and detected a wobble. This wobble, very carefully measured, confirmed that there had to be a liquid layer that separated the porous rocky core of Enceladus from its icy shell. No other internal structural configuration could explain with the same precision the dynamics of the wobble that was observed. The conclusion was that under its shell of ice, Enceladus was warm enough to sustain a global ocean. The surface near the southern pole exhibited a pattern of deep crevasses, later called "tiger stripes" and highlighted in blues and greens in the figure. This unexpectedly relatively warm and cracked terrain, reaching across Enceladus's active south pole was found to be in motion. Moved by gravitational forces, it was stretched in some places and buckled in others. To their amazement, scientists detected huge, soaring plumes of water vapor and ice particles over the warm fractures in this part of the crust. The analysis of images provided conclusive evidence that these plumes originated near the hottest spots of the tiger stripe fissures and that they sped out at about at over 1000 km per hour. During a close flyby in 2008, *Cassini*'s instruments sampled a plume directly and detected a surprising mix of volatile gases, water vapor, ice grains, hydrogen, carbon dioxide, carbon monoxide, salts and silica, as well as organic materials.

Life as we know it is thought to be possible in stable environments that offer liquid water, essential chemical elements (such as carbon, hydrogen, nitrogen, oxygen, phosphorus, and sulphur), and a source of energy (from sunlight or chemical reactions). It had been an open question as to whether conditions that might have led to life on Earth billions of years ago could exist in the solar system beyond Mars. The spacecraft *Galileo* had already shown that Jupiter's moon Europa is a place where this is the case. The discoveries by *Cassini* affirmed that such conditions also exist on Saturn's moons Titan and Enceladus. *Cassini*'s original mission was so successful that it was extended for nine additional years. Finally, in September of 2017, when *Cassini*'s fuel was entirely spent, it was sent on a programmed plunge down into Saturn's atmosphere and disappeared. It was one month shy of celebrating the 20th anniversary of its launch. While the spacecraft is no longer, the analysis of the data that *Cassini* has transmitted continues. A recent significant discovery was the detection of complex carbon-based molecules in the jets of water that emerge from the active regions of Enceladus's south pole. The study that announced the discovery asserts that "these huge molecules contain a complex network often built from hundreds of atoms" and are "the first ever detection of such complex organics coming from an extraterrestrial water-world." The study goes on to pose the important question: are these molecules the result of chemical processes only, or are they—as they are on Earth—biologically created? On Earth, the technology to answer this question already exists, so that "the next logical step is to go back to Enceladus soon with a dedicated payload and see if there is extraterrestrial life."

The two outermost planets of the solar system are Uranus and Neptune. Uranus had been observed as a moving point of light before, but it was William Herschel who identified it as a planet in 1781. The axis of rotation of Uranus was later observed to lie in the plane of its orbit so that it moves in its orbit as though it were a rolling ball. Neptune, predicted as a consequence of observed irregularities in the orbit of Uranus, was discovered in 1846. (Some details about the discovery of both Uranus and Neptune are provided in Chapter 1F.) So far, NASA's *Voyager 2* is the only spacecraft to have visited Uranus and Neptune. Most of what we now know about the two planets was gathered by its instruments during flybys that lasted only a few hours. Both of the planets are cold, dark, and very windy. Each has an atmosphere consisting mostly of hydrogen and helium, with trace amounts of methane, water, and ammonia, that gradually merge below the surface into a soup of a dense, icy, liquid mix of these same chemicals that accounts for about two-thirds of the planet's mass. The methane gas in their atmospheres gives both Uranus and Neptune their bluish color. See Figure 2.19. A solid hot core, thought to be about Earth-sized, forms each planet's center.

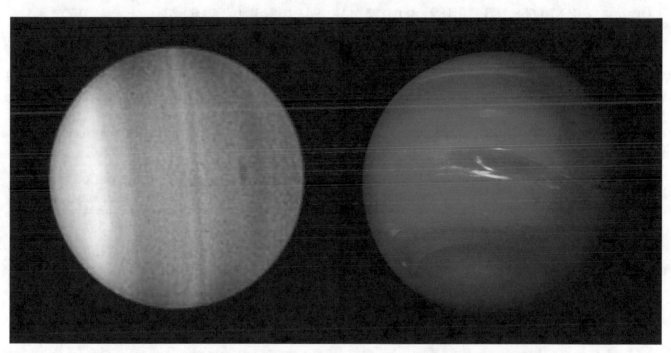

Figure 2.19. *Voyager 2* is the only spacecraft ever to cruise to Uranus and Neptune. It was near Uranus in January of 1986 and near Neptune in August of 1989. *Voyager 2*'s image of Neptune (on the right) shows the great dark spot and its companion bright smudge. Image credit: NASA/JPL-Caltech. But *Voyager 2*'s image of Uranus depicts a completely featureless bluish-green sphere. The image of Uranus shown here (on the left) is grainier, but shows some of the planet's atmospheric bands and a faint dark spot. Taken by the Advanced Camera for Surveys of *Hubble*'s space telescope in August of 2006, it is a composite of many separate exposures that use filters to capture some near-infrared wavelengths. Image credit: NASA/JPL-Caltech, Space Telescope Science Institute, and ESA

Voyager 2 has informed us that Uranus is surrounded by over a dozen faint rings and that it has more than two dozen moons swirling around it. The largest moons, Titania and Oberon (discovered by William Herschel in 1787), Umbriel, Ariel, and Miranda, are far larger than the rest. Four of these five moons (and almost all the others) are named after characters in Shakespeare's plays. *Voyager 2* observed that Neptune is surrounded by half a dozen rings and over a dozen circulating moons. Of Neptune's moons, Triton is by far the largest. Discovered in 1846 (and later named after the son of Poseidon—the Neptune of the Greek pantheon), it makes up more than 99% of the mass of Neptune's moon system. *Voyager 2* spotted a number of active geysers on Triton within the polar cap. Heated by the Sun, they eject plumes to heights of up to 8 km. The fact that *Voyager 2* detected only a few impact craters on Triton's surface confirmed the effects of its ongoing geological activity.

Some of the important quantitative information that the *Voyager, Galileo, Juno,* and *Cassini*

Table 2.4. Orbital data from NASA/JPL websites, e.g., https://solarsystem.nasa.gov/planet-compare/. The orbit periods are the sidereal periods (defined in Chapter 1G) in Earth years of 365.25 days, and "angle of orbit plane to Earth's" refers to the angle between the orbit plane of the planet and the orbit plane of Earth (defined in Chapter 6N).

Planet	semimajor axis in km	in au	eccentricity	orbit period in years	average speed in km/sec	angle of orbit plane to Earth's
Jupiter	778,340,821	5.203	0.04838624	11.862651	13.056	1.31°
Saturn	1,426,666,422	9.537	0.05386179	29.447498	9.639	2.49°
Uranus	2,870,658,186	19.189	0.04725744	84.016846	6.873	0.77°
Neptune	4,498,396,441	30.070	0.00859048	164.79132	5.435	1.77°

spacecraft have sent back about Jupiter, Saturn, Uranus, and Neptune and their many moons is collected in Tables 2.4, 2.5, and 2.6. Table 2.4 provides the basic orbital data for the four outer planets.

Table 2.5. The table lists the sizes and orbital data for the largest moons of each of the outer planets as well as Earth's moon from JPL/NASA websites, e.g., https://ssd.jpl.nasa.gov/?sat_elem.

Planet	moon	diameter in km	semimajor axis in km	eccentricity	orbit period in Earth days
Earth	Moon	3,475	384,400	0.0554	27.322
Jupiter	Ganymede	5,262	1,070,400	0.0013	7.155
	Callisto	4,820	1,882,700	0.0074	16.689
	Io	3,643	421,800	0.0041	1.769
	Europa	3,122	671,100	0.0094	3.551
Saturn	Titan	5,151	1,221,930	0.0288	15.945
	Rhea	1,527	527,108	0.0013	4.518
	Iapetus	1,469	3,560,820	0.0286	79.321
	Dione	1,123	377,396	0.0022	2.737
	Tethys	1,062	294,619	0.0001	1.888
	Enceladus	504	238,020	0.0045	1.370
	Mimas	396	185,539	0.0196	0.942
Uranus	Titania	1,577	435,910	0.0011	8.076
	Oberon	1,523	583,520	0.0014	13.463
	Umbriel	1,169	266,300	0.0039	4.144
	Ariel	1,158	191,020	0.0012	2.520
	Miranda	472	129,900	0.0013	1.413
Neptune	Triton	2,705	354,759	0.00002	5.877

The data in the column "angle of orbit plane to Earth's" of Table 2.4 together with the data in the same column of Table 2.1 tell us that all the planets orbit close to the *ecliptic*, the Earth's orbital plane. Mercury's orbit deviates the most. Its plane is separated from Earth's by 7 degrees. Similarly, the "eccentricity" columns of these two tables tell us that all planetary orbits are very nearly circles. Again, Mercury deviates the most. With its eccentricity of about 0.2, its orbit is the most elliptical. Table 2.5 provides data for the sizes and orbits of the planets' major moons. The list includes all the moons in the solar system with a diameter of over 1000 km. Table 2.6 finally, lists the sizes and masses of the four outer planets. Such accurate values for GM are obtained by

Table 2.6. These data come from NASA/JPL websites and W. M. Folkner, J. G. Williams, D. H. Boggs, R. S. Park, and P. Kuchynka, The Planetary and Lunar Ephemerides DE430 and DE431, *The Interplanetary Network Progress Report*, vol 42–196, February 15, 2014, https://ipnpr.jpl.nasa.gov/progress_report/42-196/196C.pdf (see Table 8).

Planet	diameter in km	precise: GM in m^3/sec^2	estimate: mass M in kg
Jupiter	142,984	$1.26712765 \times 10^{17}$	1.90×10^{27}
Saturn	120,536	$3.79405852 \times 10^{16}$	5.68×10^{26}
Uranus	51,118	$5.79454860 \times 10^{15}$	8.68×10^{25}
Neptune	49,528	$6.83652710 \times 10^{15}$	1.02×10^{26}

inserting orbital data as well as flyby data from spacecraft into Newton's version of Kepler's third law. The corresponding estimates for the masses M are derived by inserting a standard value for G. For these, see Chapter 1H.

As was already observed in the context of the inner planets, the outer planets and the moon systems around them change their basic quantitative characteristics very slightly over time. So the numerical parameters of the tables are continually remeasured. The values that are presented for them in the literature often differ a little from one listing to the next.

We have seen that the inner planets Mercury, Venus, Earth, and Mars are small, dense, metal-rich, and that they all orbit relatively near the Sun. They are the *rocky or terrestrial planets*. The outer planets, Jupiter, Saturn, Uranus, and Neptune are large, less dense, hydrogen-rich, and they orbit much farther from the Sun. They are the *gas giants*. While each of the outer planets have dozens of moons in orbit around it, the inner planets have a total of only three (that of Earth and the two small ones of Mars). All the moons listed had been discovered by the 1850s with the exception of Uranus's moon Miranda (discovered in 1948). The moons of all the planets—with one exception—are much less massive than the planets that they orbit around. The exception is our Moon. The ratio of its mass to Earth's mass is close to 1 to 81 (as we learned in the paragraph *The Mass of the Moon* in the Problems and Discussions section of Chapter 1). The fact that 1 au $\approx 1.5 \times 10^8$ km tells us that the planets Mercury, Venus, Mars, Jupiter, Saturn, Uranus, and Neptune are, respectively, about 0.39, 0.72, 1.5, 5.2, 9.5, 19.1, and 30 au from the Sun. This gives a better sense of the relative distances of the orbits of the planets from each other. All planets circle the Sun in the same direction as the Earth. If you were to imagine yourself high above Earth's northern hemisphere, you would see all the planets revolving counterclockwise around the Sun.

2D. About Asteroids. As already described in Chapter 1F, the solar system began as a hot cloud of gas and dust. Over large spans of time, gravity bound these materials together into larger and larger clumps to form the planets and their moons. Asteroids are the scattered debris of rocky remnants left over from this process. Our solar system contains millions of asteroids, most of which are in orbit within the *Main Asteroid Belt* between Mars and Jupiter. Over a time span of hundreds of millions of years many main belt asteroids were thrown out of their orbits when they passed near massive planets such as Jupiter and Saturn. Some were hurled in the direction of the Sun and others away from it. As a consequence, there are asteroids all over the solar system. The sizes of asteroids range from a few meters to hundreds of kilometers across. On rare occasions they impact Earth. When they do, they can cause extensive damage. A crater off the coast of the Yucatan Peninsula in Mexico is the record of a massive impact that led to the extinction of the dinosaurs 65 million years ago. Meteor Crater in Arizona was made by an impact about 50,000 years ago. Studies of Earth's geologic history tell us that about once every few thousand years an object the size of a football field smashes into Earth's surface. In recent decades there has been an organized effort to detect and track asteroids. The asteroids that have received much attention are the *Near Earth Asteroids* (NEAs), those with an orbit that brings them within 0.3 au or 45 million km of Earth's orbit. Around a thousand NEAs with a diameter of at least 1 kilometer have been identified. The *Potentially Hazardous Asteroids* (PHAs) have received even more scrutiny. The PHAs are those whose paths bring them to within 0.05 au or 7.5 million kilometers of Earth's orbit. This distance corresponds to about 20 times the average distance of 384,000 km from Earth to the Moon.

A few asteroids have been studied up close by orbiting spacecraft. The spacecraft *Dawn* was launched in 2007 to study Vesta and Ceres, two of the largest and most massive. It was the goal of the mission to focus the craft's sophisticated instrumentation on a study of these bodies in an effort to better understand the early moments of the formation of the solar system. This included an investigation of the question as to why their developments took such different paths. By July of 2011, *Dawn* had reached Vesta and had begun its 14 months in orbit around it. One surprising finding was that some of the materials found on Vesta's surface did not originate there. There are carbon-rich elements and minerals containing water molecules that were most likely delivered by impacts with debris coming from farther out in the solar system. Another observation of *Dawn* was that Vesta had developed a layered, onion-like structure—similar to that of the inner planets— consisting of a rocky outer crust, a metal inner core, and a rocky mantel between them. During the formation of such a layered structure, Vesta must have been in a hot and molten state. This would have allowed gravity to pull heavier materials into the interior to form the core. Vesta is the only known object from the early days of the solar system that has such a layered composition. By late 2012, *Dawn* had left Vesta behind and was heading for Ceres. It entered orbit around Ceres in March of 2015. The precise data gathered about *Dawn*'s orbits made it possible to make accurate determinations of the mass of Ceres (and earlier about the mass of Vesta). Tracking the slight deflections in *Dawn*'s flight path and measuring the variable gravitational field gave scientists information about the evolution of the surface as well as the internal structure of Ceres. *Dawn*'s camera made a careful record of its pockmarked surface. The evidence suggests that Ceres's crust is a mixture of ice, salts, and rock. Its surface reveals multiple features formed by flowing materials.

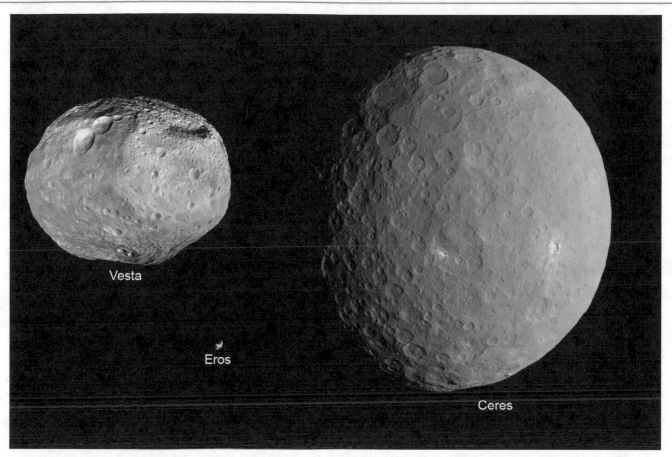

Figure 2.20. *Dawn* obtained this image of Vesta from a distance of 5,200 km in July of 2011 and that of Ceres from a distance of 13,600 km in May of 2015. The much smaller asteroid Eros was visited by the spacecraft *NEAR-Shoemaker* in February of 2000. Image credit: NASA/JPL-Caltech/UCLA/DLR(the German Aerospace Center).

This includes volcano-shaped mountains believed to have been created by the expulsion of molten ice. *Dawn*'s instruments have also detected organic materials on Ceres and discovered evidence of chemical activity. The widespread presence of minerals containing water and the emission of water vapor from its surface tell scientists that Ceres may have had a vast ocean in the past. They think that what remains of this former ocean is now mostly frozen and bound up within the crust. At the end of its mission, *Dawn* is to be placed in a stable orbit around Ceres in order to prevent a crash landing that would contaminate its surface. Figure 2.20 shows Ceres and Vesta as well as the asteroid Eros correctly in terms of their relative sizes. Only the cameras of spacecraft in orbit or in flyby have captured images of asteroids and other small bodies of the solar system with similar accuracy and resolution.

Table 2.7 lists a few of the most significant asteroids. Since asteroids are named and numbered in order of their discovery, those numbered 1 through 4 were the first four to be discovered. Chapter 1F narrates some of the history of their discovery. Those numbered 1, 2, 4, and 10 are the four largest and most massive. The asteroid 45 Eugenia was found to have two satellites. The 433rd asteroid to be discovered was the peanut-shaped rock 433 Eros. See Figures 2.20 and 2.21. It has a length of 33 km and is one of the largest of the NEAs. It was studied by the spacecraft *NEAR* (Near Earth Asteroid Rendezvous) that orbited and then landed on the asteroid in the years 2000-01. The

Table 2.7. From NASA's Asteroid Fact Sheet https://nssdc.gsfc.nasa.gov/planetary/factsheet/asteroidfact.html and more recent studies. Ceres is close to spherical, but the others are irregularly shaped and for them "diameter" refers to the length of the longest axis. For the precise meaning of "angle of orbit plane to Earth's" see Chapter 6N.

Asteroid	diameter in km	mass in kg	semimajor axis of orbit in au	orbit period in years	eccen- tricity	angle of orbit plane to Earth's
1 Ceres	965	9.39×10^{20}	2.768	4.60	0.0758	10.59°
2 Pallas	582	2.05×10^{20}	2.772	4.61	0.2310	34.84°
3 Juno	234	2.00×10^{19}	2.670	4.36	0.2563	12.99°
4 Vesta	569	2.59×10^{20}	2.362	3.63	0.0889	7.14°
10 Hygeia	530	8.67×10^{19}	3.142	5.57	0.1146	3.84°
45 Eugenia	215	6.10×10^{18}	2.721	4.49	0.0835	6.60°
433 Eros	33	6.69×10^{15}	1.458	1.76	0.2227	10.83°
4179 Toutatis	4.5	1.73×10^{13}	2.534	3.98	0.6294	0.45°
99942 Apophis	0.33	6.1×10^{10}	0.922	0.89	0.1911	3.33°
101955 Bennu	0.50	7.8×10^{10}	1.126	1.20	0.2037	6.03°
162173 Ryugu	0.90	4.5×10^{11}	1.190	1.30	0.1902	5.88°

NEAR mission will be studied in detail in Chapter 6. The asteroid 4179 Toutatis is the largest of the known *Potentially Hazardous Asteroids* (PHAs). It came to within 0.0104 au (or 1.5 million km) of Earth in 2004. The orbit of Toutatis is well understood and the probability that its trajectory will intersect the Earth's orbit in the next six centuries is essentially zero. There is a PHA that poses a much greater threat. Observations indicate that the asteroid 99942 Apophis is about 330 meters wide with a mass of roughly 50 billion kilograms. Trajectory studies tell us that it will buzz Earth on

Figure 2.21. This view of 433 Eros is a mosaic of six images taken by the *NEAR* spacecraft from a distance of about 200 km in February of 2000. Image credit: *NEAR* Project, NLR, JHU APL, Goddard SVS, NASA.

April 13, 2029 and come to within 30,000 km to 36,000 km of our planet. Since this corresponds to about three Earth diameters this will be a very near miss! The Japanese Space Agency JAXA sent the craft *Hayabusa-2* (the name comes from the Japanese word for Peregrine falcon) into space in December 2014 to explore and collect samples from the PHA 162173 Ryugu. The probe reached its destination in June 2018. In April 2019 it blasted the asteroid with an explosive charge in order to

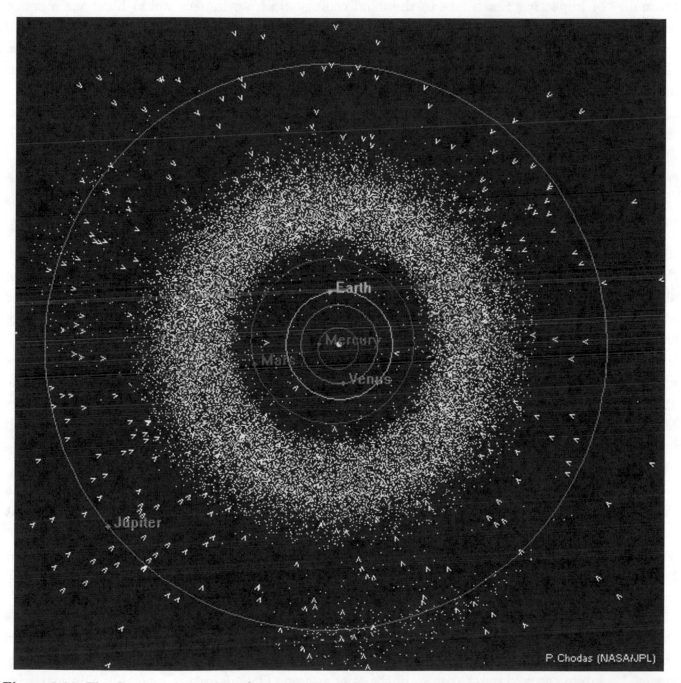

Figure 2.22. This diagram of the inner solar system viewed from above the Earth's orbital plane shows the positions of all numbered asteroids (almost all of them in the Main Asteroid Belt) and all numbered comets on January 1, 2018. Asteroids are represented by yellow dots and comets by sunward-pointing wedges. The orbits of the inner planets and Jupiter are also shown. Image credit: P. Chodas/NASA/JPL-Caltech. From https://ssd.jpl.nasa.gov/ go to SITE MAP and then to Solar System Diagrams.

loosen and expose materials from below its surface. *Hayabusa-2* landed on Ryugu in July 2019 to gather up some of the debris. The plan is to return these materials to Earth by December 2020. The hope is that their analysis will shed light on the mysteries of the birth of the solar system. The asteroid 101955 Bennu is a coal-black PHA. Every six years its orbit brings it to within about 300,000 km from Earth. It has a relatively high probability—a 1-in-2700 chance is a preliminary estimate—of impacting Earth late in the 22nd century. Such an impact would blow out a crater nearly 5 km wide and 400 meters deep and collapse buildings and level trees up to 50 km away. In September of 2016 the spacecraft *OSIRIS-REx*—the name is derived from key words of its mission: Origins, Spectral Interpretation, Resource Identification, Security - Regolith (the dusty surface layer of an asteroid) Explorer—was launched to study Bennu's physical and geologic characteristics. In December 2018, the craft went into a tight circular orbit around the asteroid. Sharp images taken since have led to the discovery that Bennu is ejecting streams of particles from its surface. *OSIRIS-REx* is scheduled to approach the surface of the asteroid and scoop up a small sample of the asteroid's surface materials. These are to be returned to Earth for study in September 2023.

The fact that a number of asteroids have been found to contain precious metals and minerals raises the prospect of mining operations in space by robotic and possibly manned spacecraft.

2E. About Comets. We now turn to the study of comets, another group of small travelers in the solar system. Both asteroids and comets were formed early in the history of the solar system about 4.5 billion years ago. Asteroids are made up of metals and rocky material like the inner planets. Comets were formed in colder regions, farther from the Sun. They consist of a *nucleus* of compactified ice, dust, and rocky materials and are often referred to as dirty snowballs. When a comet's path takes it to the inner solar system, the Sun warms this frozen composite. The gases and particles that this releases bubble into a misty cloud that surrounds the nucleus with a fragile atmosphere called the *coma*. The pressure from the solar wind, the Sun's radiation, stretches a part of this cloud of ejected particles into a *tail* of dust and gas. A comet's nucleus is generally less than a few tens of kilometers across, but its coma can extend for thousands and even millions of kilometers. Figure 2.22 gives a sense of the location of the asteroids and comets within Jupiter's orbit.

In August of 2014, the astronomer Terry Lovejoy discovered a comet from his observatory in Brisbane, Australia. The comet was photographed in early February 2015 just a few days after it passed perihelion at a distance of 1.29 au or 193,000,000 km from the Sun. When it was close to perihelion the comet was seen in the night sky with its coma nearly the size of a full Moon (but less bright) and its tail stretching faintly for eight Moon diameters. The chemical compounds vaporizing from the comet gave its coma a striking green hue. The image of Figure 2.23 shows separate tails of gases and dust, the shorter tail of gases below the longer tail of dust. The Sun's radiation makes the tail of gases glow and the Sun's light illuminates the tail of dust. (Barely visible near the end of the tail is the spiral galaxy NGC 891 depicted in Figure 1.30.) The comet has now begun its lonely voyage back into the cold outer regions of the solar system. The analysis of its highly elliptical orbit tells astronomers that it will not return to our skies for at least another 13,000 years.

The comet discovered by Robert McNaught in 2006 presented a spectacular scene over the skies of the southern hemisphere near the time it reached perihelion. See Figure 2.24. It became the brightest comet in 40 years. Its tail was seen dispersed over a huge swath of sky.

Comets have been studied and tracked during their passage through the inner solar system for some time now. A comet that is in an elliptical orbit around the Sun with a period of less than 200 years is known as a *short-period* comet. In their official designation such comets are catalogued with a number and the letter P. The number records the order in which they have been identified.

Figure 2.23. This image of the comet that Terry Lovejoy discovered was taken with a telescope in February 2015 by astrophotographer Damian Peach from England's south coast. Many thanks to him for permitting its use.

For example, the designation 1P/Halley for Halley's comet (see Chapter 1B) tells us that it was the first to be identified as a short-period comet and 2P/Encke names the second comet found to have a predictable short-period orbit. Halley and Encke are the astronomers who calculated the orbits. When a short-period comet can no longer be detected—comets can break up, disintegrate, and disappear—the letter D replaces the P in its designation. The comet 3D/Biela was first recorded in 1772 and identified as periodic in 1826 by the Austrian army officer von Biela. It has not been seen since 1852 and is believed to have disintegrated. Comets that were determined to have periods greater than 200 years are catalogued with the letter C along with the year they were first observed, followed by a second letter and a number. The second letter identifies the first or second half of the month in the year of the discovery, and the number (mostly 1, but also 2, or 3) tells us that the comet was the first, second, or third such comet to be detected (in the given year and period of the month). The designations of the comets depicted in Figures 2.23 and 2.24 as C/2014 Q2 and

Figure 2.24. This striking picture of the comet discovered by Robert McNaught in 2006 was taken at sunset from the Paranal Observatory in Chile in January 2007. The comet is setting over the Pacific Ocean, its majestic spreading tail backlit by the Sun. Image credit: S. Deiries/European Space Observatory.

C/2006 P1 tell us that the were discovered in September and August, respectively, of the given year.

The label 67P/Churyumov-Gerasimenko refers to the 67th short-period comet to have been identified and recognizes its two discoverers. It was the focus of study of the *Rosetta* mission of the European Space Agency (ESA). Launched in March 2004, it took the *Rosetta* spacecraft ten years to reach the comet and to join it on its journey around the Sun. The craft was inserted into orbit around the comet in September 2014. The images that its cameras captured show a duck-shaped nucleus consisting of two connected parts of about 4 km and 2.5 km in length. See Figure 2.25. The robotic lander that *Rosetta* carried was ejected from the craft in November 2014 and became the first probe to touch down on a comet's nucleus. The high-resolution cameras of both the spacecraft and the probe observed substantial changes on the comet's surface, especially during the time the comet was close to perihelion when the Sun's gravitational force on it was greatest. A fracture in the neck region was observed to grow in size; boulders tens of meters wide were seen to be displaced by many tens of meters; and an outburst of dust and gas was observed that led to the collapse of a cliff over 100 meters high exposing the comet's icy interior. The crafts' instruments detected over a dozen solid organic compounds on the surface of the comet's nucleus. They also detected gases streaming away from the nucleus. Along with water vapor, these included carbon monoxide, carbon dioxide and gases containing nitrogen and sulphur. The mission ended in September of 2016 with *Rosetta's* controlled crash-landing on the comet.

Table 2.8 is a listing of some of the brightest comets and a few of the most interesting. In some

Figure 2.25. The comet 67P/Churyumov-Gerasimenko imaged by Rosetta's narrow-angle camera in August 2014 from a distance of 285 km. Image credit: ESA/MPS/OSIRIS Team.

cases, the table provides the name of the discoverer or discoverers (below the identifying label) except in the case of C/1680 V1 where Newton's study of the comet is recognized. (See Chapter 1B). The comets C/1577 V1, C/1618 W1, C/1664 W1, C/1680 V1, C/1843 D1, and C/1882 R1 were all celebrated as "great comets of the year." Some of the most spectacular comets ever observed have been "sun-grazers," meaning that they came very close to the Sun during their perihelion passage. The first sun-grazing comet to receive a lot of attention was the comet that Newton tracked and that his contemporary, the Dutchman Lieve Verschuier painted with its spectacular tail sweeping across the sky. See Figure 1.8. A look at the perihelion distances that the table provides tells us that the comets C/1843 D1, C/1882 R1, and C/1965 S1 were sun-grazers as well. They are, when at perihelion, the fastest objects in the solar system. (See the discussion in the paragraph *Sungrazing Comets and their Speeds* of the Problems and Discussions section of this chapter.) The comets C/1975 V1, C/1995 O1, and C/1996 B2 were also "great comets" in the years 1976, 1997, and 1996 when they were at their brightest in the sky. The fact that the eccentricities of the comets C/1980 E1 and C/2006 P1 are greater than one indicates that they were traveling through the inner solar system along hyperbolic trajectories.

Table 2.8 requires some comments. The first concerns the angle between the plane of a comet's orbit and the plane of Earth's orbit. Consider Encke's comet for example. That the angle between the orbital planes is given as 3.3° means that if the the plane of the comet's orbit is revolved around its line of intersection with Earth's orbital plane by 3.3° so that the planes coincide, then Encke and Earth would orbit in this common plane in the same direction. Turning to a case where the angle is greater than 90°, we'll consider Halley's comet where the angle between the two orbital planes is listed as 162.3°. Here the understanding is that after Halley's orbital plane is revolved by $180° - 162.3° = 17.7°$ around its line of intersection with Earth's orbital plane so that the planes coincide, then Halley's motion in this common orbital plane is in the direction opposite to Earth's. The next comment about the data in Table 2.8 is the fact that a comet's eccentricity together with

Table 2.8. Data taken from NASA/JPL websites, https://nssdc.gsfc.nasa.gov/planetary/factsheet/cometfact.html and the JPL Small-Body Database Browser https://ssd.jpl.nasa.gov/sbdb.cgi. Note that the table omits the orbital periods of the comets Newton and West, both specified as having elliptical orbits with eccentricities very close to 1.

Comet	data from the years	eccen-tricity ε	perihelion distance q in au	semimajor axis a in au	aphelion distance in au	orbital period T in years[4]	angle of orbit plane to Earth's[5]
C/1577 V1[1] (Tycho Brahe)	1577-78	1.0	0.17750	–	–	–	104.9°
1P/Halley	1835–	0.967143	0.585978	17.83	35.08	75.32	162.3°
C/1618 W1[1]	1618-19	1.0	0.38954	–	–	–	37.2°
C/1664 W1[1]	1664-65	1.0	1.02553	–	–	–	158.7°
C/1680 V1 (Newton)	1680-81	0.999986	0.00622	444.43	888.85	–	60.7°
2P/Encke	2009-17	0.848320	0.33599	2.22	4.09	3.30	11.8°
3D/Biela	1832	0.751299	0.87907	3.53	6.19	6.65	13.2°
C/1843 D1	1843	0.999914	0.00553	64.27	128.53	513.00	144.4°
C/1882 R1[2]	1882-83	0.999899	0.00775	76.73	153.46	669.00	142.0°
C/1965 S1 (Ikeya-Seki)[3]	1965-66	0.999915	0.00779	91.60	183.19	880.00	141.9°
67P/Churyumov-Gerasimenko	1995–	0.640582	1.24529	3.46	5.68	6.45	7.0°
73P/Schwassmann-Wachmann	2016-17	0.685567	0.972190	3.09	5.21	5.44	11.2°
C/1975 V1 (West)	1975-76	0.999971	0.19663	6780.21	13,560.22	–	43.1°
C/1980 E1 (Bowell)	1980-86	1.057732	3.36394	58.27	–	–	3.9°
C/1995 O1 (Hale-Bopp)	2005-13	0.994961	0.91741	182.05	363.19	2456.41	89.2°
C/1996 B2 (Hyakutake)	1996	0.999899	0.23023	2272.08	4543.93	108,303.74	124.9°
C/2006 P1 (McNaught)	2006-07	1.000019	0.17074	8953.85	–	–	77.8°
C/2014 Q2 (Lovejoy)	2014-16	0.997773	1.29036	579.38	1157.46	13,946.02	80.3°

1) The value 1 for the eccentricity of the (parabolic) orbit is based on the original observations and is thus speculative. 2) The comet C/1882 R1 broke into several fragments during its last perihelion. 3) The comet C/1965 S1 broke into three smaller comets at its last perihelion. 4) The orbital period in the elliptical cases refers to the perihelion period. 5) The meaning of the angle between the comet's and Earth's orbital plane is given in the comments about this table.

the perihelion distance (of columns 3 and 4) determine its semimajor axis, and (if the orbit is elliptical) its aphelion distance and its orbital period (columns 5, 6, and 7). This can be seen by using facts from Chapter 1C. In the elliptical case, with the eccentricity $\varepsilon < 1$ and the perihelion distance q given, the semimajor axis a satisfies the equality $q = a(1-\varepsilon)$. So $a = q(1-\varepsilon)^{-1}$ is determined by ε and q. Since the aphelion distance is $a(1+\varepsilon) = q(\frac{1+\varepsilon}{1-\varepsilon})$, it too is determined by ε and q. By applying Newton's version $GM = \frac{4\pi^2 a^3}{T^2}$ of Kepler's third law (with M the mass of the Sun), we get that the period is $T = (\frac{4\pi^2 a^3}{GM})^{\frac{1}{2}}$. Since a is determined by ε and q and GM (with M the mass of the Sun is known—see Chapter 1H), T is determined by a and hence also by ε and q. In terms of the entries for columns 5, 6, and 7 in the hyperbolic case $\varepsilon > 1$, only the semimajor axis is defined. Chapter 1C tells us that it is given by $a = q(\varepsilon-1)^{-1}$. The point is that the entries in Table 2.8 are arrived at as follows: those in columns 3, 4, and 8 are the consequence of careful observations during the passage of the comet through the inner solar system, and those in columns 5, 6, and 7 can be derived from the entries in columns 3 and 4.

Example 2.1. For the comet C/1843 D1, $\varepsilon = 0.999914$ and $q = 0.00553$ au, so that the semimajor axis is $a = q(1-\varepsilon)^{-1} = 0.00553(0.000086)^{-1} \approx 64.30$ au and the aphelion distance is $q(\frac{1+\varepsilon}{1-\varepsilon}) = 0.00553\frac{1.999914}{0.000086} \approx 128.60$ au. Since 1 au $= 149{,}597{,}870.7$ km $\approx 1.49598 \times 10^{11}$ m, and $GM = 1.32712 \times 10^{20}$ m^3/sec^2 (from Chapter 1H), we get that the period of the comet's orbit is

$$T = \left(\tfrac{4\pi^2 a^3}{GM}\right)^{\frac{1}{2}} \approx 2\pi \left(\tfrac{(6.43023 \cdot 1.49598 \times 10^{12})^3}{1.32712 \times 10^{20}}\right)^{\frac{1}{2}} \approx 2\pi \left(\tfrac{8.90137 \times 10^{38}}{1.32712 \times 10^{20}}\right)^{\frac{1}{2}} \approx 1.62725 \times 10^{10} \text{ sec.}$$

Since 1 year has 365.25 days and hence $365.25(86{,}400) \approx 3.15576 \times 10^7$ seconds (see Chapter 1G), the period of the comet is equal to $\frac{1.62723}{3.15576} \times 10^3 \approx 516$ years. The small discrepancies between these values and those of Table 2.8 are explained by the fact that the computations of the entries in the table (provided by the JPL Small-Body Database Browser) rely on a more accurate value for q.

The fact is that the orbits of many comets are stable and predictable. The orbits of the short-period comets that Halley and Encke observed are examples. But there are comets that have experienced extreme changes in their trajectories after their encounters with the gravitational forces of the Sun (near perihelion) or with one of the massive outer planets (on close approaches). There is little doubt that this explains the disappearance of the comet 3D/Biela. The comet 73P/Schwassmann-

Figure 2.26. A *Hubble Space Telescope* image of the fragments or "string of pearls" of Shoemaker-Levy 9 taken in May 1994 when the comet was a distance of approximately 660 million km from Earth. Image credit: NASA, Weaver (JHU), T. Smith (Space Telescope Science Institute).

Wachmann, its nucleus already separated into dozens of fragments, is another comet that faces the prospect of complete disintegration. The most dramatic example is the comet D/1993 F2-Shoemaker-Levy 9, the ninth comet discovered by the pair of comet hunters Shoemaker and Levy. This comet was orbiting Jupiter when it was first observed in March 1993. The numerical analysis of the comet's motion backward in time made it possible to reconstruct its history. It had been in a short-period orbit around the Sun when, in 1970, the gravitational force of Jupiter captured it and pulled it into orbit. As the result of another too close encounter with Jupiter in July of 1992, the comet broke up into at least 21 icy fragments of up to 2 km in diameter that stretched across 1.1 million km of space (3 times the distance between Earth and the Moon). See Figure 2.26. In the summer of 1994 these fragments slammed spectacularly into Jupiter's atmosphere at speeds of about 60 km/sec. They created scars on Jupiter's surface that were visible for many months. Never before had anyone seen the impact of such large objects on a planet.

2F. Trans-Neptunian Objects. The term *Trans-Neptunian object* refers generically to any object in the solar system beyond the orbit of Neptune. Our discussion has shown that a comet that loops through the solar system in an orbit with a period of only a few years or tens of years faces a challenge. As the comet swings past the Sun near perihelion in its relatively small and tight elliptical orbit it is subjected to increased heat and heightened gravitational stresses. During each of its relatively frequent passages a little of its ice melts and some of its rocky substance tears. The occasional flyby of a larger planet add to the strain. Over long periods of time, say a few million years, the fragmentation of the nucleus and the disintegration of the comet is likely. Given the fact that the solar system is several billion years old, the implication is that short-period comets should have disappeared altogether. Yet astronomers still track several hundred of them. This was the puzzle that led the Dutch astronomer Gerard Kuiper to propose the existence of a swarm of icy objects in orbits beyond the planet Neptune. These small remnants from the dawn of our solar system continued to circle over vast stretches of time without clumping together into larger bodies. Periodically, some of these objects from what is now known as the *Kuiper belt* would be deflected by gravitational interactions toward the inner solar system and into orbits closer to the Sun. This, according to Kuiper, is the dynamic that continues to replenish the supply of short period comets.

Beginning in the late 1980s astronomers began to scan the heavens in search of the dim and icy objects beyond Neptune that Kuiper conjectured to exist. It was several years before their efforts were rewarded. In 1992 a reddish-colored speck appeared in the sights of a telescope on Mauna Kea in Hawaii to reveal a slowly moving, faint, icy body in a near circular orbit around the Sun. With a radius of about 50 au it orbited far beyond Neptune. Since then over two thousand trans-Neptunian objects have been identified. They receive a provisional designation (such as 2007 OR10) before they are named—in recent years, after deities in Greek, Roman, Polynesian, American Indian, and Inuit Eskimo mythologies—and numbered. The names Quaoar, the force of creation for a north American Indian tribe; Orcus, a god of the underworld in Roman mythology; Haumea, a Hawaiian goddess of fertility; and Makemake, the Polynesian god of creation, are examples.

Table 2.9 lists the trans-Neptunian objects of diameter of 900 km or greater discovered so far along with their orbital data. The distance data of the table suggests that the Kuiper belt extends from about 30 au to 55 au from the Sun. It is thought that there are over 100,000 such objects with a diameter of more that 100 km in addition to an estimated trillion or more comets. The discovery of the object 2003 UB313 in 2005 presented a problem. Images from both the *Hubble Space Telescope* and telescopes in Hawaii revealed that it was slightly smaller than Pluto but that it was more massive. The new object was later named Eris for the Greek goddess of discord and strife. This name turned out to be appropriate, given what happened next. With the discovery of an object similar to Pluto and the prospect of the existence of many more such objects, Pluto's status as a major planet fell into question. Should the set of planets be expanded to include all such distant rocky, icy shapes? In 2006, the International Astronomical Union defined a planet to be an object in orbit around the Sun that is not only massive enough be rounded into a spherical shape by its own gravitational force, but also massive enough to have cleared the neighborhood of its orbital path of other objects (by collision or by attracting them as moons). In a decision that continues to be controversial, Pluto was demoted to the new class of *dwarf planet*. Figure 2.20 tells

Table 2.9. Data taken from the JPL Small-Body Database Browser at https://ssd.jpl.nasa.gov/sbdb.cgi. See also http://www.mikebrownsplanets.com/2015/01/ten-years-of-eris.html.

Trans-Neptunian Object	diameter in km	perihelion distance in au	aphelion distance in au	orbit period in years	eccen-tricity	angle of orbit to Earth orbit
Pluto 1930	2380	29.66	49.31	247.74	0.2502	17.09°
Quaoar 2002 LM60	1070	41.97	45.16	287.53	0.0366	7.99°
2002 MS4	934	35.98	47.77	271.00	0.1408	17.67°
Haumea 2003 EL61	1595	35.15	51.57	285.48	0.1894	28.20°
Orcus 2004 DW	917	30.73	48.07	247.29	0.2201	20.58°
Makemake 2005 FY9	1430	38.64	52.79	309.10	0.1547	28.99°
Eris 2003 UB313	2326	37.77	97.53	556.41	0.4417	44.20°
Sedna 2003 VB12	995	76.05	899.48	10,772.69	0.8441	11.93°
2007 OR10	1535	33.18	101.10	550.18	0.5058	30.87°
2013 FY27	1113	35.83	82.19	452.34	0.3919	33.00°

tells us why the asteroid Ceres was promoted to the new category (but why the smaller asteroids were not). The Kuiper belt objects Quaoar, Haumea, Makemake, and Eris joined Pluto and Ceres in also receiving this classification. In November 2003, the team of astronomers that discovered Haumea, Makemake, and Eris announced the detection of a "far-out" trans-Neptunian object. It computed the perihelion and aphelion distances of its elliptical orbit around the Sun to be 76 au and 900 au. Later named Sedna, for an Inuit goddess who lives at the bottom of the frigid Arctic ocean, this object circles far beyond the outer radius of 55 au of the Kuiper belt and takes 10,000 years to complete a single orbit. Figure 2.27 shows how the objects listed in Table 2.9 fit into the solar system relative to the orbits of the planets and the comets Halley and Hale-Bopp.

New Horizons was the first spacecraft sent by NASA to study Pluto and other Trans-Neptunian objects. It was designed and engineered by the Johns Hopkins University Applied Physics Laboratory (APL). Its payload of scientific instruments was developed under the direction of the Southwest

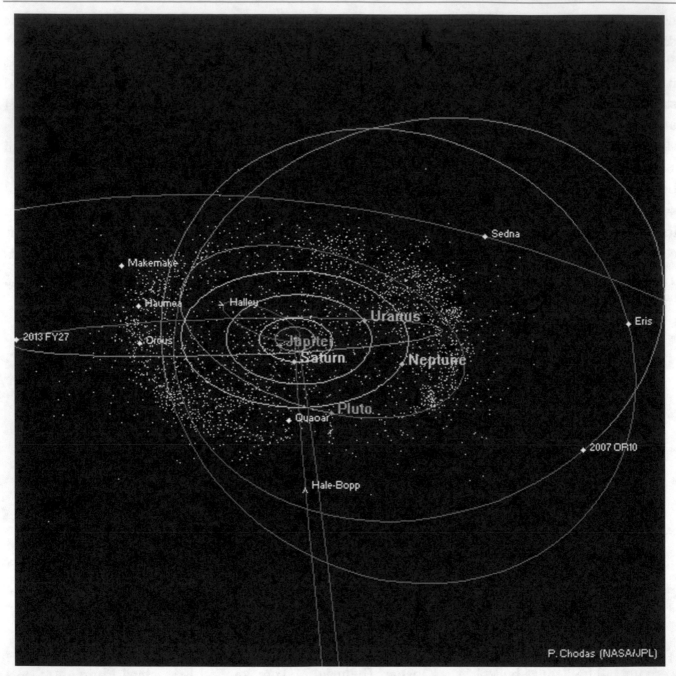

Figure 2.27. This diagram of the solar system from above the Earth's orbital plane shows the positions of significant Trans Neptunian objects as well as the comets Halley and Hale-Bopp on January 1, 2018. The orbits are drawn in two slightly different colors. The brighter violet is used for the part of the orbit above the ecliptic plane and the paler violet for the part below it. Some of the larger Trans-Neptunian objects are shown as white diamonds and many smaller ones as yellow dots. Image credit: P. Chodas/NASA/JPL-Caltech. From https://ssd.jpl.nasa.gov/ go to SITE MAP and then to Solar System Diagrams.

Research Institute (SwRI) with contributions from university laboratories and aerospace corporations. Fully fueled, the piano-sized probe weighed 478 kg at launch. The *New Horizons* spacecraft is lightweight, but robust, and designed to withstand its demanding mission to some cold and dark regions beyond our planetary system. Launched in January of 2006, the spacecraft went on a journey that would take over 9 years and cover a distance of over 5 billion kilometers before arriving to study Pluto (at the time of the launch of *New Horizons* Pluto was still a planet) and its five moons.

Pluto turned out to be much more than just a frozen ball. Its surface was found to feature remarkably diverse landforms and terrain types. The conspicuous heart-shaped region presents a striking example. Its upper part, bright and smooth, is about 1000 km across. It is a great icy plain, thought to be covered with nitrogen ice flows. The fact that it is free of impact craters means that it is geologically young. The closeup images that *New Horizons* captured (from as near as 12,500 km from Pluto's surface) show this icy plain to be bordered by a dark, heavily cratered area so dense that the craters overlap each another. Glaciers, probably composed of nitrogen ice, flow out from a mountainous area of the lower part of the heart-shaped terrain to the great, smooth icy plain. Some back-lit images have shown that Pluto has a thin, but visible atmosphere. With a

Figure 2.28. This image of Pluto (on the left) and its moon Charon (on the right) was taken by *New Horizons* in July of 2015 from a distance of about 250,000 km. Image credit: NASA / JHU APL / Southwest Research Institute.

diameter $\frac{1}{2}$ of that of Pluto, Charon is by far the largest of the moons. It too was much more complex than had been expected. Charon is moderately to heavily to cratered. Its surface is crossed by an extensive system of rifts, faults, and depressions. Probably dominated by water ice, the moon has a flat, grayish color, in marked contrast to the light reddish hue of Pluto.

The ratio of Charon's mass to Pluto's mass is large at about $\frac{1}{8}$ (compared to the $\frac{1}{81}$ for our Moon and Earth). The average distance between them is small, at about 20,000 km and the barycenter of the Pluto-Charon system lies about 1000 km above Pluto's surface. This means that Pluto and Charon are in a gravitational dance, always facing each other in the same way, both moving in elliptical loops, while making a full turn around each other every 6.4 Earth days. Figure 2.28 is a frozen snapshot of their waltz.

With its study completed, *New Horizons* left Pluto and Charon behind. Its path is taking it through a cluster of a dozen or so Kuiper belt objects discovered between 2011 and 2014. In February of 2019 it took a picture of one of them, the very strange two-lobed 2014 MU69. The goal of *New Horizons* is the observation of the surfaces and geologies of more of these icy objects.

Given how distant and faint Kuiper belt objects are, it is generally difficult to determine their basic characteristics and in particular their sizes. What aids in their study is the fact that the electromagnetic radiation that these cold objects emit is very strong in the infrared band. Infrared radiation is not visible, but it is detected as heat. When it comes from a small, distant object, the observed emitted infrared light is a better indicator of its size than the reflected visible light. The amount of visible light that is reflected by a smaller, brighter object could be the same as that reflected by a larger, but fainter one. If very small and distant, the two objects would appear to be of the same size when viewed through an optical telescope. However, the amount of infrared radiation emitted by an object depends on the size of the surface area that emits it. So the radiation that an object emits in the infrared band can provide an accurate measure of the object's size. Since Earth's atmosphere blocks much of it, infrared evidence is most effectively captured by space based telescopes. The *Spitzer Space Telescope*, launched in 2003, has been invaluable in this regard. Its highly sensitive infrared eyes have allowed astronomers to look into regions of space that are hidden from optical telescopes. This has not only included the centers of galaxies and their newly forming planetary systems but also cooler objects in deep space, and in particular the objects of the Kuiper belt. Spitzer's much more powerful successor—the *James Webb Space Telescope*—also with an infrared focus, will soon be operating in its place.

Thousands of comets originate in the Kuiper belt, but many others come from much more distant regions. At around the time that Gerard Kuiper thought deeply about the origin of comets, another Dutch astronomer Jan Oort proposed the existence of another vast swarm of icy remnants from the early history of the solar system. Now known as the *Oort cloud*, it is thought to occupy a huge region of space and to contain as many as 2 trillion objects. In terms of its distance from the Sun it is conjectured to extend from 5,000 au to 100,000 au. Recall from the paragraph *Parallax* and *Distances to Stars* of the Problems and Discussions section for Chapter 1 that one light year—the distance that light travels in one year—is equal to about 63,200 au. So the outer edge of the Oort cloud is about 1.5 light years from the Sun and extends to the outer limit of its gravitational reach. Note that the nearest star Proxima Centauri is about 4.2 light years from our Sun.

Comets originate in cold and distant reaches of the solar system. Occasionally, but over millions of years, icy balls of dust and rocks are deflected by gravitational interactions and pushed into orbits that bring them into the inner solar system as comets. The short-period comets, those with periods less than 200 years, come from the Kuiper belt. Comets with periods from several hundreds to many thousands of years have orbits with much longer semimajor axes and come from much farther away. They are the evidence for the existence of the Oort cloud. The comets C/1975 V1, C/1996 B2, and C/2006 P1 listed in Table 2.8 are likely examples. Some of the orbits of these comets were shortened by interfering gravitational forces of the planets since the time of their expulsion from the Oort cloud. For example, in April 1996 the comet C/1995 O1 (Hale-Bopp) passed within 0.77 au of Jupiter with the consequence that its orbit was shortened considerably. The possibility of similar gravitational interference on their exit from the inner solar system tells us that the long period comets listed in the table, face the possibility of significant orbital change. For instance, the comet C/1980 E1 approached the inner solar system along an orbit with a period of roughly 7 million years but an encounter with Jupiter in 1980 accelerated the comet so that it is now on an hyperbolic trajectory (with the largest known eccentricity 1.058). The possibility of gravitational interference means that the comets C/1980 E1 and C/2006 P1 currently heading out of the inner solar system on hyperbolic paths (note their eccentricities) may be destined to leave the solar system altogether.

Comets that originate in the Oort cloud can have periods of thousands or even millions of years so that they spend most of their time far from our realm of the solar system. This and the fact that the surrounding comas obscure their nuclei make them very difficult to study. The *Wide-field Infrared Survey Explorer (WISE)* spacecraft scanned the skies' infrared radiation during the years 2009 to 2011. The data that it gathered about comets has allowed scientists to "subtract" the infrared glow of the comas to estimate the sizes of their nuclei. One conclusion from this study was that there are about seven times as many long-period comets measuring at least 1 kilometer across than had been predicted previously. Another is that comets with long periods are on average up to twice as large as comets with periods of less than 20 years.

2G. The Rocket Equation.[3] To understand how the engines that drive both rockets and space-craft function, we need to turn to Newton's laws of motion. Newton's third law tells us that forces act in pairs. It says that any force has a corresponding force of equal magnitude that acts in the opposite direction. In the situation of a rocket engine, the force that pushes the craft forward is the matching force to the explosive force that drives the hot gases and exhaust particles back through the nozzle. To illustrate what is going on take a balloon, blow it up, and release it. For as long as it contains air, it is in effect a rocket engine. The force that pushes air molecules (they have mass!) out through the balloon's opening is equal in magnitude to the force that propels the balloon forward. The scientist who developed this basic idea into the first rocket engine was the American Robert Goddard(1882–1945). Regarded to be the father of modern rocket propulsion, Goddard invented, constructed, and successfully tested the first liquid fuel rockets.

For an engine to produce a steady stream of high velocity exhaust gases and particles, an ongoing controlled explosion needs to occur. But the fuels that are used (such as gasoline or kerosene) do not (and should not) combust spontaneously. In order to burn, fuels need to chemically interact with oxygen. But this introduces a complication. Jet airplanes don't need to carry their own oxidizer, since they can suck it from Earth's atmosphere. But a craft operating in the vacuum of space needs to bring its oxygen supply along for the ride. The fuel and oxidizer together are the craft's *propellant*. Most rockets and spacecraft are driven by liquid propellant engines. Such an engine is

---

[3]The last two sections of this chapter are an early introduction to the study of the flight of a spacecraft, one of the highlights—if not the highlight—of this text. They involve the calculus of derivatives, its meaning as rate of change, its application to the motion of a body, including velocity, acceleration, mass, and momentum as functions of time, and the law conservation of momentum. To pursue these two sections, a reader needs to come to an understanding of Tsiolkovsky's rocket equation but can skip its challenging derivation. In fact, a reader can skim through these two sections, get a sense of the story they tell, and return to them before turning to the last chapter of the text (that takes this story up in full).

equipped with separate fuel and oxidizer tanks that operate together with pipes, valves, a combustion chamber, a nozzle, and other plumbing. When the craft's computer sends the command, fuel and oxidizer, both highly cooled and under great pressure, are pumped into the combustion chamber. The two liquids combust spontaneously when they come into contact with each other. Together, they create the required sustained explosion that forces the residue of gases and particles to escape at high velocity through the nozzle. The exit velocity of this discharge typically ranges from 2 km to 5 km per second. Figure 2.29 is a highly simplified diagram of a liquid propellant rocket engine that illustrates what has been described. Except for the very small amount of mass that is converted by the explosion into radiated energy and resulting heat, the mass of the propellant consumed is the same as that of the corresponding mass of gases and particles expelled. (The fact

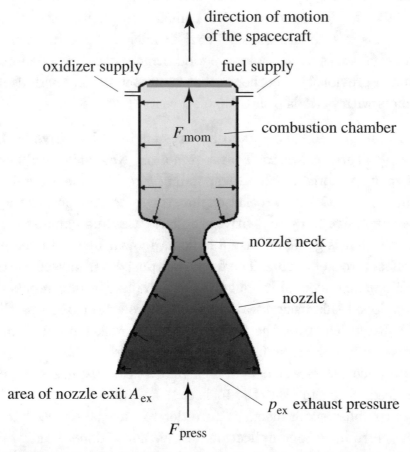

Figure 2.29. The diagram illustrates the essential elements of a liquid propellant engine but not its complexities. Both the fuel and the oxidizer are super cold liquefied gases (such as liquid hydrogen and liquid oxygen). Before they are driven into the combustion chamber, these liquids are pumped around the outside of the combustion chamber to cool it. The pumps that drive these liquids operate at extremely high pressure to overcome the pressure of the exploding propellant. With all the pipes that do the pumping and cooling, a typical liquid propellant engine looks a lot like a plumbing project gone haywire.

is that the total mass lost in this way is less that one microgram per about 40,000 liters of propellant.) So we will assume that each kilogram of rocket fuel and oxidizer comes out of the nozzle as one kilogram of hot, high-velocity gas and particles.

We'll now become technical and turn to the physics and mathematics that underly the discussion above. The first thing we need is a new formulation of Newton's second law. In the standard version of the equation $F = ma$, the magnitude F of the acting force on an object and the resulting acceleration a are both functions of time t, and the mass m is the constant of proportionality that relates them. (See Chapter 1B.) In the current situation, however, the fact that gases and particles are expelled through the engine's nozzle reduces the mass of the craft. In particular, the force that the engine generate acts on a mass that decreases as a function of time.

For the reformulation of Newton's second law, consider a body of mass m moving with velocity v. In this discussion, the term velocity refers to a numerical quantity without direction, so that it means speed. Suppose that both m and v are functions of time t, and that the motion is driven by a single force (this can be the combined effect of several forces) of magnitude F that is also a function of t. The product mv is the *momentum* of the body and Newton's second law is the assertion that

$$F = \frac{d}{dt}(mv).$$

So the magnitude of the force on the body is equal to the derivative of its momentum. Notice that in a situation where m is constant, $F = \frac{d}{dt}(mv) = m\frac{dv}{dt} = ma$ is nothing but the conventional version of Newton's second law. If the force F—the net force on the body—is zero, then $\frac{d}{dt}(mv) = 0$, so that the momentum mv is constant. This is the *law of conservation of momentum*.

Consider a craft with a liquid propellant main engine in flight in space. We'll let M be the mass of the craft and m the mass of the liquid propellant. The mass of the craft's "hardware" is $M_{\text{hard}} = M - m$. Now turn to the schematic of the standard liquid propellant rocket engine depicted in Figure 2.29. At a certain time in response to electronic signals from the craft's computer, both fuel and oxidizer are injected into the combustion chamber, where they ignite to produce a continuous, controlled explosion. We'll assume that it has reached steady state. The forces unleashed by this explosion push against the walls of the combustion chamber in all directions. The *momentum thrust* F_{mom} is the combined effect of the forces pushing against the top of the combustion chamber (shown in green in the figure). The equal and opposite force given by Newton's third law drives the mass of burned propellant out through the nozzle's neck. Let $m(t)$ be the mass of the propellant at any time t after the firing of the main engine has reached steady state. The derivative $m'(t)$ is the rate at which the propellant is consumed. This is also the rate at which the exhaust materials are expelled at the nozzle exit. Since the engine is operating at steady state, $m'(t)$ as well as the velocity v_{ex} of the exhaust (relative to the engine) are constant. An application of the reformulated version of Newton's second law tells us that

$$F_{\text{mom}} = v_{\text{ex}}\, m'(t).$$

Let A_{ex} be the area of the nozzle's exit and let p_{ex} be the pressure of the exhaust there. Since pressure times area is force, the pressure p_{ex} exerts a *pressure thrust* of

$$F_{\text{press}} = p_{\text{ex}}\, A_{\text{ex}}$$

at the nozzle's exit. The *thrust* of the main engine is the sum

$$\boxed{F = F_{\text{mom}} + F_{\text{press}} = v_{\text{ex}}\, m'(t) + p_{\text{ex}}\, A_{\text{ex}}}$$

of the momentum and pressure thrusts. The forces described above are generated internally within the craft and are not external forces on the craft.

Since space is a vacuum, the craft experiences no atmospheric drag. We'll assume that the gravitational forces on the craft are also negligible, so that there are no external forces acting on the craft. Let $M(t)$ be the mass of the craft at time t after the firing of the main engine reached steady state and notice that $M(t) = M_{\text{hard}} + m(t)$. Let's observe the craft at time t and then again at time $t + \Delta t$ soon thereafter. The difference $M(t) - M(t + \Delta t) = \Delta M$ is both the mass of the propellant that is burned during time Δt and also the mass of the exhaust materials that are expelled as a result of this burn. Let $v(t)$ be the velocity of the craft at time t and let $v_{\Delta M}$ be the (constant) velocity of the expelled mass ΔM. The reference point for these velocities depends on the mission of the craft and is typically the Earth, the Sun, any of the planets, the moon of a planet, an asteroid, or a comet. Figure 2.30 represents what has been described. Notice that $M(t) = M(t + \Delta t) + \Delta M$. Since $v(t + \Delta t)$ is the velocity of the craft at time $t + \Delta t$, the speed of

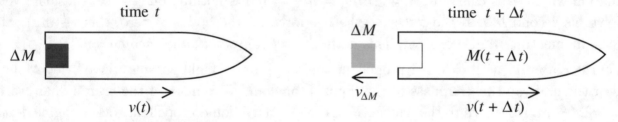

Figure 2.30. The red square represents the propellant that burns during the time interval from t to $t + \Delta t$. The beige square represents the matching exhaust materials that this propellant produces. Both have mass ΔM.

the expelled mass ΔM is equal to $v_{\Delta M} = v(t + \Delta t) \pm v_{\text{ex}}$, where v_{ex} is the speed of the exhaust materials relative to the engine. Depending on the particular situation, this speed can be less or greater than the speed of the rocket. In the situation of the figure, the craft is speeding up. In this case, the mass ΔM is forced out in a direction opposite to the motion of the craft, so that the $-$ sign applies. So $v_{\Delta M}$ can be positive or negative. In order to slow the craft down, the craft must be oriented so that the nozzle of the engine points in the direction of the craft's motion. In this case, an engine burn forces the mass ΔM out in the same direction. So the $+$ sign applies and $v_{\Delta M}$ is positive. At time t, the mass ΔM of the propellant is included within the mass of the craft. At time $t + \Delta t$, the mass ΔM consists of the corresponding exhaust materials that has separated from the craft. The momentum (velocity time mass) of the craft at time t is $v(t)M(t)$. At time $t + \Delta t$ the momentum of the craft plus exhaust materials is $v(t + \Delta t)M(t + \Delta t) + v_{\Delta M}\Delta M$. The assumption that there are no outside forces acting on the craft means that momentum is conserved, so that

$$v(t + \Delta t)M(t + \Delta t) + \big(v(t + \Delta t) \pm v_{\text{ex}}\big)\big[M(t) - M(t + \Delta t)\big] = v(t)M(t).$$

By multiplying the left side out and noticing that the terms $v(t + \Delta t)M(t + \Delta t)$ subtract off, we see that what remains on the left side is $v(t + \Delta t)M(t) \pm v_{\text{ex}}\big(M(t) - M(t + \Delta t)\big)$. By setting this expression equal to $v(t)M(t)$, it follows that $\big(v(t + \Delta t) - v(t)\big)M(t) = \pm v_{\text{ex}}\big((M(t + \Delta t) - M(t)\big)$. By dividing both sides by Δt and then by $M(t)$, we see that

$$\frac{v(t + \Delta t) - v(t)}{\Delta t} = \pm v_{\text{ex}} \frac{1}{M(t)} \frac{M(t + \Delta t) - M(t)}{\Delta t}.$$

By pushing Δt to zero on both sides and using the definition of the derivative of a function, we get

$$v'(t) = \pm v_{\text{ex}} \frac{M'(t)}{M(t)}.$$

By applying a basic property of the natural logarithm function ln and then taking antiderivatives of each side,

$$v(t) = \pm v_{\text{ex}} \ln M(t) + C,$$

where C is a constant. Let t_1 and t_2 with $t_2 > t_1$ be any two instants of time. After plugging both into the equation just derived, we get $v(t_2) - v(t_1) = \pm v_{\text{ex}}\big(\ln M(t_2) - \ln M(t_1)\big)$. Hence by two more basic properties of the natural log, $v(t_2) - v(t_1) = \pm v_{\text{ex}} \ln \frac{M(t_2)}{M(t_1)} = \pm v_{\text{ex}}\big(-\ln \frac{M(t_1)}{M(t_2)}\big)$. Therefore

$$\boxed{v(t_2) - v(t_1) = \pm v_{\text{ex}} \ln \frac{M(t_1)}{M(t_2)}}$$

This is the rocket equation first developed by the Russian Konstantin Tsiolkovsky (1857–1935), the founding father of the theoretical aspects of rocketry and spaceflight. It establishes a connection between the change in the craft's velocity and the amount of fuel consumed. Since propellant was burned, $M(t_2) < M(t_1)$, so $\frac{M(t_1)}{M(t_2)} > 1$ and hence $\ln\big(\frac{M(t_1)}{M(t_2)}\big)$ is positive. With the original understanding about the $\pm$ now reversed, in Tsiolkovsky's equation the $+$ sign applies if the burn increases the craft's speed and the $-$ sign applies if it decreases it.

Example 2.2. Suppose a spacecraft is in mid-flight to an outer planet. It has a mass of 4500 kg including the propellant. The exhaust that its main engine generates has a velocity of 3000 m/sec. In order to keep on the flight path that its mission calls for, the craft's velocity needs to be increased by 660 m/sec. After the craft is oriented correctly, its main engine begins its burn at time t_1 and reaches steady state immediately. It shuts off at time t_2 when the required increase in the craft's velocity is achieved. Letting the craft's mass at that time be $M(t_2)$ and applying Tsiolkovsky's rocket equation, we get

$$660 = v(t_2) - v(t_1) = 3000 \ln \frac{4500}{M(t_2)}.$$

Since $\ln\big(\frac{4500}{M(t_2)}\big) = \frac{660}{3000} = 0.22$, we get $\frac{4500}{M(t_2)} = e^{0.22}$. Therefore, $M(t_2) = 4500e^{-0.22} \approx 3611$ kg. It follows that the craft's main engine burned close to 889 kg of propellant during this maneuver. Since $\frac{889}{4500} \approx 0.20$, this is 20% of the total mass that the rocket had when its main engine began to fire.

Example 2.3. A rocket has placed the spacecraft of the previous example into an initial near-Earth solar orbit. The orbit is elliptical, has a semimajor axis close to 1 au, an eccentricity close to 0.02, and a perihelion period close to 1 year. Since it is in a near-Earth elliptical orbit, the paragraph *About Speeds of Objects in the Solar System* of the Problems and Discussion section of Chapter 1 informs us that the craft moves in this orbit at a speed of about 30 km/sec. The velocity increase of 0.66 km/sec for the spacecraft is small in comparison, especially given the large quantity of 889 kg of propellant necessary to achieve it.

The two examples raise a serious question. Suppose that it is the mission of a spacecraft to explore one of the outer planets of the solar system. Its journey begins on the launch pad of a three-stage rocket with the craft folded into the upper stage. The rocket's engines fire, it lifts off, the first and second stages are jettisoned after their fuel is spent, and the craft is placed into a low-Earth parking orbit. With a burn of the last stage followed by the craft's separation from it, the craft's speed can be increased to insert it into a near-Earth solar orbit or into a solar orbit that can take it to Mars. So far so good, but the much longer distance that needs to be navigated to reach an outer planet presents a problem. The fact is that a combination of technology and cost limitations make it impossible to provide the craft with a post-launch speed high enough to allow it to reach an outer planet. How then is it possible to impart to a spacecraft the kind of speed increase it needs to break out of a near-Earth or near-Mars solar orbit and to send it on its way to the more distant Jupiter, Saturn, Uranus, or Neptune? The obvious solution—to increase the craft's speed along the way by firing the spacecraft's main engine—is a non-starter because, as the two examples tell us, the amount of propellant required for such a speed increase is prohibitive. How then—given the constraints on the weight of such crafts, the size of the rockets that launch them, and the great amounts of propellant involved—can a spacecraft be sent to the outer reaches of our planetary system and beyond? The astonishing answer is that it is possible only with a flight path so designed that the spacecraft is brought into the tight vicinity of a planet (or several in succession) in such a way that the gravitational pull of the planet drags the craft along to increase its speed (and change its direction) so that it can be on its way to the destination that its mission calls for.

The diagram of Figure 2.31 illustrates what is involved. After launch, the spacecraft (labeled C in the figure) briefly orbits the Earth. By flying in the vicinity of Earth, the craft also orbits the Sun. An injection burn increases the craft's velocity and propels the craft into an expanded elliptical orbit around the Sun. This orbit is carefully directed and timed to send the craft on a course to meet another planet, labeled P in the figure. A flyby within the gravitational neighborhood of the planet P pulls the craft in the general direction of the planet's orbital path. This

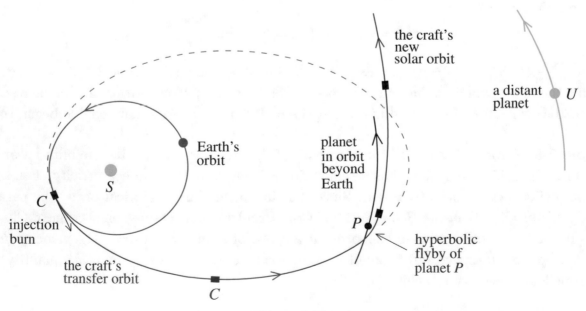

Figure 2.31

pulling action—called a *gravity assist*—changes the speed of the craft, bends its trajectory, and sends it on a rendezvous with a more distant planet, the U in the figure. With the exception of the injection burn and a few minor trajectory correction maneuvers along the way, not a single drop of the spacecraft's fuel has been burned.

The flight path of a spacecraft achieved by an engine burn with the craft in a near-Earth solar orbit that results in an expanded solar orbit that brings the craft to the vicinity of another planet (as represented by P in Figure 2.31, for instance) is an example of a *Hohmann Transfer Orbit*. This important navigational strategy was conceived by Walter Hohmann (1880–1945), a German engineer and an early pioneer of space travel. The book that he published in 1925 developed the basic orbital dynamics for spacecraft more than 30 years before the Russians first sent the world's very first space probe *Sputnik 1* into orbit around Earth.

2H. The Flight Path of *Juno*. The spacecraft *Juno* was launched from Kennedy Space Center in August 2011 on an *Atlas V* rocket that propelled it into a solar orbit. The goal of *Juno*'s mission was the study of the planet Jupiter. The preparation for an interplanetary flight to any distant destination requires an understanding of the energy that the rocket provides to the craft at its launch. Often referred to as *characteristic energy*, this is the energy required—after the spacecraft breaks free of Earth's gravity—to reshape its initial solar orbit to the desired trajectory. The characteristic energy depends on the alignment of Earth, Sun, and a target planet and hence on the projected dates of departure and arrival. The characteristic energy as calculated for *Juno* meant that the craft would have enough initial energy to be able to cruise to Mars, but that it would need to be given additional velocity along the way in order to reach Jupiter. Without it, the Sun's gravity would keep the craft bound to the inner solar system. Would *Juno*'s main engine be able to supply the extra velocity that was required?

Juno's mass at launch was 3625 kg. This included 1,280 kg of fuel and 752 kg of oxidizer for a total of 2032 kg of propellant. So 56% of the craft's mass was propellant. *Juno*'s main engine could deliver 645 newtons of thrust. At full throttle in the vacuum of space, the main engine could eject the exhaust materials through its nozzle with a velocity of $v_{ex} = 3124\,\text{m/sec}$. Figure 2.4 shows an artist's depiction of the craft, its main engine ablaze. Surely, with the amount of fuel that it had on board, this engine could provide *Juno* with the additional velocity it would need to reach Jupiter. Not so! The fuel supply—though large—would not have been sufficient.

In the case of *Juno*, the gravity assist trajectory that made its flight to Jupiter possible was both simple and surprising. After its launch, *Juno* emerged on a trajectory that brought it past Mars. About a year after launch, its main engine fired in two carefully designed burns to *slow the craft* down, trim its trajectory, and bring it back inside the orbit of Mars. This maneuver was programmed in such a way that one more year later, *Juno* eased into a gravity assist flyby of Earth that sped up the craft and sent it off to its rendezvous with Jupiter. Three years later, *Juno* was in orbit around Jupiter as planned. Refer to Figure 2.32 for a diagram of *Juno*'s flight and to

https://www.youtube.com/watch?v=sYp5p2oL51g

for a simulation. We'll now turn to the changes in the craft's velocity in terms of numerical specifics. To control its orientation—to rotate the craft, adjust its up/down and left/right attitudes—and to make minor trajectory corrections, *Juno* carries 12 small thrusters in addition to its single main engine. In February 2012, a few months after its launch, the craft fired these thrusters and burned 3 kg of propellant to undertake a minor *Trajectory Correction Maneuver* (TCM). After this orbital adjustment, *Juno* had a mass of $3625 - 3 = 3622$ kg. The two critical *Deep Space Maneuvers* DSM-1 and DSM-2 that slowed the craft followed. They occurred around the aphelion of *Juno*'s initial orbit, the first 2 days before and the second 12 days thereafter. During each maneuver, after the small thrusters pointed the craft's main engine nozzle forward in the direction of *Juno*'s flight, the main engine fired for about 30 minutes each time to reduce the craft's speed by 344 m/sec and 388 m/sec, respectively. The rocket equation of the previous section tells us how much of the craft's propellant was consumed in the process. For DSM-1, let t_1 be the instant that the engine began to fire and let t_2 be the instant it shut down. Let $M(t_1)$ and $M(t_2)$ be *Juno*'s mass at these two

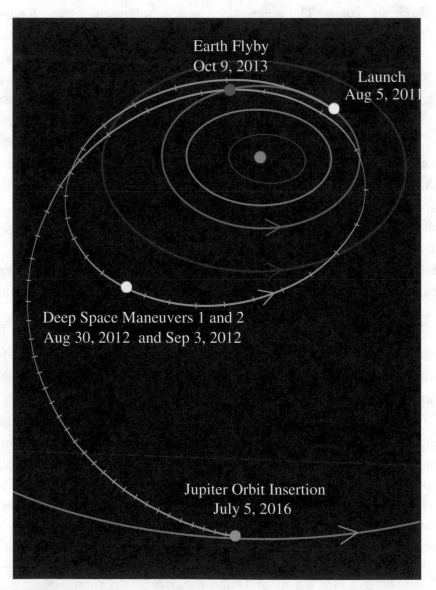

Figure 2.32. The diagram shows Earth's orbit in blue (and inside it, the orbits of Venus in gray and Mercury in purple), the orbit of Mars in red, and that of Jupiter in orange. *Juno* appears in white and its trajectory in gray. Image credit: NASA/JPL-Caltech.

instances. Since $M(t_1) = 3622$ kg, the rocket equation tells us that

$$-344 = -v_{\text{ex}} \ln\left(\frac{3622}{M(t_2)}\right) = -3124 \ln\left(\frac{3622}{M(t_2)}\right),$$

hence that $\frac{3622}{M(t_2)} = e^{\frac{344}{3124}}$ and therefore that $M(t_2) = 3622\, e^{\frac{-344}{3124}} = 3244$. So DMS-1 consumed $3622 - 3244 = 378$ kg of propellant. The same calculation for DMS-2 (with t_1 and t_2 now the start and stop times of the firing of the craft's engine for this second maneuver) shows that

$$-388 = -v_{\text{ex}} \ln\left(\frac{3244}{M(t_2)}\right) = -3124 \ln\left(\frac{3244}{M(t_2)}\right),$$

so that $\frac{3244}{M(t_2)} = e^{\frac{388}{3124}}$ and hence $M(t_2) = 3244\, e^{\frac{-388}{3124}} = 2865$. So DMS-2 consumed $3244 - 2865 = 379$ kg of propellant. Together, DSM-1 and DSM-2 used 757 kg of the 2032 kg of the propellant that *Juno* had available. This was more than 37% of the total and the trip was only barely underway.

Several more TCMs, number 5 in October 2012 (a maneuver to "clean up" the earlier DSMs), and numbers 6 and 7 in August and September of 2013 (TCM-8 was cancelled), moved *Juno* into position for a fly-by of Earth in October of 2013. This was the *Earth Gravity Assist* (EGA) in which *Juno* maneuvered to within 559 km of Earth's surface to allow Earth's gravitational pull to drag it along and to increase its speed by 7,300 m/sec. This increase sent the craft off on its rendezvous with Jupiter. Unlike the much smaller changes in the velocity that the two DSMs produced, this substantial and critical increase was accomplished without firing *Juno*'s main engine. This velocity increase (and one more TCM in November 2013 that fine-tuned the trajectory) placed *Juno* into a solar orbit that put it on course to intercept Jupiter. On July 5th 2016 with its mass at 2825 kg, *Juno*'s main engine—again pointing in *Juno*'s direction of flight—fired a third and final time for about 35 minutes. Operating in Jupiter's intense radiation environment, this burn slowed the craft by 542 m/sec and put *Juno* into orbit around the planet. *Jupiter Orbit Insertion* (JOI) had been achieved. By applying the rocket equation as before, we see that

$$-542 = -v_{\text{ex}} \ln\left(\frac{2825}{M(t_2)}\right) = -3124 \ln\left(\frac{2825}{M(t_2)}\right)$$

and hence that $\frac{2825}{M(t_2)} = e^{\frac{542}{3124}}$ and $M(t_2) = 2825\, e^{\frac{-542}{3124}} = 2375$ kg. The JOI had burned $2825 - 2375 = 450$ kg of the propellant.

Example 2.4. During DSM-1, *Juno*'s main engine fired for 30 minutes and burned 378 kg of propellant. Estimate the rate at which *Juno*'s main engine expelled its exhaust mass during this maneuver and use this information to show that the momentum thrust F_{mom} that the main engine generated was about 656 newtons.

The initial orbit around Jupiter was highly elliptical with a period of $53\frac{1}{2}$ days. When problems with the engine's valves were detected, the *Period Reduction Maneuver* (PRM) designed to tighten *Juno*'s orbit and to reduce its period to 14 days was canceled. *Juno* would remain in its 53-day orbit for the remainder of its mission. Soaring over Jupiter's swirling, bands of clouds as close as 4,100 km (nearly 10 times closer than any previous mission) with its scientific instruments fully operational, it gathered the amazing images and information already described in section 2C. The website

https://www.nasa.gov/mission_pages/juno/main/index.html

presents an ongoing, up-to-date account of *Juno*'s discoveries.

Table 2.10 tells us that the main engine burns of DMS-1 and DSM-2 lasted about $\frac{1}{2}$ hour each. We'll assess the duration of *Juno*'s Earth flyby with a comparison of the magnitudes of the gravitational forces of the Earth and Sun on the craft. Letting m be the mass of *Juno*, M_E and M_S the masses of the Earth and Sun, and applying Newton's law of universal gravitation, we get that the ratio of these magnitudes is

$$\frac{mGM_E}{d_E^2} \bigg/ \frac{mGM_S}{d_S^2} = \frac{GM_E}{GM_S} \cdot \left(\frac{d_S}{d_E}\right)^2,$$

where d_E and d_S are the distances from *Juno* to the centers of the Earth and the Sun, respectively. Data in Chapter 1H tells us that $GM_E \approx 3.986 \times 10^{14}$ m^3/sec^2 and $GM_S \approx 1.327 \times 10^{20}$ m^3/sec^2.

Table 2.10

date	event	main engine burn duration	$\Delta v^{1)}$ m/sec	propellant burned$^{2)}$ in kg	mass of craft in kg before/after the event
30 Aug 2012	DSM-1	30 min	-344	378	3622/3244
14 Sep 2012	DSM-2	30 min	-388	379	3244/2865
9 Oct 2013	EGA	–	7,300	0	2865
5 July 2016	JOI	35 min	542	450	2825/2375

1) Δv refers to the change in velocity in meters per second that the maneuver produced. The information in this column comes from NASA/JPL websites via

    http://spaceflight101.com/juno/juno-mission-trajectory-design/
    http://spaceflight101.com/juno/juno-joi-data/
    http://spaceflight101.com/juno/spacecraft-information/ and
    http://spaceflight101.com/juno/mission-updates/

2) These entries were computed with the rocket equation. They correspond closely to those that the websites provide.

Since *Juno* was about as far from the Sun as Earth during the flyby, we'll take $d_S = 1.496 \times 10^8$ km. We know that *Juno* was 559 km above Earth's surface at the time of closest approach to Earth, 8700 km above Earth's surface 18 minutes later, and that it was 6.81×10^6 km from Earth 13 hours and 39 minutes after its closest approach. Using $r_E = 6378$ km for Earth's radius and the values $d_0 = 559 + 6378 = 6937$ km, $d_1 = 8700 + 6378 = 15{,}078$ km, and $d_2 = 6.81 \times 10^6$ km, respectively, for the distance d_E, we get the corresponding values

$$\frac{3.986 \times 10^{14}}{1.327 \times 10^{20}} \left(\frac{1.496 \times 10^8}{6937}\right)^2 \approx 1400, \quad \frac{3.986 \times 10^{14}}{1.327 \times 10^{20}} \left(\frac{1.496 \times 10^8}{15{,}078}\right)^2 \approx 300, \text{ and } \frac{3.986 \times 10^{14}}{1.327 \times 10^{20}} \left(\frac{1.496 \times 10^8}{6.81 \times 10^6}\right)^2 \approx 0.0014$$

for the ratio of the gravitational force of the Earth on *Juno* to that of the Sun. Notice that while the Earth's gravitational pull on *Juno* was dominant during the initial minutes after the flyby, a few hours later the pull of the Sun returned to dominance and the Earth's pull became negligible.

It took *Juno* about 5 years to travel from its initial solar orbit to Jupiter and through its first few orbits around the planet. During this entire time—with the exception of the launch, the few hours that *Juno*'s thrusters and main engine fired (for orbit corrections and *Juno*'s orbit insertion around Jupiter) and the few hours it took for *Juno*'s flyby of Earth—the space craft was subject to a single, dominant centripetal force. This was the gravitational pull of the Sun and, after the orbit insertion, the gravitational pull of Jupiter. Since evidence tells us that gravitational forces are subject to inverse square laws, it follows from Conclusion C of Chapter 1D that with the exception of just a few hours, *Juno*'s flight path was some conic section. Since parabolas and hyperbolas were not involved (the craft was never on a trajectory to leave the solar system), this was always an ellipse. The focal points were the center of the Sun until orbit insertion and the center of Jupiter thereafter. After each of the trajectory changing maneuvers already described, the parameters of the ellipses—its eccentricity, semimajor axis, its focal point, its plane, and its orientation—changed. In sum, the flight path of *Juno* from the Sun, to Jupiter, to its orbits around it consisted of a sequence of ellipses that were connected by the short, more complex stretches of flight during which *Juno* was propelled by more than one significant force. (During the craft's flyby of Earth when Earth's gravity was dominant, *Juno*'s path was a hyperbola that had the center of Earth as the focal point.)

The fact is that in almost all situations, missions of spacecraft take years to complete and that, outside of just a few hours, spacecraft are driven in their flight paths by a single force— a single, dominant gravitational force. Any segment of the path so determined falls into a single plane. It lies on a conic section and can be studied with elementary one variable calculus. During each brief intervening time period a craft is also propelled by its thrusters, main engine, or a second gravitational force of a planet or moon. These short stretches connect the sequence of conic sections. They are too complex to be informed by elementary analysis.

2I. Problems and Discussions. This section begins with problems that call for the application of some results and conclusions of Chapter 1, for example Newton's version of Kepler's third law and the speed formulas for an elliptical orbit. Take the value of the gravitational constant G to be $G = 6.67384 \times 10^{-11} \frac{\text{m}^3}{\text{kg} \cdot \text{sec}^2}$. (See Chapter 1D and Chapter 1H.)

1. The Race to the Moon. In the 1960s Russia—then the dominant country of the Soviet Union of Communist states—and the United States became locked in an intense race to send humans to the Moon. It was perceived to be a test of superiority between the Soviet totalitarian system and the democratic system of the free world. More ominously, in terms of rocket development and payload delivery, this race also had military implications.

The first man-made object to reach the Moon was the unmanned Russian probe *Luna 2* with a hard landing on its surface in September 1959. The far side of the Moon was first photographed in October 1959 by the followup mission *Luna 3*. In these early years, Russian efforts to explore the Moon met with several successes, but American attempts were mostly failures. In the race to the Moon, America was clearly behind. In an effort to waken the competitive and patriotic spirits of his country, President John F. Kennedy famously proclaimed to a Joint Session of Congress on May 25, 1961, "I believe that this nation should commit itself to achieving the goal, before this decade is out, of landing a man on the Moon and returning him safely to the Earth."

Russia would remain in the lead for some time, but the U.S. was catching up. The soft landing in February 1963 of Russia's *Luna 9* and its transmission of photographs from the lunar surface, was matched by the American *Surveyor 1* probe four months later. The successful placement of the artificial satellite *Luna 10* into orbit around the Moon by Russia in April 1966, was duplicated when America sent *Lunar Orbiter 1* into a lunar orbit a little over four months later.

Problem 2.1. The Russian spacecraft *Luna 10* was the first probe to be placed into orbit around the Moon. The craft was launched in March of 1966 into an orbit around Earth and injected on a trajectory towards the Moon from its Earth-orbiting platform. At a distance of 8000 km from the Moon, *Luna 10* rotated to point its main engine forward for an engine burn that slowed the craft by 0.64 km/s. This maneuver put *Luna 10* into a lunar orbit on April 3, 1966. Its distance from the Moon's surface ranged from 350 km to 1017 km. The semimajor axis of the orbit was 2413 km and its period was 178.05 minutes. Use this information to provide the estimate of 7.28×10^{22} kg for the mass of the Moon. [Hint: Be aware that you are provided with some extraneous information.]

With their *Luna 17* mission in November 1970 the Soviets sent the unmanned *Lunokhod 1* rover (the name is Russian for *moonwalker*) to a soft landing on the lunar surface. It was the first human-made vehicle to move freely on a body of the solar system (other than Earth). *Lunokhod 1* explored the Moon for 321 days, traveling a total distance of $10\frac{1}{2}$ km. It returned thousands of television pictures and over 200 high-resolution panoramic images. Along the way, it analyzed the Moon's soil. Several more Russian *Luna* missions followed. One of them sent the rover *Lunokhod 2* to the Moon in 1973. It rolled over 40 km of terrain, including hilly uplands and long depressions, returning many more television pictures and panoramic images. Later, two more *Luna* probes collected samples of lunar soil and to brought them back to Earth.

The United States was beginning to respond to President Kennedy's call. The powerful *Saturn V* rocket was designed and developed. The *Apollo Program* for sending men to the Moon was conceived and planned. A *Saturn V* would carry an *Apollo* spacecraft, consisting of a cone-shaped *Command Module* that carried a crew of three astronauts, and an attached cylindrical *Service Module* that provided the electrical power and the propulsion, into orbit around Earth. From there, propelled by its rocket motor, the *Command/Service Module* would travel to the Moon and go into orbit around it. The Service Module was to carry a *Lunar Excursion Module* that would bring two astronauts to the Moon's surface, while the third would remain at the controls of the *Command/Service Module*. Once the two astronauts would complete their tasks, the *Lunar Excursion Module* would return them to the *Command Module*. The *Lunar Excursion Module* was then to be discarded and the *Command/Service Module* would start its return to Earth. Just before *Apollo*'s reentry into Earth's atmosphere, the *Service Module* was to be disconnected from the *Command Module* and allowed to burn up in the atmosphere. The *Command Module* with the three astronauts would descend by parachute to a safe splashdown in the Pacific Ocean near one of the island chains east of Australia.

The hardware of this concept was tested successfully in 1968 and 1969. The flights of *Apollo 8* and *Apollo 10* tested the *Command/Service Module* in orbit around the Moon, and *Apollo 9* tested the *Lunar Excursion Module* while in orbit around Earth. Finally, on July 20, 1969, the *Apollo 11* mission landed two astronauts for an exploration of the Moon and returned all three of its astronauts safely to Earth. The BBC celebrated the 50 anniversary of this event on July 11, 2019 with the story *Apollo 11: 'The greatest single broadcast in television history'*. See

https://www.bbc.com/news/world-us-canada-48857752

Five more successful *Apollo* expeditions in the years from 1969 to 1972 sent 10 more astronauts to the surface of the Moon. The last three missions carried lunar rovers that the astronauts drove over the Moon's dusty terrain. The Russians were unable to match this astonishing and historic American effort and the race to the Moon was over.

Problem 2.2. While on its way to the Moon, *Apollo 13* experienced an explosion of one of the fuel tanks of its *Service Module* that incapacitated the module. The command center in Houston, in a miraculous display of "seat-of-the pants" ingenuity, found a way to use the *Lunar Excursion Module* as a lifeline to return the three crew members safely to Earth. The video

$$\text{https://www.youtube.com/watch?v=sJ3Q3kL7jcA}$$

entitled "Houston, we have a Problem" tells this dramatic story that gripped the world.

 2. Exploring Venus and Mars. While the race to the Moon provided most of the drama, Russia and the U.S. also competed in efforts to explore Earth's planetary neighbors Venus and Mars.

 With its *Venera Program* of missions in the 1960s, 70s, and early 80s, Russia concentrated primarily on Venus. In August 1970, *Venera 7* became the first spacecraft to execute a soft landing on another planet. Seven more *Venera* spacecraft landed probes on Venus. They returned the first photographs of its terrain, measured its surface temperature, and analyzed its soil and rocks. The last two *Venera* probes orbited Venus, deployed penetrating radar to peer through Venus's dense cloud cover, and mapped a part of its northern hemisphere. The thick atmosphere of Venus, consisting mainly of carbon dioxide with clouds of droplets of sulfuric acid, traps heat in a runaway greenhouse effect. With surface temperatures hot enough to melt lead, Venus is the hottest planet in our solar system. The extreme heat put the *Venera* landers and their instrumentation under great stress. None of them survived for more than 2 hours.

 The first missions of the American *Mariner Program* in the years 1962 to 1968 also focused on Venus, but met with mixed success. In July 1965, after seven months of interplanetary flight, *Mariner 4* flew past Mars and returned high quality photographs and scientific data. Later *Mariner* probes between 1969 and 1973 also performed flybys of Mars. The spacecraft *Mariner 9* was launched in May 1971, and was inserted into orbit around Mars in November 1971 (see Table 2.11). The first

Table 2.11. Orbital data for *Mariner 9*'s initial orbit around Mars and the two corrected orbits that followed. The period—the duration from one periapsis to the next—is listed in Earth days.

Mariner 9 orbits of Mars in 1971	semimajor axis in km	eccentricity	orbit period in hours	angle of orbit plane to Mars equator
initial orbit Nov 14 to Nov 16	13055	0.63	12.62	64.6°
post trim 1 orbit Nov 16 to Dec 31	12631	0.62	11.97	64.8°
post trim 2 orbit Dec 31 to	12647	0.60	11.99	64.4°

man-made object to orbit another planet, it photographed the moons of Mars, mapped 70 percent of the planet's surface, and returned data showing that Mars was geologically and meteorologically active.

Problem 2.3. Compute the constant GM for M the mass of Mars for each of the three different orbits.

Problem 2.4. Use information from the Problems and Discussions section of Chapter 1 to compute the maximum and minimum speeds of *Mariner 9* for each of its three different orbits.

The spacecraft of NASA's two *Viking* missions continued the exploration of Mars from 1975 to 1982. Both carried an orbiter and a lander. After each craft entered into orbit around Mars, the landers separated and descended to the planet's surface. The orbiter of *Viking 1* circled Mars for four years, concluding its mission in August 1980. The lander of *Viking 1* was the first probe to land safely on the surface of Mars. It transmitted data and images of its terrain until November 1982.

The missions *Pioneer 10* and *11* were the first spacecraft to fly through the asteroid belt to the outer planets. *Pioneer 10* arrived at Jupiter in December 1973 and studied its atmosphere and satellites. *Pioneer 11* reached Jupiter a year later. It provided the first observations of Jupiter's polar regions and sent back clear images of its Great Red Spot. *Pioneer 11* continued on to Saturn, reaching it in September 1979. It discovered a small moon, an additional ring, and traveled under its ring plane to return striking pictures of Saturn's rings. It also studied Saturn's largest moon Titan.

The investigation of the solar system has not only increased in pace and reach ever since—see the discussions in earlier sections of this chapter—it has also become international. Successful missions undertaken by the European, Japanese, Chinese, and Indian space agencies have added sophisticated studies—for instance those of asteroids and comets as described in this chapter—to the ongoing explorations by the United States and Russia.

3. Videos and Images about the Exploration of the Solar System. The videos of the websites below provide visual illustrations and important background information for a number of the discussions in this chapter. In case the address of a website has become inactive, it should be possible to use keywords from the context or from its URL address to search and find an updated or related version of the site.

Problem 2.5. Two of the most amazing and productive missions into the solar system were the flights of the *Voyagers*. The videos

https://www.jpl.nasa.gov/video/details.php?id=1514 and

https://www.youtube.com/watch?v=YAnxt1YPWbk

provide an overview of what they achieved.

Problem 2.6. The video https://www.youtube.com/watch?v=0vl0FXPBWnQ captures several launches of the American *Space Shuttle* and https://www.youtube.com/watch?v=--X9zfgZtS0 tells the story of the shuttle flight that sent astronauts on a repair mission to correct the flawed mirror of the *Hubble Space Telescope*.

Problem 2.7. Use the orbital data of Table 2.2 for the moons Phobos and Deimos to compute the mass of Mars in kilograms. (You should get 6.42409×10^{23} kg and 6.39988×10^{23} kg, respectively.)

A number of the videos in the listings below come from the website

https://www.jpl.nasa.gov/video/

that presents a large gallery of JPL-produced videos about the solar system and the spacecraft that have been sent to explore it. To access one of the JPL videos go the particular URL and download a suitable version (e.g., Webm Format) to play it.

Problem 2.8. The videos listed below are brief descriptions of space missions to the Inner Planets. You are invited to explore them.

https://www.jpl.nasa.gov/video/details.php?id=1477 (Overview of the Mars Missions)

https://www.jpl.nasa.gov/video/details.php?id=1518 (*Curiosity*'s Mars Panorama)

https://www.youtube.com/watch?v=NXbCNAIIAxw (Amazing Mars)

https://www.youtube.com/watch?v=zqhK8dA7iO8 (5 years of *Curiosity*)

https://www.jpl.nasa.gov/video/details.php?id=1237 (*Curiosity* watches Phobos pass Deimos)

https://www.youtube.com/watch?v=yzqbN6z8ncc (A Look at Venus)

https://www.youtube.com/watch?v=POLvR56lKjU (*MESSENGER'S* Flight Path to Mercury)

https://www.youtube.com/watch?v=hDrSK3yrGM4 (*MESSENGER* Tells us about Mercury)

Problem 2.9. The videos listed below are brief descriptions of missions to the Outer Planets and their moons. You are invited to fly along.

https://www.youtube.com/watch?v=NNHfoNIiZ8Y (About Jupiter and *Juno*)

https://www.youtube.com/watch?v=kZS4UsOHmLE (*Juno*'s flyby of Earth)

https://www.nasa.gov/mission_pages/juno/earthflyby.html#. (*Juno*'s Flyby of Earth)

https://www.youtube.com/watch?v=ZCxZkf1aVUM (Jupiter by JunoCam in 2017 and 2018)

https://www.jpl.nasa.gov/video/details.php?id=1383 (About Jupiter's Moon Europa)

https://www.missionjuno.swri.edu (About Jupiter and *Juno*)

https://www.nasa.gov/mission_pages/cassini/main/index.html (*Cassini* at Saturn)

https://www.jpl.nasa.gov/video/details.php?id=1458 (*Cassini* and the Secrets of Enceladus)

https://www.jpl.nasa.gov/video/details.php?id=1451(*Cassini* Cruises Saturn's Rings)

https://www.jpl.nasa.gov/video/details.php?id=1466 (About *Cassini*'s Grand Finale)

Problem 2.10. The videos listed are brief descriptions of some asteroids, comets, and the Kuiper belt object Pluto. You are invited to explore them.

https://www.youtube.com/watch?v=mQHfGP5kpr8 (Ten Interesting Asteroids)

https://www.nasa.gov/osiris-rex (About the asteroid Bennu)

https://www.youtube.com/watch?v=F9ihB52Kr3A (About Comets, especially Halley)

https://www.nasa.gov/mission_pages/newhorizons/main/index.html (The *New Horizons* mission)

4. Sungrazing Comets and their Speeds. Recall from Chapter 1G that the semimajor axis a of Earth's orbit around the Sun is very nearly equal to 1 au and that the Earth's orbital period T is very nearly equal to 1 year. So for the Earth, the ratio $\frac{a^3}{T^2}$ is approximately equal to 1 in the units au and year. It follows from Kepler's third law that for any body in orbit around the Sun (planets, asteroids, and periodic comets) that $\frac{a^3}{T^2} \approx 1$, where a is the semimajor axis in au and T the period in years of the body under consideration.

The study of speed in the upcoming context is more meaningful in km/sec rather than au/year. Since 1 au = 149,597,870.7 km and 1 year = 365.25 days = 31,557,600 sec,

$$1\,\text{au/year} \;=\; \frac{149{,}597{,}870.7 \text{ km}}{1 \text{ year}} \times \frac{1 \text{ year}}{31{,}557{,}600 \text{ sec}} \;\approx\; 4.74\,\text{km/sec}.$$

Recall from the paragraph *About Speeds of Objects in the Solar System* of the Problems and Discussions section for Chapter 1 that the maximal speed of any body in an elliptical orbit occurs at perihelion and is equal to $v_{\max} = \frac{2\pi a}{T}\sqrt{\frac{1+\varepsilon}{1-\varepsilon}}$, where a and T are (as before) the semimajor axis and period of the orbit and ε is the eccentricity.

Problem 2.11. Consider any object in an elliptical orbit around the Sun. Use the discussion above to show that

$$v_{\max} \approx 2\pi \sqrt{\frac{1+\varepsilon}{a(1-\varepsilon)}} < \frac{2\pi \cdot \sqrt{2}}{\sqrt{a(1-\varepsilon)}},$$

where $v_{\max}$ is expressed in au/year, and a and $a(1-\varepsilon)$ are the semimajor axis and perihelion distance in au, respectively.

One conclusion that can be drawn from this result is that the term $\frac{2\pi \cdot \sqrt{2}}{\sqrt{a(1-\varepsilon)}}$ with $a(1-\varepsilon)$ the minimal possible perihelion distance in au, is the maximal speed limit in au/year for all objects in an elliptical orbit around the Sun. Astronomers believe that a larger comet has a chance to survive a close encounter with the Sun if its flyby brings it no closer to the Sun's surface than about 25,000 km. Since the average radius of the Sun is about 695,000 km, this means that in order to survive, a comet needs to have a perihelion distance of at least $695{,}000 + 25{,}000 = 720{,}000$ km or 0.0048 au. (We will focus our discussion on comets, but we'll assume that it applies to asteroids as well.)

Problem 2.12. Show that the maximum speed of an object in an elliptical orbit around the Sun cannot exceed 128.25 au/year or about 608 km/sec.

Turn to Chapter 1C and consider an ellipse with semimajor axis a and semiminor axis b. Figure 1.9 tells us that $b = \sqrt{a^2 - a^2\varepsilon^2} = a\sqrt{1-\varepsilon^2}$, where ε is the eccentricity of the ellipse. An ellipse is flat whenever b is small compared to a, and hence when ε is close to 1. The closer ε is to 1, the smaller b is relative to a, and the flatter the ellipse. This implies, in view of the conclusion of Problem 2.11, that the maximum speed of a comet with a very flat elliptical orbit is tightly approximated by $\frac{2\pi\cdot\sqrt{2}}{\sqrt{a(1-\varepsilon)}}$. It follows that among such comets, those with the smallest perihelion distances $a(1-\varepsilon)$ have the greatest maximum speeds. Table 2.8 confirms that the orbits of most of the historic comets are very flat (see the eccentricity data) and that several of them have very small perihelion distances. Comets that orbit the Sun tightly are known as "sungrazing" comets. Since their orbits are generally very flat, they are—when at perihelion—the fastest-moving objects in the solar system.

Some of the most spectacular comets ever observed have been sungrazers. The first sungrazing comet to receive a lot of attention was the Great Comet of 1680. We saw in Chapter 1B that Newton tracked it carefully. The painting by the Dutchman Lieve Verschuier (who was there to observe it himself) of Figure 1.8 informs us that its tail swept across the sky in a spectacular arch. It shows Newton's Dutch contemporaries observing the comet with great interest. Some of them are seen pointing cross-staffs—simple devices for measuring angles—skyward. The comet passed about 235,000 km above the Sun's surface, so that its perihelion distance was close to $695,000 + 235,000 = 930,000$ km or 0.0062 au. Unlike some other sungrazers, Newton's comet survived its close encounter with the Sun.

Problem 2.13. Use the information developed above to show that the maximum speed that the Great Comet of 1680 attained was around 535 km/sec.

During its journey through the inner solar system the Great Comet of 1843 was widely seen at daytime and described as "an elongated white cloud." Observations provided a tight perihelion distance of 820,000 km or 0.0055 au and an orbital eccentricity of $\varepsilon = 0.999914$. After perihelion, the comet diminished in brightness but its tail grew enormously, eventually attaining a length of 320 million km or over 2 au. The next super comet to come along was the Great Comet of 1882. First spotted by a group of Italian sailors in the Southern Hemisphere, it brightened dramatically as it approached its rendezvous with the Sun and became visible in broad daylight. Orbital data implied that its perihelion distance was 1,150,000 km or about 0.0077 au and its orbital eccentricity 0.999907. During its perihelion passage the nucleus broke into several parts, but the comet emerged from behind the Sun and was described as a "blazing star." In the days and weeks that followed, its tail continued to shine brilliantly. The brightest comet of the 20th century was discovered in the fall of 1965 only a little over a month before its perihelion passage by the two Japanese amateur astronomers Ikeya and Seki. The comet was glowing in the sky "ten times brighter than the Full Moon." Before perihelion, its nucleus was observed to have broken up. The two new nuclei emerged with slightly different orbits, both with eccentricity 0.999918 and perihelion distance 1,170,000 km or 0.0078 au. After it passed the Sun, its 120 million km long tail dominated the morning sky.

Problem 2.14. Estimate the maximum speeds of the great comets of 1843, 1882, and 1965 in km/sec.

Terry Lovejoy, the astronomer who discovered the comet C/2014 Q2 (see Figure 2.23), had spotted another comet three years earlier in November 2011. The elliptical orbit of the comet (later designated C/2011 W3) was observed to have a perihelion distance of 0.0055 au and an eccentricity of 0.99993. At perihelion on the 16th of December 2011, it passed close to the Sun's surface and was not expected to survive the severe conditions, such as extreme gravitational stresses and the exposure to temperatures of more than one million degrees Celsius, that it would encounter during its trip through the corona (the Sun's atmosphere). However, the comet emerged from the corona intact. The incredible video

https://phys.org/news/2012-03-comet-lovejoy-survive-sun.html

taken by NASA's Solar Dynamics Observatory (see Figure 1.27) caught the comet scurrying through the Sun's corona. While it survived its perihelion passage, it emerged with a degraded nucleus.

Problem 2.15. Show that the speed of the comet C/2011 W3 at perihelion was approximately 120 au/year and hence about 568 km/sec.

The narrative of this paragraph has not considered the complicating factor that the ellipses of all objects in orbit around the Sun do not have the center of the Sun but the barycenter of the solar system as focal point (see Chapter 1H). A determination of the maximum speeds of the sungrazing comets that takes this into account—see Figure 1.28 in this regard—would require a detour probably not scenic enough for the aims of this fundamental text. Instead, we'll make a comment about the barycenters of the planetary systems.

5. The Barycenters of the Planetary Systems. We saw in Chapter 1E in reference to the motion of the Earth and Moon that it is the center of mass of the Earth-Moon system—the barycenter of the system—that sits at a focal point of the elliptical orbit of the Moon around the Earth and also of the elliptical wobble that the Earth undergoes. Given that man-made Earth satellites have relatively little mass, this barycenter is also a focal point of their elliptical orbits around Earth. It is this barycenter that is in an elliptical orbit around the Sun. In the same way, it is the barycenters of the other planet-moon systems that orbit the Sun along elliptical paths. And, as pointed out in Chapter 1I, it is the barycenter of the entire solar system rather than the center of mass of the Sun that is at a focal point of the elliptical solar orbits of these systems. Figure 1.28 gives a sense of the position of this barycenter relative to the center of mass of the Sun.

To continue this discussion, we'll consider data about the planet-moon systems. Table 2.12 provides information about the sizes and masses of the planets and the mass of the most massive moon for each system of moons. We see from the table that the Sun is over a 1000 times more massive than Jupiter and that Jupiter is by far the most massive planet. The Sun is over 3000 times more massive than Saturn and almost 20,000 times more massive than Neptune, the second and third most massive of the planets. The table also informs us that Mercury and Venus have no moons and that the largest of the two moons of Mars is small. For any of the planet-moon systems, let B be the barycenter and C the center of mass of the planet. The barycenter of the system depends on the location of the moons of the planet at a given time. It follows that the barycenter moves and that the distance between B and C varies over time. Let d_{BC} be the maximum distance between B and C over the time period from the year 2000 to the year 2050. The last two columns of the table provide estimates for d_{BC} both in terms of kilometers and the radius of the planet. The

Table 2.12. From NASA and NAIF fact sheets, 2013.

System	average radius of central body in km	mass of central body in kg	most massive moon in kg	d_{BC} in km	d_{BC} in radii
Sun	695,508	1.9885×10^{30}	–	1,378,196	1.98
Mercury	2,439.7	3.3010×10^{23}	0	0	0
Venus	6,051.8	4.8676×10^{24}	0	0	0
Earth	6,371.0	5.9726×10^{24}	7.4×10^{22}	4942	0.8
Mars	3,389.5	6.4174×10^{23}	1.06×10^{16}	0.002	≈ 0
Jupiter	69,911	1.8983×10^{27}	1.48×10^{23}	220	0.003
Saturn	58,232	5.6836×10^{26}	1.34×10^{23}	312	0.005
Uranus	25,362	8.6816×10^{25}	3.5×10^{21}	43	0.00017
Neptune	24,622	1.0242×10^{26}	2.14×10^{22}	74	0.003

last column informs us that the location of the barycenters of all the planetary systems remain within the body of the planet (at least for the fifty years in question). Figure 1.28 tells us by contrast, that the barycenter of the solar system spent about 60% of the 50 years from 1945 to 1995 outside the body of the Sun. We see from the last column of the table that the only planetary system for which the distance between the barycenter and the center of mass of the planet is substantial is the Earth-Moon system. For all other planets this distance is reasonably small, so that the distinction between barycenter and center of the central body is of lesser relevance.

6. Dawn's Ion Propulsion Engine. Recall from section 2D that in 2007 the spacecraft *Dawn* was sent on a mission to study the large asteroids Vesta and Ceres. One unique aspect of *Dawn* is the revolutionary propulsion system of its main engine. The basic principle behind it is still Newton's third law, namely that the force generated is the "equal and opposite" of the force with which the engine expels particles in the opposite direction. However in *Dawn's* engine, the expelled particles are not the exhaust mass produced by the explosion of a propellant in a combustion chamber. Instead, they are atoms of the gas xenon that are given a positive charge and turned into xenon ions via the bombardment of xenon gas with high energy electrons. These positively charged xenon ions are pulled into a grid that is negatively charged. They are accelerated and shot out in the form of a streaming ion beam with a velocity of up to 150,000 km per hour. The exhaust velocity of the ions in the beam is determined by the voltage applied. The required electric power is drawn from the craft's solar panels. Xenon gas is ideal for this use. It is easily ionized, chemically inert, and has a high storage density.

Whereas a typical liquid propellant engine generates a thrust of 500 or more newtons, the maximum thrust produced by *Dawn's* ion engine is only 0.091 newtons. A stack of four pennies in the palm of your hand pushes with a greater force. At full throttle, it would take the *Dawn* spacecraft four days to accelerate from zero to 100 km per hour. But by firing for about 50,000 hours (over 5 years) over the duration of its mission in the zero gravity, frictionless, environment of space, the small effect of the thrust of its engine provided *Dawn* with the additional velocity it needed to reach orbit around

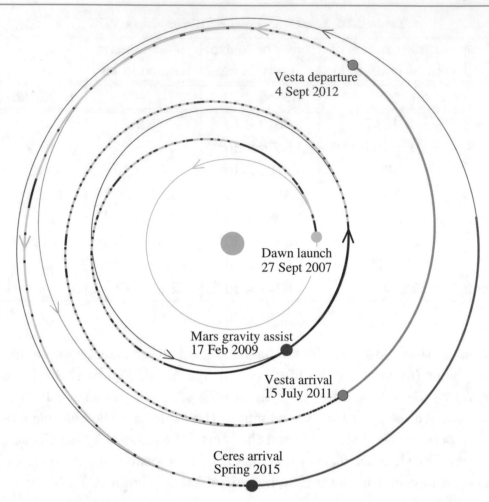

Figure 2.33. The color coding shows the part of *Dawn*'s flight driven by its ion propulsion engine in blue (with underlying black dots and streaks) and the gravity-driven part of the flight in black. Here too, gravity assist is an important component of the flight path. Image credit: NASA/JPL-Caltech.

Vesta, to spiral to lower altitudes over Vesta, to cruise to Ceres, and to spiral to a low altitude orbit around Ceres. Figure 2.33 depicts *Dawn*'s flight path. By firing over a long period of time, *Dawn*'s small engine achieved trajectory changes equivalent to those produced by liquid propellant engines that fired intermittently in short bursts of only minutes in duration. The video

https://vimeo.com/117835245

presents a simulation of *Dawn*'s orbit.

The importance of the rectangular, or Cartesian coordinate system to the disciplines of geometry, trigonometry, and calculus cannot be overstated. These disciplines depend on the interplay between geometry and algebra that the Cartesian coordinate system makes possible. However, this system is not the only one that makes this connection. We will now discuss a coordinate system, the *Polar Coordinate System* originally introduced by Isaac Newton, that provides a better framework for some mathematical investigations. This is a system that identifies a location in a plane with respect to a fixed reference point, rather than an intersecting pair of perpendicular lines. It is therefore tailor-made for the analysis of the gravitational force produced by a central body and the trajectories of objects that move in response to it. It is therefore ideal for the study of the orbits of the planets, asteroids, and comets around the Sun, the moons around the planets, as well as the flight of spacecraft.

This chapter presents the essential aspects of polar coordinates and its mathematics in detail. It studies polar functions and their graphs with a special focus on the conic sections. Full details are provided and the interplay between Cartesian and polar equations is explored. The derivative of a polar function and its geometric meaning are developed in full. The definite integral of polar functions is studied in the context of the lengths of polar graphs and the areas that they encircle. The geometric interpretation of the derivative leads to the definition of the equiangular spiral. These are discovered to play a role in the geometry of spiral galaxies in general and our Milky Way galaxy in particular.

3A. The Unit Circle and Trigonometry. We'll assume that the reader is familiar with the basics of trigonometry, but we will recall some of the fundamental definitions and concepts. Let θ be any real number. Assume first that $\theta \geq 0$ and consider a segment of length θ as depicted in Figure 3.1. Let A be the right endpoint of the segment and label the left endpoint by P_θ. This notation tells us that when distance is measured from A, the location of the left endpoint is determined by the

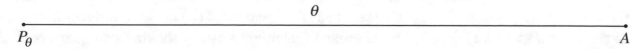

Figure 3.1

A. J. Hahn (ed.), *Basic Calculus of Planetary Orbits and Interplanetary Flight*,
https://doi.org/10.1007/978-3-030-24868-0_3

length θ. Now take a Cartesian xy-plane and consider the circle of radius 1 and center the origin O. This circle is known as the *unit circle*. See Figure 3.2. Measure off the distance θ along the perimeter of the unit circle: start at the point $(1,0)$ and proceed in a *counterclockwise* direction until the entire distance θ is measured off and the left endpoint P_θ is placed on the circle. Think of the segment as a string of length θ, place its right end A at the point $(1,0)$ and wind it counterclockwise (possibly many times) around the circle until the left endpoint P_θ lands. This is illustrated in the figure. What if θ is a negative real number? Then $-\theta$ is positive, and we'll let the segment of Figure 3.1 have length $-\theta$. To place the point P_θ on the circle in this case, measure off the length $-\theta$ on the perimeter. As before, start at $A = (1,0)$, but this time go in the *clockwise* direction to measure off the segment and to locate the point P_θ.

We can now interpret any real number θ as an angle. Consider the segment from O to $(1,0)$. It is shown in green in Figure 3.2. Keep the end at O fixed, but let the segment be free to rotate around it. The number θ interpreted as angle is the opening generated by letting the free end of the segment follow around the perimeter of the circle, counterclockwise if $\theta \geq 0$ and clockwise if $\theta < 0$, possibly many times, until it reaches the point P_θ. This opening is the angle that corresponds to

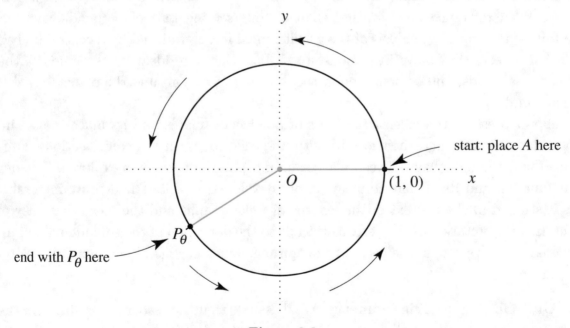

Figure 3.2

the number θ. The number θ is known as the *radian measure* of the angle.

Let's look at some basic examples of the concepts that were just introduced. Since the radius of the unit circle is 1, its circumference is 2π. For the angle that corresponds to the real number $\theta = \frac{\pi}{2}$, start at $(1,0)$, go around the perimeter of the first quarter of the circle counterclockwise, and stop at the point $P_{\frac{\pi}{2}} = (0,1)$. For $\theta = \pi$, go around (counterclockwise) the first two quarters of the circle and stop at $P_\pi = (-1,0)$. For $\theta = \frac{3\pi}{2}$, go around the first three quarters to $P_{\frac{3\pi}{2}} = (0,-1)$. For $\theta = 2\pi = \frac{4\pi}{2}$, go around all four quarters to $P_\theta = P_{2\pi} = A = (1,0)$. For the angle that corresponds to $\theta = -2\pi$, start at $(1,0)$ and go around the perimeter of the circle clockwise for a distance of 2π

to end up back at $(1,0)$. To measure off $\theta = -\pi$, go around one-half the circle (clockwise) and stop at $P_{-\pi} = (-1,0)$, For $\theta = -\frac{\pi}{2}$, start at $(1,0)$, go around a quarter circle (clockwise) and stop at the point $P_{-\frac{\pi}{2}} = (0,-1)$, and so on.

What about degrees? The fact that the angle of π radians corresponds to $180°$ means that an angle of 1 radian is equal to $\left(\frac{180}{\pi}\right)° \approx 57.30°$. Therefore an angle of θ radians has $\theta \cdot \frac{180}{\pi}$ degrees. So for example, an angle of 100π radians is equal to $18,000°$, an angle of 10π radians is equal to $1800°$, an angle of 0.1π radians has $18°$, and an angle of 0.01π radians has $1.8°$. In the other direction, an angle of 1 degree is equal to $\frac{\pi}{180}$ in radians. So an angle of θ degrees is equal to $\theta \cdot \frac{\pi}{180}$ in radians. For example, the angle $10°$ has $10 \cdot \frac{\pi}{180} \approx 0.175$ radians and $100°$ is equal to $100 \cdot \frac{\pi}{180} \approx 1.745$ radians.

We now turn to the definitions of the trigonometric quantities $\sin\theta$ and $\cos\theta$ for any real number θ. Continue to consider the unit circle of Figure 3.2. For any real number θ, locate the point P_θ. Let the coordinates of P_θ be x and y and define

$$\cos\theta = x \quad \text{and} \quad \sin\theta = y$$

as illustrated in Figure 3.3. Since $P_{\frac{\pi}{2}} = (0,1)$, we see that $\cos\frac{\pi}{2} = 0$ and $\sin\frac{\pi}{2} = 1$. Since

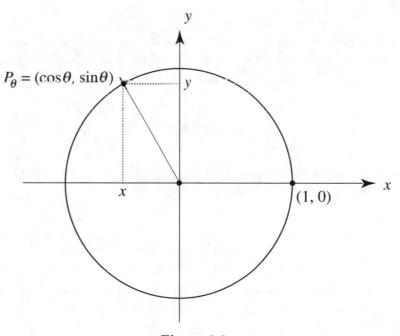

Figure 3.3

$P_\pi = (-1,0)$, we get $\cos\pi = -1$ and $\sin\pi = 0$. Finally, $P_{\frac{3\pi}{2}} = (0,-1)$ implies that $\cos\frac{3\pi}{2} = 0$ and $\sin\frac{3\pi}{2} = -1$.

Let θ be an angle given in degrees. The sine and cosine of θ are equal to the sine and cosine of the radian equivalent $\theta \cdot \frac{\pi}{180}$ of θ. So with θ in degrees, $\sin\theta = \sin\left(\theta \cdot \frac{\pi}{180}\right)$ and $\cos\theta = \cos\left(\theta \cdot \frac{\pi}{180}\right)$.

Example 3.1. Consider a typical angle θ with $0 < \theta < \frac{\pi}{2}$. Draw the point P_θ into a copy of Figure 3.3 and drop a perpendicular from this point to the x-axis to form a right triangle that has hypotenuse equal to 1. Let a and b be the horizontal and vertical sides of this triangle and conclude that $\cos\theta = a$ and $\sin\theta = b$.

Example 3.2. Use an equilateral triangle of side 1 and an isosceles right triangle with hypotenuse 1 to show that

$$\cos\frac{\pi}{6} = \frac{\sqrt{3}}{2} \text{ and } \sin\frac{\pi}{6} = \frac{1}{2}, \ \cos\frac{\pi}{3} = \frac{1}{2} \text{ and } \sin\frac{\pi}{3} = \frac{\sqrt{3}}{2}, \text{ and } \cos\frac{\pi}{4} = \frac{1}{\sqrt{2}} \text{ and } \sin\frac{\pi}{4} = \frac{1}{\sqrt{2}}.$$

For any real number θ the point $P_\theta = (\cos\theta, \sin\theta)$ is on the unit circle. So it satisfies the equation $x^2 + y^2 = 1$, and therefore

$$\sin^2\theta + \cos^2\theta = 1.$$

Refer to Figure 3.3 once more. By observing the moving point P_θ, notice that as θ varies from 0 to $\frac{\pi}{2}$, $\sin\theta$ varies from 0 to 1, and as θ moves from $\frac{\pi}{2}$ to π, $\sin\theta$ varies from 1 to 0, and so on. Plotting the various points $(\theta, \sin\theta)$ provides the graph of the sine function $f(\theta) = \sin\theta$. Doing a

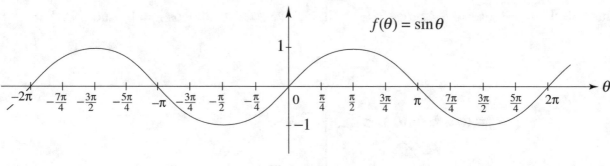

Figure 3.4

similar thing for the cosine, gives us the graph of the function $g(\theta) = \cos\theta$. See Figures 3.4 and 3.5.

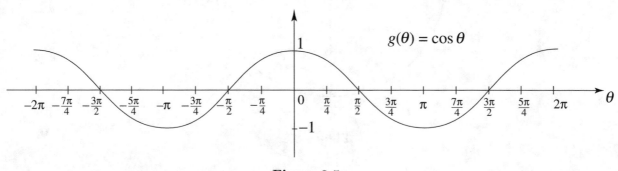

Figure 3.5

Let θ be any real number and observe that P_θ and $P_{(\theta+2\pi)}$ end up being the same point. Therefore

$$\sin(\theta + 2\pi) = \sin\theta \quad \text{and} \quad \cos(\theta + 2\pi) = \cos\theta.$$

Consider any θ as well as $-\theta$. A comparison of the points P_θ and $P_{-\theta}$ (see Figure 3.6a, for example) tells us that they have the same x-coordinates, and that the y-coordinate of one is the negative of

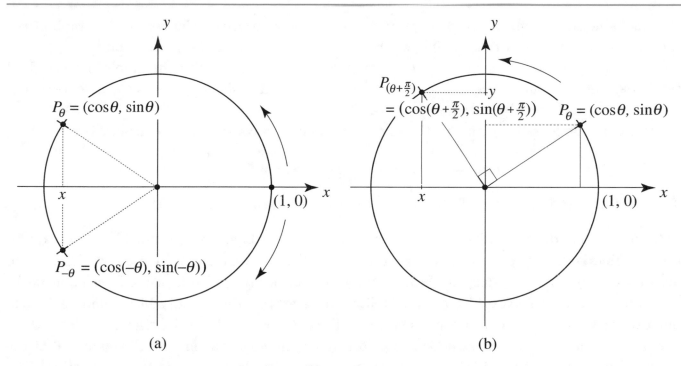

Figure 3.6

the y-coordinate of the other. It follows that

$$\cos(-\theta) = \cos\theta \ \text{ and } \ \sin(-\theta) = -\sin\theta.$$

Next, consider any θ and also $\theta + \frac{\pi}{2}$ as well as the corresponding points P_θ and $P_{(\theta+\frac{\pi}{2})}$. Figure 3.6b illustrates a typical situation. A careful study of the figure tells us that the triangle determined by P_θ and its coordinates is similar to the triangle determined by $P_{(\theta+\frac{\pi}{2})}$ and its coordinates x and y. It follows from this and the figure that $\cos(\theta + \frac{\pi}{2}) = x = -\sin\theta$ and that $\sin(\theta + \frac{\pi}{2}) = y = \cos\theta$. Thus the identities

$$\cos(\theta + \frac{\pi}{2}) = -\sin\theta \ \text{ and } \ \sin(\theta + \frac{\pi}{2}) = \cos\theta$$

are consequence of the study of Figure 3.6b.

Example 3.3. Verify the identities $\cos(\pi - \theta) = -\cos\theta$ and $\sin(\pi - \theta) = \sin\theta$ in two ways. First by applying identities already derived. Then again by making use of a diagram similar to those in Figure 3.6. Why are these formulas valid with $-\pi$ in place of π? (Consider applying the identities $\sin(-\theta) = -\sin\theta$ and $\cos(-\theta) = \cos\theta$.)

Let's have a brief look at the tangent function

$$\tan\theta = \frac{\sin\theta}{\cos\theta}.$$

Suppose first that $0 \le \theta \le \frac{\pi}{2}$. If $\theta = 0$, then $\tan\theta = 0$. The graphs of the sine and cosine tell us that as θ increases, the sine increases and the cosine decreases. Since both are positive, $\tan\theta$ increases. When θ is close to $\frac{\pi}{2}$, the sine is close to 1 and the cosine is close to 0, so $\tan\theta$ is very large. If $\theta = \frac{\pi}{2}$, then $\cos\theta = 0$, so that $\tan\theta$ is not defined. When $-\frac{\pi}{2} \le \theta \le 0$, the situation is similar. Since the sine is negative and the cosine is positive, $\tan\theta$ is now negative. The rest of the graph of the tangent is simply a repetition of the pattern for $-\frac{\pi}{2} \leq \theta \leq \frac{\pi}{2}$. Use the information that you now have to sketch the graph of $h(\theta) = \tan\theta$. Compare what you drew with the graph in a standard textbook. Let θ be any angle. Since $\sin(\theta + \pi) = -\sin\theta$ and $\cos(\theta + \pi) = -\cos\theta$, it follows that $\tan(\theta + \pi) = \tan\theta$. Similarly, $\tan(-\theta) = -\tan\theta$.

The trig functions that remain—the secant, cosecant, and cotangent—are defined by

$$\sec\theta = \frac{1}{\cos\theta}, \ \csc\theta = \frac{1}{\sin\theta}, \ \text{and} \ \cot\theta = \frac{1}{\tan\theta}.$$

The functions $\csc\theta$ and $\cot\theta$ will be of little relevance in the discussions of this text.

3B. Polar Coordinates. The Cartesian or rectangular coordinate system provides a way to represent points and curves in a plane in terms of numbers and equations. We'll now describe an alternative way for doing this, the *polar coordinate system*. Start with a plane, fix a point on it, and call it the *origin*. Next, fix a straight line that starts at the origin. Take a unit of length, mark off points on the line at distances one, two, three, and so on, from the origin, and label them $1, 2, 3, \ldots$. Complete this construction to a positive real number system with the number 0 at the origin. This numbered line is the *polar axis*. The origin, also labeled O, is the *polar origin*. The

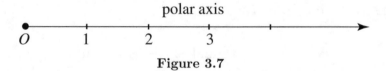

Figure 3.7

polar axis is customarily drawn horizontally and directed from the origin O to the right. See Figure 3.7.

Let P be any point in the plane and draw the segment OP. Let r be the length of this segment and let θ be the angle in radians that it makes with the polar axis. See Figure 3.8. Following our script, we regard θ to be positive if it is measured off counterclockwise and negative if it is measured

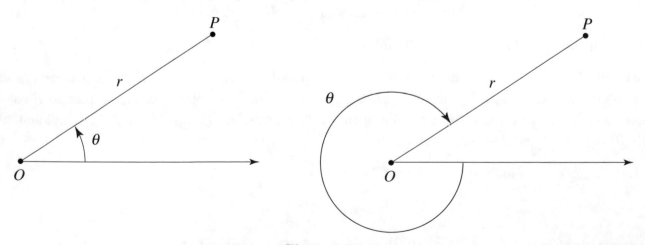

Figure 3.8

in the clockwise direction. Either way, the point P is determined by the ordered pair (r, θ) of real numbers. In each case, the numbers r and θ are called *polar coordinates* of P. This procedure can be reversed. Namely, to any pair of real numbers (r, θ) there corresponds a point in the plane. For a positive r, this is illustrated in Figures 3.9a, b, and c. If r is negative, then the understanding is that the point that corresponds to a pair (r, θ) with a negative r is obtained by taking the arrow,

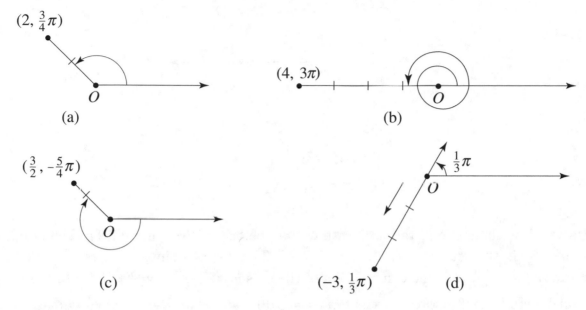

Figure 3.9

or ray, that θ determines and marking off the distance $|r|$ in the direction opposite to that of the ray, in other words along the ray $\theta + \pi$. Figure 3.9d provides an example. One difference between the Cartesian coordinate system and the polar coordinate system as just described is the fact that a point P can be represented in many (indeed infinitely many) ways as a pair (r, θ). Figure 3.10

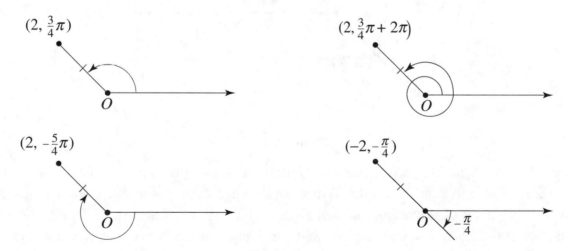

Figure 3.10

shows how four different pairs of polar coordinates all determine the same point.

3C. Polar Functions and their Graphs. We are given a plane equipped with a polar coordinate system. The *graph* of an equation involving the variables r and θ is the set of all points (r, θ) in the plane whose coordinates satisfy the equation.

Consider the polar equation $r^2 = \theta$ for instance, and notice that the points $(0, 0), (1, 1), (-1, 1),$ $(\frac{\pi}{2}, \frac{\pi^2}{4}), (-\frac{\pi}{2}, \frac{\pi^2}{4})$ are all on its graph. The graph of the equation $r = 5$ is the set of all (r, θ) with

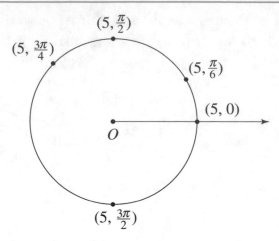

Figure 3.11

$r = 5$. Its graph is the circle of radius 5 with center the origin. See Figure 3.11. The graph of the equation $\theta = \frac{5\pi}{4}$ is the set of all (r, θ) with $\theta = \frac{5\pi}{4}$. This set of points consists of the entire line through the ray $\theta = \frac{5\pi}{4}$. It is the combination of the rays $\theta = \frac{\pi}{4}$ and $\theta = \frac{5\pi}{4}$. Refer to Figure 3.12.

We will be studying functions of the form $r = f(\theta)$ that relate the polar coordinates r and θ.

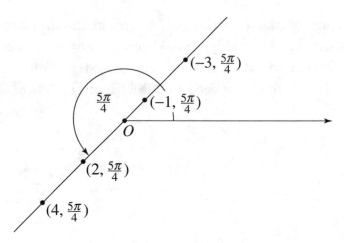

Figure 3.12

For a given θ in its domain, such a function f determines exactly one r. The constant function $r = f(\theta) = 5$ is a very simple example. When considering such polar functions as well as their graphs, it will often be of advantage to make use of Cartesian information. To set the stage for the transfer of such information, place an xy-coordinate system on top of the polar coordinate system in such a way that the two origins coincide, the polar axis coincides with the positive x-axis, the negative x-axis is obtained by extension to the other side of the origin O, and the y-axis is perpendicular to the x-axis at O. Suppose that this has been done. Let P be any point in the plane and let (r, θ) be any pair of polar coordinates for P. What are the x- and y-coordinates of P? Consider Figure 3.13. The figure illustrates a situation where both θ and r are negative. It shows the point P, the polar coordinate θ, and the unit circle. The distance OP is equal to $-r$. From the definitions of $\cos \theta$ and $\sin \theta$, we know that $(\cos \theta, \sin \theta)$ is the point on the unit circle that θ determines. The point P_1 on the unit circle has x- and y-coordinates $-\cos \theta$ and $-\sin \theta$. Because

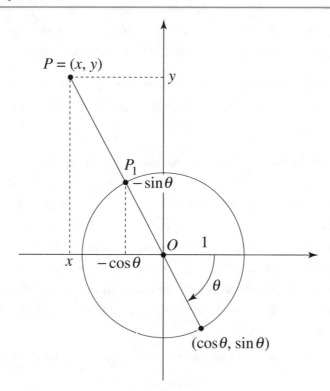

Figure 3.13

the triangles determined by P and P_1, the vertical dashed lines, and the negative x-axis are similar, we see that $\frac{1}{-r} = \frac{\cos\theta}{-x}$, and in the same way, that $\frac{1}{-r} = \frac{-\sin\theta}{y}$. Therefore,

$$x = r\cos\theta \text{ and } y = r\sin\theta.$$

It follows from these equalities that

$$r^2 = x^2 + y^2 \text{ and } \tan\theta = \frac{y}{x}.$$

These relationships hold regardless of the location of the point P and regardless of the choice of its polar coordinates. For instance, the fact that the polar coordinates $(-r, \theta + \pi)$ represent the same point as (r, θ) means that the corresponding Cartesian coordinates of the point are also the same. Therefore it must be the case that

$$x = r\cos\theta = -r\cos(\theta + \pi) \quad \text{and} \quad y = r\sin\theta = -r\sin(\theta + \pi).$$

But this follows directly from the basic trig identities $\sin(\theta + \pi) = -\sin\theta$ and $\cos(\theta + \pi) = -\cos\theta$.

We have seen that if (r, θ) is any set of polar coordinates of P, then the corresponding Cartesian coordinates for P can be computed directly from the equations $x = r\cos\theta$ and $y = r\sin\theta$. What if instead we are given Cartesian coordinates (x, y) for P and wish to determine a set of polar coordinates? If P is on the y-axis, then $P = (0, y)$ so that $(y, \frac{\pi}{2})$ is a set of polar coordinates for P. If P is not on the y-axis, then θ can be chosen to satisfy $-\frac{\pi}{2} < \theta < \frac{\pi}{2}$. See Figure 3.13 for instance (but check other cases as well). Because $\tan\theta = \frac{y}{x}$ (check this for P in each of the four quadrants) and $-\frac{\pi}{2} < \theta < \frac{\pi}{2}$, this θ is given by the definition of the inverse tangent function as

$$\theta = \tan^{-1}\frac{y}{x}.$$

The corresponding coordinate r satisfies $r^2 = x^2 + y^2$. So either $r = \sqrt{x^2 + y^2}$ or $r = -\sqrt{x^2 + y^2}$. The location of P in the plane determines which of these two possibilities applies.

Example 3.4. Determine the Cartesian coordinates for the point with polar coordinates $(3, 7)$. Then determine polar coordinates for the Cartesian point $(-4, -5)$. In each case, use a calculator to find decimal approximations of the coordinates.

Let's apply what we have learned by analyzing the polar function $r = f(\theta) = \sin\theta$ and its polar graph. Study the Cartesian graph of $\sin\theta$ from Figure 3.4 and observe: the graph of the sin function is increasing and concave down over the interval $0 \leq \theta \leq \frac{\pi}{2}$, decreasing and concave down over $\frac{\pi}{2} \leq \theta \leq \pi$, decreasing and concave up over $\pi \leq \theta \leq \frac{3\pi}{2}$, and so on. So the essential properties of the sin function have a uniform description over each of these intervals. The same is true for the analogous intervals in the negative direction. In order to understand the polar graph of the polar function $r = \sin\theta$, we will therefore consider θ over the interval $0 \leq \theta \leq \frac{\pi}{2}$, then over $\frac{\pi}{2} \leq \theta \leq \pi$, then over $\pi \leq \theta \leq \frac{3\pi}{2}$, and so forth. The information in Table 3.1 can be read off directly from the Cartesian graph of Figure 3.4.

Now to the polar graph of $f(\theta) = \sin\theta$. Notice that as the ray determined by θ rotates from

Table 3.1

1	2	3	4	5	6	7	8
$0 \leq \theta \leq \frac{\pi}{2}$	$\frac{\pi}{2} \leq \theta \leq \pi$	$\pi \leq \theta \leq \frac{3\pi}{2}$	$\frac{3\pi}{2} \leq \theta \leq 2\pi$	$0 \geq \theta \geq -\frac{\pi}{2}$	$-\frac{\pi}{2} \geq \theta \geq -\pi$	$-\pi \geq \theta \geq -\frac{3\pi}{2}$	$-\frac{3\pi}{2} \geq \theta - 2\pi$
$0 \xrightarrow{\sin\theta} 1$	$1 \xrightarrow{\sin\theta} 0$	$0 \xrightarrow{\sin\theta} -1$	$-1 \xrightarrow{\sin\theta} 0$	$0 \xrightarrow{\sin\theta} -1$	$-1 \xrightarrow{\sin\theta} 0$	$0 \xrightarrow{\sin\theta} 1$	$1 \xrightarrow{\sin\theta} 0$

$\theta = 0$ to $\theta = \frac{\pi}{2}$, the corresponding $r = \sin\theta$ stretches from $r = 0$ to $r = 1$. So the polar graph for this range of θ is an arc that curves from the point $(0, 0)$ to the point $(1, \frac{\pi}{2})$. As the ray continues its swing from $\theta = \frac{\pi}{2}$ to $\theta = \pi$, r moves from $r = 1$ to $r = 0$. The corresponding polar graph is an

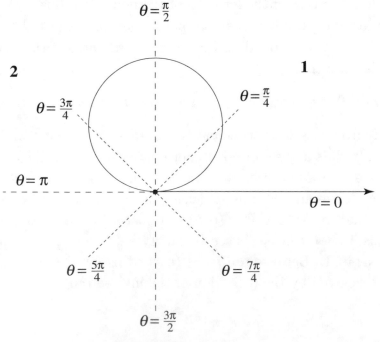

Figure 3.14

arc from $(1, \frac{\pi}{2})$ back down to $(0,0)$. It follows that as θ varies from 0 to π the polar graph loops from $(0,0)$ back to $(0,0)$. As sketched in Figure 3.14, the loop is a circle. But is it a circle?

To answer, we place an xy-coordinate system over the polar coordinate system as already described. Let P be a point in the plane. Let (x,y) be the Cartesian coordinates of P and (r,θ) any set of polar coordinates of P. If $P = (r,\theta)$ satisfies $r = \sin\theta$, then $r^2 = r\sin\theta$. In view of the equalities $r^2 = x^2 + y^2$ and $y = r\sin\theta$, we get $x^2 + y^2 = y$. So $x^2 + y^2 - y = 0$ is a Cartesian version of the equation $r = \sin\theta$. After completing squares in y, $x^2 + y^2 - y + (\frac{1}{2})^2 - (\frac{1}{2})^2 = 0$ and hence $x^2 + (y - \frac{1}{2})^2 = (\frac{1}{2})^2$. This confirms that the loop in Figure 3.14 is a circle with center $(0, \frac{1}{2})$ in the Cartesian xy-coordinates or $(\frac{1}{2}, \frac{\pi}{2})$ in polar coordinates. The radius of the circle is $\frac{1}{2}$.

Example 3.5. Do we arrive at the complete graph of $f(\theta) = \sin\theta$ by considering only the θ in the interval $0 \le \theta \le \pi$? Or is there more? Is the discussion that identifies the graph as the circle $x^2 + (y - \frac{1}{2})^2 = (\frac{1}{2})^2$ relevant to this question? Where does the graph of the function $f(\theta) = \sin\theta$ fall for θ in the intervals of columns **3** and **4** of Table 3.1. What about columns **5**, **6**, **7**, and **8**?

Example 3.6. Consider the polar function $r = f(\theta) = \tan\theta$ for $-\frac{\pi}{2} < \theta < \frac{\pi}{2}$. Use a calculator to plot the points corresponding to $\theta = 0, \frac{\pi}{6}, \frac{\pi}{4}, \frac{\pi}{3}, \frac{5\pi}{12}, \frac{11\pi}{24}, \frac{23\pi}{48}$, and finally $\frac{\pi}{2}$.

Example 3.7. Consider the polar function $r = f(\theta) = \tan\theta$. Let $0 \le \theta < \frac{\pi}{2}$ and convert $r = \tan\theta$ into an equation in the coordinates x and y. Use it to show that $x = \frac{\frac{y}{x}}{\sqrt{1 + (\frac{y}{x})^2}}$. Conclude that the graph of $r = \tan\theta$ for $0 < \theta < \frac{\pi}{2}$ lies between the vertical lines $x = 0$ and $x = 1$. Show that as $\theta \to \frac{\pi}{2}$, the graph of $r = \tan\theta$ approaches the line $x = 1$. Do a similar analysis for $-\frac{\pi}{2} < \theta \le 0$. Then sketch the polar graph of the equation $r = \tan\theta$ for $-\frac{\pi}{2} < \theta < \frac{\pi}{2}$.

The approach to the study of a polar function illustrated above is this: place a Cartesian xy-coordinate system on top of the polar coordinate system, convert the equation in r and θ given by the function into an equation in x and y, and make use of the information this equation provides.

The next section uses basic facts about the ellipse, parabola and hyperbola developed in Chapter 1C. It also makes use of the following basic observation. Take an equation in x and y

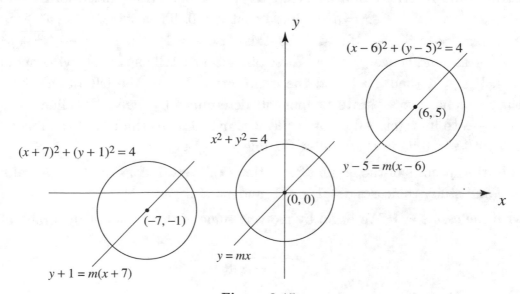

Figure 3.15

and consider its Cartesian graph. Suppose that the graph is shifted or *translated* in the xy-plane vertically and/or horizontally in such a way that it is not rotated in the process. Can an equation for such a shifted graph be identified? The matter of shifting graphs is best illustrated by observing what happens with circles and lines. Start with the circle $x^2 + y^2 = 4$ of radius 2 and center the origin $(0,0)$. Replacing x by $x - 6$ and y by $y - 5$ in this equation gives the equation $(x-6)^2 + (y-5)^2 = 4$. This circle also has radius 2, but its center has been shifted to the point $(6,5)$. In the same way, if we replace x and y by $x + 7$ and $y + 1$, respectively, then the center of the circle is shifted from $(0,0)$ to $(-7,-1)$. See Figure 3.15. The point-slope form of the equation of a line tells us that the equation of a line gotten by translating the line $y = mx$ is obtained in the same way. These considerations apply to any graph. For example, the graph of $y - 2 = (x+3)^2$ has the same shape and orientation as the graph of $y = x^2$. It is obtained by translating the graph of $y = x^2$ in such a way that the origin $(0,0)$ ends up at the point $(-3,2)$. These examples tell us that if in an equation in x and y, x is replaced by $x - c$ and y by $y - d$, then the graph is translated by c units horizontally and d units vertically. The directions of the translations (up, down, left, right) are determined by the signs (positive or negative) of the constants c and d.

3D. The Conic Sections in Polar Coordinates. A plane is provided with both a polar and a Cartesian xy-coordinate system. The origins coincide, the positive x-axis is the polar axis, and the y-axis is perpendicular to the x-axis at the origin O.

Let $\varepsilon \geq 0$ and $d > 0$ be constants. This section will examine the graph of the polar function

$$(*) \qquad\qquad r = f(\theta) = \frac{d}{1 + \varepsilon \cos \theta}.$$

The strategy described in the previous section—namely study of Cartesian versions of this equation—will play a decisive role.

We'll begin with some general observations. Suppose that $\varepsilon \leq 1$. Since $-1 \leq \cos \theta$, we see that $-1 \leq -\varepsilon \leq \varepsilon \cos \theta$, and therefore that $1 + \varepsilon \cos \theta \geq 0$. So if $\varepsilon \leq 1$, then $r > 0$ whenever r is defined (whenever $1 + \varepsilon \cos \theta > 0$). If $\varepsilon > 1$, then both $r > 0$ and $r < 0$ are possible. Consider $\theta = 0$ and $\theta = \pi$, for instance. Now return to equation $(*)$ for any $\varepsilon \geq 0$, but write it in the form $r + \varepsilon r \cos \theta = d$. Because $x = r \cos \theta$, we get $r + \varepsilon x = d$. Suppose that $r > 0$. Then $r = \sqrt{x^2 + y^2} \geq x$, and hence $d = r + \varepsilon x \geq x + \varepsilon x$. So $x \leq \frac{d}{1+\varepsilon}$. If $r < 0$ (in this case $\varepsilon > 1$), then $r = -\sqrt{x^2 + y^2} \leq -x$. So $d = r + \varepsilon x \leq -x + \varepsilon x$, and hence $\frac{d}{\varepsilon - 1} \leq x$. These observations tell us the following about the graph of equation $(*)$. If $\varepsilon \leq 1$, then $r > 0$ and the entire graph lies to the left of the line $x = \frac{d}{1+\varepsilon}$. If $\varepsilon > 1$, then the graph has two separate components determined by the vertical line $x = \frac{d}{\varepsilon + 1}$ and the vertical line $x = \frac{d}{\varepsilon - 1}$ to its right. All points (r, θ) with $r > 0$ lie to the left of the line $x = \frac{d}{\varepsilon + 1}$, and all points (r, θ) with $r < 0$ lie to the right of the line $x = \frac{d}{\varepsilon - 1}$.

Let's turn to the specifics of the graph of equation $(*)$. We will consider the equivalent equation $r + \varepsilon r \cos \theta = d$ and study the cases $\varepsilon = 1, \varepsilon < 1$, and $\varepsilon > 1$ separately.

i. Start with the case $\varepsilon = 1$. We begin by plotting some of the points of the graph of

$$r = \frac{d}{1 + \cos \theta}.$$

The values of $1 + \cos\theta$ for $\theta = 0, \frac{\pi}{4}, \frac{\pi}{2}$, and $\frac{3\pi}{4}$ are $1 + 1 = 2, 1 + \frac{\sqrt{2}}{2}, 1 + 0 = 1$, and $1 - \frac{\sqrt{2}}{2}$, respectively. This implies, for the same sequence of θ, that $r = \frac{d}{2}, r \approx 0.6d, r = d$, and $r \approx 3.4d$. The fact that $\cos(-\theta) = \cos\theta$ provides all the points of the graph of $r = \frac{d}{1+\cos\theta}$ that are plotted in Figure 3.16a. This cluster of points and the fact that $r = f(\theta)$ is not defined for $\theta = \pm\pi$ suggests that we are dealing with a parabola that has a vertical directrix to the right of the point $x = \frac{d}{2}$. Could this be so? Since $r > 0$, we get that $\sqrt{x^2 + y^2} + x = d$ is the Cartesian version of the equation $r + r\cos\theta = d$ that we are studying. Now consider the parabola with focal point the origin O and directrix the vertical line $x = d$, and refer to Figure 3.16b. A point $P = (x, y)$ is on this parabola precisely when its distance to the origin O is equal to its distance to the line $x = d$. Notice that this is the case precisely when $\sqrt{(x-0)^2 + (y-0)^2} = \sqrt{x^2 + y^2}$ is equal to $-x + d$. So the Cartesian equation of this parabola is $\sqrt{x^2 + y^2} + x = d$. Since this is also the Cartesian version of the equation

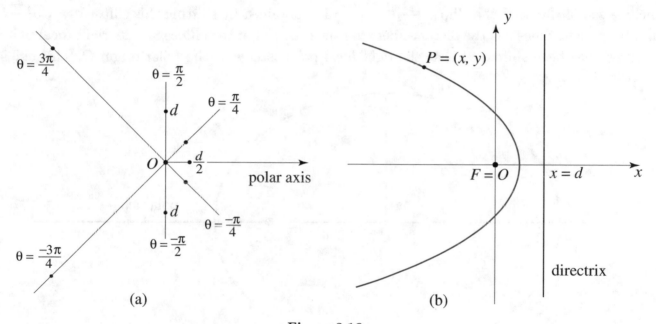

$$(a) \qquad\qquad\qquad (b)$$

Figure 3.16

$r + r\cos\theta = d$, it follows that the graph of equation $(*)$ with $\varepsilon = 1$ is the parabola with focal point O and directrix $x = d$. It is sketched in red in Figure 3.16b.

We'll take the cases $\varepsilon < 1$ and $\varepsilon > 1$ together, at least initially. Transform the Cartesian version $\pm\sqrt{x^2 + y^2} + \varepsilon x = d$ of $r + \varepsilon r\cos\theta = d$ step by step (one of them completes a square) as follows:

$$x^2 + y^2 = d^2 - 2\varepsilon dx + \varepsilon^2 x^2$$

$$(1 - \varepsilon^2)x^2 + 2\varepsilon dx + y^2 = d^2$$

$$x^2 + \frac{2\varepsilon d}{1-\varepsilon^2}x + \frac{y^2}{1-\varepsilon^2} = \frac{d^2}{1-\varepsilon^2}$$

$$x^2 + \frac{2\varepsilon d}{1-\varepsilon^2}x + \frac{\varepsilon^2 d^2}{(1-\varepsilon^2)^2} + \frac{y^2}{1-\varepsilon^2} = \frac{d^2}{1-\varepsilon^2} + \frac{\varepsilon^2 d^2}{(1-\varepsilon^2)^2}$$

$$\left(x + \frac{\varepsilon d}{1-\varepsilon^2}\right)^2 + \frac{y^2}{1-\varepsilon^2} = \frac{(1-\varepsilon^2)d^2 + \varepsilon^2 d^2}{(1-\varepsilon^2)^2}$$

$$\left(x + \tfrac{\varepsilon d}{1-\varepsilon^2}\right)^2 + \tfrac{y^2}{1-\varepsilon^2} = \left(\tfrac{d}{1-\varepsilon^2}\right)^2, \text{ and finally}$$

$$\frac{(x + \frac{\varepsilon d}{1-\varepsilon^2})^2}{(\frac{d}{1-\varepsilon^2})^2} + \frac{y^2}{\frac{d^2}{1-\varepsilon^2}} = 1.$$

ii. Suppose $\varepsilon < 1$. So $1 - \varepsilon^2 > 0$. Put $a = \frac{d}{1-\varepsilon^2}$, $b = \frac{d}{\sqrt{1-\varepsilon^2}}$, and let $c = \sqrt{a^2 - b^2}$. Since $1 - \varepsilon^2 \leq 1$, we see that $\sqrt{1-\varepsilon^2} \geq 1 - \varepsilon^2$, and hence that $a \geq b$. Since $a^2 - b^2 = \frac{d^2}{(1-\varepsilon^2)^2} - \frac{d^2}{1-\varepsilon^2} = \frac{d^2 - (1-\varepsilon^2)d^2}{(1-\varepsilon^2)^2} = \frac{\varepsilon^2 d^2}{(1-\varepsilon^2)^2}$, it follows that $c = \frac{\varepsilon d}{1-\varepsilon^2} = \varepsilon a$. After a substitution, we get the equation $\frac{(x+c)^2}{a^2} + \frac{y^2}{b^2} = 1$. Now turn to Chapter 1C. Recall that the equation $\frac{x^2}{a^2} + \frac{y^2}{b^2} = 1$ represents an ellipse with semimajor axis $a = \frac{d}{1-\varepsilon^2}$, semiminor axis $b = \frac{d}{\sqrt{1-\varepsilon^2}}$, and center the origin O. It follows from the remarks that conclude section 3C that the ellipse $\frac{(x+c)^2}{a^2} + \frac{y^2}{b^2} = 1$ is obtained by shifting this ellipse $c = \sqrt{a^2 - b^2}$ units to the left. Since c is the distance between the center C of the ellipse and its right focal point, the ellipse has been shifted so that the right focal point is now at the polar origin O. This ellipse

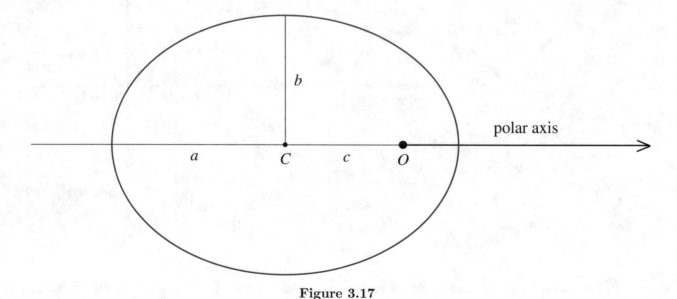

Figure 3.17

is sketched in Figure 3.17. The description of the graph of equation $(*)$ in the situation $\varepsilon < 1$ is complete. Since $c = \varepsilon a$, it follows that the ε of equation $(*)$ is the eccentricity $\varepsilon = \frac{a}{c}$ of the ellipse. The focal points are the origin O and (in either Cartesian or polar coordinates) the point $(-2c, 0)$.

iii. Suppose $\varepsilon > 1$. So $\varepsilon^2 - 1 > 0$. Let $a = \frac{d}{\varepsilon^2-1}$, $b = \frac{d}{\sqrt{\varepsilon^2-1}}$, and set $c = \sqrt{a^2 + b^2}$. Because $a^2 + b^2 = \frac{d^2}{(\varepsilon^2-1)^2} + \frac{d^2}{\varepsilon^2-1} = \frac{d^2 + (\varepsilon^2-1)d^2}{(\varepsilon^2-1)^2} = \frac{\varepsilon^2 d^2}{(\varepsilon^2-1)^2}$, we get $c = \frac{\varepsilon d}{\varepsilon^2-1}$. This time, after substituting carefully, the earlier equation becomes $\frac{(x-c)^2}{a^2} - \frac{y^2}{b^2} = 1$. Return to Chapter 1C. Consider the hyperbola $\frac{x^2}{a^2} - \frac{y^2}{b^2} = 1$ with semimajor axis a and semiminor axis b. We know that its asymptotes intersect at the origin O and that its focal points are both c units from O. It follows from the discussion that concludes section 3C that the graph of $\frac{(x-c)^2}{a^2} - \frac{y^2}{b^2} = 1$, and hence the graph of equation $(*)$, is obtained by shifting this hyperbola c units to the right. This shifted hyperbola has its left focal point

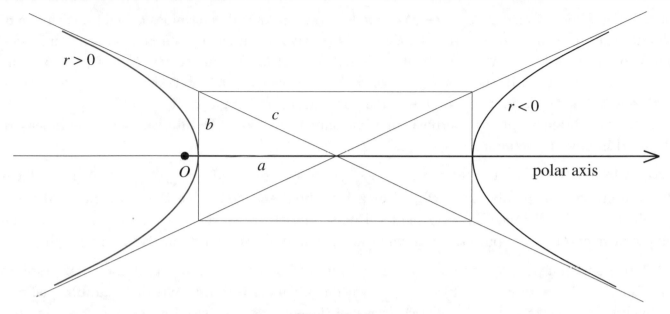

Figure 3.18

at the polar origin O. Its graph is sketched in Figure 3.18. It follows from the equalities defining a, b, and c that the ε of equation $(*)$ is the eccentricity $\varepsilon = \frac{c}{a}$ of the hyperbola.

We conclude the hyperbolic case of the function $r = f(\theta) = \frac{d}{1+\varepsilon\cos\theta}$ with a look at the way the graph of Figure 3.18 is traced out as θ varies. Figure 3.19 shows the two asymptotes in green. The two dashed green lines emanate from the polar origin O. Each is parallel to one of the asymptotes. The angles between the asymptotes and the polar axis are denoted by φ. Figure 3.19 tells us that for any θ satisfying either $0 \leq \theta < \pi - \varphi$ or $-\pi + \varphi < \theta < 0$, the ray that θ determines intersects one of the asymptotes on the left side and hence the left branch of the hyperbola. As θ varies over the interval from $-(\pi - \varphi)$ to $\pi - \varphi$, the point (r, θ) on the graph of the function traces out the entire left branch of the hyperbola from bottom to top. Both the interval and this branch are depicted

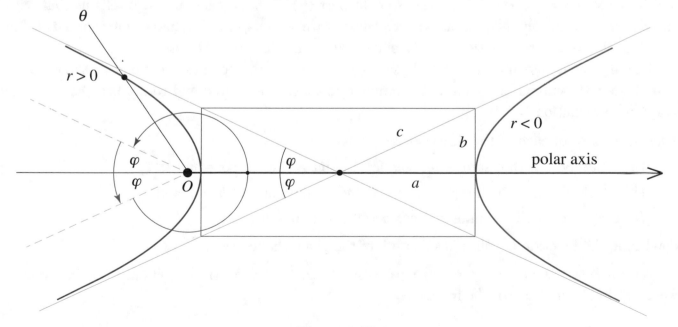

Figure 3.19

in red. For θ equal to $\pi - \varphi$ or $-(\pi - \varphi) = -\pi + \varphi$, the ray that θ determines coincides with one of the dashed green lines. In either case, the ray is parallel to an asymptote and does not intersect the hyperbola. It follows that the function is not defined for θ equal to $\pi - \varphi$ or $-(\pi - \varphi)$. One more look at Figure 3.19 tells us that $r = f(\theta)$ is negative over the interval of $\pi - \varphi < \theta < \pi + \varphi$ and that as θ varies from $\pi - \varphi$ to $\pi + \varphi$, the points (r, θ) on the graph of $r = f(\theta) = \frac{d}{1 + \varepsilon \cos \theta}$ trace out the right branch of the hyperbola from bottom to top. Both the interval and this branch are depicted in blue in the figure.

Example 3.8. Consider the functions $r = \frac{3}{1 + \frac{2}{5} \cos \theta}$, $r = \frac{4}{1 + \cos \theta}$, and $r = \frac{5}{1 + 3 \cos \theta}$. In each case identify the graph as a parabola, an ellipse, or a hyperbola and determine the eccentricity. If it's a parabola, use the distance between the focal point and the directrix to sketch the graph. For an ellipse or hyperbola, compute the semimajor and semiminor axes and then sketch the graph.

A plane with a polar coordinate system continues to be given. We saw that the polar graph of any function of the form $r = f(\theta) = \frac{d}{1 + \varepsilon \cos \theta}$, with $d > 0$ and $\varepsilon \geq 0$ constants, is a parabola, ellipse, or hyperbola. The constant ε is the eccentricity of the conic section. We'll now see, conversely, that any given conic section in the plane can be shifted and rotated so that it coincides with the graph of such a function.

If the conic section is a parabola, let d be the distance between the focal point and the directrix. Shift the given parabola so that the focal point is at the polar origin, and rotate it so that the directrix is vertical and to the right of the focal point. The parabola now coincides with that in Figure 3.16b. It is therefore the polar graph of the function $r = \frac{d}{1 + \cos \theta}$.

Let the given conic section be an ellipse with semimajor and semiminor axes a and b. Shift the ellipse, so that one of its focal points is at the polar origin, and then rotate it so that it is positioned like the ellipse in Figure 3.17. Refer to part (ii) of the discussion above and consider the equations $a = \frac{d}{1 - \varepsilon^2}$ and $b = \frac{d}{\sqrt{1 - \varepsilon^2}}$ in the variables $d > 0$ and $\varepsilon \geq 0$. They can be solved for d and ε as follows. Since $b^2 = \frac{d^2}{1 - \varepsilon^2}$, we get $ad = b^2$ and hence $d = \frac{b^2}{a}$. Since $1 - \varepsilon^2 = \frac{d}{a} = \frac{b^2}{a^2}$, we get $\varepsilon^2 = 1 - \frac{b^2}{a^2} = \frac{a^2 - b^2}{a^2}$, and hence that $\varepsilon = \frac{1}{a}\sqrt{a^2 - b^2}$. Another look at part (ii) of the discussion above tells us that with this d and ε, the graph of equation (∗) is an ellipse with semimajor axis a and semiminor axis b and that this ellipse coincides with the given ellipse (after the shift and the rotation).

The argument just used for the ellipse applies with only minor modifications to show that a hyperbola with semimajor axis a and semiminor axis b can be shifted and rotated so that it is the graph of an equation (∗).

Example 3.9. In each of the three cases,

  (a) a parabola with distance between focal point and directrix equal to 7,

  (b) an ellipse with semimajor axis $a = 6$ and semiminor axis $b = 4$, and

  (c) a hyperbola with semimajor axis $a = 6$ and semiminor axis $b = 4$,

determine the function (∗) that has a graph of the given shape.

With a view toward the upcoming application to the motion of the objects in the solar system, we'll conclude our study of the function

$$r = f(\theta) = \frac{d}{1 + \varepsilon \cos \theta}$$

by highlighting relevant information about its graph and by specifying its domain of definition. We'll rely on our earlier discussion and in particular on the graphs of Figures 3.16b, 3.17, and 3.18.

The point on the graph closest to the focal point at the polar origin O is the *periapsis*. The distance between it and O is the *periapsis distance*. A look at the graphs tells us that the periapsis distance is $r = f(0) = \frac{d}{1+\varepsilon \cos 0} = \frac{d}{1+\varepsilon}$. The length of the segment through the focal point, perpendicular to the polar axis, and bounded on both sides by the graph is the *latus rectum*. See Figure 1.13b. Another look at the graphs informs us that the latus rectum is equal to $2f(\frac{\pi}{2}) = 2(\frac{d}{1+\varepsilon \cos \frac{\pi}{2}}) = 2d$. With regard to the domain of the function, we'll deal with the elliptical, parabolic, and hyperbolic cases separately.

A. The elliptical case $\varepsilon < 1$. Here $1 + \varepsilon \cos \theta > 0$ and hence $1 + \varepsilon \cos \theta \neq 0$ for all θ. In this case, we take the domain of the function $r = f(\theta) = r = \frac{d}{1+\varepsilon \cos \theta}$ to be the full set of real numbers θ. For any stretch of increasing θ, the ray from the origin that θ determines rotates counterclockwise, so that the point (r, θ) moves counterclockwise on the ellipse of Figure 3.17. As θ varies over the domain of the function, the point will trace out the ellipse again and again. Recall that

$$a = \tfrac{d}{1-\varepsilon^2} \text{ and } b = \tfrac{d}{\sqrt{1-\varepsilon^2}}$$

are the semimajor and semiminor axes, respectively, and that $c = \frac{\varepsilon d}{1-\varepsilon^2} = a\varepsilon$. The periapsis distance is equal to $a - c = a(1 - \varepsilon) = \frac{d(1-\varepsilon)}{1-\varepsilon^2} = \frac{d}{1+\varepsilon}$. Notice that the periapsis of the graph is reached at $\theta = 0, \pm 2\pi, \pm 4\pi, \dots$. Over any interval of the form $2k\pi \leq \theta \leq 2k\pi + 2\pi = 2\pi(k+1)$ or $-2\pi(k+1) = -2k\pi - 2\pi \leq \theta \leq 2k\pi$ with $k \geq 0$ an integer, the point (r, θ) on the graph moves from one periapsis to the next. Since $d = \frac{d^2}{1-\varepsilon^2} \cdot \frac{1-\varepsilon^2}{d} = \frac{b^2}{a}$, the latus rectum in the elliptical case is equal to $2d = \frac{2b^2}{a}$.

B. The parabolic case $\varepsilon = 1$. Now $1 + \cos \theta \geq 0$ for all θ. The function is not defined when $\cos \theta = -1$ and hence for $\theta = \pm \pi, \pm 3\pi, \pm 5\pi, \dots$. Let's consider the interval $-\pi < \theta < \pi$. Figure 3.5 tells us that as θ varies from 0 to π, $\cos \theta$ varies from 1 to -1, so that the values $r = f(\theta) = \frac{d}{1+\cos \theta}$ start at $r = \frac{d}{2}$ and become larger and larger without bound. In the process, the point (r, θ) traces out the entire upper part of the parabola of Figure 3.16b. As θ varies from 0 to $-\pi$, $\cos \theta$ also varies from 1 to -1. Again, the values $r = f(\theta) = \frac{d}{1+\cos \theta}$ start at $r = \frac{d}{2}$ and become larger and larger without bound. As this occurs, the point (r, θ) traces out the entire lower part of the parabola. It follows that as θ varies over $-\pi < \theta < \pi$, the point (r, θ) traces out the entire parabola in a counterclockwise way. The function $r = f(\theta) = \frac{d}{1+\cos \theta}$ behaves in a similar way over the intervals $\pi < \theta < 3\pi$, $-3\pi < \theta < -\pi$, $3\pi < \theta < 5\pi$, $-5\pi < \theta < -3\pi$, and so on. However, we will specify the domain of the function $r = f(\theta) = \frac{d}{1+\varepsilon \cos \theta}$ in the parabolic case to be restricted to $-\pi < \theta < \pi$. Periapsis occurs only at $\theta = 0$. We have already see that the periapsis distance is $f(0) = \frac{d}{2}$ and that the latus rectum is equal to $2f(\frac{\pi}{2}) = 2d$.

C. The hyperbolic case $\varepsilon > 1$. With regard to the graph of the function in Figure 3.18,

$$a = \tfrac{d}{\varepsilon^2-1} \text{ and } b = \tfrac{d}{\sqrt{\varepsilon^2-1}}$$

are the semimajor and semiminor axes, respectively, and $c = \frac{\varepsilon d}{\varepsilon^2 - 1} = a\varepsilon$. The periapsis distance is equal to $c - a = a(\varepsilon - 1) = \frac{d(\varepsilon-1)}{\varepsilon^2-1} = \frac{d}{\varepsilon+1}$. Since $d = \frac{d^2}{\varepsilon^2-1} \cdot \frac{\varepsilon^2-1}{d} = \frac{b^2}{a}$, the latus rectum is equal to $2\frac{b^2}{a}$ in the hyperbolic case. The domain of the function $r = f(\theta) = \frac{d}{1+\varepsilon \cos \theta}$ in the hyperbolic situation is restricted to the interval $-(\pi - \varphi) < \theta < \pi - \varphi$, where φ is given by $\varphi = \tan^{-1} \frac{b}{a} = \tan^{-1}\left(\frac{d}{\sqrt{\varepsilon^2-1}} \cdot \frac{\varepsilon^2-1}{d} \right) = \tan^{-1} \sqrt{\varepsilon^2 - 1}$. To see why, refer to Figure 3.19 and the discussion of the figure. Recall in particular that when θ varies over this interval, the point (r, θ) traces out the entire left branch of the hyperbola of the figure exactly once in a counterclockwise way. Periapsis occurs at $\theta = 0$.

For $\theta = \pm(\pi - \varphi)$ the term $\frac{d}{1+\varepsilon \cos \theta}$ is not defined. To check this, note first that $\tan \varphi = \frac{b}{a}$. By elementary properties of the sine and cosine, $\frac{\sqrt{1-\cos^2 \varphi}}{\cos \varphi} = \frac{\sin \varphi}{\cos \varphi} = \tan \varphi = \frac{b}{a} = \frac{\varepsilon^2-1}{\sqrt{\varepsilon^2-1}} = \sqrt{\varepsilon^2 - 1}$. So $\frac{1-\cos^2 \varphi}{\cos^2 \varphi} = \varepsilon^2 - 1$, and hence $\frac{1}{\cos^2 \varphi} = \varepsilon^2$. It follows that $\cos \varphi = \frac{1}{\varepsilon}$. Properties of the cosine tell us that $\cos(\pi - \varphi) = \cos(\varphi - \pi) = -\cos \varphi = -\frac{1}{\varepsilon}$, confirming that $1 + \varepsilon \cos(\pm(\pi - \varphi)) = 0$.

Example 3.10. Determine the domains for the function $r = f(\theta) = \frac{d}{1+\varepsilon \cos \theta}$ in each of the three cases of Example 3.8. Find the periapsis distance and the latus rectum. In each case, compute numerical approximations of the relevant constants.

3E. The Derivative of a Polar Function. In this section it is assumed that the reader is familiar with the derivative of a function and its elementary properties, including the basic rules of differentiation (the product, quotient, and chain rules).

We'll start by recalling without proof, basic facts about the derivatives of the trig functions. The derivative of $f(\theta) = \sin \theta$ is $f'(\theta) = \cos \theta$ and the derivative of $f(\theta) = \cos \theta$ is $f'(\theta) = -\sin \theta$. An application of the quotient rule to $f(\theta) = \tan \theta = \frac{\sin \theta}{\cos \theta}$ tells us that

$$f'(\theta) = \frac{\cos \theta \cdot \cos \theta - (\sin \theta)(-\sin \theta)}{\cos^2 \theta} = \frac{1}{\cos^2 \theta} = \sec^2 \theta,$$

and by applying the chain rule to $f(\theta) = \sec \theta = (\cos \theta)^{-1}$, we get

$$f'(\theta) = (-1)(\cos \theta)^{-2} \cdot (-\sin \theta) = \frac{\sin \theta}{\cos \theta} \cdot \frac{1}{\cos \theta} = (\tan \theta)(\sec \theta).$$

Let $r = f(\theta)$ be a function in polar coordinates and let

$$f'(\theta) = \lim_{\Delta\theta \to 0} \frac{f(\theta + \Delta\theta) - f(\theta)}{\Delta\theta}$$

be its derivative. We know that the derivative of a function in Cartesian coordinates measures the slope of the tangent line to the graph of the function. We will now see that the derivative of a function in polar coordinates is also related to the tangent of the polar graph. But the connection is more complicated.

A portion of a graph of a typical function $r = f(\theta)$ is sketched in Figure 3.20. On occasion, we will restrict the domain to an interval of angles $\theta_1 \le \theta \le \theta_2$ as indicated in the figure. Suppose that $f'(\theta) > 0$ over such an interval. We know that this means that the function $r = f(\theta)$ increases with increasing θ over $\theta_1 \le \theta \le \theta_2$. So $r = f(\theta)$ grows as the ray that θ determines rotates from θ_1 to θ_2.

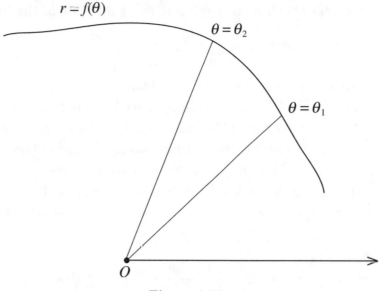

$r = f(\theta)$

$\theta = \theta_2$

$\theta = \theta_1$

O

Figure 3.20

In a similar way, if $f'(\theta) < 0$ over the interval, then $r = f(\theta)$ decreases as θ rotates from θ_1 to θ_2.

Let $P = (f(\theta), \theta)$ be any point on the graph of $f(\theta)$. Assume that P is not the polar origin O, so that $f(\theta) \neq 0$. Let γ be the angle measured in the counterclockwise direction from the tangent at $P = (f(\theta), \theta)$ to the segment from O to $P = (f(\theta), \theta)$. See Figure 3.21. Observe that $0 \leq \gamma < \pi$.

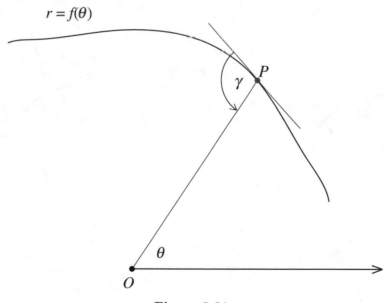

$r = f(\theta)$

γ

P

θ

O

Figure 3.21

Since γ depends on the point P and hence on θ, γ is a function $\gamma = \gamma(\theta)$ of θ. For the function $r = f(\theta) = 5$ for instance, $f'(\theta) = 0$ for all θ and, because the graph of the function is a circle with center the origin, $\gamma(\theta) = \frac{\pi}{2}$ for all θ. For a function $r = f(\theta)$, the inequalities $f'(\theta) > 0$ and $\gamma(\theta) > \frac{\pi}{2}$ both tell us that $r = f(\theta)$ is increasing at θ, and $f'(\theta) < 0$ and $\gamma(\theta) < \frac{\pi}{2}$ both tell us that $r = f(\theta)$ is decreasing at θ. This suggests that there might be an explicit connection between $f'(\theta)$ and $\gamma(\theta)$.

Let's begin the exploration of such a connection with an analysis of the limit

$$f'(\theta) = \lim_{\Delta\theta \to 0} \frac{f(\theta + \Delta\theta) - f(\theta)}{\Delta\theta}$$

that defines the derivative of the function $f(\theta)$. To facilitate our discussion, we'll assume that $f(\theta) \geq 0$ (but the conclusions hold without this assumption). Consider the point $(f(\theta + \Delta\theta), \theta + \Delta\theta)$ for a small $\Delta\theta$ and draw the segment from O to this point into Figure 3.21. Add into the figure—in red—the circular arc with center O and radius $f(\theta)$ between the rays determined by θ and $\theta + \Delta\theta$. These additions and the segment of length $f(\theta + \Delta\theta) - f(\theta)$ from the arc to the graph are shown in Figure 3.22a. We'll call the curving triangle with the red circular base the *beak* at P. Let the length of the circular arc be Δs, and observe that the radian measure of $\Delta\theta$ is equal to $\Delta\theta = \frac{\Delta s}{f(\theta)}$. So $\frac{1}{\Delta\theta} = \frac{1}{\Delta s} \cdot f(\theta)$ and after a substitution,

$$f'(\theta) = \lim_{\Delta\theta \to 0} \frac{f(\theta + \Delta\theta) - f(\theta)}{\Delta s} \cdot f(\theta).$$

In order to understand $f'(\theta)$, we need to come to grips with $\lim_{\Delta\theta \to 0} \frac{f(\theta+\Delta\theta)-f(\theta)}{\Delta s}$. Put in the tangent line *to the graph* of $r = f(\theta)$ at P in green, and let A be the point of intersection of the tangent with the ray determined by $\theta + \Delta\theta$. Also put in the tangent line *to the circle* at P again in green, and let B be the point of intersection of this tangent and the same ray. The two tangent lines and the segment connecting A and B form the triangle $\triangle APB$. We call it the *triangle* at P. It is

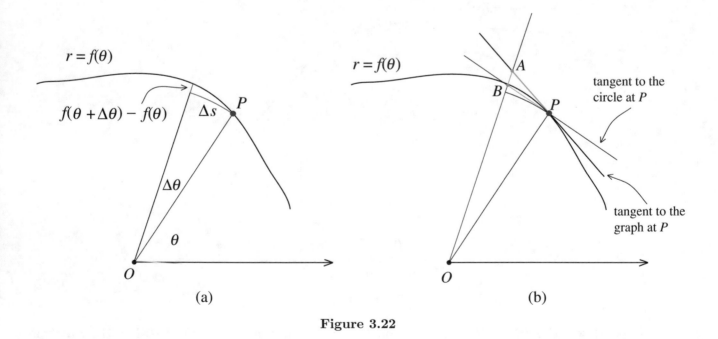

Figure 3.22

shown in Figure 3.22b. The diagram of Figure 3.23 is a composite of the two diagrams of Figure 3.22.

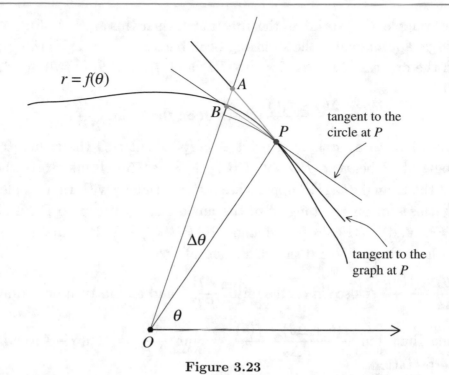

Figure 3.23

It shows both the beak at P and the triangle $\triangle APB$ at P. We'll now push $\Delta\theta$ to 0 and investigate

$$\lim_{\Delta\theta \to 0} \frac{f(\theta + \Delta\theta) - f(\theta)}{\Delta s}.$$

What happens as $\Delta\theta$ shrinks to 0 is illustrated in Figure 3.24. It is a blowup of the central part of Figure 3.23. As $\Delta\theta$ is pushed to 0, the segment OBA rotates toward the segment OP. Both the

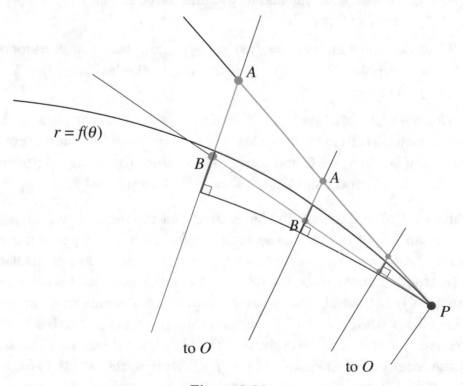

Figure 3.24

beak at P and the triangle at P shrink in the direction of their tips at P. As Figure 3.24 illustrates, the shrinking triangle approximates the shrinking beak better and better as the gap between OBA and OP closes. In the process, Δs gets closer to BP and $f(\theta + \Delta\theta) - f(\theta)$ to AB. Therefore, as $\Delta\theta$ is pushed to 0,

$$\frac{f(\theta + \Delta\theta) - f(\theta)}{\Delta s} \quad \text{closes in on the ratio} \quad \frac{AB}{BP}.$$

Because the tangent line to a circle at a point is perpendicular to the radius to that point, we know that the angle at P between PO and PB is $\frac{\pi}{2}$. So as $\Delta\theta$ shrinks to 0, the angle $\angle PBA$ approaches $\frac{\pi}{2}$, and the triangle $\triangle APB$ approaches a right triangle with right angle at B. It follows that the ratio $\frac{AB}{BP}$ closes in on the tangent of the angle $\angle APB$. Refer to Figure 3.24 once more. Because $\angle APO = \gamma$ and $\angle BPO = \frac{\pi}{2}$, the angle $\angle APB = \gamma - \frac{\pi}{2}$. By putting together what has been observed, we have demonstrated that as $\Delta\theta$ shrinks to 0,

$$\frac{f(\theta + \Delta\theta) - f(\theta)}{\Delta s} \quad \text{closes in on the ratio} \quad \frac{AB}{BP} \quad \text{and this in turn on} \quad \tan(\gamma - \tfrac{\pi}{2}).$$

So we have verified that $\displaystyle \lim_{\Delta\theta \to 0} \frac{f(\theta + \Delta\theta) - f(\theta)}{\Delta s} = \lim_{\Delta\theta \to 0} \frac{AB}{PB} = \tan(\gamma - \tfrac{\pi}{2})$ and have arrived at the geometric interpretation

$$\boxed{f'(\theta) = f(\theta) \cdot \tan(\gamma(\theta) - \tfrac{\pi}{2})}$$

of the derivative $f'(\theta)$ of a function $r = f(\theta)$ for any θ with $f(\theta) \neq 0$.

Example 3.11. Let $c > 0$ be a constant, and investigate the equation $f'(\theta) = f(\theta) \cdot \tan(\gamma(\theta) - \frac{\pi}{2})$ for the circle $r = f(\theta) = c$ of radius c.

Example 3.12. Consider the function $r = f(\theta) = \frac{1}{\cos\theta}$. Show that the corresponding Cartesian equation is $x = 1$. For any point $P = (f(\theta), \theta)$ on the graph, check that $\gamma(\theta) - \frac{\pi}{2} = \theta$. Use this to confirm that $f'(\theta) = f(\theta) \cdot \tan(\gamma(\theta) - \frac{\pi}{2})$.

Example 3.13. Consider the function $r = f(\theta) = \sin\theta$. Its graph is the circle of Figure 3.14. Let $P = (f(\theta), \theta)$ be any point on the circle. Show that $\gamma(\theta) - \frac{\pi}{2} = \frac{\pi}{2} - \theta$ (by using a property of isosceles triangles) and that $\sin(\frac{\pi}{2} - \theta) = \cos\theta$ and $\cos(\frac{\pi}{2} - \theta) = \sin\theta$ (by using trig formulas developed earlier in this chapter). Confirm the identity $f'(\theta) = f(\theta) \cdot \tan(\gamma(\theta) - \frac{\pi}{2})$.

3F. The Lengths of Polar Curves. In this section and the next, it will be assumed that the reader has an understanding of the basic aspects of integral calculus, in particular the definition of the definite integral, the fundamental theorem of calculus, as well as the substitution method.

We begin by extracting some more information from the analysis that verified the equality $f'(\theta) = f(\theta) \cdot \tan(\gamma - \frac{\pi}{2})$. Let a and b be constants with $a < b$. Assume that the function $r = f(\theta)$ is differentiable for all θ with $a < \theta < b$ (this means that the limit at the beginning of section 3E exists for these θ) and that $f(\theta)$ and its derivative $f'(\theta)$ are both continuous for all $a \leq \theta \leq b$. The focus will be on the length L of the graph of $r = f(\theta)$ between the points $(f(a), a)$ and $(f(b), b)$. We'll suppose that $f(\theta) \geq 0$ (but note that our conclusions hold without this assumption) and apply the strategy of integral calculus.

Let n be a large number and consider a set of numbers

$$a = \theta_0 < \theta_1 < \theta_2 < \cdots < \theta_i < \theta_{i+1} < \cdots < \theta_{n-1} < \theta_n = b$$

that divide the angle $b - a$ into n angles of equal size $\frac{b-a}{n} = d\theta$. Each θ_i is an angle in radian measure and the difference between consecutive angles is $d\theta$. The rays determined by $\theta_0, \ldots, \theta_{n-1}, \theta_n$ divide the graph of $r = f(\theta)$ between the points $(f(a), a)$ and $(f(b), b)$ into n pieces. Letting the lengths of these pieces be $L_0, \ldots L_{n-1}$, we get that $L = L_0 + \cdots + L_{n-1}$. Turn to Figure 3.25. Let $\theta = \theta_i$ be any of the angles selected and note that $\theta_{i+1} = \theta + d\theta$. The point $P = (f(\theta), \theta)$, the segment of the graph of length L_i, and the arc of a circle of radius $f(\theta)$ are all shown in red in the figure. We will now derive an approximation of L_i. Return to Figure 3.23 and its explanation and add the tangents at P of both the graph of $r = f(\theta)$ and the circular arc to Figure 3.25 in green. The earlier $\Delta\theta$ and Δs are now written as $d\theta$ and ds. Suppose that $d\theta$ is extremely small. A look at Figures 3.23 and 3.24 tells us that L_i is essentially equal to the length of the segment AP on the tangent to the graph at P. Recall from the earlier discussion that ΔABP is essentially a right triangle with hypothenuse AP and hence that $AP \approx \sqrt{AB^2 + BP^2}$. Return to this earlier discussion once more and observe that $BP \approx ds$ and $AB \approx f(\theta + d\theta) - f(\theta)$.

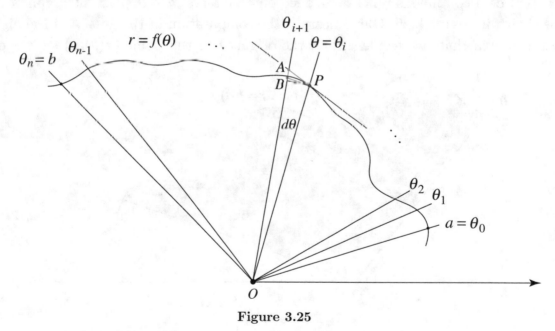

Figure 3.25

Because $d\theta = \frac{ds}{f(\theta)}$, it follows that $BP \approx ds = f(\theta)d\theta$. Because $d\theta$ is very small, $\frac{f(\theta+d\theta)-f(\theta)}{d\theta} \approx f'(\theta)$ and hence

$$AB \approx f(\theta + d\theta) - f(\theta) \approx f'(\theta)\, d\theta.$$

Therefore, $L_i \approx AP \approx \sqrt{BP^2 + AB^2} \approx \sqrt{f(\theta)^2(d\theta)^2 + f'(\theta)^2(d\theta)^2} = \sqrt{f(\theta)^2 + f'(\theta)^2}\, d\theta$. Because $L = L_1 + \cdots + L_n$, it is a consequence of the definition of the definite integral that the length L of the graph of $r = f(\theta)$ between the points $(f(a), a)$ and $(f(b), b)$ is

$$\boxed{L = \int_a^b \sqrt{f(\theta)^2 + f'(\theta)^2}\, d\theta}$$

Example 3.14. Consider a circle of radius c and an arc of length s on the circle. Put the circle into a plane with a polar coordinate system so that its center is at the polar origin O. Let the end points of the arc be given by the rays $\theta = a$ and $\theta = b$, where $0 < a < b$. Express s as a definite integral. Evaluate the integral to show that $s = c(b - a)$.

Example 3.15. Use the arc length formula to determine the length of the graph of $r = \sin\theta$ from $\theta = \frac{\pi}{4}$ to $\theta = \frac{3\pi}{4}$, then from $\theta = 0$ to $\theta = \pi$, and finally from $\theta = 0$ to $\theta = 2\pi$. Check your answers by referring to the graph of $f(\theta) = \sin\theta$ in Figure 3.14.

Example 3.16. Use the arc length formula and fact that the Cartesian equation of the polar equation $r = \frac{1}{\sin\theta}$ is the line $y = 1$ to evaluate the integral $\displaystyle\int_{\frac{\pi}{4}}^{\frac{3\pi}{4}} \frac{1}{\sin^2\theta}\, d\theta$.

3G. Areas in Polar Coordinates. This section studies the area of a region determined by the graph of a function in polar coordinates. The conclusions will play an important role in Chapter 4 in the analysis of the motion of a point-mass driven by a centripetal force. See Chapter 1D for the context.

Let $r = f(\theta)$ be a continuous function defined on an interval $a \le \theta \le b$. The graph of a typical situation is shown in Figure 3.26. Our concern is the computation of the area A of the highlighted region. It is the area that the segment from the origin O to the point $(f(\theta), \theta)$ sweeps out as it

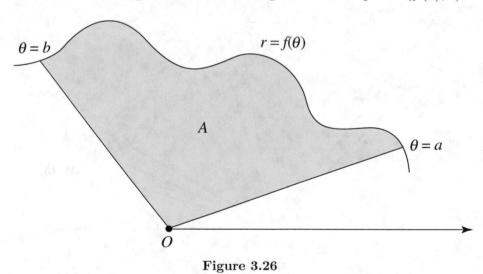

Figure 3.26

rotates from the ray $\theta = a$ to $\theta = b$. In general, there can be an overlap of areas, possibly multiple times, and each of these overlaps is counted. If $b - a \le 2\pi$, then there is no overlap. Once again we'll facilitate our argument by assuming that $f(\theta) \ge 0$ (but the conclusions hold without this assumption).

As before, let n be a very large number and chop the angle $b - a$ into n equal pieces. So

$$a = \theta_0 < \theta_1 < \cdots < \theta_{i-1} < \theta_i < \cdots < \theta_{n-1} < \theta_n = b$$

with $\theta_{i+1} - \theta_i = \frac{b-a}{n} = d\theta$ for $0 \le i \le n - 1$. Observe that the rays determined by $\theta_0, \ldots, \theta_n$ divide the area A into n pie-shaped regions as illustrated in Figure 3.27. Let $A_0, \ldots, A_i, \ldots, A_{n-1}$ denote their respective areas and observe that $A = A_0 + \cdots + A_i + \cdots + A_{n-1}$. Let $\theta = \theta_i$ be any of the

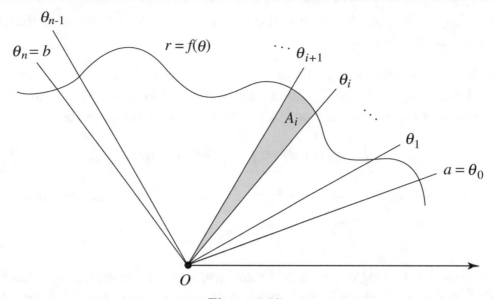

Figure 3.27

angles selected and draw a circular arc with center the origin O and radius $f(\theta)$ from the ray determined by θ to the ray given by $\theta_{i+1} = \theta + d\theta$. Consider the circular sector determined by this arc and the two rays. Figure 3.28 shows the circular sector and the relevant part of the graph of the function. Since $f(\theta)$ is the radius of the circular arc, the area of the circular sector is $\frac{1}{2}f(\theta)^2 d\theta$.

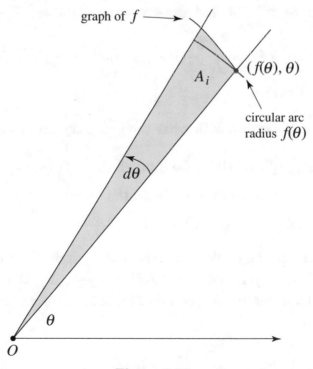

Figure 3.28

Because $d\theta$ is very small, the area A_i is essentially equal to the area of the sector. The fact that $A = A_1 + \cdots + A_{n-1}$ in combination with the definition of the definite integral informs us that

$$A = \int_a^b \tfrac{1}{2}f(\theta)^2 \, d\theta$$

Example 3.17. Apply the area formula to the function $r = f(\theta) = 5$ to show that the area of a quarter circle of radius 5 is $\frac{25}{4}\pi$.

The two examples that follow test the area formula in situations where its conclusion can be determined by other means. Consider the function $r = f(\theta) = c$ for $0 \le \theta \le 4\pi$ and $c > 0$ a constant. Its graph is the circle of radius c traced out twice. So it must be the case that

$$\int_0^{4\pi} \tfrac{1}{2} f(\theta)^2 d\theta = \int_0^{4\pi} \tfrac{1}{2} c^2 d\theta = 2(\pi c^2).$$

Because $G(\theta) = \frac{1}{2} c^2 \theta$ is an antiderivative of $g(\theta) = \frac{1}{2} c^2$, the fundamental theorem of calculus confirms that

$$\int_0^{4\pi} \tfrac{1}{2} f(\theta)^2 d\theta = \tfrac{1}{2} c^2 \theta \, \Big|_0^{4\pi} = 2\pi c^2.$$

Consider the area A of the region bounded by the graph of the function $f(\theta) = \sin\theta$ and the rays $\theta = \frac{\pi}{4}$ and $\theta = \frac{3\pi}{4}$. Refer to Figure 3.14 and notice that this region consists of a half-circle of radius $\frac{1}{2}$ plus a triangle of height $\frac{1}{2}$ and base 1. So A is equal to $\frac{1}{2}\pi(\frac{1}{2})^2 + \frac{1}{4} = \frac{\pi}{8} + \frac{1}{4}$. The area formula applied to this situation tells us that

$$A = \int_{\frac{\pi}{4}}^{\frac{3\pi}{4}} \tfrac{1}{2} \sin^2\theta \, d\theta \,.$$

Does this provide the same result? The half-angle formula $\sin^2\theta = \frac{1 - \cos 2\theta}{2}$ informs us that

$$A = \int_{\frac{\pi}{4}}^{\frac{3\pi}{4}} \tfrac{1}{2} \sin^2\theta \, d\theta = \int_{\frac{\pi}{4}}^{\frac{3\pi}{4}} \tfrac{1}{4}(1 - \cos 2\theta) d\theta = (\tfrac{1}{4}\theta - \tfrac{1}{8}\sin 2\theta) \, \Big|_{\frac{\pi}{4}}^{\frac{3\pi}{4}}$$
$$= (\tfrac{3\pi}{16} + \tfrac{1}{8}) - (\tfrac{\pi}{16} - \tfrac{1}{8}) = \tfrac{\pi}{8} + \tfrac{1}{4}.$$

Again, the result of the formula agrees with what the geometry provided.

Example 3.18. Study the graph of the polar function $r = f(\theta) = \sin\theta$ in Figure 3.14 over the interval $0 \le \theta \le \pi$. Use your observations to evaluate the integral $\int_0^\pi \tfrac{1}{2} \sin^2\theta \, d\theta$. Then evaluate the integral again by using the half-angle formula $\sin^2\theta = \frac{1 - \cos 2\theta}{2}$.

Example 3.19. Use facts from the earlier analysis of the parabola in polar coordinates to verify that the upper half of the parabola $r = f(\theta) = \frac{2}{1 + \cos\theta}$ is the graph of the function $y = \sqrt{4 - 4x}$ in Cartesian coordinates. Apply this fact (and integration by substitution) to show that $\int_0^{\frac{\pi}{2}} \frac{2}{(1 + \cos\theta)^2} \, d\theta = \frac{4}{3}$.

Example 3.20. The graph of the function $r = f(\theta) = \frac{4}{1 + \frac{1}{5}\cos\theta}$ is an ellipse. Use results from the earlier analysis of the ellipse in polar coordinates to find its semimajor axis a and semiminor axis b. Then use the fact that the area of this ellipse is $ab\pi$ to show that $\int_0^\pi \frac{8}{(1 + \frac{1}{5}\cos\theta)^2} \, d\theta = (\frac{5}{\sqrt{6}})^3 \pi$.

Return to Figure 3.26, and assume that the graph of $r = f(\theta)$ is traced out by a point P moving counterclockwise around O. So the angle θ that determines the position $(f(\theta), \theta)$ of the point is an increasing function $\theta = \theta(t)$ of time t. Suppose that P is at $(f(a), a)$ at time t_1 and that it is at $(f(b), b)$ at time t_2. So $\theta(t_1) = a$ and $\theta(t_2) = b$. Figure 3.29 captures the added information. Applying the method of integration by substitution with $\theta = \theta(t)$ to the polar area formula shows that the area that the segment OP traces out over the time interval $[t_1, t_2]$ is equal to

$$\int_{t_1}^{t_2} \tfrac{1}{2} f(\theta(t))^2 \, \theta'(t) \, dt \;=\; \int_{\theta(t_1)}^{\theta(t_2)} \tfrac{1}{2} f(\theta)^2 \, d\theta \;=\; \int_a^b \tfrac{1}{2} f(\theta)^2 \, d\theta \;=\; A.$$

Suppose that P moves along the circle $f(\theta) = \sin\theta$ of Figure 3.14. Confirm that as θ varies from $\theta = 0$ to $\theta = \pi$, the entire area $\pi(\tfrac{1}{2})^2 = \tfrac{1}{4}\pi \approx 0.7854$ of the circle is traced out. Assume that the point starts at $O = (f(0), 0)$ at time $t = 0$ and moves counterclockwise around the circle at a constant angular speed of $\theta'(t) = \tfrac{1}{2}$ radians per second. Since $\theta(0) = 0$, it follows that $\theta(t) = \tfrac{1}{2}t$

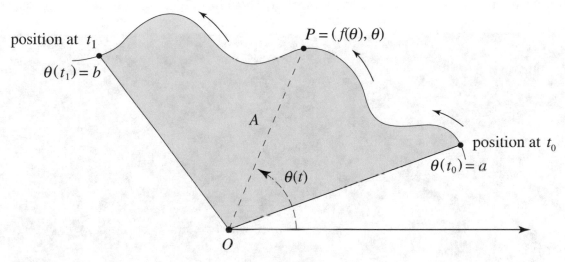

Figure 3.29

radians for any $t \geq 0$. At time $t = 4$ seconds, P has reached the point $(f(\theta(4)), \theta(4)) = (f(2), 2)$. By the above formula, the area A traced out by the segment OP during the time from $t = 0$ to $t = 4$ is

$$A \;=\; \int_0^4 \tfrac{1}{2}(\sin\theta(t))^2 \, \theta'(t) \, dt \;=\; \int_0^2 \tfrac{1}{2} \sin^2\theta \, d\theta.$$

Using $\sin^2\theta = \frac{1 - \cos 2\theta}{2}$, we get $A = \tfrac{1}{4}\big(\theta - \tfrac{1}{2}\sin 2\theta\big)\Big|_0^2 = \tfrac{1}{2} - \tfrac{1}{8}\sin 4 \approx 0.5000 + 0.0946 = 0.5946$. Since $\frac{0.5946}{0.7854} \approx 0.7571$, the segment sweeps out about 76% of the area of the circle during the 4 seconds.

3H. Spiral Galaxies and Equiangular Spirals. We observed in Chapter 1J that the spiraling arms of galaxy M51 consist of dust and pressurized hydrogen and that they are the galaxy's factories of new, hot stars. The Eagle Nebula, Messier designation M16, is such a new star factory in our own spiral galaxy, the Milky Way. It is a huge collection of clouds of hydrogen gas and dust. Depicted in Figure 3.30, the Eagle Nebula is approximately 70 light years by 55 light years across. Gravitational interactions put the clouds under pressure and contract them. With enough gas contracting and collapsing, nuclear reactions are ignited, and the compact clouds are transformed into bright stars. This kind of process created our Sun about 5 billion years ago.

Surrounding the center of a spiral galaxy is the galaxy's bulge, an enormous group of older, yellow and red stars. There is strong evidence that a spiral galaxy has an incredibly dense, tightly packed mass of matter at the center of its bulge. Such a mass exerts a gravitational force so strong that not even light can escape from it. Such masses are therefore invisible and are called *black holes*. Until recently, the evidence for their existence has been indirect. The fact that some stars are seen to be in rapid revolution around seemingly nothing at all, told us that there are massive invisible objects that exert powerful gravitational forces. Incredibly, a global network of telescopes calibrated to act together, has now captured an image of a black hole (more accurately, an image of the dark spot defined by the visible matter surrounding it) in a distant galaxy. See

https://solarsystem.nasa.gov/resources/2319/first-image-of-a-black-hole/

Everything in the disk of a spiral galaxy, including the galaxy's multitude of stars and its masses of dust and gases, revolves in elliptical orbits around the galaxy's center of mass as Kepler's second

Figure 3.30. The Eagle Nebula, also known as M16, is a several million year old star cluster (this is very young by astrophysical standards) surrounded by clouds of hydrogen dust and glowing gas. The image shows dusty columns that are light-years in length. Under the force of gravity, they collapse and contract to form stars. Many thanks to astrophotographer Russell Croman for capturing the image and permitting its use.

law predicts. As is the case in our solar system, objects closer to the center of mass move at greater speeds in their orbits than those farther away. (Refer to Tables 2.1 and 2.4 for instance.) This fact in combination with the gravitational interactions between the orbiting masses within the galaxy causes the individual elliptical orbits to rotate, so that they change their orientations and align themselves as shown in Figure 3.31. Observe from Figure 3.31b that these rotations create spiral shaped regions of matter of greater density. The hydrogen gas and dust in such regions are under gravitational pressure and form the kind of star-creating clouds described earlier. Because new stars are brightest, this accounts for the fact that in spiral galaxies there are prominent threads of bright stars

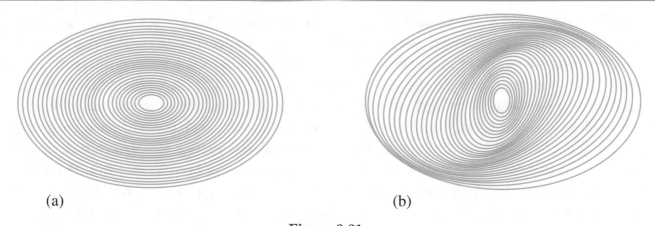

(a)   (b)

Figure 3.31

that wind their way through the galaxy's spiral arms.

The spiral galaxy with Messier designation M106 is 23.5 million light years from our own Milky Way and about 80,000 light years across. Figure 3.32 displays its dark dust lanes with their clusters

Figure 3.32. The spiral arms of galaxy M106 and the many young blue stars that course through them are captured by this composite portrait that used both Hubble exposures and images from ground-based telescopes. Image credit: NASA/ESA Hubble Space Telescope, and the Hubble Legacy Archive, a collaboration of the Space Telescope Science Institute (STScI/NASA), the Space Telescope European Coordinating Facility (ST-ECF/ESA), and the Canadian Astronomy Data Centre (CADC/NRC/CSA).

of young, blue stars and pinkish star-forming regions. They spiral toward the bright bulge of older yellowish stars. A closer look also shows structures that are not aligned with the flow of these spirals. Seen in red hues are sweeping filaments of glowing hydrogen gas that rise from the central bulge. They are thought to be powered by the galaxy's central black hole. The black hole of M106 is massive. It is believed to be about 30 million times as massive as our Sun. (By contrast, the Milky Way's central black hole is about 4 million times more massive than the Sun.) The enormous gravitational pull of the galaxy's black hole arranges the surrounding galactic matter into an orbiting disk. The matter near the inner rim of this disk are under great pressure and become very hot. Some of this matter is sucked into the black hole, but some of it is flung back into the galaxy. These materials along with streams of radiation slam into galactic material along the way and create the reddish fibrous formations that are observed.

Observations have shown that for a majority of spiral galaxies the prominent curves of their star-forming arms follow the geometry of an *equiangular spiral*. Such spirals have the property that the angle $\gamma = \gamma(\theta)$ defined in Figure 3.21 is constant throughout the entire flow of the spiral. See Figure 3.33. In spite of the fact that galaxies are places of constant turbulence, disruptions, and

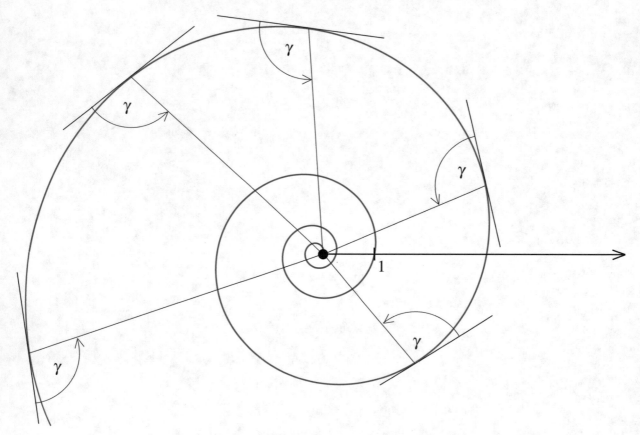

Figure 3.33

explosive events, the expanding arms of these galaxies follow such a geometry "in the large."

The seventy-fourth entry in Messier's catalogue is a formation of stars discovered in 1780 in the constellation Pisces. Through the small telescope of an amateur observer M74 appears as a faint patch of light. Today's powerful advanced telescopes have informed us that M74 is a galaxy about 32 million light-years away from Earth, that it is about 30,000 light-years across, and that it contains about 100 billion stars. Figure 3.34 shows the galaxy face-on as it is seen from Earth.

Figure 3.34. This Hubble image of M74 is a composite of exposures taken in 2003 and 2005 at visible and infrared wavelengths. A small patch used data from the Canada-France-Hawaii Telescope and the Gemini Observatory to fill in what Hubble did not observe. Image credits: NASA, ESA, and The Hubble Heritage Team (STScI/AURA).

The galaxy's two spiral arms curve outward from its bulge. They feature dark lanes of cosmic dust and are dotted with clusters of hot, young stars. Bright knots of reddish glowing gas light up the spiral arms in environments rich in star formation. M74 is a classic example of what is known as a "grand-design" spiral galaxy.

A pair of identical equiangular spirals is superimposed over the bluish lanes of the galaxy's arms. Their angle γ lies between $108°$ and $108.5°$. In spite of the turbulence that the galaxy experiences, the galaxy's expanding arms are in close alignment with the two smooth geometric spirals.

We now turn to the mathematical description of equiangular spirals. They are the graphs of differentiable polar functions of the form $r = f(\theta)$ with the property that the angle $\gamma = \gamma(\theta)$ defined in Figure 3.21 is constant. Let $r = f(\theta)$ be such a function. Since $\gamma = \gamma(\theta)$ is constant, $c = \tan(\gamma - \frac{\pi}{2})$ is a constant as well. A basic property of the derivative of a polar function developed in section 3E tells us that $f'(\theta) = f(\theta) \cdot \tan(\gamma(\theta) - \frac{\pi}{2})$. Therefore, $f(\theta)$ satisfies

$$f'(\theta) = cf(\theta).$$

Because the equation involves a function and its derivative it is called a *differential equation*. The fact that $f(\theta)$ satisfies this equation implies that

$$f(\theta) = Ae^{c\theta},$$

where A is the constant $A = f(0)$. This is easily verified. By the quotient rule,

$$\frac{d}{d\theta} \frac{f(\theta)}{e^{c\theta}} \;=\; \frac{f'(\theta)e^{c\theta} - f(\theta)\cdot c e^{c\theta}}{e^{2c\theta}} \;=\; 0.$$

So $\frac{f(\theta)}{e^{c\theta}}$ is a constant. Set it equal to A, to get $f(\theta) = Ae^{c\theta}$. Since $e^0 = 1$, $A = f(0)$, and

$$f(\theta) = f(0)e^{\tan(\gamma - \frac{\pi}{2})\theta}.$$

If $\gamma = \frac{\pi}{2}$, then $f(\theta) = f(0)e^0$. The graph is a circle of radius $f(0)$. If $\gamma > \frac{\pi}{2}$, then the graph is a spiral that expands uniformly as it turns in a counterclockwise direction. If $\gamma < \frac{\pi}{2}$, then $f(\theta)$ is a decreasing function of θ, and the graph is a spiral that contracts uniformly in a counterclockwise direction. Figure 3.33 shows the graph of the equiangular spiral

$$f(\theta) = e^{\tan(\gamma - \frac{\pi}{2})\theta}$$

with $\gamma = 99.65°$ and $\tan(\gamma - \frac{\pi}{2}) = 0.170$. For the spirals of M74, $\tan(\gamma - \frac{\pi}{2}) \approx \tan 18° \approx 0.33$. Because of the tight connection between the exponential function and logarithms, equiangular spirals are also called *logarithmic spirals*.

Let's turn to consider our own galaxy, the Milky Way. It goes without saying that, given the distances involved, it is not possible to send a spacecraft to examine our galaxy from afar. So any assessment of its structure must be carried out from within. Figure 3.35 tells us what our Milky Way looks like when viewed from the souther hemisphere. A comparison of this image with that of galaxy NGC 891 of Figure 1.30 suggests that the Milky Way is a spiral galaxy. Astronomers confirmed some time ago that this is so. We know today that the Milky Way is a disk with a diameter of about 100,000 light-years. Current estimates put the thickness of the disk at around 1,000 light-years. If we could look at the Milky Way at an angle, it would look similar to the galaxy M31 depicted in Figure 1.29. If we could view the Milky Way from a distance edge-on, what we would see would resemble the depiction of galaxy NGC 891 of Figure 1.30. Like NGC 891, our galaxy has an equator and an equatorial plane. Our Sun is just one of more than two hundred billion stars that reside in our galaxy. Positioned in one of the spiral arms, it is about 26,000 light-years from the galaxy's center and about 14 light years from the galaxy's equatorial plane. So our solar system lies close to the galaxy's equatorial plane and about half-way between its center and the disk's perimeter. Just as the Earth-Moon system is in orbit around the Sun, our solar system is in orbit around the Milky Way's center of mass. It takes our solar system one "galactic year" or between 200 and 250 million Earth years to complete one orbit.

The determination of the particulars of our galaxy's structure has been a challenge. This is not surprising, given that our solar system is tucked away in one of the galaxy's arms and that the

Figure 3.35. Our Milky Way rising skyward from the Salar de Atacama salt flat in northern Chile above the lights of a nearby town. Image credit to and copyright held by Alex Tudorica of the University of Bonn. His permission to use this image is gratefully acknowledged.

probing eyes of our instruments need to peer through or around huge clusters of dark interstellar clouds along the galactic plane. Even basic questions have been difficult to answer. The study of infrared images from NASA's *Spitzer Space Telescope* has revealed that our galaxy has two major spiraling arms instead of four, as was previously thought. In addition, there are two minor arms that emanate from the ends of the bulge to rise between the two major arms.

In December 2013, the European Space Agency (ESA) sent the spacecraft *Gaia* into orbit around the a gravitational equilibrium point of the Sun-Earth system. The craft has been studying our galaxy from this vantage point ever since. With its two powerful telescopes *Gaia* has been making accurate measurements of the position, movement, and brightness of more than a billion stars of our galaxy (about 1% of the total). During its lifetime, *Gaia* will have observed each of its one billion stars about 70 times, providing a record of the position, velocity, and brightness of each of these stars over time. The main goal of the mission is to use the information about these billion stars to make the largest, most precise three-dimensional map of our galaxy. As an important step in this direction, the ESA released its second "galactic census" of detailed, high-precision data about our galaxy's stellar objects in April of 2018. The image of Figure 3.36 is a result of the data that *Gaia*

Figure 3.36. This image of the Milky Way is not a photograph but the projection of a three dimensional graph of the locations of the stars detected by the *Gaia* spacecraft onto a plane. Brighter regions indicate higher concentrations of stars. Darker regions correspond to dense, interstellar clouds of gas and dust that absorb starlight along the line of sight. The plane of the Milky Way, where most of the galaxys stars reside, runs horizontally across the center. Image credit: ESA/Gaia − CC BY-SA 3.0 IGO https://creativecommons.org/licenses/by-sa/3.0/igo/.

captured. The two bright objects in the lower right of the image are the large and small Magellanic Clouds, two dwarf galaxies orbiting the Milky Way.

Given that the data being accumulated by *Gaia* are unprecedented in terms of their accuracy and breadth, it seems likely that the analysis of the complete record will revolutionize our understanding of the formation, evolution, and structure of our galaxy. In particular, a combination of the distance and brightness information of the stars of our galaxy should add clarity to our understanding of its spiral structure. Another recent study, relying on a careful assessment of the location of certain types of standard candles (a class of objects whose distances can be computed by comparing their observed brightness with their known luminosity) has shown that the disk of the Milky way is slightly curved. The glowing strip that Figure 3.36 depicts has the shape of a stretched ∼.

There is no question that we have learned many astonishing things about our planet, our solar system, our galaxy, and the universe as a whole, since the Scientific Revolution. It is also clear that the advances in technology have greatly speeded up the learning curve. Making this point is a central purpose of the first three chapters of this text. On the other hand, we are also finding that there are large areas of the jigsaw puzzle of the universe that we don't understand at all.

Consider the galaxies depicted in Figures 3.32 and 3.34 for instance. While they are depicted as static things, they are (of course) in motion. We saw in the explanation of Figure 3.31 that the farther a star is from a galaxy's center, the slower it will move in its orbit around the galactic center. So the stars at the edges of a spiral galaxy should travel much more slowly than those in or near its bulge. But recent observations have shown that stars seem to orbit at more or less the same speed regardless of where they are in the galactic disk. How is one to explain this apparent contradiction? The explanation is that the stars of a galaxy, especially those near the edge, are subject to the gravitational effects of invisible masses of matter that permeate the galaxy. Even though they do not have an answer to the question of what such *dark matter* actually is, astrophysicists do have a sense of how much there is in the universe and how it is distributed. The amount of such dark matter can be estimated by comparing a galaxy's motion as observed, against the motion that is calculated under the assumption that the matter of the galaxy that is seen is all there is. What makes this possible is *gravitational lensing*. The fact is that the gravitational force exerted by matter in the universe, especially a large clump of matter such as a galaxy, bends the electromagnetic radiation that passes near it. This is so far all matter, including dark matter. When light from a very distant source is bent by a galaxy that lies between an observer on Earth and the source, the light-bending galaxy acts like a lens. The source appears highly distorted and multiple images of it can be observed around the lensing galaxy. The analysis of this effect tells us about the total mass of the light-bending galaxy. By subtracting the mass of its stars, dust, and gas, it is possible to estimate the properties and the amount of its dark matter. A team of over 400 scientists using this approach in ongoing studies have measured the shapes of 26 million galaxies to directly map the patterns of dark matter over billions of light-years. One of the conclusions is that 27% of everything in the universe is dark matter.

And the plot thickens. In the early 1990s, one thing was fairly certain about the expansion of the universe. Since the universe is full of matter and the attractive force of gravity pulls matter together, astrophysicists were assuming that the attractive force of gravity would slow down the expansion of the universe over time. But then came the observations via the *Hubble Space Telescope* of very distant supernovas that showed that the expansion was actually speeding up. One scientist likened the finding to throwing a set of keys up in the air expecting them to fall back down—only to see them fly straight up toward the ceiling. No one expected that the expansion of the universe would be accelerating and no one as yet knows how to explain this phenomenon. However, the solution has been given a name. It is called *dark energy*. We know how much dark energy there is because we know how it affects the universe's expansion. It turns out that roughly 68% of the universe is dark energy and hence that dark matter and dark energy comprise 95% of the total matter and energy of the universe. It follows that all ordinary matter and energy observed in the universe with all of our sophisticated telescopes and space probes—detecting electromagnetic radiation of all frequencies, ranging from gamma ray, ultraviolet, visible, infrared, radio waves, and beyond—adds up to about 5% of the total.

3I.   Problems and Discussions. It is the primary purpose of this section to provide the reader with opportunities to think about the mathematics that this chapter sets out and to engage the problem solving methods that it presents.

1. Points and Equations in Polar and Cartesian Coordinates. These problems list points and equations in polar coordinates and consider their Cartesian equivalents. Conversely, they list points and equations in Cartesian coordinates and ask about their polar versions. The graphs of polar equations and functions in polar coordinates are considered as well.

Problem 3.1. For each of the points given in polar coordinates below, find the unique corresponding Cartesian coordinates. Use a calculator to determine numerical versions of the coordinates.

   i. $(0, \frac{\pi}{3})$

   ii. $(5, \frac{\pi}{6})$

   iii. $(7, \frac{5\pi}{4})$

   iv. $(-6, -\frac{9\pi}{4})$

   v. $(7, 10)$

   vi. $(-3, -20)$

Problem 3.2. For each of the given points in Cartesian coordinates, find the only corresponding polar coordinates (r, θ) with $-\frac{\pi}{2} < \theta \le \frac{\pi}{2}$. Use a calculator to do so. Then find two more sets of polar coordinates for each point.

   i. $(4, 5)$

   ii. $(-2, 10)$

   iii. $(-7, -5)$

   iv. $(6, -15)$

Problem 3.3. Express each of the Cartesian equations below in polar coordinates. Write the answer in the form $r = f(\theta)$ when possible.

   i. $2x + 3y = 4$

   ii. $x^2 + y^2 = 4y$

   iii. $x^2 + y^2 = x(x^2 - 3y^2)$

Problem 3.4. Write each of the polar equations as an equation in Cartesian coordinates.

   i. $r = 5$

   ii. $r = 3\cos\theta$

   iii. $\tan\theta = 6$

   iv. $r = 2\sin\theta\tan\theta$

Problem 3.5. Sketch the graph of each equation. For a complete understanding of the graph, it may be necessary to convert the equation to Cartesian coordinates.

   i. $r = -6$

   ii. $\theta = -\frac{8\pi}{6}$

   iii. $r = 4\sin\theta$

   iv. $r(\sin\theta + \cos\theta) = 1$

Problem 3.6. Produce a table for $\cos\theta$ that is analogous to Table 3.1. Sketch a graph of $r = \cos\theta$. Confirm that the graph is a circle with center $(\frac{1}{2}, 0)$ (in either polar and Cartesian coordinates) and radius $\frac{1}{2}$.

Problem 3.7. Any line in the xy-plane has an equation of the form $ax + by + c = 0$, where a, b, and c are constants. Find a polar function $r = f(\theta)$ that has this line as its graph.

Problem 3.8. Sketch the graphs of the equations below. Do so by making use of the Cartesian versions of the equations.

   i. $r = \dfrac{2}{1 + \cos\theta}$ and $r = \dfrac{6}{1 + \cos\theta}$.

   ii. $r = \dfrac{2}{1 + \frac{1}{5}\cos\theta}$ and $r = \dfrac{5}{1 + \frac{1}{2}\cos\theta}$.

   iii. $r = \dfrac{3}{1 + 2\cos\theta}$ and $r = \dfrac{\frac{1}{2}}{1 + 5\cos\theta}$.

Problem 3.9. In each case determine the equation $(*)$ (refer to Section 3D) with the property that its graph satisfies the listed condition.

   i. a parabola with distance between focus and directrix equal to 10.

   ii. an ellipse with semimajor axis $a = 8$ and semiminor axis $b = 5$.

   iii. an hyperbola with semimajor axis $a = 8$ and semiminor axis $b = 5$.

    2. Conic Sections in Polar Coordinates. Let C be any conic section. A part of C is shown in Figure 3.37 below. Place a polar axis in such a way that O is at a focus, the polar axis lies on the focal axis of the conic section, and the polar axis points outward as shown. Problems 3.10 to 3.14 have this figure as their starting point. All problems involve equation $(*)$ of Section 3D.

Problem 3.10. Let C be a parabola. In equation $(*)$ set $\varepsilon = 1$ and take d to be the distance between the focus and the directrix of C. Why is the graph of $(*)$ identical to the parabola C?

Problem 3.11. Let C be the ellipse with semimajor axis 7 and semiminor axis 4. Determine ε and d so that the graph of $(*)$ is identical to the ellipse C.

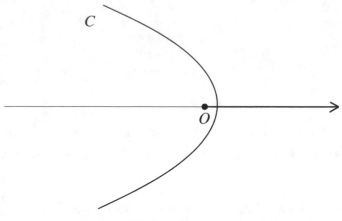

Figure 3.37

Problem 3.12. Let C be the hyperbola with semimajor axis $a = 5$ and semiminor axis $b = 3$. Determine ε and d so that the left branch of the graph of $(*)$ is the hyperbola C.

Problem 3.13. Let a and b be positive constants with $a \geq b$. Suppose that C is the ellipse with semimajor axis a and semiminor axis b. For what d and ε is the graph of $(*)$ identical to the ellipse C?

Problem 3.14. Let a and b be positive constants. Suppose that C is a hyperbola with semimajor axis a and semiminor axis b. For what d and ε is the left branch of the graph of $(*)$ identical to the hyperbola C?

Problem 3.15. Project: Let $\varepsilon \geq 0$ and $d > 0$ be constants. Modify the study of $r = \frac{d}{1+\varepsilon \cos \theta}$ to analyze the polar equation $r = \frac{d}{1+\varepsilon \sin \theta}$ and its graph. [Hint: In terms of the final results, the only difference turns out to be the orientation of the conic section. To get a sense of the difference, compare the graph of $r = \cos \theta$ (see Problem 3.6) with that of $r = \sin \theta$.] The functions $r = f(\theta) = \frac{d}{1-\varepsilon \cos \theta}$ and $r = f(\theta) = \frac{d}{1+\varepsilon \sin \theta}$ also represent all the conic sections that have the origin as a focal point after they are shifted and rotated. Their study is also very similar to the study of equation $(*)$.

3. Derivatives of Polar Functions. The next set of problems deal with derivatives of polar functions. We'll start with the function

$$r = f(\theta) = \frac{d}{1 + \varepsilon \cos \theta}$$

with $r > 0$ and θ restricted to $-\pi \leq \theta \leq \pi$.

Problem 3.16. Compute the derivative of the function $r = f(\theta)$. Show that $r = f(\theta)$ is decreasing for $\theta \leq 0$ and increasing for $\theta \geq 0$, so that $r = f(\theta)$ reaches its minimum value $\frac{d}{1+\varepsilon}$ when $\theta = 0$. Discuss the question of the existence of the maximum value of $r = f(\theta)$.

Problem 3.17. Consider the function $f(\theta) = \cos \theta$ and its graph (from Problem 3.6). Study the graph of f over each of the intervals $0 \leq \theta \leq \frac{\pi}{2}$, $\frac{\pi}{2} \leq \theta \leq \pi$, $\pi \leq \theta \leq \frac{3\pi}{2}$, and $\frac{3\pi}{2} \leq \theta \leq 2\pi$ to confirm that $r = f(\theta)$ respectively, decreases, decreases, increases, and increases over these intervals. Then compare this to the behavior of $f'(\theta) = -\sin \theta$ over each of the same intervals.

Problem 3.18. Consider the polar function $f(\theta) = \frac{1}{\sin\theta}$. Sketch its graph after converting it to Cartesian coordinates. Show that $\gamma(\theta) = \theta$ and then verify the equality $f'(\theta) = f(\theta) \cdot \tan(\gamma(\theta) - \frac{\pi}{2})$. [Hint: The equality $f'(\theta) = f(\theta) \cdot \tan(\gamma(\theta) - \frac{\pi}{2})$ relies on the trig identities $\cos(\theta - \frac{\pi}{2}) = \sin\theta$ and $\sin(\theta - \frac{\pi}{2}) = -\cos\theta$. Check these identities first.]

Problem 3.19. Check the equality $f'(\theta) = f(\theta) \cdot \tan(\gamma(\theta) - \frac{\pi}{2})$ for the polar function $f(\theta) = \cos\theta$. [Hint: By a property of isosceles triangles, $\gamma(\theta) = \frac{\pi}{2} - \theta$.]

Problem 3.20. Consider the line $r = f(\theta) = \frac{1}{\sin\theta - \cos\theta}$. Sketch the graph of this function and show for any point (r, θ) on the graph that $\gamma(\theta) = \theta - \frac{\pi}{4}$. It follows that $\tan(\gamma(\theta) - \frac{\pi}{2}) = \tan(\theta - \frac{3\pi}{4})$. Use the formulas for the sine and cosine of the sum of two angles, to show that $\tan(\theta - \frac{3\pi}{4}) = \frac{\sin\theta + \cos\theta}{\cos\theta - \sin\theta}$. Then verify the formula $f'(\theta) = f(\theta) \cdot \tan(\gamma - \frac{\pi}{2})$.

Problem 3.21. Consider Figure 3.22a. Suppose that P is a point on the graph such that the graph of $r = f(\theta)$ lies below the red circular arc to the left of P. Sketch such a situation and go through the derivation of the equality $f'(\theta) = f(\theta) \cdot \tan(\gamma - \frac{\pi}{2})$ to verify that it is valid in this case as well.

Consider a function $r = f(\theta)$ that is increasing and continuous. The fact that r increases with increasing θ, tells us that as the ray determined by θ rotates counterclockwise, the corresponding $r = f(\theta)$ increases. So the graph of $r = f(\theta)$ is a spiral that opens in a counterclockwise way. (The spiral may "wobble" in the sense that the increase in r may vary from smaller to larger, back to smaller, and so on.)

Problem 3.22. Let $c > 0$ and consider the function $r = f(\theta) = c\theta$. It's graph is an *Archimedean spiral*. Sketch it from $\theta = 0$ to $\theta = 2\pi$. Study the rate of expansion of this spiral by analyzing the equality $\frac{f'(\theta)}{f(\theta)} = \tan(\gamma(\theta) - \frac{\pi}{2})$. What can you say about γ for small positive θ? For large positive θ?

Problem 3.23. It was shown in the last part of the chapter that the graph of a function $r = f(\theta)$ with the property that the angle $\gamma(\theta) = \gamma$ is constant is an equiangular spiral and that the function has the form $f(\theta) = f(0)e^{\tan(\gamma - \frac{\pi}{2})\theta}$. Verify the equality $f'(\theta) = f(\theta) \cdot \tan(\gamma - \frac{\pi}{2})$ for such a function.

4. Definite Integrals of Polar Functions. This segment considers the definite integrals of polar functions that arise as the lengths of various polar curves and the areas bounded by them.

Problem 3.24. Use an integral formula for the length of a polar graph to determine the length of the graph of $r = f(\theta) = \sin\theta$ from $\theta = 0$ to $\theta = \pi$, then from $\theta = 0$ to $\theta = 2\pi$, and finally from $\theta = 0$ to $\theta = 3\pi$. Explain your answers by referring to the graph of $r = \sin\theta$ in Figure 3.14.

Problem 3.25. Use the graph of the polar function $r = f(\theta) = \sin\theta$ (again refer to Figure 3.14) to evaluate the integrals $\int_0^\pi \frac{1}{2}\sin^2\theta\, d\theta$ and $\int_0^{2\pi} \frac{1}{2}\sin^2\theta\, d\theta$. Then evaluate the integrals again, this time directly, by using the half-angle formula $\sin^2\theta = \frac{1 - \cos 2\theta}{2}$.

Problem 3.26. Consider the graph of the function $r = f(\theta) = \frac{1}{\sin \theta}$ over the interval $\frac{\pi}{4} \leq \theta \leq \frac{3\pi}{4}$. Set up an integral that represents the length of the graph and another that represents the area bounded by the graph and the lines $\theta = \frac{\pi}{4}$ and $\theta = \frac{3\pi}{4}$. Use the fact that the Cartesian equation of the polar equation $r = \frac{1}{\sin \theta}$ is the line $y = 1$ to compute these integrals.

Problem 3.27. Consider the circle $(x - 1)^2 + (y - 1)^2 = 2$. Find a polar function $r = f(\theta)$ that has this circle as its graph. Use the graph of the circle to evaluate the integrals $\int_0^{\frac{\pi}{2}} \frac{1}{2} f(\theta)^2 \, d\theta$ and $\int_0^{\frac{\pi}{2}} \sqrt{f(\theta)^2 + f'(\theta)^2} \, d\theta$.

Problem 3.28. Study the graph of the polar function $r = f(\theta) = \frac{4}{\cos \theta}$ over the interval $0 \leq \theta < \frac{\pi}{2}$ by using Cartesian coordinates. Use your conclusions to evaluate the integral $\int_0^{\frac{\pi}{3}} \frac{8}{\cos^2 \theta} \, d\theta$. Do so by interpreting the integral as an area and then again by interpreting it as a length.

Problem 3.29. Study the graph of the function $r = f(\theta) = \frac{3}{\sin \theta + 2 \cos \theta}$ by using Cartesian coordinates. Use your study to evaluate the integrals $\int_0^{\frac{\pi}{2}} \frac{1}{2} f(\theta)^2 \, d\theta$ and $\int_0^{\frac{\pi}{2}} \sqrt{f(\theta)^2 + f'(\theta)} \, d\theta$.

Problem 3.30. Consider the graph of the function $r = f(\theta) = \frac{2}{1 + \cos \theta}$. What does the integral $\int_0^{\frac{\pi}{2}} \frac{2}{(1 + \cos \theta)^2} \, d\theta$ represent? Convert the function $r = f(\theta)$ to a function in Cartesian coordinates and evaluate the integral.

Problem 3.31. Consider the graph of the function $r = f(\theta) = \frac{4}{1 + \frac{1}{3} \cos \theta}$. Use the formula $A = ab\pi$ for the area of an ellipse with semimajor axis a and semiminor axis b to evaluate $\int_0^{\pi} \frac{8}{(1 + \frac{1}{3} \cos \theta)^2} \, d\theta$.

Problem 3.32. Study the solutions of Problems 3.25, 3.28, 3.30, and 3.31 and then evaluate the integrals $\int_0^{\pi} \sin^2 x \, dx$, $\int_0^{\frac{\pi}{3}} \frac{1}{\cos^2 x} \, dx$, $\int_0^{\frac{\pi}{2}} \frac{1}{(1 + \cos x)^2} \, dx$, and $\int_0^{2\pi} \frac{1}{(1 + \frac{1}{3} \cos x)^2} \, dx$.

Problem 3.33. Consider the equiangular spiral $f(\theta) = \frac{1}{7} e^{\frac{\theta}{\sqrt{3}}}$ over the interval $[0, 2\pi]$. What is the constant angle γ for this spiral? Determine the length of the spiral and the area that it encloses.

Problem 3.34. Consider the equiangular spiral given by the function $r = f(\theta) = \frac{1}{4} e^{\tan(\gamma - \frac{\pi}{2})\theta}$ with $\gamma = \frac{3\pi}{4}$. Find the length of the spiral from $\theta = 0$ to $\theta = 4\pi$ and also the area enclosed by this part of the spiral. [The second part of the problem requires care.] To get a sense of what is involved, use the millimeter as the unit of length, plot the points on the spiral corresponding to θ equal to $0, \frac{\pi}{2}, \pi, \frac{3\pi}{2}$, and 2π and sketch the spiral over $0 \leq \theta \leq 2\pi$. At this point abandon scale and draw in a rough graph of the spiral over the interval $2\pi \leq \theta \leq 4\pi$.

Problem 3.35. Study Figure 3.25. Show that the length of the red circular arc from P to B is $|f(\theta)|\,d\theta$. Since this approximates the length of the graph between the two rays, we see (as in the discussion that resulted in the polar area formula) that the length L of the graph of $r = f(\theta)$ between $\theta = a$ and $\theta = b$ is equal to

$$L = \int_a^b |f(\theta)|\,d\theta.$$

What about this argument and its conclusion? Why do we know that this formula is wrong? What is the correct formula? Can you sketch a graph of a polar function where (in reference to Figure 3.25) the difference between the length of the graph of the function and that of the corresponding red circular arc (both between two rays from the origin that are close together) is substantial? For example, consider the equiangular spiral $r = f(\theta) = \frac{1}{4}e^{\tan(\gamma - \frac{\pi}{2})\theta}$ with $\gamma = \frac{3\pi}{4}$ over the interval $0 \le \theta \le 2\pi$. Using the millimeter as unit of length, plot the points of the spiral corresponding to θ equal to $0, \frac{\pi}{2}, \pi, \frac{3\pi}{2}$, and 2π, and experiment. Compute the correct and incorrect integrals and compare their values.

5. The Spiral and the Nautilus.
By way of a very wide but beautiful digression, consider the fact that spirals arise in the shell structure of a number of aquatic animals. The pearly nautilus is an example of an ocean animal with a smoothly spiraling shell. The shell of a grown nautilus is about 25 cm in diameter and has about 30 chambers. The animal lives in the successively developing outermost chamber. The chambers are connected by a tube that lets the nautilus adjust the pressure of the gases within the shell. This ability allows it to change the depth at which it is swimming. A bottom feeder, it uses its tentacles (up to about 100 in number) for capturing prey. It lives at 300 meter depths and is most commonly observed in its natural habitat through the television cameras of deep diving submersibles. The spiral of the nautilus is essentially equiangular. Figure 3.38 shows the cross section of the shell of a nautilus together with an equiangular spiral that follows the contours of the cross section rather closely. The constant angle γ of this spiral is $\gamma \approx 99.65°$.

Problem 3.36. Turn to the image of the shell of the nautilus in Figure 3.38 and place a polar axis as shown. Let the rightmost point of intersection of the spiral and the polar axis correspond to $\theta = 0$. Let s be the distance from this point to the origin O at the eye of the spiral. Remarks already made tell us that the spiral of the image of the shell is closely approximated by the graph of the function $f(\theta) = f(0)\,e^{0.17\theta} = se^{0.17\theta}$. What aspects of the geometry of the shell do the two definite integrals

$$\int_{-\frac{13\pi}{2}}^{\frac{\pi}{3}} \sqrt{f(\theta)^2 + f'(\theta)^2}\,d\theta \text{ and } \int_{-\frac{3\pi}{2}}^{\frac{\pi}{3}} \frac{1}{2}f(\theta)^2\,d\theta$$

represent at least approximately? Show that the values of these integrals are approximately equal to $6.94s$ and $1.80s^2$, respectively. Do these values appear to be consistent with Figure 3.38?

The combination of biological and mathematical factors by which the nautilus constructs one chamber of its shell after the other in such a way that an equiangular spiral emerges does not appear

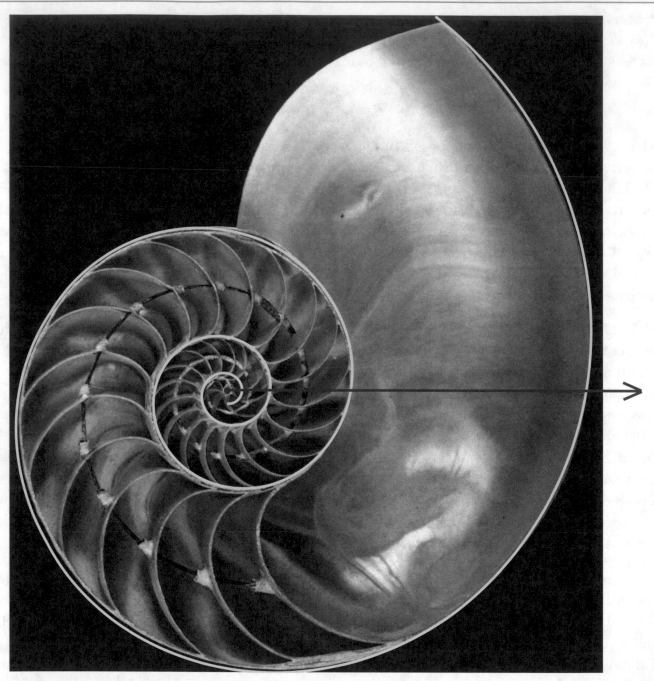

Figure 3.38. Image credit to Rutherford Platt, in *Mathematics*, Life Science Library, Time Inc., New York, 1963. Permission of use granted by Time Inc.

to be fully understood. A question that a different forum might address is this: What is the mechanism that regulates the shell's programmed sense to preserve its equiangular shape?

Centripetal Force and Resulting Trajectories

Newton's 1687 treatise *Principia Mathematica*—its central aspects have already been described in Chapter 1D—had been a miracle. It was a miracle for the synthesizing and penetrating way in which it combined basic physical laws and novel mathematical analyses to explain and confirm the conclusions about the orbits of the planets that Kepler had reached by relying on observational data alone. Newton provided a definitive solution to a problem that had occupied many of the best scientific minds since the time the early Babylonian and Greek astronomers and philosophers first began to think about the question of how the heavens worked over 3000 years ago. Whereas Kepler discovered his three laws of planetary motion with painstaking observations, Newton came to recognize the deeper underlying reality: all three of Kepler's laws of planetary motion rest on a combination of mathematical analysis, basic laws of motion, and the law of universal gravitation, the assumption—for which there is broad and conclusive evidence—that the magnitude of the gravitational force between two bodies in the universe is proportional to the product of the masses involved and the inverse of the square of the distance between them.

The focus of Newton's treatise is on the abstract study of a centripetal force acting on a point mass. One of his basic assumptions and strategies is as surprising as it is novel. It interprets and models such a force not as the continuous, smoothly pulling action that characterizes gravity, but as a rapidly repetitive "machine-gun-style" sequence of deflecting "pops." This approach triangularized the geometry and simplified the analysis. By letting the time between successive pops go to zero, Newton was able to draw out the correct relationships between a smoothly varying force and the geometry of the orbit. The link that establishes the connection between the abstract study of a centripetal force acting on a point-mass and the application to a gravitational force acting on a spherical body, is Newton's realization that his conclusions apply not only to a point-masses but also to spheres (that have their matter distributed in a radially homogeneous way).

It is the purpose of this chapter to derive afresh and in complete detail the central conclusions of Newton's analysis about the connection between a gravitational force and the trajectory of the object that it drives. Unlike Newton's treatment with its reliance on novel uses of geometric methods and its assumption that a centripetal force acts in intermittent bursts, the analysis of this chapter applies the calculus of polar functions to a smoothly acting centripetal force.

4A. A Basic Study of Forces. In mathematics, a *vector* is a quantity that has both a magnitude and a direction. A vector is represented by an arrow with the direction of the arrow indicating the direction of the vector and the length of the arrow reflecting its magnitude. The examples

© Alexander J. Hahn 2020
A. J. Hahn (ed.), *Basic Calculus of Planetary Orbits and Interplanetary Flight*,
https://doi.org/10.1007/978-3-030-24868-0_4

that are relevant in this text are velocity, acceleration, and force. We will illustrate how vector quantities behave with a look at forces. A completely analogous discussion applies to velocity and acceleration. Suppose that a unit of length and a unit of force are given. (This could be the centimeter and the newton, or the inch and the pound, and so on. See Chapter 1G.) Given a force, the vector representing it points in the direction of the force and has length equal to the numerical magnitude of the force. The most basic law of forces is the parallelogram law. It tells us that if two forces F_1 and F_2 act at the same point P, then the combined effect, or *resultant*, of the two forces—both direction and magnitude—is the force F determined by the vector given by the diagonal of the parallelogram that the point P and the vectors of F_1 and F_2 provide. What is involved is illustrated in Figure 4.1a. Figure 4.1b tells us that the resultant can also be obtained by shifting the tail of one of the vectors to the tip of the other. In discussions of forces our notation

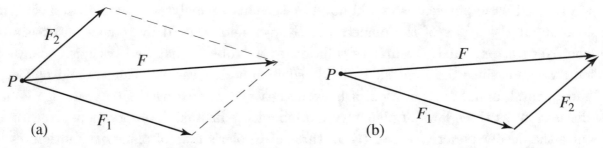

Figure 4.1

will play a dual role. In the statement "the forces F_1 and F_2 act at the same point" the symbols F_1 and F_2 are used simply to identify the forces. However, we will simultaneously use the symbols F_1 and F_2 to represent the magnitudes of these forces. An example of this practice follows.

Let's look at the parallelogram law quantitatively. In Figure 4.2a, θ is the angle between the two forces F_1 and F_2 and φ is the angle $\pi - \theta$. It follows by applying the law of cosines to the

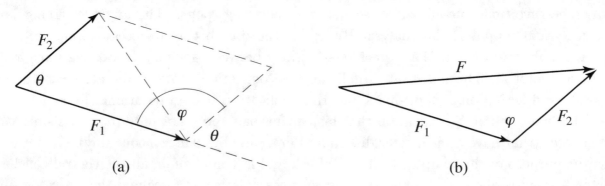

Figure 4.2

triangle of Figure 4.2b and using Example 3.3, that the magnitude of the resultant of F_1 and F_2 is

$$F = \left(F_1^2 + F_2^2 - 2F_1F_2\cos(\pi - \theta)\right)^{\frac{1}{2}} = \left(F_1^2 + F_2^2 + 2F_1F_2\cos\theta\right)^{\frac{1}{2}}.$$

Example 4.1. Let F_1 and F_2 be two forces acting at a point. Suppose that their magnitudes are 3 and 4 newtons respectively and that the angle between them is $50°$. Show that the magnitude of the resultant is $F \approx 6.36$ newtons.

The parallelogram law can also be used to separate a force into components. Let a force F be given and choose any parallelogram that has the arrow representing F as a diagonal. See Figure 4.3a. We'll let φ_1 and φ_2 be the respective angles between the diagonal of the parallelogram and its two sides. Refer to Figure 4.3b and observe that the parallelogram provides vectors F_1 and F_2 that have F as their resultant. The vector F_1 is the *component of F in the direction of φ_1* and the vector F_2 is the *component of F in the direction of φ_2*. By applying the law of sines to the upper triangle of Figure 4.3b, we get

$$\frac{\sin \varphi_2}{F_1} = \frac{\sin \varphi_1}{F_2} = \frac{\sin(\pi - (\varphi_1 + \varphi_2))}{F} = \frac{\sin(\varphi_1 + \varphi_2)}{F}.$$

The last equality makes use of Example 3.3. These equalities tell us that if F and the angles φ_1

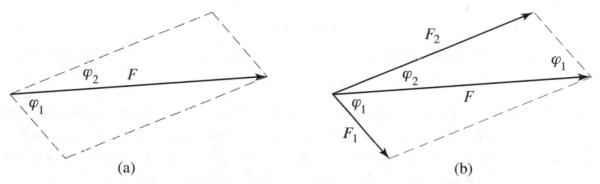

(a)   (b)

Figure 4.3

and φ_2 are given, then the magnitudes of the two components F_1 and F_2 are

$$F_1 = F \frac{\sin \varphi_2}{\sin(\varphi_1 + \varphi_2)} \quad \text{and} \quad F_2 = F \frac{\sin \varphi_1}{\sin(\varphi_1 + \varphi_2)}.$$

Example 4.2. Let F_1 and F_2 be the components of a force F of 15 newtons in the directions of the angles $\varphi_1 = 20°$ and $\varphi_2 = 40°$, respectively. Show that $F_1 \approx 11.13$ and $F_2 \approx 5.92$ newtons.

There are infinitely many parallelograms that fulfill the diagonal requirement of the diagram of Figure 4.3a. The most important is the rectangle with horizontal and vertical sides. The components it determines are the *horizontal* and *vertical* components of the force F. Let F_1 act vertically and F_2 horizontally. Since $\varphi_1 + \varphi_2 = \frac{\pi}{2}$ in this case, we see that with the notation of Figure 4.3b, the magnitudes of the vertical and horizontal components of F are equal to $F_1 = F \cos \varphi_1$ and $F_2 = F \cos \varphi_2$, respectively, in terms of the angles φ_1 and φ_2 that they make with the resultant F.

Before we turn to a detailed mathematical study of centripetal force we'll look at an example. Take a string and tie an object P of weight W to one end. Hold the other end of the string fixed at H and twirl the object P so that the angle that the segment HP makes with the vertical is constant. It follows that P moves along a fixed circle in a horizontal plane. Figure 4.4a shows the vector F_H that represents the pull of the string on P as well as its horizontal and vertical components. Because P does not move vertically, the vertical component F_1 balances the downward pull of the weight W. So $F_1 = W$. It follows that the combined effect of F_H and the weight W is the horizontal component

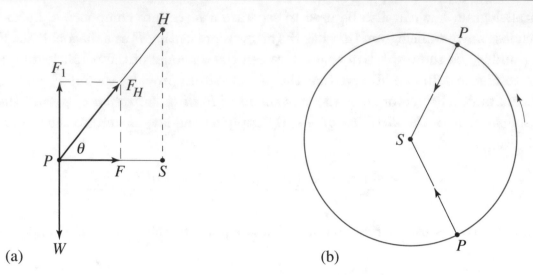

(a)   (b)

Figure 4.4

F pointing in the direction of the center S of the circle. This force is therefore a centripetal force. See Figure 4.4b.

Example 4.3. Suppose that the string PH has length 1.6 meters and that it pulls with a force of 9 newtons. Let the radius SP of the circle be 0.7 meters. Show that $\cos\theta = \frac{0.7}{1.6}$ and $\sin\theta = \frac{\sqrt{2.07}}{1.6}$. Conclude that $F = \frac{0.7}{1.6} \cdot 9 \approx 3.94$ newtons and that $W = \frac{\sqrt{2.07}}{1.6} \cdot 9 \approx 8.09$ newtons. Let m be the mass of W. Take $g = 9.8$ m/sec^2 and show that $m = \frac{\sqrt{2.07}}{1.6} \cdot \frac{9}{9.8} \approx 0.83$ kilograms.

4B. The Mathematics of a Moving Point. Consider a point moving in the Cartesian xy-plane. Take a stopwatch and start observing the point at time $t = 0$. At any elapsed time $t \geq 0$, the point will be at some position in the plane. This position varies over time, so that its x-coordinate and its y-coordinate vary with time. The two coordinates are therefore functions t that we'll denote by $x(t)$ and $y(t)$, respectively. Figure 4.5 shows a typical situation. Observe that if the functions

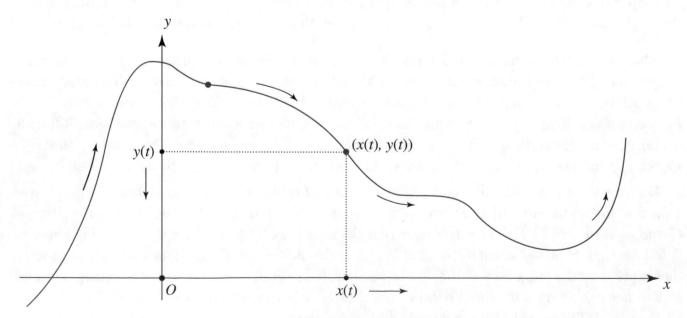

Figure 4.5

$x(t)$ and $y(t)$ are both understood, then the motion of the point is understood. Therefore the mathematics of the motion of a point in the plane (and also in space) reduces to that of a point moving along a coordinate line.

Suppose a point is moving on a coordinate axis. We'll assume that it moves smoothly without any sudden jumps, fits, and restarts. Begin to observe it at time $t = 0$. Let $p(t)$ be the function that specifies its coordinate at any time $t \geq 0$. It is the *position function* of the point. See Figure 4.6. The units of distance and time can be meters and seconds, kilometers and hours, and so forth, as

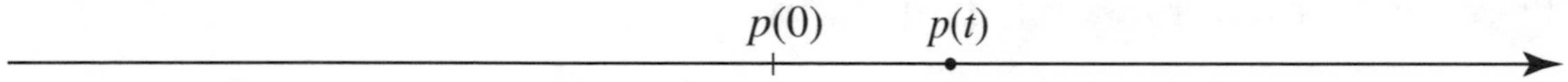

$$p(0) \qquad p(t)$$

Figure 4.6

a specific context would provide them.

Fix a time t and a time interval Δt. Our discussion assumes that Δt is positive, so that the instant $t + \Delta t$ follows t. (With only slight modifications, it is also valid for negative Δt and hence with $t + \Delta t$ preceding t.) In the term

$$\frac{p(t + \Delta t) - p(t)}{\Delta t}$$

$p(t + \Delta t) - p(t)$ is the change in the position of the point and Δt is the time it takes for this change to take place. The ratio $\frac{p(t+\Delta t)-p(t)}{\Delta t}$ is the average rate of change in the position, or *the average velocity*, of the point during the time interval from t to $t + \Delta t$. By pushing Δt to zero we get the derivative

$$p'(t) = \lim_{\Delta t \to 0} \frac{p(t + \Delta t) - p(t)}{\Delta t}$$

of $p(t)$, namely the rate at which the point's position $p(t)$ changes at the instant t. It is the point's *velocity at the instant t*. The assumption that the point moves smoothly tells us that the function $p(t)$ is differentiable and hence that this limit exists. By allowing t to vary, we get the velocity function $v(t) = p'(t)$ of the moving point. Take a given time t and turn to Figure 4.6. Note that if $v(t) = p'(t) > 0$, then $p(t)$ is increasing at t, so that the point is moving to the right. If $v(t) = p'(t) < 0$, then $p'(t)$ is decreasing at t and the point is moving to the left. And if $v(t) = p'(t) = 0$, the velocity at t is zero and the point has come to a stop.

The *speed* of the point at time t is the absolute value $|v(t)|$ of the velocity. By telling us when a point moves to the left and to the right, the velocity function $v(t) = p'(t)$ incorporates information about both the speed and the direction of its motion.

Example 4.4. Suppose that $p(t)$ is given by $p(t) = t^2 - 4t + 3$. Show that in this case, the point moves to the left for $t < 2$, to the right for $t > 2$, and stops at $t = 2$.

Let $v(t)$ be the velocity function of a point moving along a coordinate axis. Fix a time t and a time interval Δt. Our discussion again assumes that Δt is positive, so that the instant $t + \Delta t$

follows t. (With slight modifications, it is also valid for negative Δt and hence with $t + \Delta t$ preceding t.) The term

$$\frac{v(t + \Delta t) - v(t)}{\Delta t}$$

has the change $v(t + \Delta t) - v(t)$ in the velocity of the point in its numerator and the time Δt that it takes for this change to take place in its denominator. The ratio $\frac{v(t+\Delta t)-v(t)}{\Delta t}$ is the average rate of change in the velocity, or *the average acceleration*, of the point during the time interval from t to $t + \Delta t$. By pushing Δt to zero we get the derivative

$$v'(t) = \lim_{\Delta t \to 0} \frac{v(t + \Delta t) - v(t)}{\Delta t}$$

of $v(t)$, namely the rate at which the point's velocity $v(t)$ changes at the instant t. It is the point's *acceleration at the instant t*. The assumption that the point moves smoothly and in particular that its velocity varies smoothly tells us that the function $v(t)$ is differentiable and hence that this limit exists. By allowing t to vary, we get the acceleration function $a(t) = v'(t)$ of the moving point. Notice that $a(t) = p''(t)$ is the second derivative of the position function $p(t)$. Another notation that we'll use for $p''(t)$ is $\frac{d^2 p}{dt^2}$ and analogously for other second derivatives. What unit is acceleration measured in? For example, if distance is in meters and time in seconds, then velocity is given in meters per second, so that acceleration is expressed in the unit (meters per second) per second, or meters per second squared.

Recall that in mathematical studies of physical objects, a *point-mass* is a point that has non-zero mass. Assume that the moving point is a point-mass with mass m. Then by Newton's second law of motion, $F(t) = ma(t)$ is the force that acts on the point at any time t. Consider a given time t. Note that if $a(t) = v'(t) > 0$, then $v(t)$ is increasing at time t. In this case, $F(t)$ acts in the positive direction and pushes the point to the right at time t. If $a(t) = v'(t) < 0$, then $v(t)$ is decreasing at time t. Now $F(t)$ acts in the negative direction and pushes the point to the left at time t. If $v'(t) = 0$, then the velocity does not change at time t. This means that the force $F(t)$ is zero at that time.

Example 4.5. Let the position function of a moving point be $p(t) = \frac{3}{4}t^4 - 13t^3 + 75t^2 - 168t - 5$ for any time $t \geq 0$. Show that its velocity is $v(t) = p'(t) = 3(t^3 - 13t^2 + 50t - 56)$ and that the initial position and initial velocity of the point are $p(0) = -5$ and $v(0) = -168$. A substitution tells us that $v(2) = 3(8 - 52 + 100 - 56) = 0$. So the point stops at time $t = 2$. Since 2 is a root of the polynomial $v(t)$, we know that $(t - 2)$ divides $v(t)$. Check that $v(t) = 3(t - 2)(t^2 - 11t + 28)$ to confirm this. Apply the quadratic formula to $t^2 - 11t + 28 = 0$ to show that

$$t = \tfrac{11 \pm \sqrt{121 - 112}}{2} = 4 \text{ or } 7,$$

and hence that $v(4) = 0$ and $v(7) = 0$. Show that $v(t) = 3(t - 2)(t - 4)(t - 7)$ and that the point moves as follows. For $0 \leq t < 2$, observe that $v(t) < 0$, so that the point moves to the left. After it stops at $t = 2$, we see that $v(t) > 0$ for $2 < t < 4$, so that the point moves to the right over this time interval. At $t = 4$ it stops again, and since $v(t) < 0$ for $4 < t < 7$, it then moves to the left again. After its final stop at $t = 7$, $v(t) > 0$ so that it moves to the right thereafter. Suppose next that the point is a point-mass of mass m and show that the force acting on it is $F(t) = ma(t) = mv'(t) = 3m(3t^2 - 26t + 50)$. Use the quadratic formula again to show that $F(t) = 0$ for

$$t = \frac{26 \pm \sqrt{26^2 - 4 \cdot 150}}{6} = \frac{13 \pm \sqrt{19}}{3} \approx 2.88 \text{ or } 5.79.$$

Check that $F(t) = 9m(t^2 - \frac{26}{3}t + \frac{50}{3}) = 9m(t - \frac{13-\sqrt{19}}{3})(t - \frac{13+\sqrt{19}}{3})$. Conclude that for $0 \le t < \frac{13-\sqrt{19}}{3}$, the force is positive and acts to the right; when $\frac{13-\sqrt{19}}{3} < t < \frac{13+\sqrt{19}}{3}$, the force is negative and acts to the left; and when $t > \frac{13+\sqrt{19}}{3}$ it acts to the right again. Is this consistent with the earlier description of the motion of the point?

For a point moving along a coordinatized line, the functions $p(t)$, $v(t) = p'(t)$, and $a(t) = v'(t)$ for the position, velocity, and acceleration all have a directional aspect. When, for a given t, $p(t)$, $v(t)$, or $a(t)$ is positive, then the position of the point is on the positive side of the axis, the point moves in the direction of the positive axis, or its positive acceleration implies an increase in the velocity. Analogously, when $p(t)$, or $v(t)$, or $a(t)$ is negative for a given t, then the point is positioned on the negative axis, moves in the negative direction, or its negative acceleration implies a decrease in its velocity. The speed $|v(t)|$, on the other hand, always satisfies $|v(t)| \ge 0$ and does not have a directional aspect.

Now that we understand the mathematics of a point moving on a coordinate line, we can return to the motion of a point in the plane. Suppose that a point moves smoothly along a curve in the xy-plane. Turn back to Figure 4.5 for instance. For any time t (as measured by a stopwatch for example) we let $x(t)$ and $y(t)$ be the x- and y-coordinates of the position of the point. We will assume the motion to be smooth, so that both $x(t)$ and $y(t)$ are differentiable functions of t. By applying what we already learned, we know that $x'(t)$ and $y'(t)$ are the respective velocities of the x- and y-coordinates of the point. Think of $x(t)$ and $y(t)$ as the positions of the shadows of the point on the x-axis and y-axis that light sources parallel to the two axes produce. The derivatives $x'(t)$ and $y'(t)$ are the respective velocities of these shadows.

For a fixed time t and a time interval Δt, the velocities of the x- and y-coordinates of the point at time t are given by

$$x'(t) = \lim_{\Delta t \to 0} \frac{x(t + \Delta t) - x(t)}{\Delta t} \quad \text{and} \quad y'(t) = \lim_{\Delta t \to 0} \frac{y(t + \Delta t) - y(t)}{\Delta t}.$$

It should be possible to determine the velocity of the moving point in the plane in terms of the velocities $x'(t)$ and $y'(t)$. And it is! Consider the point at the two times t and $t + \Delta t$. Figure 4.7a shows the point at the two instances t and $t + \Delta t$. (In the figure, Δt is taken to be positive, but this is not essential to the argument.) For a small elapsed time Δt, the two positions of the point are near each other, so that by the distance formula of plane coordinate geometry (it relies on the Pythagorean theorem), the distance traveled by the point is approximately equal to

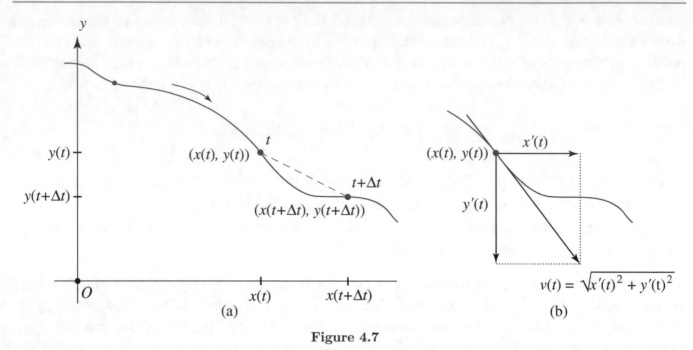

Figure 4.7

$$\sqrt{\left[x(t+\Delta t)-x(t)\right]^2 + \left[y(t+\Delta t)-y(t)\right]^2}.$$

The smaller the Δt, the closer the points are to each other, and the better this approximation is. It follows that the average speed of the point during the time interval from t to $t + \Delta t$ is approximately equal to

$$\frac{\sqrt{[x(t+\Delta t)-x(t)]^2 + [y(t+\Delta t)-y(t)]^2}}{\Delta t}.$$

Since its numerator is necessarily positive (or zero) this expression provides no information about the direction of the motion, so that it represents average speed rather than average velocity. After some algebra (pull the Δt under the radical as $(\Delta t)^2$, and distribute it over the two parts of the sum), this is equal to

$$\sqrt{\left[\frac{x(t+\Delta t)-x(t)}{\Delta t}\right]^2 + \left[\frac{y(t+\Delta t)-y(t)}{\Delta t}\right]^2}.$$

Taking the limit $\lim_{\Delta t \to 0}$ of this term we obtain that the speed of the point at time t is equal to

$$v(t) = \sqrt{x'(t)^2 + y'(t)^2}.$$

In the last step of the derivation of this formula, the operation $\lim_{\Delta t \to 0}$ needed to be moved past both the square root and the square. This a legitimate maneuver that depends on the continuity of the square root and the squaring functions. (But this is a technical point that we will not take up.)

We know that $x'(t)$ is positive when the shadow on the x-axis moves to the right and negative when it moves to the left. In the same way, the shadow on the y-axis moves up if $y'(t)$ is positive and down if $y'(t)$ is negative. So both $x'(t)$ and $y'(t)$ contain information about the direction of the motion and both represent velocity. But the term $\sqrt{x'(t)^2 + y'(t)^2}$ is always positive (or zero), includes no information about direction, and represents speed. Subsequent discussions will generally distinguish between velocity and speed, but will often use the notation v or $v(t)$ for both.

A look at Figure 4.7b confirms that the direction of the motion of the point at time t is determined by the tangent line to the curve at the point $(x(t), y(t))$ and the sign of the terms $x'(t)$ and $y'(t)$ (since they tell us about the left/right and up/down of the motion). Another look at Figure 4.7 shows that the slope of the dashed segment is $\frac{y(t+\Delta t)-y(t)}{x(t+\Delta t)-x(t)}$. It follows that the slope of the tangent to the path of the moving point at $(x(t), y(t))$ is

$$\lim_{\Delta t \to 0} \frac{y(t + \Delta t) - y(t)}{x(t + \Delta t) - x(t)} = \lim_{\Delta t \to 0} \frac{\frac{y(t+\Delta t)-y(t)}{\Delta t}}{\frac{x(t+\Delta t)-x(t)}{\Delta t}} = \frac{\lim_{\Delta t \to 0} \frac{y(t+\Delta t)-y(t)}{\Delta t}}{\lim_{\Delta t \to 0} \frac{x(t+\Delta t)-x(t)}{\Delta t}} = \frac{y'(t)}{x'(t)} .$$

So at any point $(x(t), y(t))$ the direction and speed of the motion of the point are both determined by $x'(t)$ and $y'(t)$. The direction is tangential to the path and the speed is $v(t) = \sqrt{x'(t)^2 + y'(t)^2}$.

Let's suppose that our moving point is a point-mass of mass m and that its velocities $x'(t)$ and $y'(t)$ vary smoothly. So $x'(t)$ and $y'(t)$ are differentiable functions of t and the forces on the point in the x- and y-directions are

$$mx''(t) \quad \text{and} \quad my''(t)$$

at any time t. Since force is a vector quantity, the resultant force $F(t)$ on the point-mass is given by the parallelogram law as shown in Figure 4.8a. Figure 4.8b shows the force $F(t)$ resolved into

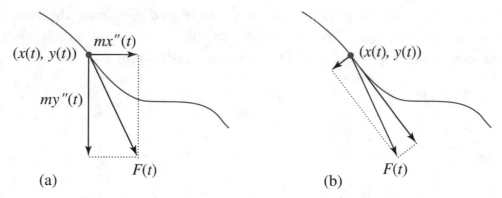

Figure 4.8

two significant components. One of them—tangential to the point's path—accelerates the point along its path. The other—perpendicular to the tangent—bends the path of the point.

Example 4.6. Consider a point-mass P of mass $m = 1$ moving in an xy-plane. Let its position at any time $t \geq -2$ be given by $x(t) = t$ and $y(t) = t^2$. Since $y(t) = x(t)^2$, the point moves on the parabola $y = x^2$. Its initial position is $(-2, 4)$. Discuss the motions of the x- and y-coordinates of P, show that it moves as illustrated in Figure 4.9a, and that its speed at any time t is $v(t) = \sqrt{1 + 4t^2}$. Deduce that the acceleration of the point P along its path is equal to $\frac{4t}{\sqrt{1+4t^2}}$. Show that the magnitude of the force $F(t)$ on P is constant and equal to 2. Consider Figure 4.8a and conclude that the force acts vertically in the upward direction. Why is this force not centripetal? Figure 4.9b shows the force $F(t)$ resolved into its tangential component and the component perpendicular to the tangent

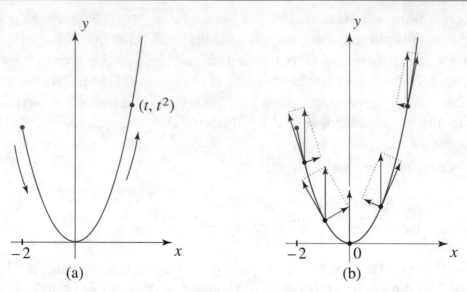

Figure 4.9

at five different instances. Why is the magnitude of the tangential component of the force equal to $\frac{|4t|}{\sqrt{1+4t^2}}$? Study Figure 4.9b and show that the magnitude of the component of the force perpendicular to the tangent is equal to $\frac{2}{\sqrt{1+4t^2}}$. Discuss the role that the two components play in determining the motion of the point as it moves from $(-2, 4)$ to $(0, 0)$ and beyond.

Example 4.7. Let the position of a point-mass P moving in an xy-plane be given by the equations $x(t) = d \cos \omega t$ and $y(t) = d \sin \omega t$, where t is time with $t \geq 0$, and d and ω are positive constants. Show that the point moves on the circle $x^2 + y^2 = d^2$ of radius d. Compute the velocities of the x- and y-coordinates of P. Show that P starts at the point $(d, 0)$ and moves counterclockwise around the circle at the constant speed $v(t) = d\omega$. So P covers a distance of $(d\omega)t$ during time t and the

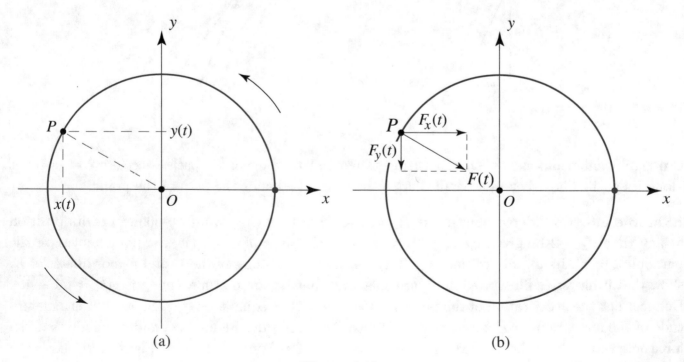

Figure 4.10

period—the time for one complete revolution—is $T = \frac{2\pi}{\omega}$. Let m be the mass of P and let $F_x(t)$ and $F_y(t)$ be the forces on P in the x- and y-directions. They are given by

$$F_x(t) = mx''(t) = -m\omega^2(d\cos\omega t) \quad \text{and} \quad F_y(t) = my''(t) = -m\omega^2(d\sin\omega t),$$

respectively. The magnitude of the resultant of these forces is

$$F(t) = \sqrt{F_x(t)^2 + F_y(t)^2} = md\omega^2\sqrt{\cos^2\omega t + \sin^2\omega t} = md\omega^2.$$

Refer to Figure 4.10b and check that when $x(t) > 0$ the force $F_x(t)$ acts to the left, and when $x(t) < 0$ the force $F_x(t)$ acts to the right. Is there a similar relationship between $y(t)$ and $F_y(t)$? Show that at any time t, the slope of the slanting segment in Figure 4.10a that connects the point P and the origin O is the same as the slope of the segment in Figure 4.10b that determines the direction of the resultant of the forces $F_x(t)$ and $F_y(t)$. Conclude that the resultant of the forces $F_x(t)$ and $F_y(t)$ is a centripetal force acting on P in the direction of the origin O.

4C. Centripetal Force in Cartesian Coordinates. A point-mass P of mass m is in motion in a plane. It is acted on by a single force, a centripetal force with center in the same plane. Place an xy-coordinate system into the plane so that the center of force is at the origin O. Let x and y be the coordinates of P. The magnitude F of the force depends on the location (x, y) of P. Click your stopwatch at time $t = 0$, and suppose that P is in a typical position at an elapsed time $t > 0$ later. Both x and y vary with time t, so that both $x = x(t)$ and $y = y(t)$ are functions of time t.

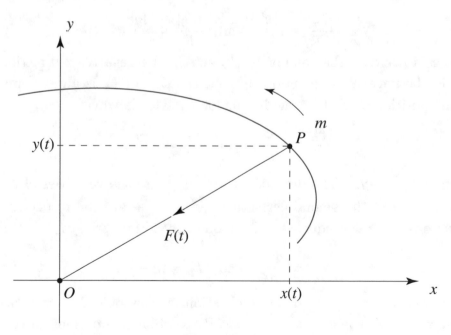

Figure 4.11

The magnitude of the force is also a function $F = F(t)$ of time t. Figure 4.11 captures what has been described.

We will assume that F acts smoothly and consequently that P moves smoothly—P does not zigzag sharply like a butterfly or a bat. In particular, we assume that $F(t)$, $x(t)$, and $y(t)$ are differentiable functions of t and that the derivatives $\frac{dx}{dt} = x'(t)$ and $\frac{dy}{dt} = y'(t)$ are differentiable as

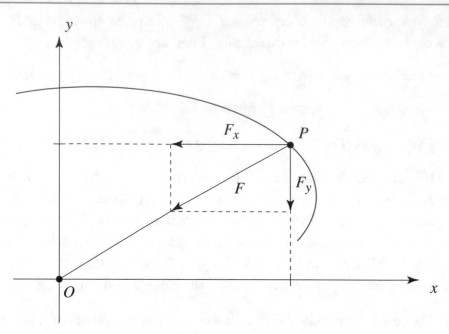

Figure 4.12

well. Let r be the distance from P to O. Let F_x and F_y be the components of F in the x- and y-directions, respectively. The quantities r, F_x, and F_y are also differentiable functions of t. Figure 4.12 illustrates the relationship between F, F_x and F_y. By the parallelogram law of forces and similar triangles,

$$\frac{|F_x|}{F} = \frac{|x|}{r} \quad \text{and} \quad \frac{|F_y|}{F} = \frac{|y|}{r} \,.$$

Notice that when x is positive, the component F_x acts in the negative x-direction, and when y is positive, F_y acts in the negative y-direction. In fact, the sign of F_x is always opposite that of the x-coordinate of the position, and the same is true for F_y. This is why

$$\frac{F_x}{F} = \frac{-x}{r} \quad \text{and} \quad \frac{F_y}{F} = \frac{-y}{r} \,.$$

So $rF_x = -xF$ and $rF_y = -yF$. The derivatives $\frac{dx}{dt}$ and $\frac{dy}{dt}$ are the velocities of P in the x- and y-directions, respectively, and the second derivatives $\frac{d^2x}{dt^2}$ and $\frac{d^2y}{dt^2}$ are the respective accelerations in the x- and y-directions. As a consequence of Newton's second law, we get

$$F_x = m\,\frac{d^2x}{dt^2} \quad \text{and} \quad F_y = m\,\frac{d^2y}{dt^2} \,.$$

The physics of the matter—it involved the application of Newton's $F = ma$ in both the x- and y-directions—is now over. *The rest is mathematics!* By combining equations already derived, we get

$$mr\,\frac{d^2x}{dt^2} = -Fx \quad \text{and} \quad mr\,\frac{d^2y}{dt^2} = -Fy$$

and therefore that

$$mr\left(x\frac{d^2y}{dt^2} - y\frac{d^2x}{dt^2} \right) \;=\; mr\,\frac{d^2y}{dt^2}\,x - mr\,\frac{d^2x}{dt^2}\,y \;=\; -Fyx + Fxy \;=\; 0.$$

It follows that

$$y \, \frac{d^2x}{dt^2} \;=\; x \, \frac{d^2y}{dt^2}.$$

(Note that this also holds when $r = 0$, because then both $x = 0$ and $y = 0$.)

Consider the difference $x \cdot \frac{dy}{dt} - y \cdot \frac{dx}{dt}$. By the product rule, the derivative of this difference is

$$\frac{d}{dt}\left(x \cdot \frac{dy}{dt} - y \cdot \frac{dx}{dt}\right) = \frac{d}{dt}\left(x \cdot \frac{dy}{dt}\right) - \frac{d}{dt}\left(y \cdot \frac{dx}{dt}\right)$$

$$= \left(\frac{dx}{dt} \cdot \frac{dy}{dt} + x \cdot \frac{d^2y}{dt^2}\right) - \left(\frac{dy}{dt} \cdot \frac{dx}{dt} + y \cdot \frac{d^2x}{dt^2}\right)$$

$$= \frac{dx}{dt} \cdot \frac{dy}{dt} - \frac{dy}{dt} \cdot \frac{dx}{dt} + x \cdot \frac{d^2y}{dt^2} - y \cdot \frac{d^2x}{dt^2}.$$

In view of the equality $y \frac{d^2x}{dt^2} = x \frac{d^2y}{dt^2}$, it follows that $\frac{d}{dt}(y \cdot \frac{dx}{dt} - x \cdot \frac{dy}{dt}) = 0$. We can therefore conclude that

$$x \cdot \frac{dy}{dt} \;-\; y \cdot \frac{dx}{dt} \;=\; c,$$

where c is a constant. This fact contains essential information about the motion of P. To extract it, we will now "go polar."

Example 4.8. Turn to the centripetal force that Example 4.7 illustrates. Show that the constant c above is equal to $d^2\omega$.

4D. Going Polar. The Cartesian part of our discussion is done. The task now will be to transfer the results that were obtained—in particular, the equality $x \cdot \frac{dy}{dt} - y \cdot \frac{dx}{dt} = c$ and the two force equations—into the context of polar coordinates. So now regard the origin O and the positive x-axis to be a polar coordinate system.

Let (r, θ), with $r > 0$ the distance from O to the point-mass P, be polar coordinates of P. The assumption $r > 0$ ensures that P does not crash into the center of force O. Let $r = f(\theta)$ be a polar

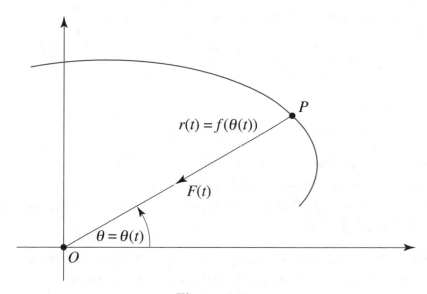

Figure 4.13

function that has the orbit of the point P as its graph. Since $\theta = \theta(t)$ is a function of the elapsed time t, $r = r(t) = f(\theta(t))$ is a function of t. The magnitude of the centripetal force is determined by the location of P and hence by θ and $r = f(\theta)$. This magnitude is therefore also a function of θ. Combining this function with $\theta = \theta(t)$ provides its connection with the force function $F(t)$. Refer to Figure 4.13. Given our general smoothness assumptions, the function $r = f(\theta)$ of θ, as well as the functions $r(t)$ and $\theta(t)$ of t, all have first and second derivatives. (The fact that the polar coordinate θ of the moving point P is a differentiable and hence a continuous function of t means that it does not jump from θ to, say, $\theta + 2\pi$ from one instant to the next.)

Because we have P moving in a counterclockwise direction, it is now convenient to organize things as follows: let $t = 0$ be an instant at which P crosses the polar axis. Then $t > 0$ is the elapsed time thereafter, and $t < 0$ refers to the time before. (In televised launches of NASA space missions, it is common to hear announcements such as "t equals minus 73 seconds and counting.") We will take $\theta(t) > 0$ for $t > 0$, $\theta(t) = 0$ for $t = 0$, and $\theta(t) < 0$ for $t < 0$.

Recall the relationships

$$x = r \cos \theta \quad \text{and} \quad y = r \sin \theta$$

between the polar and Cartesian coordinates from Chapter 3C. We will use them to rewrite the equality $x \cdot \frac{dy}{dt} - y \cdot \frac{dx}{dt} = c$ in terms of r and θ. By the product rule,

$$\frac{dx}{dt} = \frac{dr}{dt} \cos \theta + r \frac{d}{dt}(\cos \theta).$$

Since $\theta = \theta(t)$, we get by the chain rule that $\frac{d}{dt}(\cos \theta) = -\sin \theta \cdot \theta'(t)$ and hence that

$$\frac{dx}{dt} = \frac{dr}{dt} \cos \theta - r \sin \theta \cdot \frac{d\theta}{dt}.$$

In exactly the same way,

$$\frac{dy}{dt} = \frac{dr}{dt} \sin \theta + r \cos \theta \cdot \frac{d\theta}{dt}.$$

By combining the equations $x = r \cos \theta$ and $y = r \sin \theta$ with those just derived, we get

$$x \cdot \frac{dy}{dt} - y \cdot \frac{dx}{dt} = (r \cos \theta)\left(\frac{dr}{dt} \sin \theta + r \cos \theta \cdot \frac{d\theta}{dt}\right) - (r \sin \theta)\left(\frac{dr}{dt} \cos \theta - r \sin \theta \cdot \frac{d\theta}{dt}\right)$$

$$= (r^2 \cos^2 \theta)\frac{d\theta}{dt} + (r^2 \sin^2 \theta)\frac{d\theta}{dt} = r^2(\sin^2 \theta + \cos^2 \theta)\frac{d\theta}{dt}$$

$$= r^2 \frac{d\theta}{dt}.$$

We have therefore verified that $r(t)^2 \frac{d\theta}{dt} = r(t)^2 \theta'(t) = c$.

Example 4.9. Think for a moment about the equation $r(t)^2 \theta'(t) = c$. What is the meaning of $\theta'(t)$? Consider P at two different times t_1 and t_2 in its orbit, and suppose that the distance from P to O is much greater at t_1 than at t_2. Use this assumption to compare $\theta'(t_1)$ and $\theta'(t_2)$ and describe the implications for the motion of P.

Continue to consider P at two different times t_1 and t_2 in its orbit. Suppose that $t_1 \leq t_2$, and let A be the area that is swept out by the segment OP during the time interval $[t_1, t_2]$. Let $\theta(t_1) = a$ and $\theta(t_2) = b$. Figure 4.14 illustrates what has been described. Turn to Chapter 3G and note that

$$A = \int_a^b \tfrac{1}{2} f(\theta)^2 \, d\theta.$$

Because $r(t) = f(\theta(t))$, the substitutions $\theta = \theta(t)$ and $d\theta = \theta'(t)dt$ together with the discussion that concludes Chapter 3G tell us that

$$A \;=\; \int_a^b \tfrac{1}{2} f(\theta)^2 \, d\theta \;=\; \int_{t_1}^{t_2} \tfrac{1}{2} f(\theta(t))^2 \, \theta'(t)dt \;=\; \int_{t_1}^{t_2} \tfrac{1}{2} r(t)^2 \theta'(t)dt \;=\; \int_{t_1}^{t_2} \tfrac{1}{2} c \, dt \;=\; \tfrac{1}{2} ct \Big|_{t_1}^{t_2} = \tfrac{1}{2} c(t_2 - t_1).$$

Putting $\tfrac{1}{2} c = \kappa$, we see that the area A swept out by P is equal to κ times the time $t_2 - t_1$ that

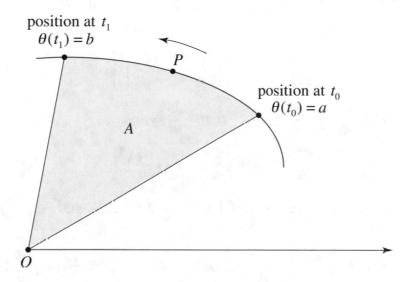

Figure 4.14

it takes to sweep it out. Taking $t_1 = 0$ and t_2 the elapsed time t, establishes the equality

$$A = \kappa t$$

for the area A traced out by the segment PO from 0 to t. This is Kepler's second law for the orbit of a point-mass propelled by a centripetal force. Newton had proved it by completely different means. (Refer to the discussion in Chapter 1D.) The constant κ is the Kepler constant of the orbit of P. Inserting $c = 2\kappa$ into the equality $r(t)^2 \frac{d\theta}{dt} = r(t)^2 \theta'(t) = c$, tells us that

$$r^2 \frac{d\theta}{dt} = 2\kappa.$$

One more order of business is the conversion of the two Cartesian force equations

$$mr \frac{d^2x}{dt^2} = -Fx \quad \text{and} \quad mr \frac{d^2y}{dt^2} = -Fy$$

of the previous section into a single force equation in polar form. Because $r(t)^2 \frac{d\theta}{dt}$ is a constant, $\frac{d}{dt}\big(r(t)^2 \frac{d\theta}{dt}\big) = 0$. By an application of the chain and product rules, $2r(t)r'(t) \cdot \frac{d\theta}{dt} + r(t)^2 \cdot \frac{d^2\theta}{dt^2} = 0$, and hence

$$2 \frac{dr}{dt} \cdot \frac{d\theta}{dt} \; + \; r \cdot \frac{d^2\theta}{dt^2} = 0.$$

Recall from above that

$$\frac{dx}{dt} = \frac{dr}{dt}\cos\theta - r\sin\theta \cdot \frac{d\theta}{dt}.$$

By differentiating this equation (use the product and chain rules several times), we get that

$$\frac{d^2x}{dt^2} = \left(\frac{d^2r}{dt^2}\cos\theta - \frac{dr}{dt}\sin\theta \cdot \frac{d\theta}{dt}\right) - \frac{dr}{dt}\sin\theta \cdot \frac{d\theta}{dt} - r\left(\cos\theta \cdot \frac{d\theta}{dt} \cdot \frac{d\theta}{dt} + \sin\theta \cdot \frac{d^2\theta}{dt^2}\right)$$

$$= \frac{d^2r}{dt^2}\cos\theta - 2\frac{dr}{dt}\sin\theta \cdot \frac{d\theta}{dt} - r\cos\theta \cdot \left(\frac{d\theta}{dt}\right)^2 - r\sin\theta \cdot \frac{d^2\theta}{dt^2}$$

$$= \frac{d^2r}{dt^2}\cos\theta - r\cos\theta \cdot \left(\frac{d\theta}{dt}\right)^2 - \sin\theta\left(2\frac{dr}{dt} \cdot \frac{d\theta}{dt} + r \cdot \frac{d^2\theta}{dt^2}\right).$$

Since the last term is equal to 0,

$$\frac{d^2x}{dt^2} = \cos\theta\left[\frac{d^2r}{dt^2} - r \cdot \left(\frac{d\theta}{dt}\right)^2\right].$$

Doing the same thing with $\dfrac{dy}{dt} = \dfrac{dr}{dt}\sin\theta + r\cos\theta \cdot \dfrac{d\theta}{dt}$, gives us

$$\frac{d^2y}{dt^2} = \sin\theta\left[\frac{d^2r}{dt^2} - r \cdot \left(\frac{d\theta}{dt}\right)^2\right].$$

From the two Cartesian force equations $-mr\frac{d^2x}{dt^2} = Fx = F(r\cos\theta)$ and $-mr\frac{d^2y}{dt^2} = Fy = F(r\sin\theta)$ of section 4C, we get that

$$F\cos\theta = m\cos\theta\left[r \cdot \left(\frac{d\theta}{dt}\right)^2 - \frac{d^2r}{dt^2}\right] \quad \text{and} \quad F\sin\theta = m\sin\theta\left[r \cdot \left(\frac{d\theta}{dt}\right)^2 - \frac{d^2r}{dt^2}\right].$$

A look at the graphs of Figures 3.4 and 3.5 in Chapter 3A tells us that the terms $\sin\theta$ and $\cos\theta$ are never simultaneously equal to 0. Therefore by canceling and then inserting the equality $\frac{d\theta}{dt} = \frac{2\kappa}{r^2}$, we get

$$F = m\left[r \cdot \left(\frac{d\theta}{dt}\right)^2 - \frac{d^2r}{dt^2}\right] = m\left[\frac{4\kappa^2}{r(t)^3} - \frac{d^2r}{dt^2}\right].$$

The initial important goals of our study of centripetal force have now been reached. They are the two equations

$$r^2(t)\frac{d\theta}{dt} = 2\kappa \quad \text{and} \quad F(t) = m\left[\frac{4\kappa^2}{r(t)^3} - \frac{d^2r}{dt^2}\right],$$

a version of Kepler's second law and the *centripetal force equation*, respectively.

It will be of advantage to recast this *first form* of the centripetal force equation. Let $g(\theta) = \frac{1}{f(\theta)}$. So $r(t) = f(\theta(t)) = g(\theta(t))^{-1}$. By the chain rule and the formula $\frac{d\theta}{dt} = \frac{2\kappa}{r(t)^2}$,

$$\frac{dr}{dt} = -g(\theta(t))^{-2} \cdot \frac{d}{dt}g(\theta(t)) = -\frac{1}{g(\theta(t))^2}\,g'(\theta(t))\,\theta'(t)$$

$$= -\frac{1}{g(\theta(t))^2}\frac{2\kappa}{r(t)^2}\,g'(\theta(t)) = -2\kappa g'(\theta(t)).$$

By another application of the chain rule,

$$\frac{d^2r}{dt^2} = -2\kappa g''(\theta(t))\,\theta'(t) = -2\kappa g''(\theta(t))\frac{2\kappa}{r(t)^2} = -4\kappa^2 g''(\theta(t))\frac{1}{r(t)^2} = -4\kappa^2 g''(\theta(t))g(\theta(t))^2.$$

By substituting $\frac{d^2 r}{dt^2} = -4\kappa^2 g''(\theta(t)) g(\theta(t))^2$ and $r(t)^{-3} = g(\theta(t))^3$ into $F(t) = m\left[\frac{4\kappa^2}{r(t)^3} - \frac{d^2 r}{dt^2}\right]$, we arrive at the *second form*

$$F(t) = 4m\kappa^2 g(\theta(t))^2 \big[g(\theta(t)) + g''(\theta(t)) \big]$$

of the centripetal force equation.

Let's step back and summarize the results that were derived and the assumptions that were made. A point-mass P of mass m is regarded to be propelled by a centripetal force and to move smoothly in its orbit. A polar coordinate system is chosen in the plane of the orbit with the origin O placed at the center of force. The polar function $r = f(\theta)$ expresses the distance r of P from O in terms of the polar angle θ of the changing position of P. As a consequence, the polar graph of this function describes the orbit of P. Let t be the elapsed time from some fixed instant. The fact that P moves, means that both the positional angle $\theta = \theta(t)$ and the distance $r(t) = f(\theta(t))$ are functions of time t. It was demonstrated above that the motion of P satisfies $r^2 \frac{d\theta}{dt} = 2\kappa$, where κ is Kepler's constant of the orbit, and that the magnitude $F(t)$ of the centripetal force acting on P is related to the distance $r(t)$ from P to O in the explicit way the centripetal force equation prescribes.

4E. From Conic Section to Inverse Square Law. We begin this section by assuming that the orbit of P—note that as yet no assumptions have been made about the geometry of the orbit—is a conic section, either an ellipse, a parabola, or a hyperbola, and that the center of force is at a focal point. A review of the analysis of the equation

$$r = \frac{d}{1 + \varepsilon \cos\theta}$$

in Chapter 3D tells us that the polar coordinate system with the origin O at the center of force—see the summary above—can be placed in such a way that O is at a focal point of the conic section and the orbit of P is the graph of the polar function

$$r = f(\theta) = \frac{d}{1 + \varepsilon \cos\theta}$$

where $r > 0$, $\varepsilon \geq 0$ is the eccentricity of the conic section, and $d > 0$ is a constant. If the orbit is an ellipse, then $0 \leq \varepsilon < 1$; if it is a parabola, then $\varepsilon = 1$; and if it is a hyperbola, then $\varepsilon > 1$.

Differentiate the equation

$$r(t) = d(1 + \varepsilon \cos\theta(t))^{-1}$$

and use the fact that $\frac{d\theta}{dt} = \frac{2\kappa}{r(t)^2}$ to get

$$\frac{dr}{dt} = -d(1 + \varepsilon \cos\theta(t))^{-2}\left(-\varepsilon \sin\theta(t) \cdot \frac{d\theta}{dt}\right) = \varepsilon d(1 + \varepsilon \cos\theta(t))^{-2}(\sin\theta(t))2\kappa r(t)^{-2}.$$

After substituting $(1 + \varepsilon \cos\theta(t))^{-2} = \frac{r(t)^2}{d^2}$, this becomes

$$\frac{dr}{dt} = \frac{2\varepsilon\kappa}{d} \sin\theta(t).$$

Differentiating once more, we get

$$\frac{d^2r}{dt^2} = \frac{2\varepsilon\kappa}{d}\left(\cos\theta(t) \cdot \frac{d\theta}{dt}\right) = \frac{2\varepsilon\kappa}{d}(\cos\theta(t))\frac{2\kappa}{r(t)^2} = \frac{4\kappa^2\varepsilon}{d}\cos\theta(t)\frac{1}{r(t)^2}.$$

Substituting this into the first form of the centripetal force equation of the previous section, we obtain

$$F(t) = m\left[\frac{4\kappa^2}{r(t)^3} - \frac{d^2r}{dt^2}\right] = m\left[\frac{4\kappa^2}{r(t)^3} - \frac{4\kappa^2\varepsilon}{d}\cos\theta(t)\frac{1}{r(t)^2}\right] = 4m\kappa^2\left[\frac{1}{r(t)} - \frac{\varepsilon}{d}\cos\theta(t)\right]\frac{1}{r(t)^2}.$$

Because $\frac{1}{r(t)} = \frac{1+\varepsilon\cos\theta(t)}{d} = \frac{1}{d} + \frac{\varepsilon}{d}\cos\theta(t)$, we see that $F(t) = \frac{4m\kappa^2}{d}\frac{1}{r(t)^2}$. Since $L = 2d$ is the latus rectum of the conic section (this is pointed out in the latter part of Chapter 3D),

$$F(t) = \frac{8m\kappa^2}{L} \cdot \frac{1}{r(t)^2},$$

where m is the mass of the point-mass, κ and L are the Kepler constant and latus rectum of the orbit, and $r(t)$ is the distance from the object to the point of origin of the centripetal force.

The conclusion above was established by Newton with an argument that regards the centripetal force as acting "machine-gun" style on the point-mass P. (See Chapter 1D.) The derivation detailed here with its assumption of a smoothly acting centripetal force and its use of today's calculus has the advantage that it works for elliptical, parabolic, and hyperbolic orbits all at once.

It is a direct consequence of our discussion that if a point-mass P is propelled by a centripetal force and *if it has an orbit that is either an ellipse, a parabola, or a hyperbola, with the center of force at a focal point,* then *the magnitude of the force is proportional to the inverse of the square of the distance between P and this focal point.*

Suppose that the orbit is an ellipse with semimajor and semiminor axes a and b. That the area of this ellipse is $ab\pi$ is a consequence of integral calculus. To see this, position the ellipse so that its equation is $\frac{x^2}{a^2} + \frac{y^2}{b^2} = 1$. Solving for y^2, we get $y^2 = b^2(1 - \frac{x^2}{a^2}) = \frac{b^2}{a^2}(a^2 - x^2)$. So $y = \frac{b}{a}\sqrt{a^2 - x^2}$ is a function that has the upper half of the ellipse as its graph. It follows that the area of the upper half of the ellipse is equal to

$$\int_{-a}^{a} \frac{b}{a}\sqrt{a^2 - x^2}\, dx = \frac{b}{a}\int_{-a}^{a}\sqrt{a^2 - x^2}\, dx.$$

Taking $b = a$ in the situation just discussed tells us that $\int_{-a}^{a}\sqrt{a^2 - x^2}\, dx$ is the area of the upper half of the circle of radius a. Since this area is $\frac{1}{2}\pi a^2$, we can conclude that

$$\int_{-a}^{a} \frac{b}{a}\sqrt{a^2 - x^2}\, dx = \frac{b}{a}\int_{-a}^{a}\sqrt{a^2 - x^2}\, dx = \frac{b}{a}\frac{1}{2}\pi a^2 = \frac{1}{2}ab\pi$$

is the area of the upper half of the ellipse with semimajor axis a and semiminor axis b. So the area of the full ellipse is $ab\pi$. It follows that $\kappa = \frac{ab\pi}{T}$ where T is the period of the orbit. Since the latus rectum is $L = \frac{2b^2}{a}$ (as pointed out in Chapter 3D), we get the elliptical version of the inverse square law

$$F(t) = \frac{8m\kappa^2}{L} \cdot \frac{1}{r(t)^2} = 8m\frac{a^2b^2\pi^2}{T^2}\frac{a}{2b^2} \cdot \frac{1}{r(t)^2} = \frac{4\pi^2a^3m}{T^2} \cdot \frac{1}{r(t)^2}.$$

Example 4.10. Let's illustrate what has been developed in the context of Example 4.7, the point-mass driven by a centripetal force around the circle of radius d. In this situation, $r = f(\theta) = d$,

$\theta(t) = \omega t$, and $r(t) = f(\theta(t)) = d$. By Example 4.8, Kepler's constant is $\kappa = \frac{1}{2}d^2\omega$. The basic equality $r^2(t)\frac{d\theta}{dt} = 2\kappa$ translates to $d^2\omega = 2\kappa$ and the centripetal force equation becomes

$$F(t) = m\left[\frac{4\kappa^2}{r(t)^3} - \frac{d^2r}{dt^2}\right] = m\left[\frac{4\cdot\frac{1}{4}d^4\omega^2\kappa^2}{d^3} - 0\right] = md\omega^2.$$

Since the orbit is a circle and the force is centripetal directed to its center (this is also the focal point), we know from the result derived in this section that the force satisfies an inverse square law, and in particular that the formula $F(t) = \frac{4\pi^2 a^3 m}{T^2} \cdot \frac{1}{r(t)^2}$ applies. And it does. Since $T = \frac{2\pi}{\omega}$ and a and $r(t)$ are both equal to d, we get

$$F(t) = \frac{4\pi^2 a^3 m}{T^2} \cdot \frac{1}{r(t)^2} = \frac{4\pi^2 d^3 m}{(\frac{2\pi}{\omega})^2} \cdot \frac{1}{d^2} = md\omega^2.$$

While $F(t) = md\omega^2$ does not look like an inverse square law, the equality above tells us that it is.

4F. From Inverse Square Law to Conic Section. We now go in the other direction and show that the orbit of a point-mass P that is propelled by a centripetal force that satisfies an inverse square law is an ellipse, a parabola, or a hyperbola, and that the center of force is at a focal point of the orbit.

Let's first clarify what it means for a centripetal force to satisfy an inverse square law. It means that the magnitude $F(t)$ of the force on the point-mass P at any time t is given by an equation of the form

$$F(t) = C\frac{m}{r(t)^2},$$

where $r(t)$ is the distance between P and the center of force, m is the mass of P, and $C > 0$ is a constant. Letting $g(\theta) = \frac{1}{r(\theta)}$ and combining $F(t) = C\frac{m}{r(t)^2} = Cmg(\theta(t))^2$ with the second form of the centripetal force equation, we get

$$4m\kappa^2 g(\theta(t))^2\left[g(\theta(t)) + g''(\theta(t))\right] = Cmg(\theta(t))^2.$$

Dividing through by $4m\kappa^2 \cdot g(\theta(t))^2$, gives us

$$g(\theta(t)) + g''(\theta(t)) = \frac{C}{4\kappa^2}.$$

Because we are interested in the shape of the orbit, namely the precise form of the function $r = f(\theta) = \frac{1}{g(\theta)}$, we now ignore the fact that θ is a function of t, and consider the equation $g(\theta) + g''(\theta) = \frac{C}{4\kappa^2}$. This equation is an example of a *second order differential equation*. The fact that all of its solutions can be found explicitly, determines $g(\theta)$ and shows that the orbit of P has the required properties. Put $h(\theta) = g(\theta) - \frac{C}{4\kappa^2}$, and notice that the function $h(\theta)$ satisfies the equation

$$h''(\theta) + h(\theta) = 0.$$

We will see that this implies in turn that $h(\theta)$ has the form $h(\theta) = A\sin\theta + B\cos\theta$ for some constants A and B. This can be verified with some trickery that is rather like pulling a mathematical rabbit out of a hat. Let $h(\theta)$ be any function that satisfies the equation and define the two functions $A(\theta)$ and $B(\theta)$ as follows:

$$A(\theta) = (\cos\theta)h(\theta) - (\sin\theta)h'(\theta) \quad \text{and} \quad B(\theta) = (\sin\theta)h(\theta) + (\cos\theta)h'(\theta).$$

The fact that $\sin^2\theta + \cos^2\theta = 1$, quickly implies that $h(\theta) = A(\theta)(\cos\theta) + B(\theta)(\sin\theta)$. If we can show that the functions $A(\theta)$ and $B(\theta)$ are both constant and set $A = A(\theta)$ and $B = B(\theta)$, then $h(\theta) = A\sin\theta + B\cos\theta$ as required. Since $h''(\theta) + h(\theta) = 0$,

$$A'(\theta) = [(-\sin\theta)h(\theta) + (\cos\theta)h'(\theta)] - [(\cos\theta)h'(\theta) + (\sin\theta)h''(\theta)] = 0$$

and hence $A(\theta)$ is a constant. A similar computation shows that $B'(\theta) = 0$ and hence that $B(\theta)$ is also constant. So we have verified that

$$h(\theta) = A\cos\theta + B\sin\theta,$$

where A and B are constants. As a consequence,

$$g(\theta) = A\sin\theta + B\cos\theta + \frac{C}{4\kappa^2}.$$

We will now suppose that there is a point of "closest approach" for P, namely, an angle θ at which the distance $r = f(\theta)$ from P to O has a local minimum. (This is always so in the case of an object driven by a real centripetal force in a real orbit or flyby.) Rotate the polar axis, while keeping the center of force at the origin O, so that this local minimum occurs at $\theta = 0$. We next need to obtain some information about the constants A and B. The effort to get it is clarified by the Cartesian graph of $g(\theta) = A\sin\theta + B\cos\theta + \frac{C}{4\kappa^2}$. (What is meant by Cartesian graph in this context is illustrated in Figures 3.4 and 3.5 for $\sin\theta$ and $\cos\theta$.) Because $g(\theta) = \frac{1}{f(\theta)}$, the function $g(\theta)$ has a local maximum at $\theta = 0$. So by a basic theorem of elementary calculus, $g'(0) = 0$. Since $g'(\theta) = A\cos\theta - B\sin\theta$, it follows that $A = 0$. So $g'(\theta) = -B\sin\theta$ and $g''(\theta) = -B\cos\theta$. Assume for a moment that B is negative. Because $g''(0) = -B > 0$, the second derivative test of elementary calculus would tell us that $g(\theta)$ has a local minimum at $\theta = 0$. But for a nonzero B, $g(\theta) = B\cos\theta + \frac{C}{4\kappa^2}$ cannot have both a local maximum and a local minimum at $\theta = 0$. Therefore $B < 0$ cannot be, and hence $B \geq 0$. In view of the fact that $f(\theta) = \frac{1}{g(\theta)} = \frac{1}{B\cos\theta + \frac{C}{4\kappa^2}}$, we now get

$$f(\theta) = \frac{1}{\frac{C}{4\kappa^2}\left(1 + \frac{4\kappa^2 B}{C}\cos\theta\right)} = \frac{\frac{4\kappa^2}{C}}{1 + \frac{4\kappa^2 B}{C}\cos\theta}.$$

Since this polar function has the form $f(\theta) = \frac{d}{1 + \varepsilon\cos\theta}$ with $d = \frac{4\kappa^2}{C} > 0$ and $\varepsilon = \frac{4\kappa^2 B}{C} \geq 0$, we know from the study in Chapter 3D that its graph is a conic section with eccentricity ε and focal point the origin O. It is also shown at the end of Chapter 3D that $\frac{4\kappa^2}{C} = d = \frac{L}{2}$, where L is the latus rectum of the orbit.

We have verified that *if the magnitude of the centripetal force acting on P satisfies an inverse square law,* then *the orbit of P is an ellipse, a parabola, or a hyperbola with the center of force at a focal point.*

From $\frac{4\kappa^2}{C} = \frac{L}{2}$, we get $C = \frac{8\kappa^2}{L}$, so that the equality $F(t) = C\frac{m}{r(t)^2}$ brings us back to the earlier force formula

$$F(t) = \frac{8m\kappa^2}{L} \cdot \frac{1}{r(t)^2}.$$

4G. Summary of Newton's Theory. This chapter considered the abstract situation of a centripetal force with center of force O that propels a point-mass P along a path—the orbit or trajectory of P—that lies in a plane. Refer to Figure 4.15. The motion is timed starting at $t = 0$. The elapsed

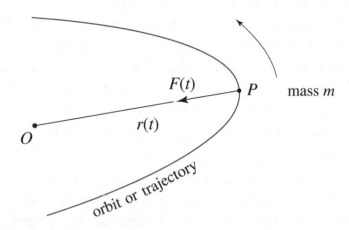

Figure 4.15

time thereafter is denoted by t. Our discussion established the following conclusions:

Conclusion A. The segment OP sweeps out equal areas in equal times. In particular, there is a constant κ with the property that if $A(t)$ is the area swept out by the segment OP during elapsed time t, then $\frac{A(t)}{t} = \kappa$. This property is Kepler's second law and the constant κ is Kepler's constant.

Conclusion B. If the orbit of P is either an ellipse, a parabola, or a hyperbola with the center of force O at a focus, then the magnitude $F(t)$ of the centripetal force is given by the equation

$$F(t) = \frac{8\kappa^2 m}{L} \cdot \frac{1}{r(t)^2},$$

where m is the mass of the point-mass, L is the latus rectum of the orbit, and $r(t)$ is the distance between P and O. If the orbit is an ellipse with semimajor axis a and period T, then

$$F(t) = \frac{4\pi^2 a^3 m}{T^2} \cdot \frac{1}{r(t)^2}.$$

Conclusion C. If throughout the orbit of P, the magnitude $F(t)$ of the centripetal force is given by an inverse square law, in other words, by an equation of the form $F(t) = Cm\frac{1}{r(t)^2}$, where m is the mass of P and $C > 0$ is a constant, then the orbit of P is either an ellipse, a parabola, or a hyperbola, the center of force O is at a focal point, and $C = \frac{8\kappa^2}{L}$, where κ is Kepler's constant and L is the latus rectum of the orbit.

It has become tradition in the mathematical sciences to follow axiomatic approaches. Definitive theories in these disciplines are often cast in the following form: certain basic underlying laws or principles, often referred to as axioms or postulates, are taken as starting point, and all other relevant propositions are deduced from these by the force of logic and mathematics alone. The paradigm of such axiomatic approaches is the development of plane geometry in Euclid's *Elements*.

Newton's theory of planetary motion can be cast in this form. Start with Newton's three basic laws of motion (see Chapter 1B) and his *Law of Universal Gravitation*. Recall that the law of universal gravitation is the assertion that any two point-masses in the universe attract each other with a force given by the formula $F = G \frac{m_1 \cdot m_2}{r^2}$, where m_1 and m_2 are their masses, r is the distance between them, and G is a *universal constant*. Observations over many years, confirmed by trajectory data of man-made satellites and spacecraft, have provided solid evidence that the law of universal gravitation is valid anywhere in the solar system. Since the Sun is much more massive than any of the planets (we saw in Chapter 1I that the Sun contains over 99% of the mass of the solar system), it attracts each planet with what is essentially a centripetal force. Conclusion C tells us that *Kepler's first law* holds, namely that the orbit of a planet must be an ellipse with the Sun at a focus. (The fact that the parabola and hyperbola have infinite extent rules such trajectories out.) *Kepler's second law* for planetary orbits is provided by Conclusion A. The derivation of *Kepler's third law* in Newton's more explicit form $\frac{a^3}{T^2} = \frac{GM}{4\pi^2}$ was carried out in Chapter 1D.

There are two important issues that require further discussion. The first is the fact that our study has focused abstractly on a point-mass P moved along its orbit by a centripetal force. But does this study really apply to our solar system? Does it apply to a planet in motion around the Sun? Can it be assumed that the gravitational pull by the Sun on a planet is directed to the center of mass of the planet? A planet, after all, is not a point-mass, but a composite of a myriad of point-masses. The next section will answer these questions in the affirmative (after some reasonable assumptions are made).

A second concern has to do with Conclusion C. We saw that the study of the orbit of P around O rests on the determination of the polar function $r = f(\theta)$ that expresses the distance r of P from O in terms of the angle θ. The graph of this function describes the geometry of the orbit as an ellipse, parabola, or hyperbola, shaped in each case by orbital constants. But what about r as a function of elapsed time t? Since $r = r(t)$ is the composite of the two functions $r = f(\theta)$ and $\theta = \theta(t)$, this reduces to the determination of the angle $\theta = \theta(t)$ as a function of t. Can the function that measures the angle that the segment OP sweeps out be identified? This subtle question will be taken on in Chapter 5 for elliptical orbits and in Chapter 6 for hyperbolic and parabolic orbits.

4H. Gravity and Geometry. Newton was aware that there was a fundamental unanswered problem that stood in the way of the application of his study of centripetal force for point-masses to the situation of the gravitational attraction of bodies in the solar system and beyond. Given that the gravitational attraction of any small particle of matter on any other satisfies the inverse square law, why should it be that large massive bodies—namely, huge collectives of such particles—attract each other in the same way? Is the net force exerted by all the particles of matter of a massive sphere on all the particles of another massive sphere directed from the center of mass of one sphere to the center of mass of the other? Is the magnitude of this resultant force inversely proportional to the square of the distance between these centers of mass? Only if this is so does the study of gravitational force reduce to the situation of point-masses. This question presented a formidable challenge to Newton. Some scholars have in fact claimed that the matter was the cause (or at least one of them) for the 20-year delay between Newton's first thoughts about universal gravitation and the composition of the *Principia*.

The problem facing Newton was subtle. Consider the gravitational force F that a body B of mass M exerts on a point-mass m. (In the discussion of this section, the symbol m will refer to both the point-mass and its mass.) Then it is *not the case in general* that the magnitude F of this force is given by Newton's law of universal gravitation $F = G\frac{mM}{d^2}$, where d is the distance from the point-mass to the center of mass of B. Both the shape of the body B and the distribution of the mass within it play a critical role. Both need to be configured symmetrically for Newton's law to hold. Fortunately, the formula is correct in the important situation of a sphere that has its mass distributed in a certain radially homogeneous way. (Since the Sun, Moon, Earth, and the planets satisfy this property very nearly, Newton's law of universal gravitation is on target in these cases.) The proof of this assertion makes use (more than once) of the essential strategy of integral calculus. Slice up the body B into a very large number of smaller pieces, and determine the force with which each of the pieces acts on m. To understand how B acts on m is a matter of computing the resultant of all the smaller forces. The summation strategy of integral calculus is the tool that makes this computation possible.

Part 1. To start, take B to be a thin homogeneous circle of matter of mass M and radius r. Think of a circular loop of a thin wire. Homogeneous means that the matter of the circle is evenly distributed (in particular, there are no lumps). Suppose that the point-mass m lies on the perpendicular to the circle through its center O, at a distance c from the center. Divide the circle into a huge even number, say, $2k$, of small equal segments. Each segment is $\frac{2\pi r}{2k} = \frac{\pi r}{k}$ units long and has a mass of $M_{\text{seg}} = \frac{M}{2k}$. Refer to Figure 4.16. By the Pythagorean theorem, the distance between m

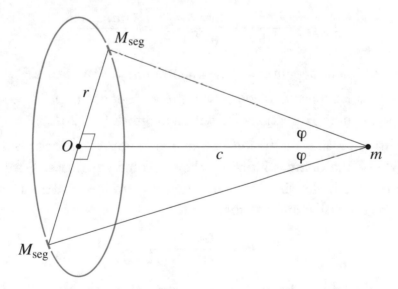

Figure 4.16

and each of the segments is equal to $\sqrt{c^2 + r^2}$. Since the segments are very small, Newton's law of universal gravitation for point-masses tells us that the magnitude of the gravitational force with which each of them attracts m is equal to

$$\frac{GmM_{\text{seg}}}{(\sqrt{c^2 + r^2})^2} = \frac{GmM_{\text{seg}}}{c^2 + r^2}.$$

Since the number of identical segments is even, the segments can be paired, as Figure 4.16 indicates, each with the one on the opposite side of the circle. Figure 4.17 depicts the vectors representing the forces of attraction that a matching pair of segments exerts on m as well as

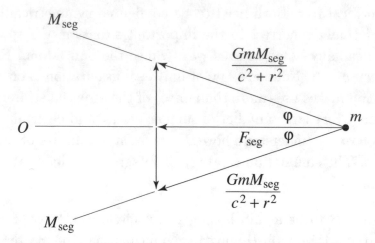

Figure 4.17

the relevant components of these forces. Notice that the two components along the line from m to O (only one of them is shown in the figure) are both equal to $F_{\text{seg}} = \frac{GmM_{\text{seg}}}{c^2+r^2} \cos\varphi$ and that the two components perpendicular to this line cancel each other out. By Figure 4.16, $\cos\varphi = \frac{c}{\sqrt{c^2+r^2}}$, so that

$$F_{\text{seg}} = \frac{GmM_{\text{seg}}}{c^2+r^2} \cdot \frac{c}{\sqrt{c^2+r^2}} = \frac{GmM_{\text{seg}}\,c}{(c^2+r^2)^{\frac{3}{2}}}.$$

The magnitudes of the two components along the line from m to O add to $2F_{\text{seg}} = 2\frac{GmM_{\text{seg}}c}{(c^2+r^2)^{\frac{3}{2}}}$. Considering the fact that the circular mass B is composed of k such opposite pairs, we see that the force of attraction of B on m points in the direction of O with a magnitude of $2kF_{\text{seg}} = \frac{Gm(2kM_{\text{seg}})c}{(c^2+r^2)^{\frac{3}{2}}} = \frac{GmMc}{(c^2+r^2)^{\frac{3}{2}}}$.

The various equalities in this discussion are in fact approximations. But when k is pushed to infinity (in the style of the definite integral) they become equalities. We have shown that the gravitational force with which the thin circular mass B of radius r attracts the point-mass m is directed to the center of the circle and has magnitude

$$F = \frac{GmMc}{(c^2+r^2)^{\frac{3}{2}}},$$

where M is the mass of B and c is the distance from the point-mass to the center of the circle. The gravitational force of attraction does point to the center of mass of B. However, the fact that r is not zero means that the magnitude of this force is *not* given by Newton's law $F = \frac{GmM}{c^2}$.

Part 2. We now turn to consider a thin homogeneous spherical shell B of radius R and mass M. Since only the surface of the sphere is included, think of B as the thin spherical skin of a ball. Let O be the shell's center. Suppose that the point-mass m lies at a distance c from O. We will assume that $c \geq R$, so that the point-mass is on or outside the shell, and analyze the gravitational force that the shell exerts on m.

Begin by slicing up the spherical shell into a large number of very thin ring-like sections. All cuts are perpendicular to the axis—placed horizontally—that connects the center O with m.

A typical ring of the spherical shell is shown in blue in Figure 4.18. (It is not drawn "very thin" in the diagram, but thick enough so that the relevant mathematics can be explained.) The angle θ with $0 \leq \theta < \pi$ determines the point Q. The angle θ also determines the right boundary of the ring, and the sliver of an angle $d\theta$ determines its thickness and (along with θ) its left boundary. The points Q and Q' lie at distances $r = R\sin\theta$ and $R\sin(\theta + d\theta)$ from the horizontal axis, respectively. Applying the definition of radian measure to the angle $d\theta$, we see that the length of the arc QQ' is $R\,d\theta$. An application of the formula for the area of a truncated cone tells us that the surface area of the blue ring is tightly approximated by $\pi\big(R\sin(\theta + d\theta) + R\sin\theta\big)R\,d\theta$. Let M_{ring} be the mass of the ring. Since the surface area of the entire spherical shell is $4\pi R^2$, the fact that mass and surface area are proportional (this is a consequence of the homogeneity) tells us that

$$\frac{M_{\mathrm{ring}}}{M} \approx \frac{\pi R^2\big(\sin(\theta + d\theta) + \sin\theta\big)\,d\theta}{4\pi R^2} = \tfrac{1}{4}\big(\sin(\theta + d\theta) + \sin\theta\big)\,d\theta.$$

Therefore, $M_{\mathrm{ring}} \approx \tfrac{1}{4}M\big(\sin(\theta + d\theta) + \sin\theta\big)\,d\theta$. Let s be the length of the segment connecting Q to the mass m. Let x be the base of the triangle with hypotenuse s. Since the distance from O to m is c, the remaining segment has length $c - x$. By applying Part 1 to the ring and the mass m, we get that the force of attraction of the ring on m is directed to the center O of the sphere and has

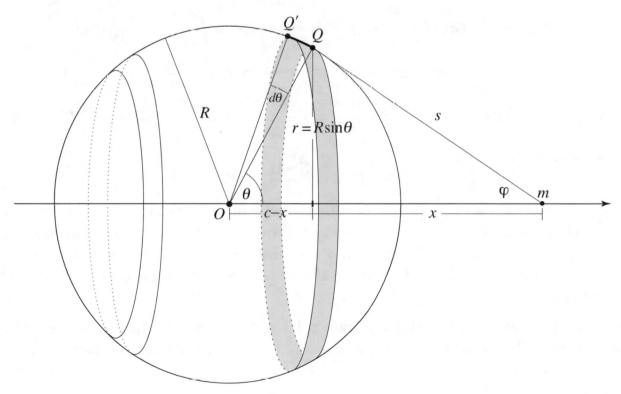

Figure 4.18

approximate magnitude

$$F_{\mathrm{ring}} \approx \frac{GmM_{\mathrm{ring}}\,x}{(x^2 + r^2)^{\frac{3}{2}}} \approx \frac{\tfrac{1}{4}GmMx\big(\sin(\theta + d\theta) + \sin\theta\big)\,d\theta}{(x^2 + r^2)^{\frac{3}{2}}} \approx \frac{\tfrac{1}{2}GmMx\sin\theta}{(x^2 + r^2)^{\frac{3}{2}}}\,d\theta.$$

(Use the fact that $d\theta$ is very small for the last approximation.) Adding this approximation over all the ring-like sections that were cut (with θ varying from 0 to π) gives us an approximation of the gravitational force F_{shell} with which the entire shell pulls on m. It is the message of integral calculus that by slicing up the shell into sections that are thinner and thinner, these approximations of F_{shell} get tighter and tighter, and that in the limit,

$$F_{\text{shell}} = \int_0^\pi \frac{\frac{1}{2}GmMx\sin\theta}{(x^2+r^2)^{\frac{3}{2}}}\,d\theta.$$

Given that $r = R\sin\theta$ and x are both functions of θ, this would appear to be a complicated integral. However, it turns out that it can be solved rather quickly by expressing the integrand in the variable $s = \sqrt{x^2+r^2}$. Since $(x^2+r^2)^{\frac{3}{2}} = s^3$, it remains to express $\sin\theta\,d\theta$ and x in terms of s. Applying the law of cosines to the angle θ and the triangle ΔQOm, we get $s^2 = R^2 + c^2 - 2Rc\cos\theta$. By differentiating, $2s\frac{ds}{d\theta} = -2Rc(-\sin\theta) = 2Rc\sin\theta$. Therefore $\sin\theta\,d\theta = \frac{1}{Rc}s\,ds$. To express x in terms of s, use the law of cosines again to get $R^2 = s^2 + c^2 - 2sc\cos\varphi$. Since $\cos\varphi = \frac{x}{s}$, this implies that $R^2 = s^2 + c^2 - 2cx$, so that $x = \frac{s^2+c^2-R^2}{2c}$. Feeding everything back into the integrand of the integral above, we get $\frac{1}{2}\frac{GmMx\sin\theta}{(x^2+r^2)^{\frac{3}{2}}}\,d\theta = \frac{1}{2}GmM \cdot \frac{s^2+c^2-R^2}{2c} \cdot \frac{1}{s^3} \cdot \frac{1}{Rc}s\,ds$. Notice that when $\theta = 0$, $s = c - R$, and when $\theta = \pi$, $s = c + R$. Therefore

$$F_{\text{shell}} = \frac{1}{4Rc^2}GmM\int_{c-R}^{c+R}\frac{s^2+c^2-R^2}{s^2}\,ds = \frac{1}{4Rc^2}GmM\int_{c-R}^{c+R}\left(1+\frac{c^2-R^2}{s^2}\right)ds.$$

Since $s - (c^2-R^2)s^{-1} = s - (c^2-R^2)\frac{1}{s}$ is an antiderivative of $1 + \frac{c^2-R^2}{s^2}$,

$$\int_{c-R}^{c+R}\left(1+\frac{c^2-R^2}{s^2}\right)ds = s - (c^2-R^2)\frac{1}{s}\bigg|_{c-R}^{c+R} = c+R-(c-R)-\left(c-R-(c+R)\right) = 4R.$$

We have therefore shown that

$$F_{\text{shell}} = \frac{GmM}{c^2}.$$

Since the spherical shell is homogeneous, O is its center of mass. The force of attraction of each ring on m points in the direction of O, so the same is true for the sum of all these forces. Since c is the distance between O and m, Newton's law of universal gravitation applies both to the direction and to the magnitude of the force of attraction of a thin homogeneous spherical shell of mass M on a point-mass m.

Part 3. We have arrived at the important point of our discussion. Using what was already established, we will now show that the gravitational force F that a sphere of mass M exerts on a point-mass m a distance c from the center of the sphere is given by Newton's law of universal gravitation

$$F = \frac{GmM}{c^2},$$

provided that the matter within the sphere is distributed in a certain symmetric way. Some assumption about the way that the matter within the sphere is distributed is surely necessary. Why? Consider a sphere made of a light material that has embedded within it a small but dense and heavy kernel of matter. Since every particle attracts every other, surely the gravitational force that the larger sphere exerts on a point-mass depends decisively on the location of the small, heavy kernel within it.

Let's turn to a sphere B of matter with radius R and center O. The assumption that we will make is this: any two small bits of matter in B that are the same distance from the center of the sphere have the same density. This assumption is met by a sphere that is composed of concentric homogeneous layers, each in the shape of a spherical shell. Think of the way an onion is structured. Since each shell is homogeneous, the discussion of Part 2 applies to it. So each shell pulls on the point-mass m in accordance with Newton's law. Our intuition should tell us therefore, that the entire sphere should pull on m in this way. Intuition is great. But a detailed argument is better.

The first thing to do is to define a density function for the sphere B. Let P be any point inside the sphere, and let S be a small sphere inside B with center P and radius s. Let m_S be the mass of S, V_S the volume of S, and consider the ratio $\frac{m_S}{V_S}$. This is the average density of the matter comprising the small sphere S. The density $\rho(P)$ at P is the limit

$$\rho(P) = \lim_{s \to 0} \frac{m_S}{V_S}.$$

If the point P is on the sphere, a similar definition works (with only a part of the sphere around P being relevant). The assumption that we will make about the matter within the sphere B is as

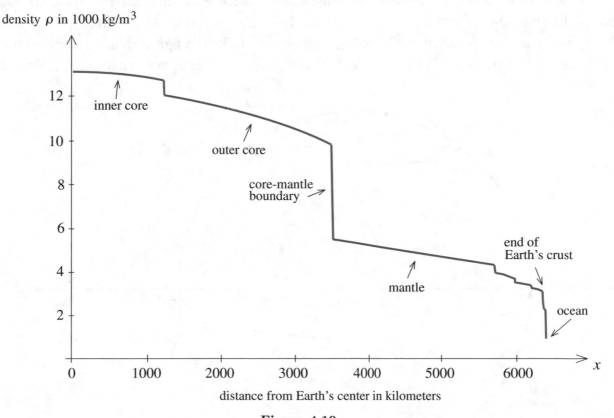

Figure 4.19

follows.

If P and Q are any two points in B that are the same distance from O, then $\rho(P) = \rho(Q)$.

We can now define a density function $\rho(x)$ for the sphere. For any x with $0 \leq x \leq R$, choose a point P in the sphere that is distance x from O, and set $\rho(x) = \rho(P)$. We will assume that $\rho(x)$ is a continuous function of x. Intuitively, a sphere that has such a density function is made up of shells—again, think of an onion—that are homogeneous.

Let's turn to our planet Earth for a moment. For Earth, the graph of the function $\rho(x)$ is sketched in Figure 4.19, with 1000 kg/m^3 the unit of density. What it tells us is that Earth has a very dense inner core with a radius of about 1250 km, a dense outer core about 2250 km thick, a less dense mantle about 2000 km thick, and that the layer consisting of the crust, water, rock, and soil of Earth's surface is relatively light and thin. The density of the water of the oceans is approximately 1000 kg/m^3.

We'll now verify Newton's law of universal gravitation for spheres with the density property that we have described. Let B be a sphere of radius R and turn to Figure 4.20. Let n be a large positive integer (large relative to R) and let $dx = \frac{R}{n}$. Choose $n-1$ points between 0 and R

$$0 = x_0 < x_1 < x_2 < \cdots < x_{i-1} < x_i < \cdots < x_{n-2} < x_{n-1} < x_n = R$$

that divide the interval $0 \leq x \leq R$ into n equal subintervals of length dx. So $x_{i+1} - x_i = dx$, for $i = 0, 1, \ldots n-1$. Figure 4.20 depicts the sphere B in black and, in blue, the spherical shell of thickness dx that the points x_i and x_{i+1} determine. (For purposes of "visibility," the segment $x_i \leq x \leq x_{i+1}$ and the spherical shell are both much thicker in the figure than in our description.) The volumes of the spheres of radii x_{i+1} and x_i are $\frac{4}{3}\pi x_{i+1}^3$ and $\frac{4}{3}\pi x_i^3$ respectively. Since dx is very

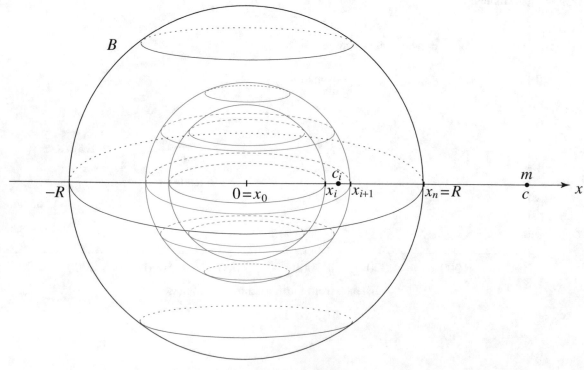

Figure 4.20

small, the volume of the blue spherical shell is

$$\tfrac{4}{3}\pi x_{i+1}^3 - \tfrac{4}{3}\pi x_i^3 = \tfrac{4}{3}\pi\big((x_i + dx)^3 - x_i^3\big) = \tfrac{4}{3}\pi\big(3x_i^2 dx + 3x_i(dx)^2 + (dx)^3\big) \approx 4\pi x_i^2 dx.$$

Take c_0 to be a point in the subinterval $x_0 \leq x \leq x_1$, c_1 a point in $x_1 \leq x \leq x_2$, ..., c_i a point in $x_i \leq x \leq x_{i+1}$, and so on. Given that the shell is thin, it follows that its density is nearly equal to $\rho(c_i)$ throughout, so that the shell is nearly homogeneous. Since density is mass divided by volume, mass is volume times density. It follows therefore that the mass M_i of the thin shell is approximately

$$M_i \approx (4\pi c_i^2 dx)\rho(c_i) = 4\pi c_i^2 \rho(c_i)dx.$$

Since c is the distance of the point-mass m from the shell's center O, Part 2 tells us that the gravitational force F_i of this shell on the point-mass m is

$$F_i \approx G\frac{mM_i}{c^2} \approx \frac{Gm}{c^2} 4\pi c_i^2 \rho(c_i)dx,$$

By doing this for each of the n shells, adding the results, and noticing that the sum $\sum\limits_{i=0}^{n-1} M_i$ of the masses of the n shells is the mass M of the sphere, we get that the force F of the entire sphere on m satisfies

$$F \approx \frac{Gm}{c^2}\sum_{i=0}^{n-1} 4\pi c_i^2 \rho(c_i)\, dx \approx \frac{Gm}{c^2}\sum_{i=0}^{n-1} M_i = \frac{GmM}{c^2}.$$

Repeating this computation again and again and letting n go to infinity we see that the corresponding $dx = \frac{R}{n}$ goes to zero. The definition of the definite integral applied to the sums $F \approx \dfrac{Gm}{c^2}\sum\limits_{i=0}^{n-1} 4\pi c_i^2 \rho(c_i)\, dx$ and $M \approx \sum\limits_{i=0}^{n-1} 4\pi c_i^2 \rho(c_i)\, dx$ tells us that

$$F = \tfrac{Gm}{c^2}\int_0^R 4\pi x^2 \rho(x)\, dx = \tfrac{GmM}{c^2},$$

and we have established that if the density of a sphere of mass M satisfies the "onion property" described above, then its force of attraction on a point-mass m that is a distance c from its center satisfies Newton's law of universal gravitation.

Suppose, finally, that two spheres A and B of matter both satisfy the onion property. We know from Part 3 that every particle of A is attracted by B in accordance with Newton's formula, where c is the distance from the particle to the center of B. From the point of view of A, therefore, the entire mass of B can be considered to be concentrated at its center. In other words, B can be considered as a point-mass. But this point-mass is attracted by A in accordance with Newton's gravitational formula. Therefore Newton's formula holds for the two spheres A and B. The gravitational force with which they attract each other is equal to G times the product of their masses divided by the square of the distance between their centers.

We have verified that Newton's law of universal gravitation is valid for spherical bodies in the universe that are radially homogeneous (as defined by the onion property). However, we saw in Chapter 2 that the smaller bodies in the solar system, in particular, the comets and asteroids (and many of the moons of the planets), are highly irregular in shape. Newton's law does not generally apply to such bodies (the paragraph *Where the Law of Universal Gravitation Fails* in the Problems and Discussions section of this chapter explores this question). The large bodies in the solar system, the Sun, the planets, and the dwarf planets, are spheres and it seems probable that they, like the Earth, satisfy the onion property. However, these bodies rotate around an axis and this has caused them to bulge at their equators. But in reference to Newton's universal gravitation, the impact of this is minimal, at least in the case of the Earth.

A mathematical model for such deformed spheres can be obtained as follows. Let a be the radius of the body at the equator and let b be the radius at a pole and note that $a \geq b$. Take the ellipse that has semimajor axis a and semiminor axis b and revolve it one complete revolution around the

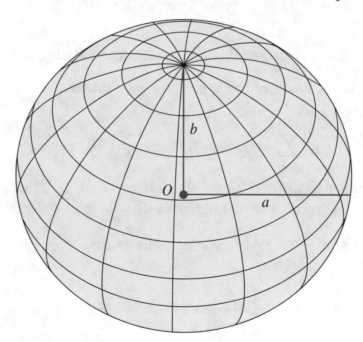

Figure 4.21

line of the minor axis. The surface obtained is known as an *ellipsoid*. See Figure 4.21. Let's compute the volume of the ellipsoid just described. Consider an *xy*-coordinate system and the ellipse $\frac{x^2}{a^2} + \frac{y^2}{b^2} = 1$, where a and b are the semimajor and semiminor axes. Solving for y, we get $y = \pm\frac{b}{a}\sqrt{a^2 - x^2}$. The upper half of the volume of the ellipsoid is the volume of revolution obtained by revolving the graph of $f(x) = \frac{b}{a}\sqrt{a^2 - x^2}$ one revolution around the *y*-axis as Figure 4.22a illustrates. The thin black strip in the figure revolves to generate a cylindrical shell of circumference $2\pi x$, height $y = \frac{b}{a}\sqrt{a^2 - x^2}$, and thickness dx. Since the volume of this shell is $2\pi x\frac{b}{a}\sqrt{a^2 - x^2}\, dx$, the summation strategy of integral calculus tells us that the volume of the upper half of the ellipsoid is

$$\int_0^a 2\pi x\frac{b}{a}\sqrt{a^2 - x^2}\, dx\,.$$

Example 4.11. The substitution $u = a^2 - x^2$ shows that the volume of the ellipsoid is $V = \frac{4}{3}a^2 b\pi$.

Figure 4.22b shows the sphere of radius b placed inside the ellipsoid. The region outside the sphere and inside the ellipsoid is the *equatorial bulge*. Its volume is $\frac{4}{3}a^2b\pi - \frac{4}{3}b^3\pi = \frac{4}{3}b(a^2 - b^2)\pi$.

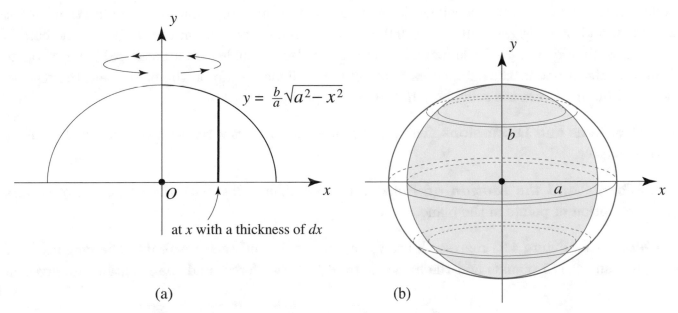

$$y = \frac{b}{a}\sqrt{a^2 - x^2}$$

at x with a thickness of dx

(a)   (b)

Figure 4.22

In the case of Earth, $a = 6378.1$ km and $b = 6356.8$ km. So the Earth's volume is

$$\frac{4}{3}\pi (6378.1 \times 10^3)^2 (6356.8 \times 10^3) \approx 1.0832 \times 10^{21} \, \text{m}^3.$$

Refer to Table 2.3 for the estimate 5.97×10^{24} kg for the Earth's volume. It follows that the Earth's average density is

$$\frac{5.97 \times 10^{24}}{1.0832 \times 10^{21}} \approx 5.51 \times 10^3 \, \text{kg/m}^3.$$

The volume of its equatorial bulge is

$$\frac{4}{3}\pi (6356.8 \times 10^3)(6378.1^2 - 6356.8^2) \times 10^6 \approx 7.22 \times 10^{18} \, \text{m}^3.$$

The density of the outermost 20 km of the Earth's crust is about 2.8×10^3 kg/m^3, so that the mass of the bulge is about

$$(7.22 \times 10^{18})(2.8 \times 10^3) \approx 2.02 \times 10^{22} \, \text{kg}.$$

It follows that the mass of the bulge is about $\frac{1}{3}$ of 1 per cent of the total mass of the Earth.

It follows that the bulge is of little consequence for the gravitational pull both by and on the Earth. On the other hand, we saw in *About the Earth-Moon System* of the Problems and Discussion section of Chapter 1, that the bulge causes small perturbations in Earth's motion over time.

The fact that the Earth bulges out at the equator means that the tip of Mount Chimborazo in Ecuador (163 kilometers south of the equator) is the point on the surface of the Earth that is farthest from the Earth's center. One could argue therefore that Mount Chimborazo (at 6,310 m above sea level) is the highest peak on Earth and not Mount Everest (at 8,850 m above sea level).

With regard to an assessment of the smoothness of the surface of the Earth, let's consider the relative height of Mount Everest if the Earth were shrunk to the size of a basketball. Let's take 6,370 km or 6,370,000 m as the average radius of the Earth and 0.122 m as the radius of a basketball. (The radius of an official basketball lies in the range of 0.120 m to 0.124 m.) This implies that Mount Everest would be $\frac{0.122}{6,370,000} \cdot 8850\,\text{m} \approx 0.00017\,\text{m} = 0.17$ millimeters high on the surface of the ball. Is this higher than one of the little mounds—officially called a pebble—on a basketball? These range from a height of about 0.33 millimeters to about 0.64 millimeters. So on the basketball, the relative height of Mount Everest is roughly half of that of a pebble.

4I. Problems and Discussions. This set of problems considers various aspects of the discussions of this chapter.

1. Forces and the Motion of Points. These problems deal with the properties of vectors and the motion of points in the plane.

Problem 4.1. Figure 4.23 represents two forces F_1 and F_2 and their resultant. The magnitude of the resultant is 115 pounds and the angles between the two forces and the resultant are given in

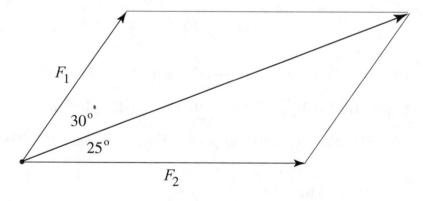

Figure 4.23

the figure. Use the law of sines to determine the magnitudes of F_1 and F_2.

Problem 4.2. Use the law of cosines to determine the magnitude of the resultant of the two vectors

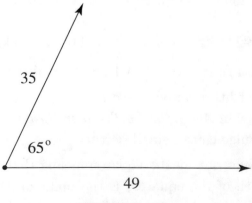

Figure 4.24

in Figure 4.24. Then use the law of sines to find the angles between the resultant and these vectors.

Problem 4.3. Consider Example 4.3 and the accompanying Figure 4.4. Given that the string has a length of 1.6 meters and that $\theta = 60°$, find the radius of the circular path of the object P. If the mass of P is $2\,\mathrm{kg}$, determine the magnitudes of the forces F_H and F.

Problem 4.4. Consider a point-mass P of mass $m = 1$ moving in an xy-plane. Let its position be given by $x(t) = \sqrt{t}$ and $y(t) = t$ for $t \geq 0$. Check that this point moves on the parabola $y = x^2$ starting at $(0,0)$. Analyze the motion of the point-mass by following what was done in Example 4.6. How does this motion compare with that of Example 4.6 for $t \geq 0$? What are the chief differences?

Problem 4.5. Suppose that a point-mass P is driven by a centripetal force in such a way that the angle $\theta(t)$ of Figure 4.13 increases at a constant rate. Show that $r(t)$ as well as $F(t)$ are constant and that the trajectory is a circle.

Problem 4.6. Suppose that a point-mass P moves along a straight line L with constant speed. Observe that P can be regarded as being subject to a centripetal force of zero magnitude centered at some point S not on L. Draw a picture of what is going on and show that the segment SP sweeps out equal areas during equal times. Let v be the speed of P and d the distance from S to L. Show that Kepler's constant is $\kappa = \frac{1}{2}vd$. Does this example contradict Newton's Conclusion C?

Problem 4.7. Suppose that a centripetal force of magnitude $F(t)$ propels a point-mass P on a trajectory that is the graph of the function $r = f(\theta) = \frac{d}{a\sin\theta + b\cos\theta}$, where a, b, and d are constants with $d \neq 0$. Let $g(\theta) = \frac{1}{f(\theta)} = \frac{1}{d}(a\sin\theta + b\cos\theta)$ and show that $g(\theta) + g''(\theta) = 0$. Conclude from the second form of the force equation that $F(t) = 0$. Review Chapter 3C and determine the graph of $r = f(\theta)$.

Problem 4.8. Consider a point-mass P of mass $m = 1$ moving in an xy-plane with position functions $x(t) = t\cos t$ and $y(t) = t\sin t$ with $t \geq 0$. Find the coordinates of the point for $t = 0, \frac{\pi}{4}, \frac{\pi}{2}, \frac{3\pi}{4}, \pi, \frac{5\pi}{4}, \frac{3\pi}{2}, \frac{7\pi}{4}$, and 2π. Analyze the point's motion by following Example 4.6. Show that the speed of the point is $v(t) = \sqrt{t^2 + 1}$. Verify that the magnitude of the force on P is $\sqrt{t^2 + 4}$. Show that the magnitudes of the tangential component of the force and the component perpendicular to it are $\frac{t}{\sqrt{t^2+1}}$ and $\frac{t^2+2}{\sqrt{t^2+1}}$, respectively. Discuss these magnitudes for t increasing from small to large.

The motion of P is best understood by going polar. Consider the functions $r = f(\theta) = \theta$ and $\theta = \theta(t) = t$. So $r(t) = f(\theta(t)) = t$. Use the connection between polar and Cartesian coordinates to show that $x(t) = r(t)\cos\theta(t) = t\cos t$ and $y(t) = r(t)\sin\theta(t) = t\sin t$. Therefore the two functions $r = f(\theta) = \theta$ and $\theta = \theta(t) = t$ represent the motion of P. The graph of $r = f(\theta) = \theta$ is an Archimedean spiral (see Problem 3.22). Since the angle is given by $\theta(t) = t$, this expanding spiral is traced out at a constant angular speed of $\theta'(t) = 1$ radian per unit time. Could the force that propels P be a centripetal force? The term $r(t)^2\theta'(t)$ provides the answer.

Problem 4.9. Let the position of a point-mass P of mass m moving in an xy-plane be given by the equations $x(t) = a\cos\omega t$ and $y(t) = b\sin\omega t$, where t is time with $t \geq 0$, and a, b, and ω are positive constants with $a \geq b$. Show that the distance between P and O at any time t is $r(t) = \sqrt{a^2\cos^2\omega t + b^2\sin^2\omega t}$. Show that the point starts at $(a, 0)$ and that it moves counterclockwise

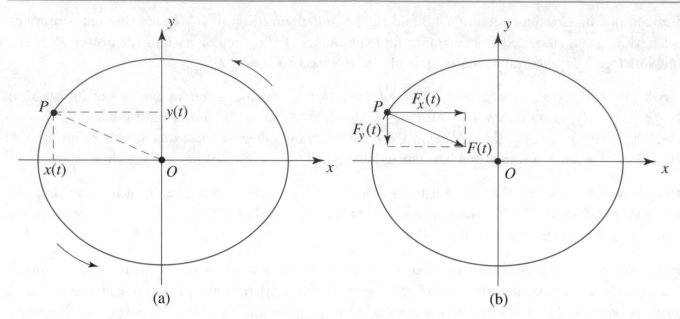

Figure 4.25

on the ellipse $\frac{x^2}{a^2} + \frac{y^2}{b^2} = 1$ with semimajor axis a and semiminor axis b. Refer to Figure 4.25 and analyze the motion of P by following what was done in Example 4.7. Conclude in particular that the force on P is a centripetal force in the direction of the origin O and that it has magnitude
$$F(t) = \sqrt{F_x(t)^2 + F_y(t)^2} = m\omega^2 \sqrt{a^2 \cos^2 \omega t + b^2 \sin^2 \omega t} = m\omega^2 r(t).$$

The fact that P starts at time $t = 0$ and completes its first revolution when t satisfies $\omega t = 2\pi$ tells us that the Kepler constant of its orbit is $\kappa = \frac{ab\omega\pi}{2\pi} = \frac{ab\omega}{2}$. Since the force on P is centripetal, $F(t)$ and $r(t)$ satisfy the force equation $F(t) = m\left[\frac{4\kappa^2}{r(t)^3} - \frac{d^2r}{dt^2}\right]$. (This can also be verified directly with a labor-intensive calculus exercise.) If $a > b$, then $r(t)$ varies. Use the equality $F(t) = m\omega^2 r(t)$ to show that $F(t)$ cannot satisfy an inverse square law. The geometric reason for this is that the center of force is the center of the ellipse and not a focal point of the ellipse. If $a = b$, then the ellipse is a circle and the center is also the focal point. In this case, $r(t) = a$ is a constant, and we saw in Example 4.10 that the centripetal force $F(t)$ does satisfy an inverse square law.

2. *Projectile Motion on Earth.* Let's go from motion in the abstract to the motion of a thrown object P near Earth's surface. We'll assume that air resistance is negligible. This is so if the projectile has a low initial speed (it's not a speeding bullet), a large enough weight (it's not a feather), and that its size is small (it's not a car). A baseball that is lobbed or a basketball of a jump shot are two examples. In such a situation, the trajectory of the projectile can be described accurately and relatively easily in mathematical terms. The discussion of section 4H allows us to suppose that the gravitational pull on P is directed to the Earth's center of mass C and that this pull satisfies an inverse square law. It follows from conclusion C of section 4H that the trajectory of P is a conic section with C at a focus. See Figure 4.26. We'll assume that the initial and terminal points of the trajectory of the projectile are close relative to the distance from P to C (which exceeds the 6356 km of Earth's polar radius). This means that the trajectory of the projectile can be studied by placing an xy-coordinate axis into the plane of the trajectory as shown in Figure 4.27 and by

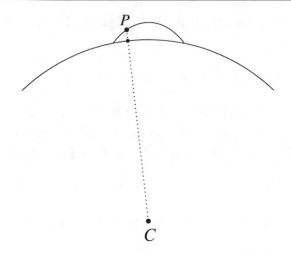

Figure 4.26

assuming that the pull of gravity on the projectile is parallel to the y-axis throughout its flight.

Let the projectile start its motion at time $t = 0$ with initial position the point $(0, y_0)$. It has an initial velocity that is represented by the vector in the figure. The length of the vector is equal to the initial speed v_0 of the projectile. The angle of the vector with respect to the horizontal is known as the *angle of elevation* or *angle of departure*. It is labeled by φ, where $0 \leq \varphi \leq \frac{\pi}{2}$. For any time $t \geq 0$, the x- and y coordinates of the position of P are $x(t)$ and $y(t)$, respectively. Note that $x(0) = 0$ and $y(0) = y_0$. For any t, the velocities of the projectile in the x- and y-directions are $x'(t)$ and $y'(t)$, respectively, and the speed of the projectile is equal to $v(t) = \sqrt{x'(t)^2 + y'(t)^2}$.

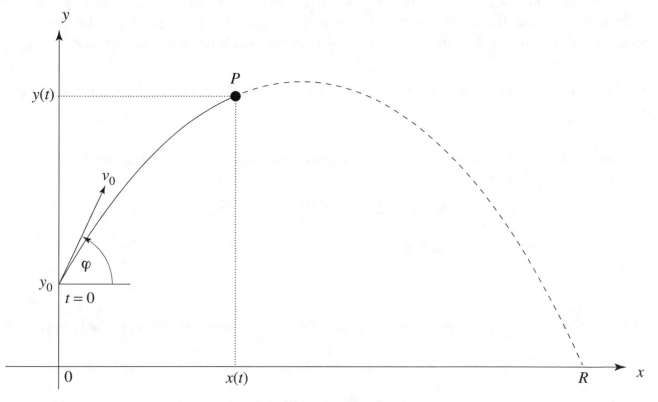

Figure 4.27

So $v_0 = \sqrt{x'(0)^2 + y'(0)^2}$ is the initial speed. Refer to Figure 4.27 again and observe that $\cos\varphi = \frac{x'(0)}{v_0}$ and $\sin\varphi = \frac{y'(0)}{v_0}$. Therefore,

$$x'(0) = v_0 \cos\varphi \quad \text{and} \quad y'(0) = v_0 \sin\varphi.$$

The only force on the projectile during its flight is gravity acting in the negative y-direction. Since the force in the x-direction is zero, the acceleration in the x-direction is also zero. Thus $x''(t) = 0$ and $x'(t)$ is constant. Since $x'(0) = v_0 \cos\varphi$ and $x(0) = 0$, it follows that

$$x'(t) = v_0 \cos\varphi \quad \text{and} \quad x(t) = (v_0 \cos\varphi)t.$$

Because gravity produces an acceleration of $-g$ in the y-direction, $y''(t) = -g$. Since $y'(0) = v_0 \sin\varphi$ and $y(0) = y_0$, we find that

$$y'(t) = -gt + v_0 \sin\varphi \quad \text{and} \quad y(t) = -\frac{g}{2}t^2 + (v_0 \sin\varphi)t + y_0.$$

From above, $t = \frac{x(t)}{v_0 \cos\varphi}$. So $t^2 = \frac{(x(t))^2}{v_0^2 \cos^2\varphi}$ and by a substitution into the expression for $y(t)$, we get

$$y(t) = \frac{-g}{2v_0^2 \cos^2\varphi}(x(t))^2 + (\tan\varphi)x(t) + y_0.$$

So the position $(x(t), y(t))$ of the projectile at any time t satisfies the equation

$$y = \left(\frac{-g}{2v_0^2 \cos^2\varphi}\right)x^2 + (\tan\varphi)x + y_0.$$

This is the equation of a parabola. Therefore, the trajectory of the projectile is a parabola, the very specific parabola that the constants v_0, y_0, and φ determine. For the rest of our discussion we'll assume that the terrain is flat and horizontal and that the x-axis lies along the ground.

Problem 4.10. Use some elementary calculus to show that the maximal height reached by the projectile is $\frac{1}{2g}v_0^2 \sin^2\varphi + y_0$.

Problem 4.11. At what time and how far downrange will the projectile hit the ground? Show that the time t_{imp} of impact is

$$t_{\text{imp}} = \frac{v_0 \sin\varphi + \sqrt{v_0^2 \sin^2\varphi + 2gy_0}}{g}$$

and that impact occurs down range at

$$x = R = x(t_{\text{imp}}) = \tfrac{v_0}{g}\cos\varphi\left[v_0 \sin\varphi + \sqrt{v_0^2 \sin^2\varphi + 2gy_0}\right].$$

Assume that $y_0 = 0$. The trig formula $\sin 2\varphi = 2\sin\varphi \cos\varphi$ provides the simplified expression

$$R = \frac{v_0^2}{g}\sin 2\varphi$$

for the range R. So when $y_0 = 0$, the maximal range is achieved for $\varphi = \frac{\pi}{4}$ and is equal to

$$R_{\max} = \frac{v_0^2}{g}.$$

If $y_0 \neq 0$, then the question of the maximal range is more complicated. The value $\varphi = \frac{\pi}{4}$ does not provide the greatest range R.

Problem 4.12. Show that the maximum value of the function

$$R(\varphi) = \tfrac{v_0}{g} \cos \varphi \left[v_0 \sin \varphi + \sqrt{v_0^2 \sin^2 \varphi + 2gy_0} \right]$$

occurs when $\sin \varphi = \frac{v_0}{\sqrt{2v_0^2 + 2gy_0}}$. [The strategy is as expected: set $R'(\varphi) = 0$ and see what you get. The algebra is a bit involved, but routine. Multiply through by the denominator $\sqrt{v_0^2 \sin^2 \varphi + 2gy_0}$ and then square both sides. Also show that $\sin \varphi = \frac{v_0}{\sqrt{2v_0^2 + 2gy_0}}$ is equivalent to $\tan \varphi = \frac{v_0}{\sqrt{v_0^2 + 2gy_0}}$.]

Problem 4.13. Show that the speed of the projectile at any time t is

$$v(t) = \sqrt{v_0^2 + g^2 t^2 - 2g(v_0 \sin \varphi)t}.$$

Determine the speed at impact as well as the angle of impact.

Problem 4.14. A baseball player throws a ball in the direction of a teammate. The gravitational constant at the location is $g = 9.8\,\mathrm{m/sec^2}$. The ball is released 1.5 meters from the ground with an initial velocity of 21.7 meters/sec and an angle of elevation of 30°. What is the maximum height that the ball reaches? How long will the ball remain in the air before it strikes the ground? What will be the ball's speed when it impacts the ground in front of the outstretched glove of the teammate?

3. Tossing the Hammer. The hammer throw is an Olympic track and field event that appears to have its origin in the Scottish/English sport of sledge hammer throwing. It has been an Olympic event for men since the Paris Olympics in 1900. For women the hammer throw was first included in the Sydney Olympics in 2000. A 16 pound, or equivalently a 7.257 kg metal ball is attached to a 3 foot $11\frac{3}{4}$ inch, or 1.215 meter, piece of wire and the wire in turn is attached to a handle. The thrower grasps the handle with both hands and swings the ball in a circular arc with extended arms. As the ball moves in a circle the thrower's body spins, both with successively increasing rotational

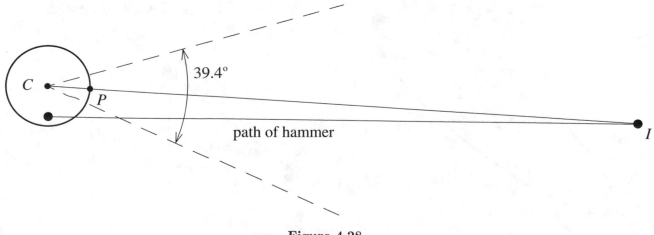

Figure 4.28

speed. The thrower attempts to achieve the longest possible throw with an optimal combination of the speed and angle of inclination of the ball at the instant he releases the handle and the ball flies off (along with the handle and the wire). For a throw to be valid, the thrower has to remain within the designated throwing circle of 7 feet in diameter throughout and the ball has to land within a specified wedge shaped sector. Both are depicted in Figure 4.28. The official distance of a valid throw is the distance from the point of impact I of the ball to the point P on the throwing circle. Again refer to Figure 4.28.

The world record for the men's hammer throw is held by the Ukranian Yuriy Sedykh, who threw 86.74 m (284 ft 7 in) at the European Track and Field Championships in Stuttgart, West Germany, on the 30th of August, 1986. Having stood for over 30 years, it is one of the oldest records in track and field. Yuriy came into the championship meet as the world record holder with 86.66 m. The four best throws in his sequence of six at this championship meet were: 85.28 m, 85.46 m, 86.74 m, and 86.68 m. His remarkable performance is captured in the video

https://www.youtube.com/watch?v=4qAE2PrCVhY

The record setting throw was recorded with a camera at 200 frames per second and analyzed by Ralph Otto and Gabriele Hommel, in their article, Hammer Throw World Record, Photo Sequence-Yuriy Sedykh, 1992. See

http://www.hammerthrow.org/wp-content/uploads/photosequences/otto_sedykh_wr.pdf

The data that this study develops include:

> The speed of the ball at release: $v_0 = 30.7$ m/sec.
> The height of the ball at release: $y_0 = 1.66$ m.
> The angle of elevation of the ball at release: $\varphi = 39.9°$.
> The final release phase of the throw: with a duration of 0.27 sec, it increased the
> ball's speed from 24.1 m/sec to 30.7 m/sec.
> Given its latitude and elevation, the gravitational constant g for Stuttgart is
> essentially the same as that of Paris. The latter is $g = 9.81$ m/sec^2.

Use these data and the formula $R = \frac{v_0}{g} \cos \varphi \left[v_0 \sin \varphi + \sqrt{v_0^2 \sin^2 \varphi + 2gy_0} \right]$ derived in Problem 4.11 for the horizontal distance that a projectile achieves in a vacuum to check that $R \approx 96.56$ m.

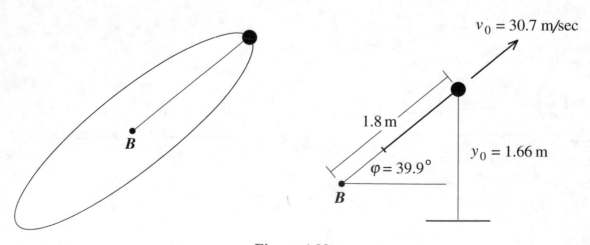

Figure 4.29

This tells us that air resistance and the retarding effect of the wire and the handle reduced this theoretical value of the record throw by $96.56 - 86.74 = 9.82$ m.

Figure 4.29 depicts the instant of the release of Yuriy's record throw. In terms of the physics of what is happening, the matter is similar to the motion of the Moon around the Earth. See Chapter 1E. In both situations, two bodies go around a common center of force, the barycenter of the system. In this last phase of Yuriy's throw both the hammer and (a part of) his torso are in orbit around a barycenter B that lies along the segment determined by his extended arms and the cable of the hammer. A study of the video and the fact that the wire is about 1.2 m long, suggests that B lies about 1.8 m from the center of the ball.

To compute the force with which the hammer pulls on the thrower just before the moment of release, we'll assume that the hammer moves with a speed of 30.7 m/sec along the very last arc of a circular orbit of radius $r = 1.8$ m. The period T of this assumed orbit satisfies $\frac{2\pi r}{T} = 30.7$. So $T = \frac{3.6\pi}{30.7} = 0.37$ sec. The formula for the centripetal force

$$F = \frac{4\pi^2 a^3 m}{T^2} \cdot \frac{1}{r_P^2}$$

applies to the current situation with $r_P = a = 1.8$ and $m = 7.257$ kg, so that the force with which the revolving ball pulls on the arms of the thrower just before release is

$$F = \frac{4\pi^2 a^3 m}{T^2} \cdot \frac{1}{r_P^2} = \frac{4\pi^2 (1.8)m}{T^2} \approx \frac{(71.06)(7.257)}{0.37^2} \approx 3767 \text{ newtons.}$$

This force is much greater than the weight of the thrower. Since Yuriy had a mass of 110 kg at the time of the throw, his weight was $9.81 \cdot 110 = 1079$ newtons. Given this difference, why didn't Yuriy fly off with the ball before he released it?

The analysis just undertaken has ignored the downward pull of the weight $9.81 \cdot 7.257 = 71.19$ newtons of the hammer. The fact that this is much smaller than the force exerted by the moving hammer makes this a reasonable assumption. It has also ignored the force of friction pushing against the soles of the throwers athletic shoes. This is difficult to assess, but the angle of the body of the thrower and the way the soles of the shoes touch the floor of the throwing circle both suggest that this too is a lesser factor.

From 1980 onward, the world record in the hammer throw has evolved as follows:

80.64 m (Yuriy Sedykh, May 16, 1980); 81.66 m (Sergey Litvinov, May 24, 1980)
81.80 m (Yuriy Sedykh, July 31, 1980); 83.98 m (Sergey Litvinov, June 4, 1982)
84.14 m (Sergey Litvinov, June 21, 1983); 86.34 m (Yuriy Sedykh, July 3, 1984)
86.66 m (Yuriy Sedykh, June 22, 1986); 86.74 m (Yuriy Sedykh, August 30, 1986)

Yuriy Sedykh and the Russian Sergey Litvinov were the greatest hammer throwers of all time. From 1976 until 1991, one or the other won the world championship as well as Olympic gold. Litvinov still holds the Olympic record with his throw of 84.80 m at the Seoul Olympics in 1988.

4. Supersized Reflecting Telescopes. There are two types of telescopes: refracting and reflecting. In a refracting telescope it is the objective lens that collects, bends, and brings to a focus the parallel rays of light from a distant object. In a reflecting telescope it is a curving primary mirror that plays this role. The world's most powerful telescopes are reflecting telescopes. The structure of most of them goes back to a design by the Scotsman James Gregory, a mathematician and astronomer working in the 17th century. The essential scheme of the Gregorian telescope is shown in Figure 4.30. The primary mirror collects the light and brings it to a focus before a second inward curving mirror reflects it back through a hole in the center of the primary mirror to form

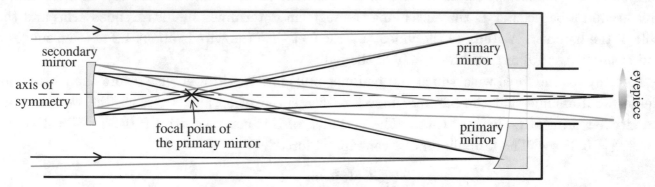

Figure 4.30. Gregorian telescope. From https://en.wikipedia.org/wiki/Gregorian_telescope#/media/File:Gregorian_telescope.svg

an image that is enlarged by the eyepiece. Reflecting telescopes have several important advantages over refracting telescopes. In a lens, the volume of glass or plastic has to have a precise geometry, be homogeneous, and be free of imperfections throughout. In an optical mirror, only the surface has to be perfectly shaped and polished. A large objective lens is heavy and can be distorted by gravity. In contrast, a large mirror can be supported by a lighter frame on the other side of its reflecting face with no gravitational sag. In other words, in terms of the critical light-collecting ability of a telescope, it is not feasible to build huge lenses, but it is feasible to build huge light-collecting mirrors.

All large telescopes built in the 20th and 21st centuries, as well as those still under construction, are reflecting telescopes, and most of these are Gregorian telescopes (or modified versions of the Gregorian design) with primary mirrors that have central holes. The largest are three super telescopes, the Giant Magellan Telescope, the Thirty Meter Telescope, and the European Extremely Large Telescope, all three scheduled to come on line in the decade of the 2020s.

The primary mirror of the Giant Magellan Telescope (GMT) will consist of a configuration of seven circular mirrors all of the same diameter. One mirror at the center (with the hole that the Gregorian design requires) is to be surrounded by six more mirrors aligned in such a way that their combined surface will lie on a parabola of revolution. All of the mirrors are to be manufactured by the Mirror Lab at the University of Arizona. We will focus on the mirror at the center to describe the manufacturing process. A cylindrical form of about 8.4 m in diameter is placed with its base in a horizontal position so that its central axis is vertical. A tight, tiled arrangement of hundreds of hexagonal boxes made of heat-resistant material forms the supporting base of the mirror during its manufacture. Chunks of the purest glass, about 20,000 kg in all, are placed into the cylinder on top of this base. See Figure 4.31. The cylinder sits inside a tub that forms the bottom half of a furnace. With the lid in place, the cylinder is enclosed in the furnace. Powerful heating elements melt the glass in the cylinder at a temperature of 1160 °C (degrees Celsius). The furnace assembly is rotated in the horizontal plane around the central axis of the cylinder at a constant rate of 5 revolutions per minute. This rotation pushes the liquid glass outward toward the rim of the cylinder. The molten glass reaches steady state with its upper surface curving from the center of the cylinder up to its edge. After the correct mirror geometry has been achieved, the cooling process begins. It takes about three months to cool the glass to room temperature. The slowness of the process ensures that the glass will not develop cracks. To preserve the geometry, the furnace assembly continues to rotate during this time. The process that has been described is called *spin casting*.

The remarkable fact is that the spin-casting process provides the upper surface of the glass, and hence the eventual mirror, with the parabolic geometry that it needs to have! A second remarkable

Figure 4.31. Casting the central mirror GMT4 for the Giant Magellan Telescope. Image credit: Ray Bertram, Richard F. Caris Mirror Lab, University of Arizona.

fact is that the verification that this is so relies on Newton's formula for the magnitude of a centripetal force.

Suppose that the glass surface has reached steady state. Take a plane through the vertical axis of the rotation. Think of this vertical plane to be fixed, and consider the curve obtained by intersecting the upper surface of the molten glass with this vertical plane. Figure 4.32 shows this plane along with an xy-coordinate system. The coordinates 4.2 and 1.15 represent the radii of the mirror and that of its central hole (both in meters). The y-axis is the axis of rotation of the furnace assembly and that of the graph of the function $y = f(x)$ that represents the surface of the molten glass. Consider a small particle of molten glass of mass m riding on this surface. Let α be the angle between the tangent line to the graph at the glass particle and the horizontal. Let F_b be the buoyant force with which the molten glass pushes against the particle. It acts perpendicularly to the tangent line. Because steady state has been reached, the vertical component V of the buoyant force is equal in magnitude to the force of gravity F_g acting on the glass particle. Let F_c be the horizontal component of the buoyant force. Extend the line of force of F_b, and draw in the angle β. Notice that $\alpha + \beta = 90°$. The angle

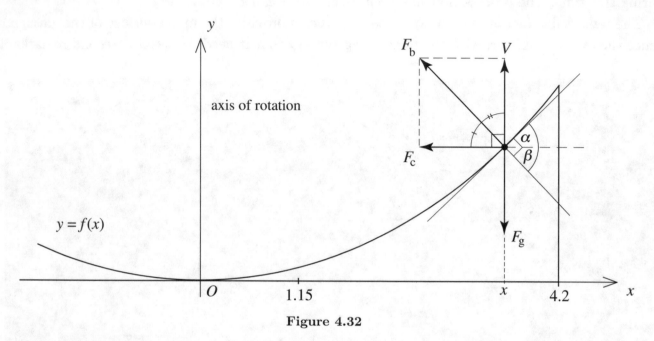

Figure 4.32

marked with the single stroke is equal to β so that the angle marked with the double stroke is equal to α. Since the tangent of this angle is equal to $\frac{F_c}{V} = \frac{F_c}{F_g}$, it follows that

$$\tan \alpha = \frac{F_c}{F_g}.$$

Suppose that the cylindrical form turns at a constant rate of τ revolutions per unit time. Since steady state has been reached, the particle moves at this rate on a fixed horizontal plane in a circle of constant radius x. The particle is kept in "orbit" on this horizontal plane by the horizontal component of the buoyant force of magnitude F_c, which acts centripetally in the direction of the axis of rotation. Since the mass m moves in a circle of radius x, we know by the circular case of the centripetal force equation (with $a = r(t) = x$) that

$$F_c = \frac{4\pi^2 x m}{T^2},$$

where T is the time of one complete revolution. If the glass particle takes the time T for 1 revolution, it completes $\frac{1}{T}$ revolutions in 1 unit of time. Therefore $\frac{1}{T} = \tau$, and it follows that

$$F_c = 4\pi^2 \tau^2 m x.$$

Because $f'(x)$ is the slope of the tangent,

$$f'(x) = \tan \alpha = \frac{F_c}{F_g} = \frac{4\pi^2 \tau^2 m x}{mg} = \frac{4\pi^2}{g} \tau^2 x.$$

Since the derivatives of $y = f(x)$ and $y = \frac{2\pi^2}{g}\tau^2 x^2$ are the same and $f(0) = 0$, it follows, by taking antiderivatives, that the upper surface of the molten glass is obtained by rotating the parabola

$$f(x) = \frac{2\pi^2}{g}\tau^2 x^2$$

one complete revolution around the y-axis. The focal point of the parabola has coordinates $\left(0, \frac{g}{8\pi^2\tau^2}\right)$. This follows from the study of the parabola $x^2 = 4cy$ in Chapter 1C. Notice that the shape of the parabola depends (in addition to π and g) only on the speed of the rotation τ and not on the density of the molten glass. So the shape is the same regardless of the liquid that is being rotated.

The constant τ controls the geometry of the mirror. For the central mirror of the GMT, the rotational speed was set at $\tau = \frac{1}{12}$ revolutions per second (the equivalent of 5 revolutions per minute). Since $g \approx 9.80$ m/s^2 (at the location of the University of Arizona in Tucson, Arizona), it follows that in meters,

$$f(x) \approx 0.014x^2.$$

The parabolic surface that has now been described only approximates the final shape of the mirror. The precision requirements on the mirror are extraordinary. After the glass mass has cooled, its parabolic surface is made smooth with diamond grinding wheels. This brings the accuracy of the surface to within about $\frac{1}{10}$ of 1 millimeter of what is needed. Finally, after polishing the glass, the necessary tolerance of less that $\frac{1}{10,000}$ of 1 millimeter is achieved. The final step—to be undertaken for each of the mirrors on location of the GMT high in the Chilean Andes—is the application of a thin, fragile, reflective aluminum coating to the glass surface. Only then will the manufacture of the mirrors be complete. The telescope is scheduled to begin operation in the year 2029.

Problem 4.15. Consider the shape of the central mirror of the Giant Magellan Telescope as it is described by Figure 4.32 and the function $f(x) \approx 0.014x^2$. Show that the depth of the mirror, namely, the vertical distance between the horizontal plane at the mirror's rim and its central hole is $f(4.2) - f(1.15) \approx 0.014[(4.2)^2 - (1.15)^2] \approx 0.23$ m. Then show that the focal point of the central mirror is

$$\frac{g}{8\pi^2\tau^2} \approx \frac{(9.80)(12^2)}{8(9.87)} \approx 17.87 \text{ m}$$

above the origin O.

We have described the manufacture of the central primary mirror of the GMT. The six mirrors surrounding the central mirror are constructed in the same way. But there is one important difference. The parabola of the central mirror determines the parabola of the entire configuration, and its central axis coincides with the focal axis of this parabola. The surfaces of the other six mirrors lie higher on the parabola. So they are off-axis and not rotationally symmetric. After the spin-cast glass forms of the six non-central mirrors emerge from the furnace, they need to be ground and polished with complex precision to receive the delicate geometry that they need to have. When complete, the seven-mirror configuration of the GMT will have a diameter of 25 m. The website

http://www.gmto.org/gallery/

provides up to date progress reports on the construction of the telescope.

There are plans to construct telescopes even larger than the GMT. One of them is the Thirty Meter Telescope (TMT) with a primary mirror of 30 m in diameter to be built on Mauna Kea in Hawaii. The design of this mirror is completely different from that of the GMT. It will be a composite of 492 individual hexagonal mirror segments that measure 1.44 m from corner to opposite corner. The hexagons, all slightly different in shape, will be carefully aligned to form the hyperbolic primary mirror of the design. (The fact that the primary and secondary mirrors are both curved and work in tandem means that the parabolic geometry is not the only option for the primary mirror.) The advantage of this approach is that the smaller mirrors are more quickly manufactured and more easily shipped to the construction site than the huge mirrors of the GMT. On the other hand, it will be much easier to control the fewer moving parts of the GMT with the necessary accuracy. The building of the TMT has faced delays. Native Hawaiians consider the proposed site of the telescope sacred and have protested its construction. The European Extremely Large Telescope (E-ELT) to be built in Chile, will have the same mirror design as the TMT. With its 798 hexagonal mirrors and a diameter of 39 m, it will be the largest of the three new super telescopes. Its mirrors are already being cast and it is scheduled to see "first light" in the year 2025.

The new large telescopes will look deeply into space to unravel the mysteries surrounding the evolution and current state of the universe (its age, galaxy formation, dark matter and dark energy, black holes, and planets that orbit distant stars). Earth-based telescopes need to compensate for atmospheric interference, such as air currents and turbulence, as well as refraction. These distort the information carried by the light from the object being observed. The powerful computers that are integrated with the optics of high-tech telescopes can measure the distortions, and the many actuators can continuously adjust the shape of the mirrors in response. Such systems are referred to as *adaptive optics*. They transform twinkling stars into clear, steady points of light. In this way, all three super telescopes will produce images that are 10 times sharper than those of the Hubble Space Telescope. (The amazingly successful Hubble with its 2.4 m mirror orbits Earth and does not have to deal with atmospheric conditions.)

5. *Where the Law of Universal Gravitation Fails.* The next two problems provide examples of bodies for which Newton's law of universal gravitation does not hold. They tell us that this law is the exception rather than the rule. For instance, it would not apply to the force with which a potato-shaped asteroid attracts an orbiting spacecraft. Of course, the law of universal gravitation along with the assumption that the spacecraft, or asteroid, or comet is a point-mass does provide accurate information in most situations where the distances involved are large.

Problem 4.16. Consider a point-mass of mass m and a thin homogeneous circular disc D of radius R and mass M. Figure 4.33 depicts D along with an x-axis that lies in the plane of the disc. The origin O is the center of D. The point-mass lies on the axis perpendicular to D through its center O at a distance c from O. The density of D is equal to $\rho = \frac{M}{\pi R^2}$. A typical circular ring of D is shown in blue. It has radius x and thickness dx. The circumference of the ring is $2\pi x$, so that it has mass $2\pi\rho x\, dx$. It follows from Part 1 of the section Gravity and Geometry that this circle attracts

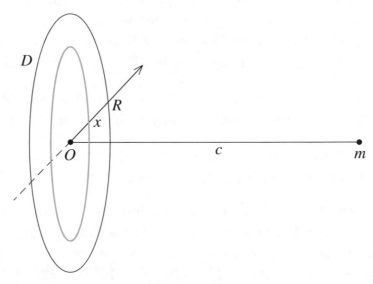

Figure 4.33

the point-mass with a force of $G\,\frac{m(2\pi\rho x\, dx)c}{(x^2+c^2)^{\frac{3}{2}}}$ directed to the center of the circle. Use the strategy of integral calculus to show that the gravitational force of the disc on the point-mass is directed to the center of the disc and has magnitude $F = G\,\frac{2mM}{R^2}\left[1 - \frac{c}{\sqrt{R^2+c^2}}\right]$.

The center of mass of the disc D is its center, so that an application of the law of universal gravitation provides the incorrect magnitude of $G\frac{mM}{c^2}$ for this attractive force.

Problem 4.17. Consider a point-mass of mass m and a homogeneous cylinder C (include the interior of the cylinder as well as its surface) of radius R, height h, and mass M. Figure 4.34 depicts C along with its central axis. The cylinder extends from its circular base at $x = 0$ to its

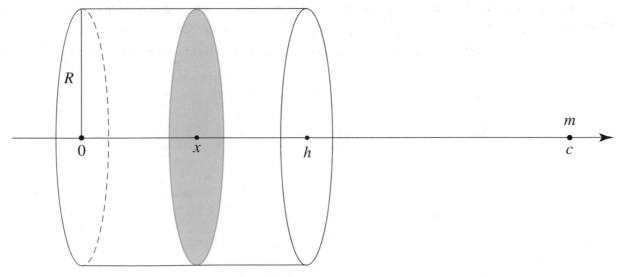

Figure 4.34

other circular boundary at $x = h$. The point-mass lies on the central axis at a distance $c \geq h$ from the base of the cylinder. The density of C is $\rho = \frac{M}{\pi R^2 h}$. A typical circular disc of C is shown in the figure in blue. It is parallel to the base, intersects the axis at x, and has a thickness of dx. The disc has a mass of $\pi R^2 dx \cdot \rho$. Use the result of Problem 4.16 to determine the gravitational force of the disc on m. Set up an integral and then solve it to show that the force of the cylinder on the point-mass is directed to the center of the cylinder and that its magnitude is $G\frac{2mM}{R^2 h}\left[h - (R^2 + c^2)^{\frac{1}{2}} + (R^2 + (c - h)^2)^{\frac{1}{2}}\right]$.

Observe therefore that Newton's law of universal gravitation does not hold for a cylinder and a point-mass, even when the point-mass is symmetrically positioned on the cylinder's central axis.

6. The Volume and Density of the Moon. The final two problems will estimate the average density of the Moon and conclude that it is much less than Earth's average density.

Problem 4.18. The Moon, like Earth, is a flattened ellipsoid. The Moon's equatorial and polar radii are 1738.1 and 1736.0 kilometers respectively. Use the conclusion of Example 4.11 to show that the volume of the Moon is closely approximated by $2.1968 \times 10^{19}\,\text{m}^3$. Show that the volume of the Moon's equatorial bulge is $\frac{4}{3}\pi(1736.0)(1738.1^2 - 1736.0^2) \times 10^9 = 5.3052 \times 19^{16}\,\text{m}^3$. This amounts to about $\frac{1}{4}$ of 1% of the Moon's total volume.

Problem 4.19. Refer to Table 2.3 of Chapter 2 for the estimate 7.35×10^{22} kg of the Moon's mass and use the result of Problem 4.18 to show that the average density of the Moon is 3.35×10^3 kg/m^3.

A study of Figure 4.19 tells us that the average density of the Moon is about the same as the density of the lightest materials of the Earth's crust. Analyses of the lunar rocks that the Apollo missions brought back have informed us that the chemical composition of these rocks differs from that of the crustal rocks of Earth. These two facts together give weight to the most widely supported theory about the origin of the Moon. This is the "large impact theory" asserting that a massive body (say, one-half the size of Earth) struck Earth. The impact threw a large amount of material, both from Earth and the impacting body, into orbit around Earth. Much of it would have come from Earth's lighter outer mass. So the large impact theory is not inconsistent with the observed differences between the densities and chemical compositions of the materials of Earth and Moon.

It's time to step back and recall some of the high points of the discussion of the previous chapters. We learned in Chapter 1 that Kepler's painstaking analysis of what Tycho Brahe had massively observed and recorded, allowed him to conclude that the planets move in accordance with what would later be called Kepler's three laws of planetary motion. These laws assert that the planets move in elliptical orbits with the Sun at a focal point, that the segment that joins each moving planet to the fixed Sun sweeps out equal areas in equal times, and that the ratio $\frac{a^3}{T^2}$, where a is the semimajor axis of an orbit and T its period, is the same for all the planets. The chapter went on to Newton's conclusions—obtained in the abstract by a combination of basic laws of motion with the mathematics of calculus—about the connection between the geometry of the path taken by a point-mass and the magnitude of the centripetal force that acts on it. Armed with these conclusions and his law of universal gravitation, Newton was able to show that Kepler's laws were logical consequences of basic laws of motion and mathematics. Chapter 2 turned to the recent history of the exploration of our solar system and the universe beyond by spacecraft and telescopes that are ever increasing in number and sophistication. The chapter presented many of the incredible images and some of the highly accurate data that they have sent back to us, with emphasis on the planets, their moons, asteroids, and comets. Chapter 3 introduced the polar coordinate system, studied polar functions, and developed their basic calculus. The primary purpose of Chapter 4 was to elaborate in detail Newton's theory of a centripetal force acting on a point-mass by relying on the calculus of polar functions. Newton's theory applies to a planet, asteroid, or comet pulled along its orbit or trajectory—we will use these two terms interchangeably—by the gravitational force of the Sun. It also applies to the orbit of a moon or man-made satellite around a planet, and to a spacecraft on a trajectory around the Sun, a planet, a moon, an asteroid, or a comet (as long as a single dominant gravitational force is involved). In all these cases, closed orbits are elliptical and open-ended curving flybys are either parabolic or hyperbolic, and the source of the attracting gravitational force is located at a focal point of the path.

What has been described provides a wealth of information about the objects in the solar system and how they move, but there are several important concerns that remain to be pursued. Chapter 5 will explore the most basic of these with a focus on the situation of elliptical orbits. Knowing the shape of the trajectory of a planet, moon, asteroid, comet, or spacecraft is fundamental, but the shape itself says nothing about the way the shape is traced out. Kepler's equal areas in equal times law does say something about the way this happens. But is it possible to compute the position of an orbiting body in precise terms by knowing only the elapsed time of its motion from some moment forward (in combination with basic orbital data)? We will see that it is. The final point on the agenda of this chapter concerns the precession of perihelion already discussed in Chapter 1G. The fact is that as a planet moves along its elliptical orbit, the orbit itself revolves very slowly around the Sun. The cause of this revolution has two aspects. One is the cumulative effect of the gravitational forces of the other planets. The second is that in any delicate, long term analysis of the motion of a planet, Newton's inverse square law of gravitational force needs to be adjusted by adding the corrective term that Einstein's theory of relativity provides. The development of a mathematical theory that predicts these effects is the last item that this chapter considers.

5A. Setting the Stage. In keeping the range of applications in mind, the motion of an object in orbit will be studied in the abstract situation of a point-mass P driven by a centripetal force in the direction of a fixed point S. This first section sets the stage for both this chapter and the next. It refers to earlier discussions for the basic notation and concepts, points to the relevant assumptions, and recalls the important facts. The key assumption is that the centripetal force satisfies an inverse square law. This implies—see Conclusion C of Chapter 4G—that the orbit of P is a conic section, either an ellipse, parabola, or hyperbola, and that S is a focal point. The point of the orbit where P is closest to S is known as *periapsis* in general, and *perihelion* if S represents the center of mass of the Sun (or more accurately, see Chapter 1I, the barycenter of the solar system), P is the center of mass of an object, and the force is the Sun's gravity. If the orbit is an ellipse, the point in the orbit where P is farthest from S is called *apoapsis* in general, and *aphelion* if S represents the Sun and the force on P is gravity.

Consider P at the instant it is at periapsis—in the case of an elliptical orbit, this can be any time at which P reaches the periapsis of its orbit—and click a stop watch. This is time $t = 0$. Beginning at this moment, the time t flows forward. We know from Chapter 4D that the area $A(t)$ that the advancing segment from S to P sweeps out during time t is given by

$$A(t) = \kappa t,$$

where κ is Kepler's constant for the orbit. We'll let m be the mass of P. At any time t, let

$$r(t) \quad \text{and} \quad F(t)$$

be the distance between P and S and the magnitude of the centripetal force acting on P. Any coherent system of units, of time, distance, mass, and force, can be used, but in this text preference will be given to the metric system MKS with its second, meter, kilogram, and newton.

Figure 5.1 captures what has been described. In the figure, Q is the periapsis position and L is the *latus rectum* of the orbit. This is the length of the segment (depicted in green in the figure) through S perpendicular to the focal axis. Newton's inverse square law—refer to Conclusion B of

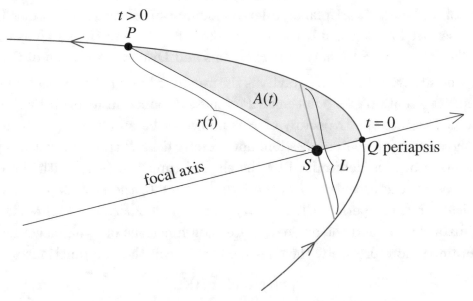

Figure 5.1

Chapter 4G—provides the formula

$$F(t) = \frac{8\kappa^2 m}{L} \cdot \frac{1}{r(t)^2}.$$

Let's turn to the situation of a body of mass M and center of mass S acting gravitationally on an object of mass m and center of mass P. Let's assume that the mass M of the body is much greater than m and that its gravitational pull is the only effective force on the object. So by the discussion of Chapter 1D, the position of the body is essentially fixed relative to the resulting motion of the object. Assume that the body is close to being spherical and that its mass is very close to being radially distributed. Suppose that this also holds for the object or else that the object is small enough to be regarded as a point-mass. From the analysis of Chapter 4H, we can conclude that the gravitational pull of the body on the object is essentially a centripetal force on the object's center of mass P in the direction of the center of mass S of the body, and that Newton's law of universal gravitation applies to it. Even though the law of universal gravitation holds only in approximation in this gravitational situation, we'll make the simplifying assumption that it holds exactly, so that if $F(t)$ is the magnitude of the gravitational pull and $r(t)$ the distance between S and P, then

$$F(t) = \frac{GmM}{r(t)^2},$$

where G is the universal gravitational constant. (See Chapter 1H for information about G.) Since this is an inverse square law, our introductory discussion and the earlier force equation applies to this gravitational situation as well. By combining the two force equations, we get the connection

$$GM = \frac{8\kappa^2}{L}$$

between the parameters κ and L of the orbit of P and the mass M of the attracting body.

What has been described applies very tightly to a planet, asteroid, or comet in orbit around the Sun, to a large moon or small object in orbit around a planet, and to a spacecraft in orbit around or hyperbolic flyby of a planet, dwarf planet, a large moon, or an asteroid. (At times the barycenters of systems of masses need to be considered. See Chapter 1E, Chapter 1I, as well as the segment *The Barycenters of the Planetary Systems* of the Problems and Discussion section of Chapter 2.)

Even though the situation involving gravity is the important and primary example, we'll continue to study the abstract situation of a point-mass P of mass m pulled in its orbit by a centripetal force directed to a point S. Since the trajectory of P is known to be an ellipse, parabola, or hyperbola, we can turn to the important related question: how exactly does P trace its trajectory out? Is it the case that the position of P is determined by the elapsed time t together with the relevant orbital constants (the mass m, the eccentricity ε, the latus rectum L, the semimajor axis a, and, in the case of an elliptical orbit, the period T)? Let $\alpha(t)$ be the angle $\angle PSQ$ of Figure 5.1 in radians and notice that the distance $r(t)$ and the angle $\alpha(t)$ are both functions of t that together determine the position of P. Framed more precisely, the question is this: can the two functions

$$r(t) \quad \text{and} \quad \alpha(t)$$

that provide the precise position of P in its orbit relative to S be determined explicitly in terms of the elapsed time t (and the relevant constants)? The connection $A(t) = \kappa t$ that Kepler's second law provides between the area that the segment PS traces out and time t, suggests that this should be possible. And it is possible! But the arguments—as we will see in this chapter for elliptical orbits and in the next chapter for hyperbolic (and parabolic) trajectories—are complicated.

Since the concern of this entire chapter will be the special case of elliptical orbits, we now assume that the orbit of the point-mass P is an ellipse and that S is one of its focal points. The basic steps involved in the determination of the functions $r(t)$ and $\alpha(t)$ go back to Kepler, but our presentation of the specifics relies on today's calculus of functions.

5B. Determining Distance and Angle. Let's begin by recalling the basics about the ellipse from Chapter 1C. In Figure 5.2, the semimajor and semiminor axis of the ellipse are a and b, respectively. The distance between the center O of the ellipse and the focal point S is $c = \sqrt{a^2 - b^2}$, so that the eccentricity is $\varepsilon = \frac{c}{a} = \frac{\sqrt{a^2 - b^2}}{a} < 1$ and $OS = a\varepsilon$. The distances of the periapsis and apoapsis positions from S are $a(1 - \varepsilon)$ and $a(1 + \varepsilon)$, and the latus rectum is $L = \frac{2b^2}{a}$. (See part A in Chapter 3D for this last fact.) In the case of a circle, $a = b$, $\sqrt{a^2 - b^2} = 0$, and $\varepsilon = 0$. Any point

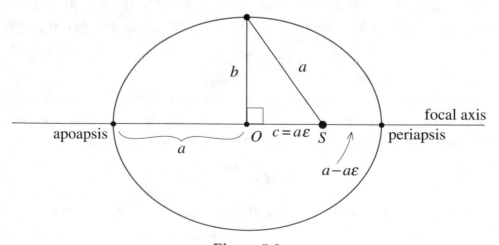

Figure 5.2

on the circle can be selected as the periapsis. The point on the circle opposite to it is the corresponding apoapsis.

The period T of the orbit of P is the time it takes for the point-mass P to move from one periapsis to the next. Note that Kepler's constant is $\kappa = \frac{ab\pi}{T}$. After substituting $\kappa^2 = \frac{(ab\pi)^2}{T^2}$ and $L = \frac{2b^2}{a}$ into the earlier force equation and the equation in the gravitational situation that relates the mass M of the pulling body to the parameters of the orbit of P, we get

$$F(t) = \frac{4\pi^2 a^3 m}{T^2} \cdot \frac{1}{r^2(t)} \quad \text{and} \quad GM = \frac{4\pi^2 a^3}{T^2}.$$

This section starts the determination of the functions $r(t)$ and $\alpha(t)$. The first thing that we'll do is place an xy-coordinate system into the orbital plane of P so that the origin is at the center of the ellipse and the focal axis coincides with the x-axis. In this xy-coordinate system, the equation of the ellipse is $\frac{x^2}{a^2} + \frac{y^2}{b^2} = 1$. Figure 5.3 shows the orbiting point-mass P in typical position at elapsed time t. As in Figure 5.1, P moves around the ellipse in a counterclockwise way. The coordinates x and y of P vary with time and are therefore functions

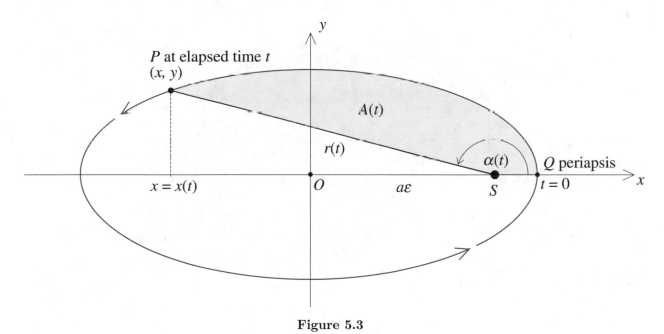

Figure 5.3

$x = x(t)$ and $y = y(t)$ of t. We know that the shaded sector of the ellipse determined by the segments SP and SQ has area $A(t) = \kappa t$. As important facilitating construction, let's surround the ellipse of Figure 5.3 with a circle of radius a and center the origin O. See Figure 5.4. The equation of the circle is $x^2 + y^2 = a^2$. It is of historical interest to note that when Kepler first studied these matters, he also surrounded the ellipse with such a circle. The point P_0 in the figure is obtained by projecting P vertically onto the circle. Its coordinates are $x = x(t)$ and $y_0 = y_0(t)$. Let $\beta = \beta(t)$ be the angle in radians determined by the points P_0, O, and Q. The area function $B(t)$ is defined by the shaded region in Figure 5.4. The formula for the area of a circular sector tells us that $B(t) = \frac{1}{2}a^2\beta(t)$. Observe that each completed orbit adds 2π to the angles α and β, and $ab\pi$ and πa^2, respectively, to the areas A and B. When it is of interest to understand the motion of P before time $t = 0$, the position of P is assigned a negative time t, where $|t|$ is the time for P to reach the periapsis at $t = 0$. The angles $\alpha(t)$ and $\beta(t)$ are measured counterclockwise when $t > 0$ and clockwise when $t < 0$.

So $\alpha(t)$ and $\beta(t)$ are positive for $t > 0$ and negative for $t < 0$. (Refer to Chapter 3A in this regard.) For $t < 0$, $A(t)$ and $B(t)$ are understood to be the negatives of the areas that the segments SP and OP_0 trace out during the motion of P from time t to $t = 0$.

We will assume that P moves smoothly along its orbit. There are no fits, stops, and starts. So $r = r(t), \alpha = \alpha(t), x = x(t), y = y(t), y_0 = y_0(t)$, and $\beta = \beta(t)$ vary smoothly and are differentiable functions of t.

The determination of the position of P and, in particular, the functions $r(t)$ and $\alpha(t)$ proceeds in three steps. Step one expresses the functions $r(t)$ and $\alpha(t)$ in terms of $\beta(t)$. Step two establishes an equation that links $\beta(t)$ and t. For a given elapsed time t, step three solves this equation for $\beta(t)$. Since t determines $\beta(t)$, and $\beta(t)$ in turn determines $r(t)$ and $\alpha(t)$, this combination of steps determines both $r(t)$ and $\alpha(t)$ in terms of t. The angle $\beta(t)$ is pivotal to the solution.

Steps one and two rely on a careful analysis of Figure 5.4. The initial focus of our discussion is on the situation depicted in the figure where P is in its first orbit. This implies in particular that t

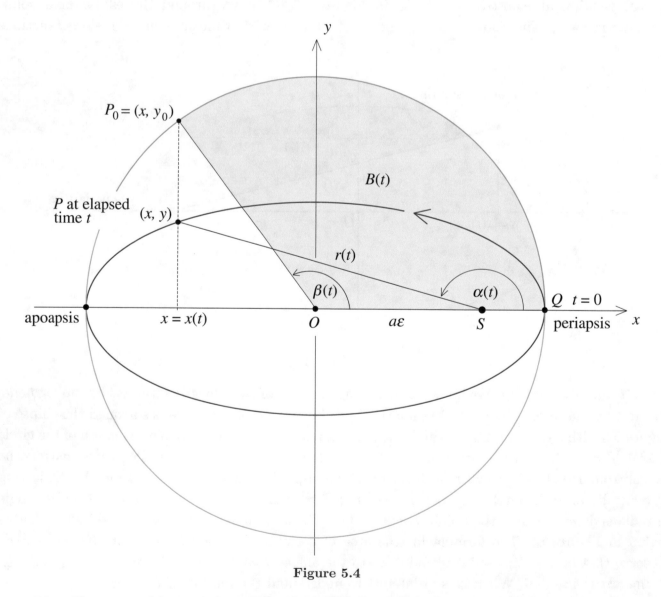

Figure 5.4

is positive. However, with routine modifications this discussion also applies to later orbits and also to the situation with negative time t.

Step one begins by relating the ellipse to the circle. Solving $\frac{x^2}{a^2} + \frac{y^2}{b^2} = 1$ for y, we get $\frac{y^2}{b^2} = 1 - \frac{x^2}{a^2} = \frac{a^2 - x^2}{a^2}$. So $y^2 = \frac{b^2}{a^2}(a^2 - x^2)$ and hence $y = \pm\frac{b}{a}\sqrt{a^2 - x^2}$. Since $x^2 + y_0^2 = a^2$, we get $y_0 = \pm\sqrt{a^2 - x^2}$. A look at Figure 5.4 tells us that P and P_0 are always on the same side of the x-axis. In other words, y and the corresponding y_0 always have the same sign. Therefore

$$y = \tfrac{b}{a}\, y_0.$$

It follows from observations about the sine and cosine in Chapter 3C that the x- and y-coordinates of the point P_0 are $x(t) = a\cos\beta(t)$ and $y_0(t) = a\sin\beta(t)$, respectively. Thus

$$\boxed{x(t) = a\cos\beta(t) \quad \text{and} \quad y(t) = b\sin\beta(t)}$$

Return to Figure 5.4. Since $x(t)$ is negative, the distance OS between the center and the focus of the ellipse is equal to $c = a\varepsilon$, and $a^2 = b^2 + c^2$, we get by applying the Pythagorean theorem, that

$$r(t)^2 = (a\varepsilon - x(t))^2 + y(t)^2 = (a\varepsilon)^2 - 2a\varepsilon x(t) + x(t)^2 + y(t)^2 = (a\varepsilon)^2 - 2a\varepsilon x(t) + x(t)^2 + \tfrac{b^2}{a^2}y_0(t)^2$$

$$= (a\varepsilon)^2 - 2a\varepsilon x(t) + x(t)^2 + \tfrac{b^2}{a^2}(a^2 - x(t)^2) = (a\varepsilon)^2 + b^2 - 2a\varepsilon x(t) + x(t)^2 - \tfrac{b^2}{a^2}x(t)^2$$

$$= a^2 - 2a\varepsilon x(t) + x(t)^2 - \tfrac{a^2 - (a\varepsilon)^2}{a^2}x(t)^2 = a^2 - 2a\varepsilon x(t) + x(t)^2 - x(t)^2 + \tfrac{(a\varepsilon)^2}{a^2}x(t)^2$$

$$= a^2 - 2a\varepsilon x(t) + \varepsilon^2 x(t)^2 = (a - \varepsilon x(t))^2.$$

Because $a \geq x(t) \geq \varepsilon x(t)$, $a - \varepsilon x(t) \geq 0$. Since $r(t) \geq 0$ and $r(t)^2 = (a - \varepsilon x(t))^2$, it follows that $r(t) = a - \varepsilon x(t)$. The substitution $x(t) = a\cos\beta(t)$ provides the equality

$$\boxed{r(t) = a\big(1 - \varepsilon\cos\beta(t)\big)}$$

As an aside, we point out that $r(t)$ is also determined by $\alpha(t)$. To see this, refer to part A of Chapter 3D for the fact that the polar equation $r = \frac{d}{1 + \varepsilon\cos\theta}$ represents an ellipse with focal point the polar origin, eccentricity ε, and semimajor axis $a = \frac{d}{(1 - \varepsilon^2)}$. This tells us that $r(t) = \frac{a(1 - \varepsilon^2)}{1 + \varepsilon\cos\alpha(t)}$. By Problem 5.6 this equality can also be derived quickly from within our current context.

The link between $\alpha(t)$ and $\beta(t)$ relies on Figure 5.4 and basic properties of the cosine. Note that

$$\cos\alpha(t) = -\cos(\pi - \alpha(t)) = -\tfrac{a\varepsilon - x(t)}{r(t)} = \tfrac{x(t) - a\varepsilon}{r(t)} = \tfrac{a\cos\beta(t) - a\varepsilon}{a(1 - \varepsilon\cos\beta(t))} = \tfrac{\cos\beta(t) - \varepsilon}{1 - \varepsilon\cos\beta(t)} \ .$$

By a double application of the standard trig identity

$$\tan^2\tfrac{\theta}{2} = \frac{1 - \cos\theta}{1 + \cos\theta}$$

(it is gotten by combining the half-angle formulas $\sin^2\frac{\theta}{2} = 1 - \cos\theta$ and $\cos^2\frac{\theta}{2} = 1 + \cos\theta$), we get

$$\tan^2\tfrac{\alpha(t)}{2} = \frac{1 - \cos\alpha(t)}{1 + \cos\alpha(t)} = \frac{1 - \frac{\cos\beta(t) - \varepsilon}{1 - \varepsilon\cos\beta(t)}}{1 + \frac{\cos\beta(t) - \varepsilon}{1 - \varepsilon\cos\beta(t)}} = \frac{\frac{1 - \varepsilon\cos\beta(t) - \cos\beta(t) + \varepsilon}{1 - \varepsilon\cos\beta(t)}}{\frac{1 - \varepsilon\cos\beta(t) + \cos\beta(t) - \varepsilon}{1 - \varepsilon\cos\beta(t)}}$$

$$= \frac{(1 + \varepsilon) - (1 + \varepsilon)\cos\beta(t)}{(1 - \varepsilon) + (1 - \varepsilon)\cos\beta(t)} = \frac{(1 + \varepsilon)(1 - \cos\beta(t))}{(1 - \varepsilon)(1 + \cos\beta(t))} = \left(\tfrac{1 + \varepsilon}{1 - \varepsilon}\right)\tan^2\tfrac{\beta(t)}{2} \ .$$

Starting at $t = 0$, follow the motion of P around the ellipse. When $\alpha(t) = 0, \pi, 2\pi, 3\pi, 4\pi, \ldots$, then $\beta(t) = 0, \pi, 2\pi, 3\pi, 4\pi, \ldots$, and when $\alpha(t)$ lies in one of the intervals $(0, \pi), (\pi, 2\pi), (2\pi, 3\pi), \ldots$, then $\beta(t)$ lies in the same interval. So if $\frac{\alpha(t)}{2}$ lies in one of the intervals $(0, \frac{\pi}{2}), (\frac{\pi}{2}, \frac{3\pi}{2}), (\frac{3\pi}{2}, 2\pi), (2\pi, \frac{5\pi}{2}), \ldots$, then $\frac{\beta(t)}{2}$ lies in this interval as well. It follows from the graph of the tangent that $\tan \frac{\alpha(t)}{2}$ and $\tan \frac{\beta(t)}{2}$ are either both positive or both negative. Since $\tan^2 \frac{\alpha(t)}{2} = \left(\frac{1+\varepsilon}{1-\varepsilon}\right) \tan^2 \frac{\beta(t)}{2}$, we can conclude that

$$\boxed{\tan \tfrac{\alpha(t)}{2} = \sqrt{\tfrac{1+\varepsilon}{1-\varepsilon}} \, \tan \tfrac{\beta(t)}{2}}$$

When $\alpha(t)$ and $\beta(t)$ are multiples of π, then neither $\tan \frac{\alpha(t)}{2}$ nor $\tan \frac{\beta(t)}{2}$ is defined. But in this case, $\alpha(t)$ and $\beta(t)$ are the same multiple of π. So $\alpha(t) = \beta(t)$. The equation derived above is *Gauss's equation*. It is named after its discoverer, the great German mathematician-astronomer Carl Friedrich Gauss (1777–1855) who had an exceptional influence on many fields of mathematics and science and is often ranked along with Archimedes and Newton as one of history's three most brilliant mathematicians. We already encountered this genius in Chapter 1F in connection with the discovery of the asteroid Ceres.

Recall that T is the period of the orbit of P. Since the elapsed times from $t = 0$ to $t = \frac{T}{2}$ and from $t = \frac{T}{2}$ to $t = T$ are the same, it follows that $A(\frac{T}{2}) + A(\frac{T}{2}) = A(T) = ab\pi$, so that $A(\frac{T}{2}) = \frac{ab\pi}{2}$. This means that P is at its apoapsis position when $t = \frac{T}{2}$, and at times $t = \frac{T}{2} + T, \frac{T}{2} + 2T, \ldots$ as well. Let k_t be the number of complete orbits that P has traced out during elapsed time $t \geq 0$ and observe that $t = t_1 + k_t T$, where $0 \leq t_1 < T$. Suppose that $0 < t_1 < \frac{T}{2}$. Observe that in this case, $\alpha(t) = \varphi(t) + 2\pi k_t$ for an angle $\varphi(t)$ with $0 < \varphi(t) < \pi$. Suppose that $\frac{T}{2} < t_1 < T$. In this case, $\pi < \alpha(t) - 2\pi k_t < 2\pi$. So $-\pi < \alpha(t) - 2\pi k_t - 2\pi < 0$, and hence $-\pi < \alpha(t) - 2\pi(k_t + 1) < 0$. With $\varphi(t) = \alpha(t) - 2\pi(k_t + 1)$, we get that $\alpha(t) = \varphi(t) + 2\pi(k_t + 1)$, where $-\pi < \varphi(t) < 0$. Now set $n_t = k_t$ when P is in motion from periapsis to apoapsis, and $n_t = k_t + 1$ when P moves from apoapsis to periapsis. So $\alpha(t) = \varphi(t) + 2\pi n_t$ in either case. From the graph of the tangent, $\tan \frac{\alpha(t)}{2} = \tan \frac{\varphi(t)}{2}$. Since $-\frac{\pi}{2} < \frac{\varphi(t)}{2} < \frac{\pi}{2}$, we know from the definition of the inverse tangent that $\tan^{-1}(\tan \frac{\alpha(t)}{2}) = \frac{\varphi(t)}{2}$. So $\alpha(t) = 2 \tan^{-1}(\tan \frac{\alpha(t)}{2}) + 2\pi n_t$. Together with Gauss's equation, this implies that

$$\boxed{\alpha(t) = 2 \tan^{-1}\left(\sqrt{\tfrac{1+\varepsilon}{1-\varepsilon}} \, \tan \tfrac{\beta(t)}{2}\right) + 2\pi n_t}$$

As noted before, when $\beta(t)$ is a multiple of π, then the term $\tan \frac{\beta(t)}{2}$ is not defined. But as we have already seen, in this case $\alpha(t) = \beta(t)$.

5C. Kepler's Equation. Kepler's equation provides the step that links t and $\beta(t)$. Its derivation uses Figure 5.4 to analyze the area $A(t)$ of Figure 5.3 for any elapsed time t.

Assume that P is in the first half of its initial orbit, and refer back to Figure 5.4. We'll start by computing the area of the circular section determined by the segments $x(t)Q$ and $x(t)P_0$. This circular section consists of the circular sector QOP_0 and the triangle $\Delta Ox(t)P_0$. Since the area of $B(t)$ is $\frac{1}{2}\beta(t)a^2$, the area of this circular section is equal to

$$\int_{x(t)}^{a} \sqrt{a^2 - x^2} \, dx = \tfrac{1}{2}\beta(t)a^2 - \tfrac{1}{2}x(t)y_0(t).$$

Notice that this holds for $x(t)$ negative (as in Figure 5.4) or positive. By multiplying through by $\frac{b}{a}$, we get that

$$\int_{x(t)}^{a} \frac{b}{a}\sqrt{a^2 - x^2}\ dx = \tfrac{1}{2}\beta(t)ab - \tfrac{1}{2}x(t)y(t)$$

is the area of the elliptical section of Figure 5.3 determined by the segments $x(t)Q$ and $x(t)P$. After subtracting the area $\frac{1}{2}(a\varepsilon - x(t))y(t)$ of the triangle $\Delta Sx(t)P$ from this elliptical section, we get that $A(t) = \frac{1}{2}\beta(t)ab - \frac{1}{2}x(t)y(t) - \frac{1}{2}(a\varepsilon - x(t))y(t) = \frac{1}{2}\beta(t)ab - \frac{1}{2}a\varepsilon y(t)$. Because $y(t) = b\sin\beta(t)$, we can conclude that

$$A(t) = \tfrac{1}{2}ab\beta(t) - \tfrac{1}{2}\varepsilon ab\sin\beta(t).$$

(Note that if $x(t) \geq a\varepsilon$, then $\frac{1}{2}(a\varepsilon - x(t))y(t)$ is negative, so that the subtraction adds the area of $\Delta Sx(t)P$ to the elliptical section. But this is exactly what needs to be done in this case.)

We'll next verify this equality without the restriction on the position of P. Let's suppose that at time t, P is in the second half of its first orbit. Let t' be the previous moment in the orbit for which $x(t') = x(t)$. Figure 5.5 considers the positions of P at the two instants t and t'. Notice that

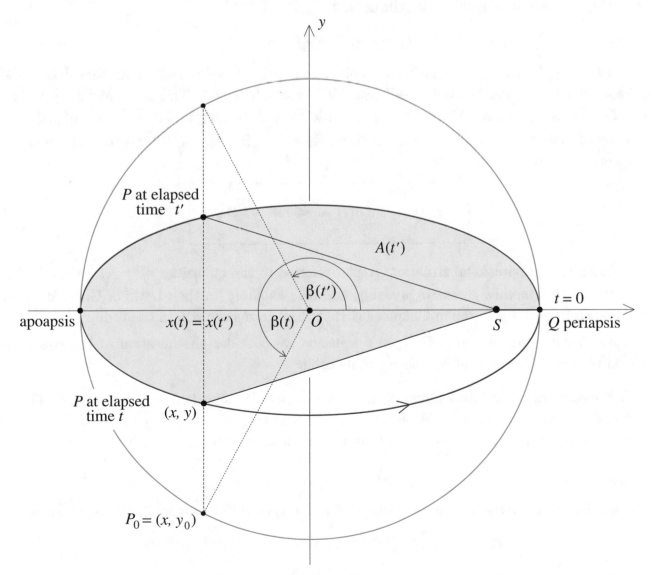

Figure 5.5

$A(t) + A(t') = ab\pi$ and that $\beta(t) + \beta(t') = 2\pi$. By applying the result already verified,

$$A(t') = \tfrac{1}{2}ab\beta(t') - \tfrac{1}{2}\varepsilon ab \sin \beta(t').$$

Since $A(t') = ab\pi - A(t)$ and $\beta(t') = 2\pi - \beta(t)$, we get by using basic properties of the sine, that

$$ab\pi - A(t) = \tfrac{1}{2}ab(2\pi - \beta(t)) - \tfrac{1}{2}\varepsilon ab \sin(2\pi - \beta(t)) = ab\pi - \tfrac{1}{2}ab\,\beta(t) + \tfrac{1}{2}\varepsilon ab \sin \beta(t).$$

Therefore as before, $A(t) = \tfrac{1}{2}ab\beta(t) - \tfrac{1}{2}\varepsilon ab \sin \beta(t)$.

Suppose next that P is in its second orbit. So the elapsed time is $t = t_1 + T$ with $0 < t_1 \leq T$, the area swept out is $A(t) = A(t_1 + T) = A(t_1) + ab\pi$, and $\beta(t) = \beta(t_1 + T) = \beta(t_1) + 2\pi$. Since P is in its first orbit at time t_1,

$$\begin{aligned} A(t) = ab\pi + A(t_1) &= ab\pi + \tfrac{1}{2}ab\,\beta(t_1) - \tfrac{1}{2}\varepsilon ab \sin \beta(t_1) \\ &= \tfrac{1}{2}ab(\beta(t_1) + 2\pi) - \tfrac{1}{2}\varepsilon ab \sin(\beta(t_1) + 2\pi) \\ &= \tfrac{1}{2}ab(\beta(t) - \tfrac{1}{2}\varepsilon ab \sin(\beta(t)). \end{aligned}$$

Doing this for every additional orbit tells us that

$$A(t) = \tfrac{1}{2}ab\beta(t) - \tfrac{1}{2}\varepsilon ab \sin \beta(t)$$

holds for any $t \geq 0$. From the definition of Kepler's constant, $\frac{A(t)}{t} = \kappa = \frac{ab\pi}{T}$. Therefore $A(t) = \frac{ab\pi \cdot t}{T}$, and hence $\tfrac{1}{2}ab\beta(t) - \tfrac{1}{2}\varepsilon ab \sin \beta(t) = \frac{ab\pi \cdot t}{T}$. So $\beta(t) - \varepsilon \sin \beta(t) = \frac{2\pi t}{T}$. This is *Kepler's equation* in celebration of its discoverer. When the centripetal force on P is the gravitational force of attraction by a body of mass M and center of mass S, then $\frac{4\pi^2}{T^2} = \frac{GM}{a^3}$. So $\frac{2\pi t}{T} = \sqrt{\frac{GM}{a^3}}\, t$, and we have arrived at the version of Kepler's equation

$$\boxed{\; \beta(t) - \varepsilon \sin \beta(t) \;=\; \frac{2\pi t}{T} \;=\; \sqrt{\frac{GM}{a^3}}\, t \;}$$

that includes this gravitational situation. Kepler referred to the quantities $\frac{2\pi t}{T} = \sqrt{\frac{GM}{a^3}}\, t$, $\beta(t)$, and $\alpha(t)$ as the *mean anomaly, eccentric anomaly,* and *true anomaly* (in their Latin or German equivalents), respectively. We will see in Chapter 6N that these terms are still in use today.

Before tackling the solution of Kepler's equation we consider the question of the speed and direction of the motion of P at any time t in its orbit.

5D. Determining Speed and Direction. Let $v(t)$ denote the speed of the point-mass P depicted in Figure 5.3 at time t in its orbit. We will use the formula $v(t) = \sqrt{x'(t)^2 + y'(t)^2}$ that was derived in Chapter 4B in combination with facts from the previous section to establish that

$$v(t) = \frac{2\pi a}{T}\sqrt{\frac{2a}{r(t)} - 1}.$$

Recall that $x(t) = a \cos \beta(t)$, $y(t) = b \sin \beta(t)$, and $r(t) = a(1 - \varepsilon \cos \beta(t))$. By the chain rule,

$$x'(t) = -(a \sin \beta(t))\beta'(t) \quad \text{and} \quad y'(t) = (b \cos \beta(t))\beta'(t),$$

so that

$$v(t)^2 = x'(t)^2 + y'(t)^2 = \left[a^2\sin^2\beta(t) + b^2\cos^2\beta(t)\right]\beta'(t)^2$$
$$= \left[a^2\sin^2\beta(t) + (a^2 - (a\varepsilon)^2)\cos^2\beta(t)\right]\beta'(t)^2 \text{ (refer to Figure 5.2)}$$
$$= \left[a^2 - (a\varepsilon)^2\cos^2\beta(t)\right]\beta'(t)^2 = \left[a^2 - a^2\varepsilon^2\cos^2\beta(t)\right]\beta'(t)^2$$
$$= a^2\left[1 - \varepsilon^2\cos^2\beta(t)\right]\beta'(t)^2 = a^2\left[\left(1 + \varepsilon\cos\beta(t)\right)\left(1 - \varepsilon\cos\beta(t)\right)\right]\beta'(t)^2.$$

Differentiate Kepler's equation $\beta(t) - \varepsilon\sin\beta(t) = \frac{2\pi t}{T}$ to get $\beta'(t) - \varepsilon\cos\beta(t)\,\beta'(t) = \frac{2\pi}{T}$. So $\beta'(t) = \frac{2\pi}{T(1-\varepsilon\cos\beta(t))}$ and hence $\beta'(t)^2 = \frac{4\pi^2}{T^2(1-\varepsilon\cos\beta(t))^2)}$. Substitute this into the expression for $v(t)^2$ just derived and cancel the term $1 - \varepsilon\cos\beta(t)$ to obtain

$$v(t)^2 = \frac{4\pi^2 a^2}{T^2} \cdot \frac{1+\varepsilon\cos\beta(t)}{1-\varepsilon\cos\beta(t)}.$$

Because $r(t) = a(1 - \varepsilon\cos\beta(t))$, we get $1 - \varepsilon\cos\beta(t) = \frac{r(t)}{a}$ and hence that $\varepsilon\cos\beta(t) = 1 - \frac{r(t)}{a}$. After two more substitutions,

$$v(t)^2 = \frac{4\pi^2 a^2}{T^2} \cdot \frac{1+(1-\frac{r(t)}{a})}{\frac{r(t)}{a}} = \frac{4\pi^2 a^2}{T^2}\left(2 - \frac{r(t)}{a}\right)\frac{a}{r(t)} = \frac{4\pi^2 a^2}{T^2}\left(\frac{2a}{r(t)} - 1\right).$$

Taking square roots finishes the verification of the formula $v(t) = \frac{2\pi a}{T}\sqrt{\frac{2a}{r(t)} - 1}$. When the centripetal force on P is the gravitational force of attraction by a body of mass M (and center of mass S), then $\frac{4\pi^2 a^2}{T^2} = \frac{GM}{a}$. So $\frac{2\pi a}{T} = \sqrt{\frac{GM}{a}}$, and therefore $v(t) = \sqrt{\frac{GM}{a}}\sqrt{\frac{2a}{r(t)} - 1} = \sqrt{GM}\sqrt{\frac{2}{r(t)} - \frac{1}{a}}$. Therefore the speed $v(t)$ of P at any time t in its orbit is equal to

$$\boxed{v(t) = \frac{2\pi a}{T}\sqrt{\frac{2a}{r(t)} - 1} = \sqrt{GM}\sqrt{\frac{2}{r(t)} - \frac{1}{a}}}$$

Example 5.1. Show that $v_{\max} = \frac{2\pi a}{T}\sqrt{\frac{1+\varepsilon}{1-\varepsilon}} = \sqrt{\frac{GM(1+\varepsilon)}{a(1-\varepsilon)}}$ and $v_{\min} = \frac{2\pi a}{T}\sqrt{\frac{1-\varepsilon}{1+\varepsilon}} = \sqrt{\frac{GM(1-\varepsilon)}{a(1+\varepsilon)}}$ are the maximum and minimum speeds of P in its orbit. The paragraph *About Speeds of Objects in the Solar System* of the Problems and Discussions section of Chapter 1 gave an elementary verification of these formulas.

It remains to study the direction of the motion of the point-mass P in its elliptical orbit. Figure 5.6 shows P in two typical positions. Let $\gamma(t)$ be the angle between the tangent to the orbit at P and the segment from P to S. The angle $\gamma(t)$ is measured counterclockwise from the tangent in the direction of the motion to the segment. So $\gamma(t)$ is always positive. Notice that at periapsis and apoapsis, $\gamma(t) = \frac{\pi}{2}$. We'll now assume that P is neither at periapsis nor apoapsis and focus on either of the two situations of the figure.

We'll let $h(t)$ be the length of the perpendicular segment from S to the tangent at P. Regard P and t to be fixed and let an additional short time Δt elapse. Let the point-mass be in position P' at time $t + \Delta t$. The area swept out by SP during the motion from P to P' is $A(t + \Delta t) - A(t)$. This area is shaded in the figure. The point P'' is the intersection of the continuation of the segment SP' with the tangent at P. Let d be the distance from P to P''. With t fixed, d is a function $d = d(\Delta t)$ of Δt. Since Δt is small, P' is close to P, so that

$$d(\Delta t) \approx \text{arc } PP' \quad \text{and} \quad A(t + \Delta t) - A(t) \approx \text{area } \Delta PP''S = \tfrac{1}{2}d(\Delta t) \cdot h(t).$$

It is apparent from the figure that the smaller the Δt is, the closer the distances PP' and PP'' are

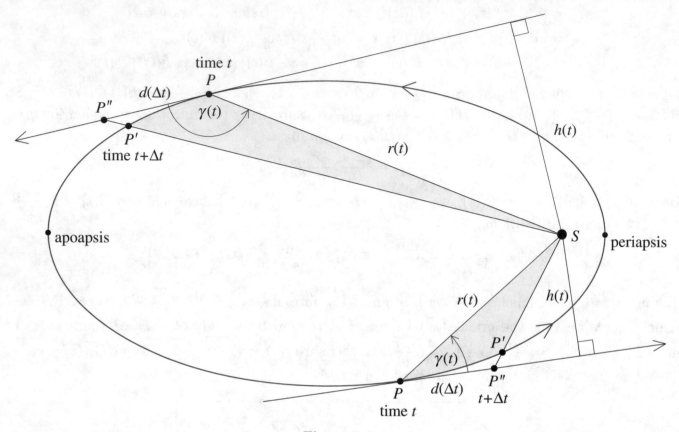

Figure 5.6

to each other, and the tighter the above approximations are. The average speed of the point during its motion from P to P' is $\frac{\text{arc } PP'}{\Delta t}$, so that the speed $v(t)$ at P is

$$v(t) = \lim_{\Delta t \to 0} \frac{\text{arc } PP'}{\Delta t} = \lim_{\Delta t \to 0} \frac{d(\Delta t)}{\Delta t}.$$

Since $\kappa = \frac{A(t+\Delta t)-A(t)}{\Delta t}$ for any Δt, we now get

$$\kappa = \lim_{\Delta t \to 0} \frac{A(t + \Delta t) - A(t)}{\Delta t} \approx \lim_{\Delta t \to 0} \frac{1}{2}\frac{d(\Delta t)}{\Delta t} \cdot h(t) = \tfrac{1}{2}v(t)\,h(t).$$

From the figure, the definition of $\gamma(t)$, and the fact that $\sin(\pi - \gamma(t)) = \sin\gamma(t)$, we see that $\sin\gamma(t) = \frac{h(t)}{r(t)}$. Therefore $\kappa = \tfrac{1}{2}v(t)r(t)\sin\gamma(t)$ and $\sin\gamma(t) = \frac{2\kappa}{r(t)v(t)}$. The equality $v(t) = \frac{2\pi a}{T}\sqrt{\frac{2a}{r(t)} - 1}$ implies that $r(t)v(t) = \frac{2\pi a r(t)}{T}\sqrt{\frac{2a}{r(t)} - 1} = \frac{2\pi a}{T}\sqrt{r(t)(2a - r(t))}$. From Figure 5.2, $b^2 = a^2 - (a\varepsilon)^2 = a^2(1 - \varepsilon^2)$, and hence $b = a\sqrt{1 - \varepsilon^2}$. Therefore $2\kappa = \frac{2ab\pi}{T} = (\frac{2a\pi}{T})b = (\frac{2a\pi}{T})(a\sqrt{1 - \varepsilon^2})$. By substituting into $\sin\gamma(t) = \frac{2\kappa}{r(t)v(t)}$, finally

$$\sin\gamma(t) = \frac{a\sqrt{1-\varepsilon^2}}{\sqrt{r(t)(2a-r(t))}}$$

A study of Figure 5.6 tells us that during the motion of P from apoapsis to periapsis $0 < \gamma(t) \leq \frac{\pi}{2}$ and during the motion of P from periapsis to apoapsis, $\frac{\pi}{2} \leq \gamma(t) < \pi$. Applying $\sin^{-1}$ to both sides of the equation for $\sin\gamma(t)$ and recalling that the inverse sine of any number between -1 and 1 needs to lie between $-\frac{\pi}{2}$ and $\frac{\pi}{2}$, we get the formulas

$$\gamma(t) = \sin^{-1}\left(\frac{a\sqrt{1-\varepsilon^2}}{\sqrt{r(t)(2a-r(t))}}\right) \quad \text{or} \quad \gamma(t) = \pi - \sin^{-1}\left(\frac{a\sqrt{1-\varepsilon^2}}{\sqrt{r(t)(2a-r(t))}}\right)$$

where the first equality applies to the motion of P from apoapsis to periapsis and the second equality to the motion of P from periapsis to apoapsis.

Example 5.2. Use the fact that $\sin^{-1}(1) = \frac{\pi}{2}$ to check that these formulas are also valid for P at periapsis or apoapsis. Suppose next that the orbit is not a circle and that $\gamma(t) = \frac{\pi}{2}$. Since $\sin\frac{\pi}{2} = 1$, it follows that $\frac{a\sqrt{1-\varepsilon^2}}{\sqrt{r(t)(2a-r(t))}} = 1$. Use the quadratic formula to show that $r(t) = a \pm a\varepsilon$. Conclude that $\gamma(t) = \frac{\pi}{2}$ occurs only at periapsis and apoapsis.

Example 5.3. Since the inverse sine is an increasing function, $\sin^{-1}\left(\frac{a\sqrt{1-\varepsilon^2}}{\sqrt{r(t)(2a-r(t))}}\right)$ reaches its minimum value when $\frac{a\sqrt{1-\varepsilon^2}}{\sqrt{r(t)(2a-r(t))}}$ is at its minimum, and hence when $\sqrt{r(t)(2a-r(t))}$ is at its maximum. Study the parabola $y = x(2a-x) = -x^2 + 2ax$ and conclude that $\sin^{-1}\left(\frac{a\sqrt{1-\varepsilon^2}}{\sqrt{r(t)(2a-r(t))}}\right)$ reaches its minimum value when $r(t) = a$. Show that this minimum value is $\sin^{-1}\sqrt{1-\varepsilon^2}$. Conclude that $\gamma(t)$ attains its minimum value $\sin^{-1}\sqrt{1-\varepsilon^2}$ when P is on approach to periapsis at a distance a from S and that $\gamma(t)$ reaches its maximum value $\pi - \sin^{-1}\sqrt{1-\varepsilon^2}$ when P is on approach to apoapsis at a distance of a from S.

5E. Solving Kepler's Equation by Successive Approximations. Consider the function $f(x) = x - \varepsilon\sin x$. This function is continuous because both x and $\varepsilon\sin x$ are. The values of $f(x)$ can be made arbitrarily large both positively and negatively by choosing x appropriately large and positive or large and negative, due to the fact that $|\varepsilon\sin x| \leq \varepsilon < 1$. Since $f(x)$ is continuous, it follows that for any real number y there is an x such that $f(x) = y$. Since $f'(x) = 1 - \varepsilon\cos x > 0$, $f(x) = x - \varepsilon\sin x$ is an increasing function, and we can conclude that for any given t, Kepler's equation has a unique solution $\beta(t)$. The trick is to find it, or at least to provide an estimate for it. This is the goal of step three. It applies a method of successive approximations to tell us for any given elapsed time t what the corresponding $\beta(t)$ is that satisfies Kepler's equation.

The successive approximations strategy that solves Kepler's equation relies on the inequality $|\sin x_1 - \sin x_2| \leq |x_1 - x_2|$ for any real numbers x_1 and x_2. To verify it we'll show that

$$|\sin x_1 - \sin x_2| < |x_1 - x_2| \quad \text{whenever } x_1 \neq x_2.$$

Consider the functions $g(x) = x - \sin x$ and $h(x) = x + \sin x$. Their derivatives are $g'(x) = 1 - \cos x$ and $h'(x) = 1 + \cos x$. Because $1 > \cos x$ except for $x = 0, \pm2\pi, \pm4\pi, \ldots$ when $\cos x = 1$, it follows that $g'(x) > 0$, except for the isolated points when $g'(0) = 0$. So $y = g(x)$ is an increasing function of x and hence $g(x_1) > g(x_2)$ whenever $x_1 > x_2$. A similar argument shows that $h(x_1) > h(x_2)$ whenever $x_1 > x_2$. To verify that

$$|\sin x_1 - \sin x_2| < |x_1 - x_2|$$

for any x_1 and x_2 with $x_1 \neq x_2$, we may take $x_1 > x_2$ (or else we can work with $x_2 > x_1$). It follows from what was established about $g(x)$ and $h(x)$ that

$$x_1 - \sin x_1 > x_2 - \sin x_2 \quad \text{and} \quad x_1 + \sin x_1 > x_2 + \sin x_2.$$

So

$$x_1 - x_2 > \sin x_1 - \sin x_2 \quad \text{and} \quad x_1 - x_2 > \sin x_2 - \sin x_1.$$

Since $|\sin x_1 - \sin x_2|$ is equal to either $\sin x_1 - \sin x_2$ or $\sin x_2 - \sin x_1$, the verification is complete.

We now solve $\beta(t) - \varepsilon \sin \beta(t) = \frac{2\pi t}{T} = \sqrt{\frac{GM}{a^3}} t$ for $\beta(t)$ by successive approximations. The application of any such method to the solution of an equation starts with an educated guess, or an informed initial stab, at a solution. This initial educated guess is then refined step by step to any desired or required degree of accuracy.

1. The first stab at $\beta(t)$ is $\beta_1 = \frac{2\pi t}{T}$. By Kepler's equation, $|\beta(t) - \beta_1| = \left|\beta(t) - \frac{2\pi t}{T}\right| = |\varepsilon \sin \beta(t)| \leq \varepsilon$. Because $\varepsilon < 1$ for any ellipse (in fact ε is usually much smaller, see Table 5.1), β_1 approximates $\beta(t)$. Notice that the first approximation β_1 is nothing but the mean anomaly $\frac{2\pi t}{T}$.

2. The approximation step: after the angle β_i has been determined (in radians), the next angle β_{i+1} is given by $\beta_{i+1} = \frac{2\pi t}{T} + \varepsilon \sin \beta_i = \beta_1 + \varepsilon \sin \beta_i$ (in radians).

Applying the approximation step (2) to $\beta_1 = \frac{2\pi t}{T}$ gives the new angle $\beta_2 = \frac{2\pi t}{T} + \varepsilon \sin \beta_1$. Repeating this with β_2, we get $\beta_3 = \frac{2\pi t}{T} + \varepsilon \sin \beta_2$. Doing this again and again, we get $\beta_4 = \frac{2\pi t}{T} + \varepsilon \sin \beta_3$, $\beta_5 = \frac{2\pi t}{T} + \varepsilon \sin \beta_4, \dots$, and $\beta_i = \frac{2\pi t}{T} + \varepsilon \sin \beta_{i-1}, \dots$. Now the question is: does the sequence

$$\beta_1, \ \beta_2, \ \beta_3, \dots, \ \beta_i, \dots$$

close in on the solution $\beta(t)$ of $\beta(t) - \varepsilon \sin \beta(t) = \frac{2\pi t}{T}$? This is indeed the case. Since $\beta(t) = \frac{2\pi t}{T} + \varepsilon \sin \beta(t)$ by Kepler's equation, it follows that

$$\beta(t) - \beta_2 = \left(\tfrac{2\pi t}{T} + \varepsilon \sin \beta(t)\right) - \left(\tfrac{2\pi t}{T} + \varepsilon \sin \beta_1\right) = \varepsilon(\sin \beta(t) - \sin \beta_1).$$

Therefore, using an inequality established earlier,

$$|\beta(t) - \beta_2| = \varepsilon|\sin \beta(t) - \sin \beta_1| \leq \varepsilon|\beta(t) - \beta_1| \leq \varepsilon^2.$$

In the same way,

$$\beta(t) - \beta_3 = \left(\tfrac{2\pi t}{T} + \varepsilon \sin \beta(t)\right) - \left(\tfrac{2\pi t}{T} + \varepsilon \sin \beta_2\right) = \varepsilon(\sin \beta(t) - \sin \beta_2),$$

so that

$$|\beta(t) - \beta_3| = \varepsilon|\sin \beta(t) - \sin \beta_2| \leq \varepsilon|\beta(t) - \beta_2| \leq \varepsilon^3.$$

Repeating this computation again and again shows that

$$|\beta(t) - \beta_4| \le \varepsilon^4, \quad |\beta(t) - \beta_5| \le \varepsilon^5, \ldots, \quad |\beta(t) - \beta_i| \le \varepsilon^i, \ldots .$$

Since $\varepsilon < 1$, the powers $\varepsilon^2, \varepsilon^3, \varepsilon^4, \ldots$ close in on zero. Therefore the distances $|\beta(t) - \beta_1|, |\beta(t) - \beta_2|$, $|\beta(t) - \beta_3|, \ldots$ between $\beta_1, \beta_2, \beta_3, \ldots$ and $\beta(t)$ close in on zero. So the numbers $\beta_1, \beta_2, \beta_3, \ldots$ close in on $\beta(t)$ as required. Since $0 \le \varepsilon < 1$ for any ellipse, this successive approximation process will always converge to $\beta(t)$.

The solution of the problem of determining the position of the point-mass P is complete: take the given elapsed time t, and solve Kepler's equation for $\beta(t)$ by the method just described. The closer the eccentricity ε is to 0, the more rapid the convergence of the sequence $\varepsilon, \varepsilon^2, \varepsilon^3, \varepsilon^4, \ldots$ to 0, and hence the more rapid the convergence of the sequence of approximations $\beta, \beta_2, \beta_3, \beta_4, \ldots$ to $\beta(t)$. Then substitute $\beta(t)$ into the equations derived in section 5B to get the corresponding $r(t)$ and $\alpha(t)$. With this in hand, turn to section 5D to compute the speed and direction of the motion of P. In this way, given the elapsed time t of the orbiting point-mass P from periapsis at $t = 0$, it is possible to determine its orbital data in terms of t and the orbital constants. The reliance on approximation methods for the computation of $\beta(t)$ raises the question as to whether there is an explicit function $\beta = f(t)$ that provides the angle β for a given time t. Such a function was defined in terms of power series by Friedrich Bessel, the astronomer who was first able to detect the stellar parallax for some near stars. (See *Parallax and Distances to Stars* of the Problems and Discussions section of Chapter 1.)

Table 5.1 provides data we need for the examples that follow next and later in the chapter. They are taken from Tables 2.1 and 2.4 in Chapter 2, but also include data for Pluto and the comet Halley.

Table 5.1. Information from NASA and JPL websites. Each orbital period in this table is the perihelion period with the year defined as 365.25 days. (See Chapter 1D.)

orbiting body	semimajor axis in km	eccentricity	orbit period in years	angle of orbit plane to Earth's	average speed in km/sec
Mercury	57,909,227	0.20563593	0.2408489	7.00°	47.362
Venus	108,209,475	0.00677672	0.6152028	3.39°	35.021
Earth	149,598,262	0.01671123	1.0000264	0.00°	29.783
Mars	227,943,824	0.0933941	1.8808645	1.85°	24.077
Jupiter	778,340,821	0.04838624	11.862757	1.31°	13.056
Saturn	1,426,666,422	0.05386179	29.447762	2.49°	9.639
Uranus	2,870,658,186	0.04725744	84.017599	0.77°	6.873
Neptune	4,498,396,441	0.00859048	164.79280	1.77°	5.435
Pluto	5,906,440,628	0.2488273	247.92287	17.14°	4.669
Halley	2,667,950,000	0.9671429	75.32	162.26°	?

Example 5.4. The eccentricities of all eight planets are small. Mercury's eccentricity $\varepsilon < 0.2057$ is the largest. But even in this case, the process described converges very quickly. Squaring $\varepsilon \approx 0.2057$ four consecutive times, we get

$$\varepsilon^2,\ \varepsilon^4,\ \varepsilon^8,\ \varepsilon^{16} < 1 \times 10^{-11}.$$

Since $|\beta(t) - \beta_{16}| < \varepsilon^{16}$, the sixteenth iteration of the approximation process provides an approximation of $\beta(t)$ that is accurate to within a tiny fraction of a radian.

Example 5.5. For a relatively large $\varepsilon < 1$, achieving good accuracy for a particular t will usually require many steps. For Halley's comet, $\varepsilon > 0.967$. By repeatedly squaring 0.967 we get,

$$\varepsilon^{128} > 0.0136,\ \varepsilon^{256} > 0.00018,\ \varepsilon^{512} > 3 \times 10^{-8} \dots .$$

So for tight accuracy, several hundred iterations might be necessary. But this is hardly a problem for a computer.

It should not come as a surprise that the mathematics in steps one to three can be used to construct a computer model of the solar system. This is done in *Computer Model of Elliptical Orbits Generated by Kepler's Equations*. Go to the website

$$\text{http://learning.nd.edu/orbital/orbital-info.html}$$

and experiment with the simulations for the inner and outer planets.

Example 5.6. We'll illustrate Kepler's equation with a look at Earth's orbit. Figure 5.7 considers Earth in position E at the "top" of its orbit, exactly halfway between its perihelion and aphelion positions. How long after perihelion will the Earth arrive there? For Earth's position at E, $\beta(t) = \frac{\pi}{2}$. Kepler's equation $\beta(t) - \varepsilon \sin \beta(t) = \frac{2\pi t}{T}$ with $\varepsilon = 0.0167$ and $T = 365.26$ days (see Chapter 1G)

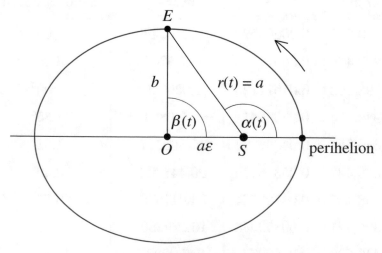

Figure 5.7

tells us that the corresponding t satisfies $t \approx \frac{365.26}{2\pi} \left(\frac{\pi}{2} - (0.0167) \cdot 1 \right) \approx 90.34$ days. So the Earth takes about 90.34 days to complete the first quarter of its orbit. We now also know that Earth takes $\frac{365.26}{2} - 90.34 \approx 92.29$ days to complete the second quarter of its orbit. This is consistent with a consequence of Kepler's equal area law, namely that Earth moves faster in its orbit when it is near perihelion than when it is farther away. We know from Figure 5.7 and Table 5.1 that the distance $r(t)$ corresponding to $\beta(t) = \frac{\pi}{2}$ is $r(t) = a \approx 149{,}598{,}000$ km. The angle $\alpha(t)$ corresponding to $\beta(t) = \frac{\pi}{2} = 90°$ is given by $\alpha(t) = 2\tan^{-1}\left(\sqrt{\frac{1+\varepsilon}{1-\varepsilon}}\,\tan\frac{\beta(t)}{2}\right) + 2n_t\pi \approx 2\tan^{-1}\left(\sqrt{\frac{1.0167}{0.9833}}\cdot 1\right) + 0 \approx 90.96°$.

The outline of the procedure that provided for the elliptical orbit of a given planet the value $\beta(t)$ for any elapsed time t, and in turn the distance $r(t)$ and the angle $\alpha(t)$, followed the analysis that Kepler carried out over 400 years ago. But the details differ. For example, instead of an approximation scheme of the sort described above, Kepler relied on his *Rudolphine Tables*. He considered the equation $\beta(t) - \varepsilon\sin\beta(t) = \frac{2\pi t}{T}$ and computed t for lots of angles $\beta(t)$ from $0°$ to $180°$ and recorded the results in the tables. To find an approximate $\beta(t)$ for a given elapsed time t, Kepler (and later astronomers) could go to the tables, locate a time close to t, and read off (and extrapolate) to find the corresponding $\beta(t)$. Example 5.6 gives a sense of the principle involved.

5F. Earth, Jupiter, and Halley. We will now illustrate the information that sections 5A through 5E provide. The examples consider the orbits of Earth, Jupiter, and Halley's comet around the Sun. We'll take the necessary data from Table 5.1 and the value $GM = 1.327124 \times 10^{11}$ km^3/s^2 for the Sun from Chapter 1H. Our computations carry an accuracy of six decimal places.

We'll begin by studying the orbiting Earth. For Earth, $a = 1.495983 \times 10^8$ km and $\varepsilon = 0.016711$. Before evaluating the distance $r(t)$, the velocity $v(t)$, and the orbital angle $\gamma(t)$ for specific times t, let's gain a sense of these values by examining their range from minimum to maximum.

From the formulas for the periapsis and apoapsis distances (see Figure 5.2), we know that

$$(1.495982 \times 10^8)(1 - 0.016712) < a(1 - \varepsilon) \le r(t) \le a(1 + \varepsilon) < (1.495984 \times 10^8)(1.016712)$$

in km and therefore that

$$1.470981 \times 10^8 < r(t) < 1.520985 \times 10^8$$

in km. By substituting the data we have into the formulas of Example 5.1 for the maximum and minimum speeds, we get the bounds

$$29.29099 < \sqrt{\frac{GM(1-\varepsilon)}{a(1+\varepsilon)}} = v_{\min} \le v(t) \le v_{\max} = \sqrt{\frac{GM(1+\varepsilon)}{a(1-\varepsilon)}} < 30.28661$$

in km/sec on the speed $v(t)$ of Earth in its orbit. Finally, turning to the orbital angle $\gamma(t)$, we find from Example 5.3 that

$$1.554084 < \sin^{-1}\sqrt{1-\varepsilon^2} = \gamma_{\min} \le \gamma(t) \le \gamma_{\max} = \pi - \sin^{-1}\sqrt{1-\varepsilon^2} < 1.587509$$

in radians for Earth in its orbit from apoapsis to periapsis and back. This is $89.04° < \gamma(t) < 90.96°$ in degrees. The Earth's near-circular orbit is the reason why $\gamma(t)$ is close to $90°$ throughout.

Data recorded by the U.S. Naval Observatory (USNO) on the website

$$\text{https://www.usno.navy.mil/USNO}$$

informs us that in 2013 perihelion occurred on Jan 01 at 11:38 pm, spring equinox on Mar 20 at 6:02 am, and summer solstice on Jun 21 at 12:04 am (all in Eastern Standard Time). It follows that in 2013, it took Earth 77.497916 days to travel from perihelion to spring equinox and 170.000046 days to move from perihelion to summer solstice. Both spring equinox and summer solstice occurred before aphelion of the year 2014 (since both times of travel are less that one-half Earth's period).

We will compute Earth's position and velocity on spring equinox of 2013 by determining $r(t), v(t)$, and $\gamma(t)$ for $t = 77.497916$ days. As first step, we'll calculate the angle $\beta(t)$ by using the approximation method of section 5E. With six-decimal-place accuracy at each step we get

$$\beta_1 = \tfrac{2\pi t}{T} \approx \tfrac{2\pi(77.497916)}{365.259636} \approx 1.333117 \,\text{radians},$$
$$\beta_2 = \beta_1 + \varepsilon \sin\beta_1 \approx 1.333117 + 0.016711(0.971887) \approx 1.349358 \,\text{radians},$$
$$\beta_3 = \beta_1 + \varepsilon \sin\beta_2 \approx 1.333117 + 0.016711(0.975583) \approx 1.349420 \,\text{radians, and}$$
$$\beta_4 = \beta_1 + \varepsilon \sin\beta_3 \approx 1.333117 + 0.016711(0.975596) \approx 1.349420 \,\text{radians}.$$

Since the process has terminated, $\beta(t) \approx \beta_4 \approx 1.349420$ radians with the required six-decimal-place accuracy. By substitution into the equations

$$r(t) = a(1 - \varepsilon \cos\beta(t)) \quad\text{and}\quad \alpha(t) = 2\tan^{-1}\!\left(\sqrt{\tfrac{1+\varepsilon}{1-\varepsilon}}\,\tan\tfrac{\beta(t)}{2}\right)$$

of section 5B (with regard to $\alpha(t)$, we'll consider Earth to be on its way from perihelion to aphelion in its initial orbit, so that $n_t = 0$), we find that

$$r(t) \approx (1.495983 \times 10^8)(1 - 0.016711(\cos 1.349420)) \approx 1.490494 \times 10^8 \,\text{km and}$$
$$\alpha(t) = 2\tan^{-1}\!\big(1.016853 \tan(0.674710)\big) \approx 2(0.682877) \approx 1.365754 \,\text{radians}.$$

So at spring equinox in 2013, $r(t)$ and $\alpha(t)$ were close to 149 million kilometers and 78.25°, respectively. Substituting into the formula $v(t) = \sqrt{\tfrac{GM}{a}}\sqrt{\tfrac{2a}{r(t)} - 1}$, we get

$$v(t) \approx \sqrt{8.871137 \times 10^2}\,\sqrt{2.007365 - 1} \approx 29.8939 \,\text{km/sec}$$

for the Earth's orbital speed. Finally, the formula $\gamma(t) = \sin^{-1}\!\left(\tfrac{a\sqrt{1-\varepsilon^2}}{\sqrt{r(t)(2a-r(t))}}\right)$ tells us that Earth's orbital angle $\gamma(t)$ was

$$\gamma(t) \approx \pi - \sin^{-1}\!\left(\tfrac{(1.495983\times10^8)\sqrt{0.999721}}{\sqrt{(1.490494\times10^8)(2(1.495983\times10^8)-(1.490494\times10^8))}}\right) \approx \pi - \sin^{-1}(0.999867) = 1.587092$$

radians or about 90.93°. After studying Example 5.3 explain why the angle $\gamma(t)$ is close to its maximum.

We turn to compute Earth's position and velocity on summer solstice of 2013 by determining $r(t), v(t)$, and $\gamma(t)$ for $t = 170.000046$ days. By the approximation method of section 5E we get

$$\beta_1 = \tfrac{2\pi t}{T} \approx \tfrac{2\pi(170.000046)}{365.259636} \approx 2.924336 \,\text{radians},$$
$$\beta_2 = \beta_1 + \varepsilon \sin\beta_1 \approx 2.924336 + 0.016711(0.215552) \approx 2.927938 \,\text{radians},$$
$$\beta_3 = \beta_1 + \varepsilon \sin\beta_2 \approx 2.924336 + 0.016711(0.212033) \approx 2.927879 \,\text{radians},$$
$$\beta_4 = \beta_1 + \varepsilon \sin\beta_3 \approx 2.924336 + 0.016711(0.212090) \approx 2.927880 \,\text{radians, and}$$
$$\beta_5 = \beta_1 + \varepsilon \sin\beta_4 \approx 2.924336 + 0.016711(0.212089) \approx 2.927880 \,\text{radians}.$$

So $\beta(t) \approx \beta_5 \approx 2.927880$ radians on summer solstice in 2013. In terms of the position and velocity of the Earth on summer solstice of 2013, check that $r(t) \approx 1.520414 \times 10^8$ km, $\alpha(t) \approx 2.931396$ radians or 167.96°, that $v(t) \approx 29.3019$ km/sec, and $\gamma(t) \approx 1.574303$ radians or very nearly 90.20°. Why was it that on summer solstice the Earth's distance $r(t)$ from the Sun was so close to its maximum, the velocity $v(t)$ so close to its minimum, and the angle $\gamma(t)$ so close to 90°? Example 5.2 provides a hint.

Example 5.7. Use the data in Table 5.1 and follow the above computations for Earth to check that Jupiter's distance from the Sun varies from about 7.41×10^8 km to about 8.16×10^8 km, that its smallest and greatest orbital speeds are $v_{\min} = 12.44$ km/sec and $v_{\max} = 13.71$ km/sec, and that the smallest angle during Jupiter's approach to the Sun is $\gamma(t) = 87.23°$. Verify that when Jupiter is 7.5×10^8 km from the Sun, it has a speed of about 13.5 km/sec. How far is Jupiter from the Sun exactly 2 years after it passes perihelion? How fast is it moving at that time?

Example 5.8. Halley's comet passed its perihelion position at $t = 0$. At the precise time t years later, it completed the first quarter of its orbit. In addition to the information in Table 5.1, refer to Figure 5.4 and use Kepler's equation to show that $t \approx 7.24$ years. Then show that Halley completed the second quarter of its orbit in about 30.42 years.

5G. Orbital Questions and Definite Integrals.[1] It is often asserted that the semimajor axis a of the elliptical orbit of a planet is the average distance of the planet from the Sun. This conclusion is usually justified by taking its maximum or aphelion distance $a(1 + \varepsilon)$ and its minimum or perihelion distance $a(1 - \varepsilon)$ and noticing that the average value of these two numbers is

$$\frac{1}{2}\big((a + a\varepsilon) + (a - a\varepsilon)\big) = a.$$

Let's think about this for a moment. Suppose you're driving along a highway from point A to point B. You start with a speed of 40 miles per hour and, say two hours later, you finish your trip with a speed of 70 miles per hour. Surely, your average speed is not necessarily $\frac{1}{2}(70 + 40) = 55$ miles per hour. But this is the essence of the above argument that a is the average value of the distance of a planet from the Sun. The problem is that the car's speed as well as the distance between the planet and the Sun are both functions that vary all along the path of travel. In both cases, it is the *average value of a function* that needs to be considered.

The definite integral provides a definition for the average value of a function. Start by considering a list of numbers, for example, 5, 3, 6, 4, 2, and 8. Their average is

$$\frac{5 + 3 + 6 + 4 + 2 + 8}{6} = \frac{28}{6} = \frac{14}{3} = 4\tfrac{2}{3}.$$

A graphical interpretation of this average is provided by Figure 5.8. The area under the graph of the function that the horizontal segments of length 1 with respective y-coordinates 5, 3, 6, 4, 2, and 8 determine is 28. So the average $4\tfrac{2}{3}$ is this area divided by its extent, namely 6, along the x-axis.

---

[1]This and the remaining sections of this chapter make use of basic methods of solving definite integrals. One of them relies on the Taylor series expression of a function, primarily the special case of the binomial series. The proofs of all the facts that play a role can be found in most any standard calculus text.

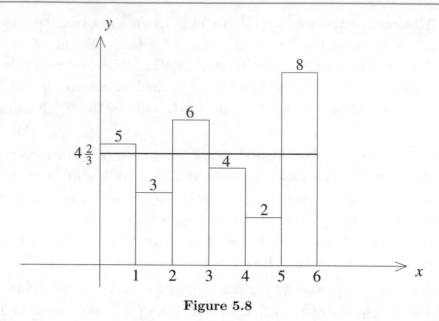

Figure 5.8

Next consider any continuous function $y = f(x)$ defined over an interval $a \leq x \leq b$ and turn to Figure 5.9. The number c in the figure is chosen to be precisely that number such that the area $c(b-a)$ of the rectangle with base $b-a$ and height c is equal to the area $\int_a^b f(x)\,dx$ under the graph of $y = f(x)$. It makes sense to say that this c is the average value of the function $y = f(x)$. So in general, we'll define *the average value of a function* $y = f(x)$ that is continuous over an interval $a \leq x \leq b$ to be

$$\frac{1}{b-a}\int_a^b f(x)\,dx.$$

Let's check this definition against ideas that we encountered in Chapter 4B. Consider a point

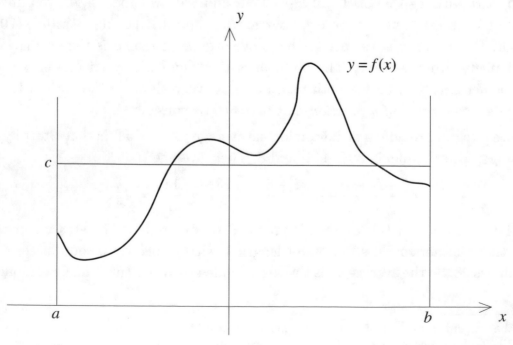

Figure 5.9

moving along a coordinate axis starting at time $t = 0$. Let $p(t)$ be its position at any time $t \geq 0$. We know that the velocity $v(t)$ and acceleration $a(t)$ of the point at any time $t \geq 0$ are given by

$$v(t) = p'(t) \text{ and } a(t) = v'(t).$$

We also know that for two instants of time t_1 and t_2 with $0 \leq t_1 < t_2$, the average velocity and the average acceleration over the time period $t_1 \leq t \leq t_2$ are

$$\frac{p(t_2) - p(t_1)}{t_2 - t_1} \quad \text{and} \quad \frac{v(t_2) - v(t_1)}{t_2 - t_1},$$

respectively. How do these averages compare to the average values

$$\frac{1}{t_2 - t_1} \int_{t_1}^{t_2} v(t)\, dt \quad \text{and} \quad \frac{1}{t_2 - t_1} \int_{t_1}^{t_2} a(t)\, dt$$

of the velocity and acceleration functions $v(t)$ and $a(t)$ over $t_1 \leq t \leq t_2$ as these were just defined? Since $p(t)$ and $v(t)$ are antiderivatives of $v(t)$ and $a(t)$, respectively, the fundamental theorem of calculus tells us that these average values are equal to

$$\frac{1}{t_2 - t_1} \cdot (p(t_2) - p(t_1)) \quad \text{and} \quad \frac{1}{t_2 - t_1} \cdot (v(t_2) - v(t_1)).$$

Therefore the two definitions of average velocity and average acceleration are the same.

We'll now turn to Figure 5.3 and the discussion in section 5B. The analysis of the varying distance r between P and S applies in particular to the situation of a planet P in orbit around the Sun S. What does the definition for the average value of a continuous function tell us about the average value of r? One challenge that this question faces is that r can be expressed as a function of several different variables. Another is that it requires a facility in dealing with definite integrals (for example the substitution method, improper integrals, and solutions that rely on the binomial series).

i.) The average value of the function $r(\beta)$. Since $r(\beta(t)) = a(1 - \varepsilon \cos \beta(t))$, it follows that $r(\beta) = a(1 - \varepsilon \cos \beta)$ is a function of the angle β. Figure 5.4 illustrates how $r(\beta)$ is determined for a given β. As β varies from 0 to 2π, the segment SP of length $r(\beta)$ traces out the entire ellipse. The average value of $r(\beta)$ over the interval $0 \leq \beta \leq 2\pi$ is

$$\frac{1}{2\pi} \int_0^{2\pi} r(\beta)\, d\beta = \frac{1}{2\pi} \int_0^{2\pi} a(1 - \varepsilon \cos \beta)\, d\beta = \frac{a}{2\pi} \left[\beta - \varepsilon \sin \beta \right]_0^{2\pi} = \frac{a}{2\pi} \cdot 2\pi = a.$$

This confirms the earlier average value of a. But the angle β does not appear to be the most meaningful variable over which to average the distance r.

The equality $r(t) = \frac{a(1 - \varepsilon^2)}{1 + \varepsilon \cos \alpha(t)}$ was established in section 5B. It follows that r is the function $r(\alpha) = \frac{a(1 - \varepsilon^2)}{1 + \varepsilon \cos \alpha}$ of the angle α.

ii.) The average value of the function $r(\alpha)$. A look at Figure 5.3 tells us that the average values of the function $r(\alpha)$ over $0 \leq \alpha \leq 2\pi$ and over $0 \leq \alpha \leq \pi$ are the same. So the average value of $r(\alpha)$ over $0 \leq \alpha \leq 2\pi$ is given by the definite integral

$$\frac{1}{\pi}\int_0^\pi r(\alpha)\, d\alpha \; = \; \frac{a(1-\varepsilon^2)}{\pi}\int_0^\pi \frac{1}{1+\varepsilon\cos\alpha}\, d\alpha.$$

The verification that this integral is equal to $b = a\sqrt{1-\varepsilon^2}$ is tricky. It requires several steps.

a. Use the substitution $u = \tan\frac{\alpha}{2}$. Consider the identities $\tan^2\frac{\alpha}{2} = \frac{1-\cos\alpha}{1+\cos\alpha}$ and $\tan\frac{\alpha}{2} = \frac{\sin\alpha}{1+\cos\alpha}$ (both are standard trig identities) and solve them for $\cos\alpha$ and $\sin\alpha$, respectively, to get $\cos\alpha = \frac{1-u^2}{1+u^2}$ and $\sin\alpha = \frac{2u}{1+u^2}$. Use these equalities to show that

$$\int \frac{1}{1+\varepsilon\cos\alpha}\, d\alpha = \int \frac{2}{u^2+1+\varepsilon(1-u^2)}\, du = \frac{1}{1+\varepsilon}\int \frac{2}{1+\left(\sqrt{\frac{1-\varepsilon}{1+\varepsilon}}u\right)^2}\, du.$$

b. Substitute $z = \sqrt{\frac{1-\varepsilon}{1+\varepsilon}}\, u$ to show that the integral of step (**a**) is equal to

$$\frac{2}{\sqrt{1-\varepsilon^2}}\int \frac{dz}{1+z^2} = \frac{2}{\sqrt{1-\varepsilon^2}}\tan^{-1}z + C = \frac{2}{\sqrt{1-\varepsilon^2}}\tan^{-1}\left(\sqrt{\frac{1-\varepsilon}{1+\varepsilon}}\tan\frac{\alpha}{2}\right) + C.$$

c. Note that the last term of (**b**) is not defined for $\alpha = \pi$. But use the strategy of improper integrals to conclude that with $0 \le \varphi < \pi$.

$$\int_0^\pi \frac{1}{1+\varepsilon\cos\alpha}\, d\alpha = \lim_{\varphi\to\pi}\int_0^\varphi \frac{1}{1+\varepsilon\cos\alpha}\, d\alpha = \lim_{\varphi\to\pi}\frac{2}{\sqrt{1-\varepsilon^2}}\tan^{-1}\left(\sqrt{\frac{1-\varepsilon}{1+\varepsilon}}\tan\frac{\varphi}{2}\right) = \frac{2}{\sqrt{1-\varepsilon^2}}\cdot\frac{\pi}{2} = \frac{\pi}{\sqrt{1-\varepsilon^2}}.$$

d. Finish the computation that shows that the average value of $r(\alpha)$ over $0 \le \alpha \le 2\pi$ is the semiminor axis $b = a\sqrt{1-\varepsilon^2}$.

Note, as an aside, that the substitution $u = \tan\frac{x}{2}$ used above is widely applicable as a method that transforms integrals involving trig functions into integrals involving polynomials.

We turn, finally, to the average of r as function of time t. The relevant interval of time is $0 \le t \le T$, where T is the period of the orbit.

iii.) The average value of the function $r(t)$. This average value is given by the integral $\frac{1}{T}\int_0^T r(t)\, dt$.

a. Use the derivative of Kepler's equation $\beta(t) - \varepsilon\sin\beta(t) = \frac{2\pi t}{T}$ as well as the equality $r(t) = a(1 - \varepsilon\cos\beta(t))$ to show that $\frac{T}{2a\pi}\beta'(t)r(t) = 1$.

b. Show that the average value of $r(t)$ over the interval $0 \le t \le T$ is

$$\frac{1}{T}\int_0^T r(t)\, dt = \frac{1}{2a\pi}\int_0^T r(t)^2\beta'(t)\, dt = \frac{a}{2\pi}\int_0^T \left(1 - \varepsilon\cos\beta(t)\right)^2\beta'(t)\, dt.$$

c. Use the substitution $u = \beta(t)$ and the formula $\cos^2 u = \frac{1}{2}(1 + \cos 2u)$ to check that this last integral is equal to

$$\frac{a}{2\pi}\left[\int_0^{2\pi} du \; - \; 2\varepsilon\int_0^{2\pi}\cos u\, du \; + \; \varepsilon^2\int_0^{2\pi}\cos^2 u\, du\right] = a\left(1 + \tfrac{1}{2}\varepsilon^2\right).$$

It follows that $\frac{1}{T}\int_0^T r(t)\, dt = a\left(1 + \tfrac{1}{2}\varepsilon^2\right)$.

So the average value of $r(t)$ as function of t over the time interval $0 \le t \le T$ is $a\left(1 + \frac{1}{2}\varepsilon^2\right)$. This largest of the three averages that were computed reflects the fact that the orbiting P moves more quickly at periapsis and more slowly at apoapsis. Therefore the larger values of $r(t)$ near apoapsis have a greater impact on this average than the smaller values of $r(t)$ near periapsis. Table 5.1 informs us that ε is small for the planets. So ε^2 is much smaller and hence the average value of $r(t)$ over one planetary orbit is close to its semimajor axis a. The computation of the integrals above involved substitutions that seemed to come out of the blue. Solving an integral is indeed often like pulling a rabbit out of a hat.

Another question that is relevant in the context of elliptical orbits concerns the distance traveled by a planet (asteroid or comet) during one complete revolution around the Sun. In more explicit terms, what is the circumference of an ellipse with semimajor axis a and semiminor axis b? The solution of this problem is also challenging. Here too we provide an outline of the important steps.

The ellipse shown in Figure 1.9 has semimajor axis a, semiminor axis b, and equation $\frac{x^2}{a^2} + \frac{y^2}{b^2} = 1$. The graph of the function $f(x) = \frac{b}{a}(a^2 - x^2)^{\frac{1}{2}}$ is the upper half of the ellipse. By the arc length formula of integral calculus applied to the upper right quarter of the ellipse, the circumference of the full ellipse is

$$4 \int_0^a \sqrt{1 + f'(x)^2}\, dx.$$

Use of the trig substitution $x = \sin\theta$ and simplifying algebra convert this definite integral to

$$4a \int_0^{\frac{\pi}{2}} \sqrt{1 - \varepsilon^2 \sin^2\theta}\, d\theta,$$

where ε is the eccentricity. (Use the equality $b^2 = a^2 - a^2\varepsilon^2$ along the way.) This is an *elliptic integral* (because it arises in the study of ellipses) that cannot be solved in closed form. This means that $\sqrt{1 - \varepsilon^2 \sin^2\theta}$ does not have an antiderivative that is given by a standard function (a function that can be expressed as a combination of algebraic, trig, logarithm, and exponential functions and the inverse functions of such functions). However, it can be solved by using the binomial series

$$(1 + x)^k = 1 + kx + \tfrac{k(k-1)}{2!}x^2 + \tfrac{k(k-1)(k-2)}{3!}x^3 + \cdots + \tfrac{k(k-1)(k-2)\cdots(k-(i-1))}{i!}x^i + \cdots$$

which is known to converge for all x with $-1 < x < 1$. The case $k = \frac{1}{2}$ provides the power series

$$\sqrt{1 + x} = 1 + \tfrac{1}{2}x + \sum_{i=2}^{\infty} \tfrac{(-1)^{i-1} 1 \cdot 3 \cdot 5 \cdots (2i-3)}{2^i i!}\, x^i.$$

By setting $x = -\varepsilon^2 \sin^2\theta$, we get the expression

$$\sqrt{1 - \varepsilon^2 \sin^2\theta} = 1 - \tfrac{\varepsilon^2 \sin^2\theta}{2} - \sum_{i=2}^{\infty} \tfrac{1 \cdot 3 \cdot 5 \cdots (2i-3)\varepsilon^{2i} \sin^{2i}\theta}{2^i i!}.$$

for the integrand that we are dealing with. This series converges for all θ because $0 \le \varepsilon^2 \sin^2\theta < 1$ for all θ. After integrating term by term and using the formula $\displaystyle\int_0^{\frac{\pi}{2}} \sin^{2i}\theta\, d\theta = \frac{1}{2} \cdot \frac{3}{4} \cdot \frac{5}{6} \cdots \frac{2i-1}{2i} \cdot \frac{\pi}{2}$ (it can be verified by combining integration by parts with the principle of mathematical induction) we obtain the formula

$$4a \int_0^{\frac{\pi}{2}} \sqrt{1 - \varepsilon^2 \sin^2 \theta} \, d\theta$$

$$= 2\pi a \left(1 - \left(\tfrac{1}{2}\right)^2 \tfrac{\varepsilon^2}{1} - \left(\tfrac{1 \cdot 3}{2 \cdot 4}\right)^2 \tfrac{\varepsilon^4}{3} - \left(\tfrac{1 \cdot 3 \cdot 5}{2 \cdot 4 \cdot 6}\right)^2 \tfrac{\varepsilon^6}{5} - \left(\tfrac{1 \cdot 3 \cdot 5 \cdot 7}{2 \cdot 4 \cdot 6 \cdot 8}\right)^2 \tfrac{\varepsilon^8}{7} - \left(\tfrac{1 \cdot 3 \cdot 5 \cdot 7 \cdot 9}{2 \cdot 4 \cdot 6 \cdot 8 \cdot 10}\right)^2 \tfrac{\varepsilon^{10}}{9} - \cdots \right)$$

$$= 2\pi a \left(1 - \tfrac{1}{4}\varepsilon^2 - \tfrac{3}{64}\varepsilon^4 - \tfrac{45}{2304}\varepsilon^6 - \tfrac{1575}{147456}\varepsilon^8 - \cdots \right)$$

for the circumference C of the ellipse with semimajor axis a and eccentricity ε. The infinite series converges quickly for a small ε but slowly for ε close to 1. The formula confirms and Figure 5.4 illustrates that C is less than the circumference $2\pi a$ of a circle of radius a.

Example 5.9. Consider the Earth's elliptical orbit around the Sun. A look at Table 5.1 tells us that we can use the approximations $a = 149,600,000$ km and $\varepsilon = 0.0167$. Since $\varepsilon^2 \approx 0.000279$ and $\varepsilon^4 \approx 0.000000078$, we can ignore the higher powers of ε to get the approximation of the circumference of Earth's ellipse

$$C \approx 2\pi (1.496 \times 10^8) \left(1 - \tfrac{1}{4}\varepsilon^2 - \tfrac{3}{64}\varepsilon^4 - \tfrac{45}{2304}\varepsilon^6 - \tfrac{1575}{147456}\varepsilon^8 - \cdots \right)$$

$$\approx 9.3996(1 - 0.00006975 - 0.00000000366) \times 10^8 \approx 9.3990 \times 10^8 \text{ km}.$$

This is only a littles less that the circumference 9.3996×10^8 km of the surrounding circle. We can test the accuracy of this estimate for C against the conclusion of section 5F that informs us that the Earth's maximum and minimum orbital speeds are 30.29 and 29.29 km/sec, respectively. Since the Earth's average speed is $\frac{C}{T}$ where $T = 1.0000264$ is its period in years. Since 1 year has $(365.25)(86,400) = 31557600$ seconds, we get that $\frac{C}{T} \approx \frac{9.3990 \times 10^8}{(1.0000264)(31557600)} = 29.78$ km/sec (in agreement with the value in Table 5.1).

Example 5.10. For Halley's comet the formula for C converges much more slowly. With $a = 2.668 \times 10^9$ km and $\varepsilon = 0.967$,

$$C \approx 2\pi (2.668 \times 10^9) \left(1 - \tfrac{1}{4}\varepsilon^2 - \tfrac{3}{64}\varepsilon^4 - \tfrac{45}{2304}\varepsilon^6 - \tfrac{1575}{147456}\varepsilon^8 - \cdots \right)$$

$$\approx 16.764(1 - 0.23377 - 0.04099 - 0.01597 - 0.00817 - 0.00481 - 0.00309 - 0.00211 - 0.00150) \times 10^9$$

$$\approx 11.5603 \times 10^9 \text{ km}.$$

Notice that the convergence is relatively slow for the circumference of Halley's orbit. Since Halley's orbital period is $T = 75.32$ years and 1 year = 31557600 seconds, the average orbital speed of Halley is

$$\frac{C}{T} \approx \frac{11.5603 \times 10^9}{75.32(3.1558 \times 10^7)} \approx 4.86 \, \text{km/sec}.$$

This result fills in the question mark in Table 5.1. An application of the speed formulas of Example 5.1 tells us that Halley's maximum and minimum orbital speeds are approximately 54.43 km/sec and 0.91 km/sec, respectively.

When it comes to the orbit of a planet, it is the gravitational pull of the Sun that plays the predominant role and gives the orbit its elliptical shape. The fact is that the Sun has an enormous mass that consists of over 99% of the total mass of the solar system. However, the small gravitational forces on a planet that the other planets of the solar system exert have an effect. We observed in

Chapter 1G that the motion of a planet around the Sun S has two aspects. At the same time that the planet moves around its elliptical orbit, the ellipse itself rotates. The Sun S remains fixed at the focal point of the ellipse as the focal axis rotates in the plane of the ellipse taking the perihelion and aphelion positions of the planet with it. In the process, the perihelion of the orbit advances in the same direction as the motion of the planet. This advance is known as the *precession of perihelion*. It is primarily the result of the gravitational tug of the other planets, but Einstein's relativistic modification of Newton's theory of gravity also plays a role. The mathematical study of these effects and the precession that they produce are the topics of the rest of this chapter.

5H. Perturbed Orbits and Precession. Let's consider a planet P in orbit around the Sun S. Let m be the mass of the planet and M_0 the mass of the Sun. By regarding S to be fixed relative to the motion of P and the masses of both S and P to be concentrated at their centers of mass, we'll assume that P is a point-mass and that the gravitational force of S on P is a centripetal force in the direction of S. By Newton's law of universal gravitation its magnitude is equal to

$$F_S(r) = \frac{GmM_0}{r^2}.$$

where r is the variable distance between S and P.

We'll now suppose that another, smaller centripetal force acts on P in the direction of S. We'll assume that its magnitude is a function $F(r)$ of r that can be positive or negative. It is positive if the force acts in the direction of S and negative if it acts in the direction opposite to S. The magnitude of the resultant of the two forces is

$$\Phi(r) = F_S(r) + F(r) = \tfrac{GmM_0}{r^2} + F(r).$$

Since $F_S(r)$ is larger than $F(r)$, $\Phi(r)$ is a centripetal force on P that acts in the direction of S. Therefore the theory of Chapter 4D applies to the force $\Phi(r)$ and the resulting trajectory of P. We will later apply our conclusions in two situations:

1. $F(r)$ is the resultant of all the gravitational forces of the other planets on P, and

2. $F(r)$ is the general relativistic correction of the Newtonian gravitational force $\frac{GmM_0}{r^2}$ on P.

Let's return to the scene of Chapter 4D. Choose a polar coordinate system that has its pole O at S. From the eccentricity data of Table 5.1, we know that the elliptical orbit of each of the eight planets is close to being a circle (only Mercury is somewhat of an outlier in this regard). We will assume that the additional force $F(r)$ perturbs the orbit of the point-mass P so that it deviates slightly from a circle with center S. Let s be the radius of this circle and specify the perturbed orbit by taking it to be the graph of the function $r = f(\theta) = s + p(\theta)$, where $p(\theta)$ is a differentiable function with the property that for all θ, $|p(\theta)|$ is much smaller than s. Click a stopwatch at time $t = 0$, and let time t flow. We'll assume that the motion of P in its orbit is smooth, and in particular that F_S, F, and Φ are differentiable functions of r and that $\theta = \theta(t)$ and $p = p(t) = p(\theta(t))$ and $r(t) = s + p(t)$ are differentiable functions of t. Since $\Phi(r)$ is centripetal in the direction of the pole $S = O$, Chapter 4D provides the equations

$$r(t)^2 \frac{d\theta}{dt} = 2\kappa \quad \text{and} \quad \Phi(r(t)) = m\left[\frac{4\kappa^2}{r(t)^3} - \frac{d^2r}{dt^2}\right],$$

where κ is the Kepler constant of the orbit of P.

Let's focus on the magnitude $\Phi(r)$ of the combined force function. We'll assume that this function has derivatives of all orders at $r = s$ (meaning that the first second, third, and so on, derivatives all exist at $r = s$) and that its Taylor series expansion centered at s converges for all values $r = s + p(t)$ with $t \geq 0$. It follows that

$$\Phi(r) = \Phi(s) + \Phi'(s)(r - s) + \frac{\Phi^{(2)}(s)}{2}(r - s)^2 + \frac{\Phi^{(3)}(s)}{3!}(r - s)^3 + \frac{\Phi^{(4)}(s)}{4!}(r - s)^4 + \ldots$$

for all $r = s + p(t)$ and all $t \geq 0$. The symbols $\Phi^{(2)}(s), \Phi^{(3)}(s), \Phi^{(4)}(s), \ldots$ denote the second, third, $\ldots$, derivatives of $\Phi(r)$ evaluated at $r = s$. It follows that

$$\Phi(r(t)) = \Phi(s) + \Phi'(s)p(t) + \frac{\Phi^{(2)}(s)}{2}p(t)^2 + \frac{\Phi^{(3)}(s)}{3!}p(t)^3 + \frac{\Phi^{(4)}(s)}{4!}p(t)^4 + \ldots$$
$$= \Phi(s) + s\Phi'(s)\frac{p(t)}{s} + \frac{s^2\Phi^{(2)}(s)}{2}\left(\frac{p(t)}{s}\right)^2 + \frac{s^3\Phi^{(3)}(s)}{3!}\left(\frac{p(t)}{s}\right)^3 + \frac{s^4\Phi^{(4)}(s)}{4!}\left(\frac{p(t)}{s}\right)^4 + \ldots.$$

Since $|p(t)|$ is much smaller than s, the term $\frac{p(t)}{s}$ is very close to zero, and we regard $\left(\frac{p(t)}{s}\right)^i$ to be negligible for all $i \geq 2$. So

$$\Phi(r(t)) = \Phi(s) + s\Phi'(s)\frac{p(t)}{s}$$

for all $t \geq 0$. Since $r(t) = s + p(t)$, we see that $\frac{d^2r}{dt^2} = \frac{d^2p}{dt^2}$. So the right side of the force equation above is equal to $m\left[\frac{4\kappa^2}{(s+p(t))^3} - \frac{d^2p}{dt^2}\right]$. Facts about the binomial series tell us that

$$(1 + x)^k = 1 + kx + \frac{k(k-1)}{2!}x^2 + \frac{k(k-1)(k-2)}{3!}x^3 + \frac{k(k-1)(k-2)(k-3)}{4!}x^4 + \ldots$$

for all $|x| < 1$, and hence that for $k = -3$,

$$\left(1 + \frac{p(t)}{s}\right)^{-3} = 1 - 3\frac{p(t)}{s} + \frac{3\cdot4}{2!}\left(\frac{p(t)}{s}\right)^2 - \frac{3\cdot4\cdot5}{3!}\left(\frac{p(t)}{s}\right)^3 + \frac{3\cdot4\cdot5\cdot6}{4!}\left(\frac{p(t)}{s}\right)^4 - \ldots.$$

Again regarding the terms $\left(\frac{p(t)}{s}\right)^i$ to be negligible for all $i \geq 2$, we get $\left(1 + \frac{p(t)}{s}\right)^{-3} = 1 - 3\frac{p(t)}{s}$, in turn $\frac{4\kappa^2}{(s+p(t))^3} = 4\kappa^2 s^{-3}\left(1 - 3\frac{p(t)}{s}\right)$, and therefore $m\left[\frac{4\kappa^2}{(s+p(t))^3} - \frac{d^2p}{dt^2}\right] = m\left[4\kappa^2 s^{-3}\left(1 - 3\frac{p(t)}{s}\right) - \frac{d^2p}{dt^2}\right]$. By combining the equalities we have,

$$\Phi(s) + \Phi'(s)p(t) = m\left[\frac{4\kappa^2}{(s+p(t))^3} - \frac{d^2p}{dt^2}\right] = m\left[4\kappa^2 s^{-3} - (4\kappa^2 s^{-3})3\frac{p(t)}{s} - \frac{d^2p(t)}{dt^2}\right]$$
$$= \frac{4m\kappa^2}{s^3} - 3\frac{4m\kappa^2}{s^3}\frac{p(t)}{s} - m\frac{d^2p(t)}{dt^2}.$$

If the planet's orbit were a circle of radius s, then $r(t) = s$ implies that $\frac{dr}{dt} = 0$ and $\frac{d^2r}{dt^2} = 0$, so that by the force formula, $\Phi(r(t)) = \Phi(s) = \frac{4m\kappa^2}{s^3}$. This fact is the reason for the choice of the initial condition $\Phi(s) = \frac{4m\kappa^2}{s^3}$. By inserting it above, we get $\Phi'(s)p(t) = -3\Phi(s)\frac{p(t)}{s} - m\frac{d^2p(t)}{dt^2}$ and

$$\frac{d^2p(t)}{dt^2} + \frac{1}{m}\left(\frac{3}{s}\Phi(s) + \Phi'(s)\right)p(t) = 0.$$

Consider the differential equation $Ay'' + By' + Cy = 0$ and observe that the equation just derived above has this form with $y = p(t)$, $A = 1$, $B = 0$, and $C = \frac{1}{m}\left(\frac{3}{s}\Phi(s) + \Phi'(s)\right)$. We'll refer to any standard text on differential equations for the solutions of such equations and the basic fact that the roots of the polynomial $Ax^2 + Bx + C = x^2 + C$ govern the outcomes. The conclusions tell us that in our current situation, there are constants D_1 and D_2 such that

1. The case $C < 0$. The polynomial $x^2 + C$ has the two real roots $\sqrt{-C}$ and $-\sqrt{-C}$, and

$$p(t) = D_1 e^{\sqrt{-C}\,t} + D_2 e^{-\sqrt{-C}\,t}.$$

2. The case $C = 0$. The polynomial $x^2 + C = x^2$ has the single root 0, and

$$p(t) = D_1 t + D_2.$$

3. The case $C > 0$. The polynomial $x^2 + C$ has the two complex roots $\sqrt{C}i$ and $-\sqrt{C}i$, and

$$p(t) = D_1 \cos \sqrt{C}\,t + D_2 \sin \sqrt{C}\,t.$$

If $D_1 \neq 0$ in either the solution $p(t) = D_1 e^{2\sqrt{-C}t} + D_2 e^{-2\sqrt{-C}t}$ of Case 1 or $p(t) = D_1 t + D_2$ in Case 2, the term $|p(t)|$ becomes larger and larger as t becomes larger and larger. If $D_1 = 0$ in Case 1 then $p(t)$ goes to zero for increasing t, and if $D_1 = 0$ in Case 2, then $p(t)$ is constant. So the orbit of P is a circle or it converges to a circle over time. Given the basic assumption that the orbit of P is a perturbed circle, it is the solution $p(t) = D_1 \cos \sqrt{C}t + D_2 \sin \sqrt{C}t$ of Case 3 that applies.

As on earlier occasions, let's assume that P is at perihelion at $t = 0$ and that the perihelion distance is $r(0) = q$. Since this is a minimum value of $r(t)$, we know that $r'(0) = 0$. The fact that $p(t) = r(t) - s$, tells us that $p(0) = q - s$. Since $p(t) = D_1 \cos \sqrt{C}t + D_2 \sin \sqrt{C}t$, we get $p'(t) = -D_1\sqrt{C} \sin \sqrt{C}t + D_2\sqrt{C} \cos \sqrt{C}t$. Since $p'(0) = r'(0) = 0$, it follows that $D_2 = 0$, so that $p(t) = D_1 \cos \sqrt{C}t$. By taking $t = 0$, we get $q - s = p(0) = D_1$. Therefore

$$p(t) = (q - s)\cos(\sqrt{C}t) \quad \text{and} \quad r(t) = s + (q - s)\cos(\sqrt{C}t).$$

From the graph of the cosine (see Figure 3.5) we know that as t flows from $t = 0$ to $t = \frac{\pi}{2\sqrt{C}}$ to $t = \frac{\pi}{\sqrt{C}}$ to $t = \frac{2\pi}{\sqrt{C}}$, the value of $\cos(\sqrt{C}t)$ goes from 1 to 0 to -1 and back to 1. Since $\cos \pi = -1$ is the largest negative value of the cosine, P reaches aphelion at $t = \frac{\pi}{\sqrt{C}}$. The aphelion distance is $r(\frac{\pi}{\sqrt{C}}) = 2s - q$. At $t = \frac{2\pi}{\sqrt{C}}$, P is back at perihelion. Since $C = \frac{1}{m}\left(\frac{3}{s}\Phi(s) + \Phi'(s)\right)$, it follows that

$$\tau = \frac{2\pi}{\sqrt{C}} = \frac{2\pi}{\sqrt{\frac{1}{m}(\frac{3}{s}\Phi(s)+\Phi'(s))}} = \frac{2\pi\sqrt{m}}{\sqrt{\frac{3}{s}\Phi(s)+\Phi'(s)}} = 2\pi\sqrt{m}\left(\tfrac{3}{s}\Phi(s) + \Phi'(s)\right)^{-\frac{1}{2}}$$

is the time it takes for P to move from one perihelion to the next.

Now let ψ be the angle that the segment PS sweeps out in going from one perihelion to the next. Recalling that $r(t)^2 \frac{d\theta}{dt} = 2\kappa$ and using the binomial series

$$(1+x)^k = 1 + kx + \tfrac{k(k-1)}{2!}\,x^2 + \tfrac{k(k-1)(k-2)}{3!}\,x^3 + \cdots$$

with $k = -2$ and $x = \frac{p(t)}{s}$, we get

$$\frac{d\theta}{dt} = 2\kappa(s + p(t))^{-2} = 2\kappa s^{-2}\left(1 + \tfrac{p(t)}{s}\right)^{-2} = \tfrac{2\kappa}{s^2}\left(1 - 2\tfrac{p(t)}{s} + 3\left(\tfrac{p(t)}{s}\right)^2 - \cdots\right).$$

Since the terms $\left(\frac{p(t)}{s}\right)^i$ are negligible for $i \geq 2$, we can take $\frac{d\theta}{dt} = \frac{2\kappa}{s^2}\left(1 - 2\frac{p(t)}{s}\right)$. The term $2\frac{p(t)}{s}$, while not negligible, is small. Because (as is shown in the paragraph *Using Definite Integrals* in the Problems and Discussions section of this chapter) the average of this term over the time τ is zero, we will take $\frac{d\theta}{dt} = \frac{2\kappa}{s^2}$ over τ. Since $\frac{d\theta}{dt}$ is the angular velocity of the revolving segment PS, it follows that $\psi = \tau\frac{2\kappa}{s^2}$. Since $\Phi(s) = \frac{4m\kappa^2}{s^3}$ and $\frac{\Phi(s)}{ms} = \frac{4\kappa^2}{s^4}$, we get $\frac{2\kappa}{s^2} = \sqrt{\frac{\Phi(s)}{ms}}$, and therefore

$$\psi = \tau \sqrt{\frac{\Phi(s)}{ms}} = 2\pi\sqrt{m}\left(\tfrac{3}{s}\Phi(s) + \Phi'(s)\right)^{-\frac{1}{2}}\sqrt{\frac{\Phi(s)}{ms}} = 2\pi\left(\tfrac{3}{s}\Phi(s) + \Phi'(s)\right)^{-\frac{1}{2}}\left(\tfrac{s}{\Phi(s)}\right)^{-\frac{1}{2}} = 2\pi\left(3 + s\,\tfrac{\Phi'(s)}{\Phi(s)}\right)^{-\frac{1}{2}}.$$

Notice that $\psi - 2\pi$ measures the advance or slippage in radians of the perihelion over one orbit. If T is the planet's sidereal period in years (refer back to Chapter 1G), then

$$\boxed{\frac{\psi - 2\pi}{T} = \frac{2\pi\left(\left(3 + s\,\frac{\Phi'(s)}{\Phi(s)}\right)^{-\frac{1}{2}} - 1\right)}{T}}$$

is the advance or slippage of the perihelion of the orbit in radians per year.

Example 5.11. Suppose that the magnitude $F(r)$ of the perturbing force is equal to zero, so that $\Phi(r) = F_S(r) = \frac{GmM_0}{r^2} = GmM_0 r^{-2}$. Show that

$$\Phi'(r) = -2GmM_0 r^{-3} = \frac{-2GmM_0}{r^3} = \frac{-2}{r}\Phi(r)$$

and hence that $\frac{r\Phi'(r)}{\Phi(r)} = -2$. Conclude that $\psi = 2\pi$ and that there is no precession of the perihelion.

5I. The Gravitational Force of one Planet on Another. The primary cause of the precession of the perihelion of a planet's orbit is the cumulative effect of the gravitational forces that the other planets exert on it. Table 5.2 tells us for each of the planets, that the rotation of its perihelion through a single degree requires hundreds of years. So the collective impact of these forces on a

Table 5.2. The approximate number of years it takes for a 1° precession of a planet's perihelion.

Mercury	Venus	Earth	Mars	Jupiter	Saturn	Uranus	Neptune
625	1756	314	222	550	183	1078	10,000

planet is extremely small, and this tells us that that the forces themselves are very weak. It will be the delicate task of this section to compute the gravitational pull on a planet P that a single planet Q in orbit beyond P exerts.[2] By considering their masses to be concentrated at their centers of

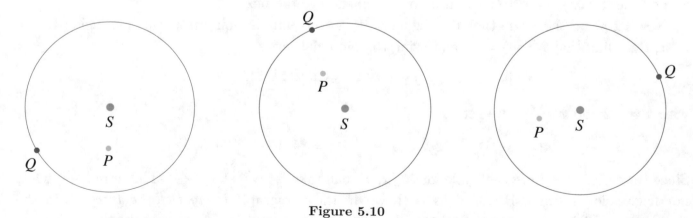

Figure 5.10

_____

[2]The primary goal is the derivation of a formula for the magnitude of this gravitational force—see the formula for $G(r)$ at the end of the section. The derivation consists of two steps. The first uses the strategy of integration to express the force as a definite integral and the second evaluates the integral. Both steps are technical and computationally lengthy.

mass, we will assume that both P and Q are point-masses.

Newton's law of universal gravitation tells us that the magnitude of the force of Q on P is inversely proportional to the square of the distance between P and Q. The fact that this distance varies considerably—see Figure 5.10—means that the magnitude of this weak force varies considerably as well. Without some simplifying assumptions, the estimate of the gravitational force of the planet Q on P as well as its long term effect on the orbit of P would hardly seem possible.

The values for the eccentricities of the planetary orbits in Table 5.1 supports the assumption that the orbit of Q is a circle with center the Sun S. The data of Table 5.2 implies that Q needs to revolve hundreds of times around this circle for its gravitational force on P to have an observable effect. In so doing, the planet Q exerts its gravitational force on P from all locations around the circle over many hundreds of years. This means that it makes sense to average this gravitational force on P over time and to assume that the mass of Q is spread uniformly throughout its circular orbit. The assumption that the circular orbit of Q is a ring with mass, will allow us to compute the magnitude of this averaged force on P and to show that it is a centripetal force in the direction away from the Sun. The case where Q orbits inside the orbit of P can be dealt with in a similar way. See the paragraph *The Perturbing Force of an Interior Planet* in the Problems and Discussions section of this chapter. In this situation the average force acts in the direction of the Sun.

Consider the force of the circular ring on P when P and S are aligned horizontally as shown in Figure 5.11(a). Suppose that in this case, the magnitude of the force can be determined and it can be shown that it acts in a direction opposite the direction from P to S. Then these same conclusions hold with P in any position (see Figures 5.11(b) and 5.11(c) for instance). To see this,

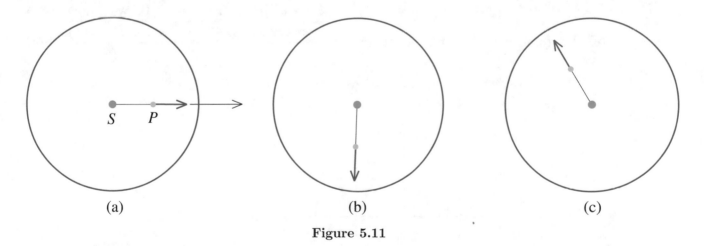

(a)   (b)   (c)

Figure 5.11

simply rotate P into the position it has in Figure 5.11(a). Therefore our study of the gravitational force on P starts with Figure 5.11(a).

i.) Setting up the Force as an Integral. Figure 5.12 shows a polar coordinate system with polar origin O and angle coordinate ϕ. The planet P is represented by a point-mass of mass m positioned at O. The point S is the center of mass of the Sun and the distance from O to S is r. The circular ring centered at S represents the mass of the planet Q. The ring has radius R, mass M, and a constant linear density of $\frac{M}{2\pi R}$. Observe that $0 < r < R$.

Our first order of business will be to determine the polar function $g(\phi)$ with $g(\phi) > 0$ that has the circle as its graph. From the triangle on the right in Figure 5.12 and the law of cosines we

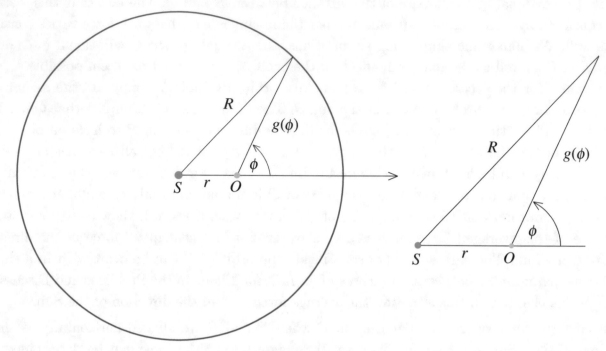

Figure 5.12

get $R^2 = r^2 + g(\phi)^2 - 2rg(\phi)\cos(\pi - \phi)$. Since $\cos(\pi - \phi) = -\cos\phi$ (see Figure 5.13), it follows that $R^2 = r^2 + g(\phi)^2 + 2rg(\phi)\cos\phi$. Since $g(\phi)^2 + (2r\cos\phi)g(\phi) + r^2 - R^2 = 0$, we know by an application of the quadratic formula, that $g(\phi) = \dfrac{-2r\cos\phi \pm \sqrt{4r^2\cos^2\phi + 4(R^2 - r^2)}}{2}$. Since

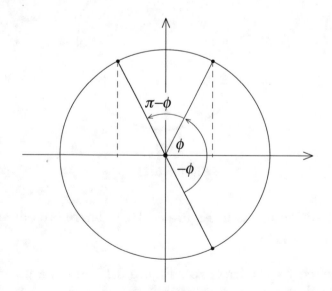

Figure 5.13

$\sqrt{4r^2\cos^2\phi + 4(R^2 - r^2)} > \sqrt{4r^2\cos^2\phi} \geq 2r\cos\phi$ and $g(\phi) > 0$, the $+$ option applies. So

$$g(\phi) = \sqrt{r^2\cos^2\phi + R^2 - r^2} - r\cos\phi = \sqrt{R^2 - r^2\sin^2\phi} - r\cos\phi.$$

Example 5.12. Figure 5.12 considers a situation with $\phi < \pi$. Show that the formula $g(\phi) = \sqrt{R^2 - r^2 \sin^2 \phi} - r \cos \phi$ is also valid for $\phi \geq \pi$.

We now turn to the computation of the gravitational force that the circular ring of mass M exerts on the point-mass P at O. Figure 5.14 considers the angles ϕ and $\phi + d\phi$ for a thin sliver of an angle $d\phi$, and puts in the two wedges that they determine. The angle $d\phi$ is taken so small that the two blue arcs—labeled 1 and 2—that it cuts from the ring can be regarded to be point-masses.

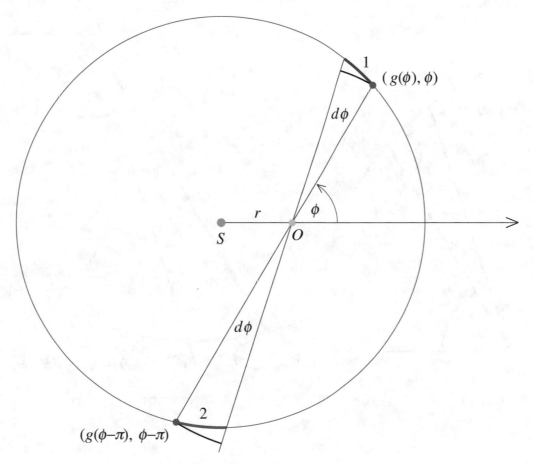

Figure 5.14

Let the lengths of these arcs be ds_1 and ds_2, respectively. Given that the linear density of the ring is $\frac{M}{2\pi R}$, these blue arcs have masses $\frac{M}{2\pi R} ds_1$ and $\frac{M}{2\pi R} ds_2$, respectively. We will now recast the expressions for these two masses.

Define the function $h(\phi)$ by $h(\phi) = g(\phi - \pi) = \sqrt{R^2 - r^2 \sin^2(\phi - \pi)} - r \cos(\phi - \pi)$. By making use of Figure 5.13, we get

$$h(\phi) = \sqrt{R^2 - r^2 \sin^2 \phi} + r \cos \phi.$$

Consider the two black circular arcs of Figure 5.14 that emanate from the two blue points $(g(\phi), \phi)$ and $(g(\phi - \pi), \phi - \pi) = (h(\phi), \phi - \pi))$. Notice that the circles on which they lie have radii $g(\phi)$ and $h(\phi) = g(\phi - \pi)$, respectively. It follows from the definition of radian measure that the lengths of the two black circular arcs are

$$g(\phi)d\phi \quad \text{and} \quad h(\phi)d\phi = g(\phi - \pi)d\phi,$$

respectively. It follows from Figure 5.15(a) and the analysis of the arc length of a polar curve in Chapter 1E (it is based on the Pythagorean theorem), that

$$ds_1 \approx \sqrt{(g(\phi)d\phi)^2 + (g(\phi+d\phi) - g(\phi))^2} = \sqrt{(g(\phi)d\phi)^2 + \frac{(g(\phi+d\phi)-g(\phi))^2}{(d\phi)^2}(d\phi^2)}$$
$$\approx \sqrt{g(\phi)^2 + g'(\phi)^2}\, d\phi\,.$$

Since $d\phi$ is very small, this approximation of ds_1 is very tight. In a similar way, by referring to

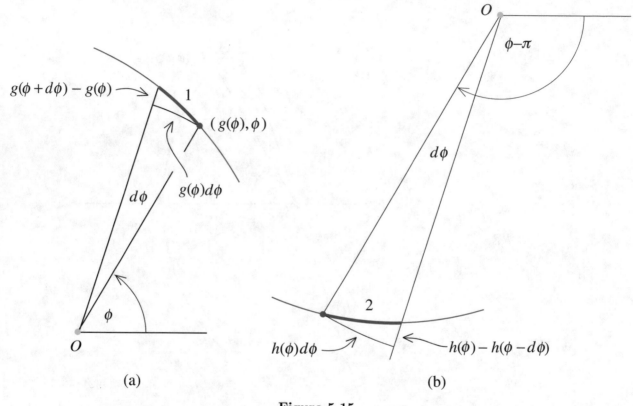

(a)   (b)

Figure 5.15

Figure 5.15(b), we get the very tight approximation

$$ds_2 \approx \sqrt{(h(\phi)d\phi)^2 + (h(\phi) - h(\phi - d\phi))^2} = \sqrt{(h(\phi)d\phi)^2 + \frac{(h(\phi-d\phi)-h(\phi))^2}{(-d\phi)^2}(-d\phi)^2}$$
$$\approx \sqrt{h(\phi)^2 + h'(\phi)^2}\, d\phi$$

of ds_2. These two approximations provide the approximations

$$\tfrac{M}{2\pi R}\, ds_1 \approx \tfrac{M}{2\pi R}\, \sqrt{g(\phi)^2 + g'(\phi)^2}\, d\phi \quad \text{and} \quad \tfrac{M}{2\pi R}\, ds_2 \approx \tfrac{M}{2\pi R}\, \sqrt{h(\phi)^2 + h'(\phi)^2}\, d\phi$$

of the masses of the blue arcs 1 and 2. By applying Newton's law of universal gravitation twice, we see that the magnitudes of the forces with which these two arcs pull on the point-mass P positioned at O are approximately equal to

$$\tfrac{GmM}{2\pi R} \cdot \frac{\sqrt{g(\phi)^2 + g'(\phi)^2}}{g(\phi)^2}\, d\phi \quad \text{and} \quad \tfrac{GmM}{2\pi R} \cdot \frac{\sqrt{h(\phi)^2 + h'(\phi)^2}}{h(\phi)^2}\, d\phi,$$

respectively. The two forces pull in opposite directions. So the magnitude of the resultant of the two forces is

$$\frac{GmM}{2\pi R} \cdot \frac{\sqrt{g(\phi)^2 + g'(\phi)^2}}{g(\phi)^2}\,d\phi - \frac{GmM}{2\pi R} \cdot \frac{\sqrt{h(\phi)^2 + h'(\phi)^2}}{h(\phi)^2}\,d\phi = \frac{GmM}{2\pi R}\left(\frac{\sqrt{g(\phi)^2 + g'(\phi)^2}}{g(\phi)^2} - \frac{\sqrt{h(\phi)^2 + h'(\phi)^2}}{h(\phi)^2}\right)d\phi.$$

This difference turns out to be positive for $-\frac{\pi}{2} < \phi < \frac{\pi}{2}$, so that the force with which arc 1 pulls on P is greater than the force with which arc 2 pulls on P. (This is so because the distance of arc 1 from O is smaller than that of arc 2, but it is not obvious because arc 2 being longer has greater mass.) The next step is the computation of

$$\frac{\sqrt{g(\phi)^2 + g'(\phi)^2}}{g(\phi)^2} - \frac{\sqrt{h(\phi)^2 + h'(\phi)^2}}{h(\phi)^2}.$$

After some elementary algebraic moves,

$$\frac{\sqrt{g(\phi)^2 + g'(\phi)^2}}{g(\phi)^2} - \frac{\sqrt{h(\phi)^2 + h'(\phi)^2}}{h(\phi)^2} = \frac{h(\phi)^2\sqrt{g(\phi)^2 + g'(\phi)^2} - g(\phi)^2\sqrt{h(\phi)^2 + h'(\phi)^2}}{g(\phi)^2 h(\phi)^2}$$

$$= \frac{h(\phi)\sqrt{h(\phi)^2 g(\phi)^2 + h(\phi)^2 g'(\phi)^2} - g(\phi)\sqrt{g(\phi)^2 h(\phi)^2 + g(\phi)^2 h'(\phi)^2}}{g(\phi)^2 h(\phi)^2}.$$

We'll focus on the terms $\sqrt{h(\phi)^2 g(\phi)^2 + h(\phi)^2 g'(\phi)^2}$ and $\sqrt{g(\phi)^2 h(\phi)^2 + g(\phi)^2 h'(\phi)^2}$. Given that

$$g(\phi) = (R^2 - r^2\sin^2\phi)^{\frac{1}{2}} - r\cos\phi \text{ and } h(\phi) = (R^2 - r^2\sin^2\phi)^{\frac{1}{2}} + r\cos\phi,$$

we get $g(\phi) \cdot h(\phi) = (R^2 - r^2\sin^2\phi) - r^2\cos^2\phi = R^2 - r^2$. So

$$h(\phi)^2 g(\phi)^2 = (g(\phi)h(\phi))^2 = (R^2 - r^2)^2.$$

By the chain rule,

$$g'(\phi) = \tfrac{1}{2}(R^2 - r^2\sin^2\phi)^{-\frac{1}{2}}(-2r^2\sin\phi\cos\phi) + r\sin\phi$$

$$= -(R^2 - r^2\sin^2\phi)^{-\frac{1}{2}}(r^2\sin\phi\cos\phi) + r\sin\phi.$$

Therefore,

$$h(\phi)g'(\phi) = -r^2\sin\phi\cos\phi + r\sin\phi(R^2 - r^2\sin^2\phi)^{\frac{1}{2}}$$

$$- (R^2 - r^2\sin^2\phi)^{-\frac{1}{2}}r^3\sin\phi\cos^2\phi + r^2\cos\phi\sin\phi$$

$$= r\sin\phi(R^2 - r^2\sin^2\phi)^{\frac{1}{2}}\left[1 - \frac{r^2\cos^2\phi}{R^2 - r^2\sin^2\phi}\right].$$

Since $1 - \frac{r^2\cos^2\phi}{R^2 - r^2\sin^2\phi} = \frac{R^2 - r^2}{R^2 - r^2\sin^2\phi}$, we finally get

$$h(\phi)g'(\phi) = \frac{(R^2 - r^2)r\sin\phi}{(R^2 - r^2\sin^2\phi)^{\frac{1}{2}}}.$$

After putting together what has been derived,

$$\sqrt{h(\phi)^2 g(\phi)^2 + h(\phi)^2 g'(\phi)^2} = \left((R^2 - r^2)^2 + \frac{(R^2 - r^2)^2 r^2\sin^2\phi}{R^2 - r^2\sin^2\phi}\right)^{\frac{1}{2}}$$

$$= (R^2 - r^2)\left[1 + \frac{r^2\sin^2\phi}{R^2 - r^2\sin^2\phi}\right]^{\frac{1}{2}}$$

$$= (R^2 - r^2)\left[\frac{R^2 - r^2\sin^2\phi + r^2\sin^2\phi}{R^2 - r^2\sin^2\phi}\right]^{\frac{1}{2}}.$$

It follows that

$$\sqrt{h(\phi)^2 g(\phi)^2 + h(\phi)^2 g'(\phi)^2} = \frac{R(R^2 - r^2)}{\sqrt{R^2 - r^2 \sin^2 \phi}},$$

so that the first part of our computation is complete. The equalities

$$h'(\phi) = \tfrac{1}{2}(R^2 - r^2 \sin^2 \phi)^{-\frac{1}{2}}(-2r^2 \sin \phi \cos \phi) - r \sin \phi$$
$$= -(R^2 - r^2 \sin^2 \phi)^{-\frac{1}{2}}(r^2 \sin \phi \cos \phi) - r \sin \phi$$

and a calculation identical to the one just carried out (except for a minus sign that gets "squared away") show that

$$\sqrt{g(\phi)^2 h(\phi)^2 + g(\phi)^2 h'(\phi)^2} = \frac{R(R^2 - r^2)}{\sqrt{R^2 - r^2 \sin^2 \phi}}.$$

By inserting these equalities into the earlier expression for $\frac{\sqrt{g(\phi)^2 + g'(\phi)^2}}{g(\phi)^2} - \frac{\sqrt{h(\phi)^2 + h'(\phi)^2}}{h(\phi)^2}$, we get

$$\frac{\sqrt{g(\phi)^2 + g'(\phi)^2}}{g(\phi)^2} - \frac{\sqrt{h(\phi)^2 + h'(\phi)^2}}{h(\phi)^2} = \frac{h(\phi)\frac{R(R^2 - r^2)}{\sqrt{R^2 - r^2 \sin^2 \phi}} - g(\phi)\frac{R(R^2 - r^2)}{\sqrt{R^2 - r^2 \sin^2 \phi}}}{(R^2 - r^2)^2}$$

$$= \frac{h(\phi)\frac{R}{\sqrt{R^2 - r^2 \sin^2 \phi}} - g(\phi)\frac{R}{\sqrt{R^2 - r^2 \sin^2 \phi}}}{R^2 - r^2}$$

$$= \frac{h(\phi) - g(\phi)}{R^2 - r^2}\frac{R}{\sqrt{R^2 - r^2 \sin^2 \phi}} = \frac{2r \cos \phi}{R^2 - r^2}\frac{R}{\sqrt{R^2 - r^2 \sin^2 \phi}}$$

$$= \frac{2r \cos \phi}{R^2 - r^2}\frac{R}{R\sqrt{1 - \frac{r^2}{R^2} \sin^2 \phi}} = \frac{1}{R^2 - r^2}\frac{2r \cos \phi}{\sqrt{1 - \frac{r^2}{R^2} \sin^2 \phi}}.$$

Substituting this result into the earlier force equation, tells us that the resultant of the pull of the two blue arcs in Figure 5.14 on the point-mass m located at O is equal to

$$\frac{GmM}{2\pi R}\frac{1}{R^2 - r^2}\frac{2r \cos \phi}{\sqrt{1 - \frac{r^2}{R^2} \sin^2 \phi}}\, d\phi = \frac{GmMr}{\pi R(R^2 - r^2)}\frac{\cos \phi}{\sqrt{1 - \frac{r^2}{R^2} \sin^2 \phi}}\, d\phi.$$

By the parallelogram law of Chapter 4A, the magnitudes of the vertical and horizontal components of this resultant are

$$\left(\frac{GmMr}{\pi R(R^2 - r^2)}\frac{\cos \phi}{\sqrt{1 - \frac{r^2}{R^2} \sin^2 \phi}}\, d\phi\right)\sin \phi \quad \text{and} \quad \left(\frac{GmMr}{\pi R(R^2 - r^2)}\frac{\cos \phi}{\sqrt{1 - \frac{r^2}{R^2} \sin^2 \phi}}\, d\phi\right)\cos \phi,$$

respectively.

A careful look at the way the two blue arcs of Figure 5.14 are paired, tells us that they sweep out the entire circle when ϕ varies from $-\frac{\pi}{2} \leq \phi \leq \frac{\pi}{2}$. Review the summation strategy that led to the curve-length and area integrals of Chapters 3F and 3G. Applying it to the sum of all the vertical components for ϕ ranging over $\frac{\pi}{2} \leq \phi \leq \frac{\pi}{2}$ tells us that the vertical component of the force with which the entire circular ring attracts the point-mass P at O is equal to the definite integral

$$\frac{GmMr}{\pi R(R^2 - r^2)}\int_{-\frac{\pi}{2}}^{\frac{\pi}{2}} \frac{\cos \phi \sin \phi}{\sqrt{1 - \frac{r^2}{R^2} \sin^2 \phi}}\, d\phi.$$

This integral is easy to solve. Let $u = \frac{r}{R}\sin \phi$. By differentiating, $du = \frac{r}{R}\cos \phi\, d\phi$. So $\sin \phi = \frac{R}{r}u$, and $\cos \phi\, d\phi = \frac{R}{r}du$. With this substitution and use of the fact that $-(1 - u^2)^{\frac{1}{2}}$ is an antiderivative of $\frac{u}{\sqrt{1 - u^2}}$, we get

$$\frac{GmMr}{\pi R(R^2-r^2)}\int_{-\frac{r}{R}}^{\frac{r}{R}}\frac{\frac{R}{r}\cdot\frac{R}{r}u}{\sqrt{1-u^2}}\,du \;=\; \frac{GmMR}{\pi r(R^2-r^2)}\int_{-\frac{r}{R}}^{\frac{r}{R}}\frac{u}{\sqrt{1-u^2}}\,du \;=\; \frac{GmMr}{\pi R(R^2-r^2)}\Big(-\big(1-(\tfrac{r}{R})^2\big)^{\frac12}+\big(1-(-\tfrac{r}{R})^2\big)^{\frac12}\Big)=0.$$

So the vertical component of the force of the circular ring on the point-mass P is zero.

Therefore the force of the entire ring acts horizontally on P in the direction opposite to S. See Figure 5.12. And the magnitude of this force on P is obtained by summing up all the horizontal components computed above. Applying the summation strategy of the definite integral once more, we see that the magnitude of the force of the entire ring on the point-mass is

$$\frac{GmMr}{\pi R(R^2-r^2)}\int_{-\frac{\pi}{2}}^{\frac{\pi}{2}}\frac{\cos^2\phi}{\sqrt{1-\frac{r^2}{R^2}\sin^2\phi}}\,d\phi\,.$$

Since $\displaystyle\int_{-\frac{\pi}{2}}^{\frac{\pi}{2}}\frac{\cos^2\phi}{\sqrt{1-\frac{r^2}{R^2}\sin^2\phi}}\,d\phi=\int_{0}^{\frac{\pi}{2}}\frac{\cos^2\phi}{\sqrt{1-\frac{r^2}{R^2}\sin^2\phi}}\,d\phi+\int_{-\frac{\pi}{2}}^{0}\frac{\cos^2\phi}{\sqrt{1-\frac{r^2}{R^2}\sin^2\phi}}\,d\phi$ and the substitution $\phi\to-\phi$

transforms the second integral into the first, $\displaystyle\int_{-\frac{\pi}{2}}^{\frac{\pi}{2}}\frac{\cos^2\phi}{\sqrt{1-\frac{r^2}{R^2}\sin^2\phi}}\,d\phi=2\int_{0}^{\frac{\pi}{2}}\frac{\cos^2\phi}{\sqrt{1-\frac{r^2}{R^2}\sin^2\phi}}\,d\phi$. Therefore our

focus is on the computation $\displaystyle\int_{0}^{\frac{\pi}{2}}\frac{\cos^2\phi}{\sqrt{1-\frac{r^2}{R^2}\sin^2\phi}}\,d\phi$.

ii.) Solving the Integral. The solution of this integral is challenging. The binomial series

$$(1-x)^{-\frac12}=1+\tfrac12\,x+\tfrac{3}{2^2\cdot2!}\,x^2+\tfrac{3\cdot5}{2^3\cdot3!}\,x^3+\tfrac{3\cdot5\cdot7}{2^4\cdot4!}\,x^4+\cdots+\tfrac{3\cdot5\cdots(2k-1)}{2^k\cdot k!}\,x^k+\cdots$$

is the key. As was already pointed out in section 5G, it converges to $(1-x)^{-\frac12}$ for all x with $|x|<1$. Since $0\le\frac{r^2}{R^2}\sin^2\phi\le\frac{r^2}{R^2}<1$,

$$(1-\tfrac{r^2}{R^2}\sin^2\phi)^{-\frac12}=1+\tfrac12\tfrac{r^2}{R^2}\sin^2\phi+\tfrac{3}{2^2\cdot2!}\big(\tfrac{r^2}{R^2}\sin^2\phi\big)^2$$
$$+\tfrac{3\cdot5}{2^3\cdot3!}\big(\tfrac{r^2}{R^2}\sin^2\phi\big)^3+\tfrac{3\cdot5\cdot7}{2^4\cdot4!}\big(\tfrac{r^2}{R^2}\sin^2\phi\big)^4+\cdots+\tfrac{3\cdot5\cdots(2k-1)}{2^k\cdot k!}\big(\tfrac{r^2}{R^2}\sin^2\phi\big)^k+\cdots$$

Multiplying this equality through by $\cos^2\phi=1-\sin^2\phi$, gives us

$$\frac{\cos^2\phi}{(1-\frac{r^2}{R^2}\sin^2\phi)^{\frac12}}=(1-\sin^2\phi)+\tfrac12\tfrac{r^2}{R^2}(\sin^2\phi)(1-\sin^2\phi)+\tfrac{3}{2^2\cdot2!}\big(\tfrac{r^2}{R^2}\big)^2(\sin^4\phi)(1-\sin^2\phi)$$

$$+\tfrac{3\cdot5}{2^3\cdot3!}\big(\tfrac{r^2}{R^2}\big)^3(\sin^6\phi)(1-\sin^2\phi)+\tfrac{3\cdot5\cdot7}{2^4\cdot4!}\big(\tfrac{r^2}{R^2}\big)^4(\sin^8\phi)(1-\sin^2\phi)+\cdots$$

$$+\tfrac{3\cdot5\cdots(2(k-1)-1)}{2^{(k-1)}\cdot(k-1)!}\big(\tfrac{r^2}{R^2}\big)^{k-1}(\sin^{2(k-1)}\phi)(1-\sin^2\phi)+\tfrac{3\cdot5\cdots(2k-1)}{2^k\cdot k!}\big(\tfrac{r^2}{R^2}\big)^k(\sin^{2k}\phi)(1-\sin^2\phi)+\cdots$$

$$=1-\Big[1-\tfrac12\tfrac{r^2}{R^2}\Big]\sin^2\phi-\Big[\tfrac12\tfrac{r^2}{R^2}-\tfrac{3}{2^2\cdot2!}\big(\tfrac{r^2}{R^2}\big)^2\Big]\sin^4\phi-\Big[\tfrac{3}{2^2\cdot2!}\big(\tfrac{r^2}{R^2}\big)^2-\tfrac{3\cdot5}{2^3\cdot3!}\big(\tfrac{r^2}{R^2}\big)^3\Big]\sin^6\phi$$

$$-\Big[\tfrac{3\cdot5}{2^3\cdot3!}\big(\tfrac{r^2}{R^2}\big)^3-\tfrac{3\cdot5\cdot7}{2^4\cdot4!}\big(\tfrac{r^2}{R^2}\big)^4\Big]\sin^8\phi-\Big[\tfrac{3\cdot5\cdot7}{2^4\cdot4!}\big(\tfrac{r^2}{R^2}\big)^4-\tfrac{3\cdot5\cdot7\cdot9}{2^5\cdot5!}\big(\tfrac{r^2}{R^2}\big)^5\Big]\sin^{10}\phi-\cdots$$

$$-\Big[\tfrac{3\cdot5\cdots(2(k-1)-1)}{2^{(k-1)}\cdot(k-1)!}\big(\tfrac{r^2}{R^2}\big)^{k-1}-\tfrac{3\cdot5\cdots(2k-1)}{2^k\cdot k!}\big(\tfrac{r^2}{R^2}\big)^k\Big]\sin^{2k}\phi-\cdots$$

$$=1-\Big[1-\tfrac12\tfrac{r^2}{R^2}\Big]\sin^2\phi-\tfrac12\tfrac{r^2}{R^2}\Big[1-\tfrac{3}{2\cdot2}\tfrac{r^2}{R^2}\Big]\sin^4\phi-\tfrac{3}{2^2\cdot2!}\big(\tfrac{r^2}{R^2}\big)^2\Big[1-\tfrac{5}{2\cdot3}\tfrac{r^2}{R^2}\Big]\sin^6\phi$$

$$-\tfrac{3\cdot5}{2^3\cdot3!}\big(\tfrac{r^2}{R^2}\big)^3\Big[1-\tfrac{7}{2\cdot4}\tfrac{r^2}{R^2}\Big]\sin^8\phi-\tfrac{3\cdot5\cdot7}{2^4\cdot4!}\big(\tfrac{r^2}{R^2}\big)^4\Big[1-\tfrac{9}{2\cdot5}\tfrac{r^2}{R^2}\Big]\sin^{10}\phi-\cdots$$

$$-\tfrac{3\cdot5\cdots(2k-3)}{2^{(k-1)}\cdot(k-1)!}\big(\tfrac{r^2}{R^2}\big)^{k-1}\Big[1-\tfrac{2k-1}{2\cdot k}\tfrac{r^2}{R^2}\Big]\sin^{2k}\phi-\cdots\,.$$

It follows that

$$\int_0^{\frac{\pi}{2}} \frac{\cos^2 \phi}{\sqrt{1 - \frac{r^2}{R^2}\sin^2 \phi}}\, d\phi$$

$$= \int_0^{\frac{\pi}{2}} 1\, d\phi - \left[1 - \frac{1}{2}\frac{r^2}{R^2}\right]\int_0^{\frac{\pi}{2}} \sin^2 \phi\, d\phi - \frac{1}{2}\frac{r^2}{R^2}\left[1 - \frac{3}{4}\frac{r^2}{R^2}\right]\int_0^{\frac{\pi}{2}} \sin^4 \phi\, d\phi - \frac{3}{2^2 \cdot 2!}\left(\frac{r^2}{R^2}\right)^2\left[1 - \frac{5}{6}\frac{r^2}{R^2}\right]\int_0^{\frac{\pi}{2}} \sin^6 \phi\, d\phi$$

$$- \frac{3 \cdot 5}{2^3 \cdot 3!}\left(\frac{r^2}{R^2}\right)^3\left[1 - \frac{7}{8}\frac{r^2}{R^2}\right]\int_0^{\frac{\pi}{2}} \sin^8 \phi\, d\phi - \frac{3 \cdot 5 \cdot 7}{2^4 \cdot 4!}\left(\frac{r^2}{R^2}\right)^4\left[1 - \frac{9}{10}\frac{r^2}{R^2}\right]\int_0^{\frac{\pi}{2}} \sin^{10} \phi\, d\phi$$

$$- \cdots - \frac{3 \cdot 5 \cdots (2k-3)}{2^{(k-1)}(k-1)!}\left(\frac{r^2}{R^2}\right)^{k-1}\left[1 - \frac{2k-1}{2k}\frac{r^2}{R^2}\right]\int_0^{\frac{\pi}{2}} \sin^{2k} \phi\, d\phi - \cdots.$$

By applying the formula $\int_0^{\frac{\pi}{2}} \sin^{2k}\phi\, d\phi = \frac{1}{2}\cdot\frac{3}{4}\cdot\frac{5}{6}\cdots\frac{2k-1}{2k}\cdot\frac{\pi}{2}$ for any $k \geq 1$ (see section 5G), we get

$$\int_0^{\frac{\pi}{2}} \frac{\cos^2 \phi}{\sqrt{1 - \frac{r^2}{R^2}\sin^2 \phi}}\, d\phi$$

$$= \frac{\pi}{2} - \left[1 - \frac{1}{2}\frac{r^2}{R^2}\right]\frac{1}{2}\frac{\pi}{2} - \frac{1}{2}\frac{r^2}{R^2}\left[1 - \frac{3}{4}\frac{r^2}{R^2}\right]\frac{1}{2}\frac{3}{4}\frac{\pi}{2} - \frac{3}{2^2\cdot2!}\left(\frac{r^2}{R^2}\right)^2\left[1 - \frac{5}{6}\frac{r^2}{R^2}\right]\frac{1}{2}\frac{3}{4}\frac{5}{6}\frac{\pi}{2} - \frac{3\cdot5}{2^3\cdot3!}\left(\frac{r^2}{R^2}\right)^3\left[1 - \frac{7}{8}\frac{r^2}{R^2}\right]\frac{1}{2}\frac{3}{4}\frac{5}{6}\frac{7}{8}\frac{\pi}{2}$$

$$- \frac{3\cdot5\cdot7}{2^4\cdot4!}\left(\frac{r^2}{R^2}\right)^4\left[1 - \frac{9}{10}\frac{r^2}{R^2}\right]\frac{1}{2}\frac{3}{4}\frac{5}{6}\frac{7}{8}\frac{9}{10}\frac{\pi}{2} - \cdots - \frac{3\cdot5\cdots(2k-3)}{2^{(k-1)}(k-1)!}\left(\frac{r^2}{R^2}\right)^{k-1}\left[1 - \frac{2k-1}{2k}\frac{r^2}{R^2}\right]\frac{1}{2}\frac{3}{4}\frac{5}{6}\cdots\frac{2k-1}{2k}\frac{\pi}{2} - \cdots$$

$$= \frac{1}{2}\pi - \frac{1}{2}\frac{1}{2}\left[1 - \frac{1}{2}\frac{r^2}{R^2}\right]\pi - \frac{1}{2^3}\frac{3}{4}\frac{r^2}{R^2}\left[1 - \frac{3}{4}\frac{r^2}{R^2}\right]\pi - \frac{3^2}{2^5(2!)^2}\frac{5}{6}\left(\frac{r^2}{R^2}\right)^2\left[1 - \frac{5}{6}\frac{r^2}{R^2}\right]\pi - \frac{3^2\cdot5^2}{2^7(3!)^2}\frac{7}{8}\left(\frac{r^2}{R^2}\right)^3\left[1 - \frac{7}{8}\frac{r^2}{R^2}\right]\pi$$

$$- \frac{3^2\cdot5^2\cdot7^2}{2^9(4!)^2}\frac{9}{10}\left(\frac{r^2}{R^2}\right)^4\left[1 - \frac{9}{10}\frac{r^2}{R^2}\right]\pi - \cdots - \frac{3^2\cdot5^2\cdots(2k-3)^2}{2^{2k-1}((k-1)!)^2}\frac{2k-1}{2k}\left(\frac{r^2}{R^2}\right)^{k-1}\left[1 - \frac{2k-1}{2k}\frac{r^2}{R^2}\right]\pi - \cdots.$$

We can now conclude that

$$\frac{GmMr}{\pi R(R^2-r^2)}\int_{-\frac{\pi}{2}}^{\frac{\pi}{2}} \frac{\cos^2 \phi}{\sqrt{1 - \frac{r^2}{R^2}\sin^2 \phi}}\, d\phi$$

$$= \frac{GmM}{R^2-r^2}\frac{r}{R}\left[1 - \frac{1}{2}\left[1 - \frac{1}{2}\frac{r^2}{R^2}\right] - \frac{1}{2^2}\frac{3}{4}\frac{r^2}{R^2}\left[1 - \frac{3}{4}\frac{r^2}{R^2}\right] - \frac{3^2}{2^4(2!)^2}\frac{5}{6}\left(\frac{r^2}{R^2}\right)^2\left[1 - \frac{5}{6}\frac{r^2}{R^2}\right] - \frac{3^2\cdot5^2}{2^6(3!)^2}\frac{7}{8}\left(\frac{r^2}{R^2}\right)^3\left[1 - \frac{7}{8}\frac{r^2}{R^2}\right]\right.$$

$$\left. - \frac{3^2\cdot5^2\cdot7^2}{2^8(4!)^2}\frac{9}{10}\left(\frac{r^2}{R^2}\right)^4\left[1 - \frac{9}{10}\frac{r^2}{R^2}\right] - \cdots - \frac{3^2\cdot5^2\cdots(2k-3)^2}{2^{2(k-1)}((k-1)!)^2}\frac{2k-1}{2k}\left(\frac{r^2}{R^2}\right)^{k-1}\left[1 - \frac{2k-1}{2k}\frac{r^2}{R^2}\right] - \cdots\right]$$

$$= \frac{GmM}{R^2-r^2}\frac{r}{R}\left[\frac{1}{2} + \frac{1}{2^2}\frac{1}{4}\left(\frac{r}{R}\right)^2 + \frac{3^2}{2^4(2!)^2}\frac{1}{6}\left(\frac{r}{R}\right)^4 + \frac{3^25^2}{2^6(3!)^2}\frac{1}{8}\left(\frac{r}{R}\right)^6 + \frac{3^25^27^2}{2^8(4!)^2}\frac{1}{10}\left(\frac{r}{R}\right)^8\right.$$

$$\left. + \frac{3^25^27^29^2}{2^{10}(5!)^2}\frac{1}{12}\left(\frac{r}{R}\right)^{10} + \cdots + \frac{3^2\cdot5^2\cdots(2k-3)^2}{2^{2k}(k!)^2}\frac{1}{2(k+1)}\left(\frac{r}{R}\right)^{2k} + \cdots\right]$$

$$= \frac{GmM}{2(R^2-r^2)}\frac{r}{R}\left[1 + \frac{1}{2^2}\frac{1}{2}\left(\frac{r}{R}\right)^2 + \frac{3^2}{(2\cdot4)^2}\frac{1}{3}\left(\frac{r}{R}\right)^4 + \frac{(3\cdot5)^2}{(2\cdot4\cdot6)^2}\frac{1}{4}\left(\frac{r}{R}\right)^6 + \frac{(3\cdot5\cdot7)^2}{(2\cdot4\cdot6\cdot8)^2}\frac{1}{5}\left(\frac{r}{R}\right)^8\right.$$

$$\left. + \frac{(3\cdot5\cdot7\cdot9)^2}{(2\cdot4\cdot6\cdot8\cdot10)^2}\frac{1}{6}\left(\frac{r}{R}\right)^{10} + \cdots + \frac{(3\cdot5\cdots(2k-1))^2}{(2\cdot4\cdots2k)^2}\frac{1}{k+1}\left(\frac{r}{R}\right)^{2k} + \cdots\right].$$

This last expression—we'll denote it by $G(r)$—is the magnitude of the gravitational force with which the ring of mass M and radius R attracts the point-mass P of mass m positioned at O at a distance $r < R$ from the ring's center. The force acts along the line SP in the direction away from S. In our applications, we'll take r to be the semimajor axes of the planet P and R to be the semimajor axis of the planet Q. With this understanding and the data of Table 5.1, the terms of the sum containing $\left(\frac{r}{R}\right)^{2k}$ with $k \geq 6$ are very small (and become smaller with increasing k). Given that approximations are our goal, we'll regard them to be negligible, so that

$$G(r) = \frac{GmM}{2(R^2-r^2)} \frac{r}{R} \left[1 + \frac{1}{2^3}\left(\frac{r}{R}\right)^2 + \frac{3}{2^6}\left(\frac{r}{R}\right)^4 + \frac{5^2}{2^{10}}\left(\frac{r}{R}\right)^6 + \frac{5\cdot7^2}{2^{14}}\left(\frac{r}{R}\right)^8 + \frac{3^3\cdot7^2}{2^{17}}\left(\frac{r}{R}\right)^{10} \right]$$

Since any gravitational force in the direction of the Sun is taken to be positive, this perturbing force on a planet by a planet exterior to its orbit is considered to be negative.

Example 5.12. Use the product rule to show that

$$G'(r) = \frac{GmM}{(R^2-r^2)^2} \frac{r^2}{R} \left[1 + \frac{1}{2^3}\left(\frac{r}{R}\right)^2 + \frac{3}{2^6}\left(\frac{r}{R}\right)^4 + \frac{5^2}{2^{10}}\left(\frac{r}{R}\right)^6 + \frac{5\cdot7^2}{2^{14}}\left(\frac{r}{R}\right)^8 + \frac{3^3\cdot7^2}{2^{17}}\left(\frac{r}{R}\right)^{10} \right]$$
$$+ \frac{GmM}{2(R^2-r^2)} \frac{1}{R} \left[1 + \frac{3}{2^3}\left(\frac{r}{R}\right)^2 + \frac{3\cdot5}{2^6}\left(\frac{r}{R}\right)^4 + \frac{5^2\cdot7}{2^{10}}\left(\frac{r}{R}\right)^6 + \frac{5\cdot7^2\cdot9}{2^{14}}\left(\frac{r}{R}\right)^8 + \frac{3^3\cdot7^2\cdot11}{2^{17}}\left(\frac{r}{R}\right)^{10} \right].$$

5J. Perihelion Precession for Mercury. The focus for the rest of the chapter will be on the planet Mercury. We will see in section 5K that the study of the precession of the perihelion of Mercury's orbit, in particular the discrepancy between the observed precession and that predicted by Newton's theory of gravity, played an important role in the history of modern astronomy.

We'll let m be the mass of Mercury and take the circle of its perturbed orbit to have radius equal to the semimajor axis a of its elliptical orbit. We'll also assume that the radii of the circular rings that represent the masses of the seven planets that pull on Mercury as they orbit the Sun are also equal to the semimajor axes of their elliptical orbits. These choices are consistent with the fact—verified in section 5G—that the time-averaged distance of any planet from the Sun is $a(1 + \frac{1}{2}\varepsilon^2)$ where a is its semimajor axis and ε its eccentricity. We'll number the planets from Venus to Neptune from 2 to 8, denote the radii of their circular rings by $R_2, \ldots, R_8$, and their masses by $M_2, \ldots, M_8$. The force with which any of these rings pulls Mercury in the direction of the Sun was studied in the previous section. Accordingly, we will denote the forces of Venus, $\ldots$, Neptune on Mercury by $G_2(a), \ldots, G_8(a)$. We'll take $a = 57.909227 \times 10^9$ m for the semimajor axis of Mercury's orbit from Table 5.1. The mass of the Sun is denoted by M_0. The value $GM_0 = 1.3271244042 \times 10^{20} \frac{\text{m}^3}{\text{sec}^2}$ as well as the values GM_i for the planets are taken from Table 8 of the JPL study listed as [5] in the References for Chapter 2. See Tables 2.3 and 2.6 in this regard.

In order to apply the formula for the angular advance or slippage of a planet's precession that was derived in section 5H to Mercury's orbit, we need to compute

$$\frac{a\Phi'(a)}{\Phi(a)},$$

where $\Phi(a) = F_S(a) - F(a) = \frac{GmM_0}{a^2} - \sum_{i=2}^{8} G_i(a)$ and $\Phi'(a) = F'_S(a) - F'(a) = \frac{-2GmM_0}{a^3} - \sum_{i=2}^{8} G'_i(a).$

The mass m appears in all terms of the formulas for $G(r)$ and $G'(r)$ at the end of the previous section so that it can be factored out from the right sides of the two expressions above. This simplifies our calculations and improves their accuracy. So we'll compute

$$\frac{1}{m}\Phi(a) = \frac{GM_0}{a^2} - \sum_{i=2}^{8} \frac{1}{m} G_i(a) \quad \text{and} \quad \frac{1}{m}\Phi'(a) = \frac{-2GM_0}{a^3} - \sum_{i=2}^{8} \frac{1}{m} G'_i(a)$$

instead. The fact that the magnitudes of the forces $G_i(a)$ are comparatively small means that the computation of the terms $\frac{1}{m}G_i(a)$ and $\frac{1}{m}G'_i(a)$ is delicate. Since accurate data is available, we'll carry out our computations with 9 significant figure accuracy.

We'll start by computing $\frac{GM_0}{a^2}$. Its value is

$$\frac{GM_0}{a^2} = \frac{1.3271244042 \times 10^{20}}{57.909227^2 \times 10^{18}} = 0.0003957456043956 \times 10^2 = 39574560.44 \times 10^{-9} \text{ N/kg.}$$

The conclusion of the study undertaken in section 5I tells us that for each of the planets from Venus to Neptune, $\frac{1}{m}G_i(a)$ is the product of $\frac{GM_i}{2(R_i^2-a^2)}\frac{a}{R_i}$ and $1 + \frac{1}{2^3}\left(\frac{a}{R_i}\right)^2 + \cdots + \frac{3^3 \cdot 7^2}{2^{17}}\left(\frac{a}{R_i}\right)^{10}$. The results of the computation[3] of these products (obtained by using the the data of Tables 2.1, 2.3, 2.4, and

Table 5.3. The computation of the terms $\frac{1}{m}G_i(a)$ for each of the seven planets orbiting outside Mercury's orbit.

planet	$\dfrac{GM_i}{2(R_i^2-a^2)}\dfrac{a}{R_i}$ in 10^{-9} N/kg	$1 + \cdots + \dfrac{3^3 \cdot 7^2}{2^{17}}\left(\dfrac{a}{R_i}\right)^{10}$	$\dfrac{1}{m}G_i(a)$ in 10^{-9} N/kg
Venus	10.40299008	1.04033766	10.82262236
Earth	4.05487822	1.01987359	4.13546321
Mars	0.11192869	1.00826980	0.11285432
Jupiter	7.82419068	1.00069338	7.82961582
Saturn	0.37894014	1.00020608	0.37901823
Uranus	0.00709528	1.00005088	0.00709564
Neptune	0.00217496	1.00002072	0.00217501

2.6) are collected in Table 5.3. By adding the entries of the last column, we get

$$\sum_{i=2}^{8}\frac{1}{m}G_i(a) = (10.82262236 + 4.13546321 + 0.11285432 + 7.82961582$$

$$+ 0.37901823 + 0.00709564 + 0.00217501) \times 10^{-9}$$

$$= 23.28884459 \times 10^{-9} \text{ N/kg.}$$

Therefore

$$\frac{1}{m}\Phi(a) = \frac{GM_0}{a^2} - \sum_{i=2}^{8}\frac{1}{m}G_i(a) = (39574560.44 - 23.28884459) \times 10^{-9}$$

$$= 39574537.15115541 \times 10^{-9}$$

$$= 3.95745372 \times 10^{-2} \text{ N/kg.}$$

---

[3]The computations that produced the data in Tables 5.3 and 5.4 are manageable when carried out with a calculator such as the one provided by https://web2.0calc.com/.

We turn next to the computation of $a\frac{1}{m}\Phi'(a) = a\left(\frac{-2GM_0}{a^3} - \sum_{i=2}^{8} \frac{1}{m}G_i'(a)\right)$. Using the fact that $GM_0 = 1.3271244042 \times 10^{20} \frac{\text{m}^3}{\text{sec}^2}$, we get

$$a\frac{2GM_0}{a^3} = \frac{2GM_0}{a^2} = \frac{2(1.3271244042)}{57.9092277^2} \times 10^2 = 0.000791491209 \times 10^2 = 7.91491209 \times 10^{-2} \text{ N/kg}.$$

By the formula of Example 5.12, for each of the planets $\frac{1}{m}G_i'(a)$ is the sum of the products

$$\frac{GM_i}{(R_i^2-a^2)^2}\frac{a^2}{R_i} \text{ by } 1 + \frac{1}{2^3}\left(\frac{a}{R_i}\right)^2 + \cdots + \frac{3^3 \cdot 7^2}{2^{17}}\left(\frac{a}{R_i}\right)^{10} \text{ and } \frac{GM_i}{2(R_i^2-a^2)}\frac{1}{R_i} \text{ by } 1 + \frac{3}{2^3}\left(\frac{a}{R_i}\right)^2 + \cdots + \frac{3^3 \cdot 7^2 \cdot 11}{2^{17}}\left(\frac{a}{R_i}\right)^{10}.$$

The results of the tedious arithmetic are presented in Table 5.4 (this arithmetic makes use of the middle column of Table 5.3 as well) with the sum of the products listed in the last column. By

Table 5.4. The computation of the terms $\frac{1}{m}G_i'(a)$ for each of the seven planets orbiting outside Mercury's orbit.

planet	$\frac{GM_i}{(R_i^2-a^2)^2}\frac{a^2}{R_i}$ in 10^{-20} N/kg·m	$\frac{GM}{2(R_i^2-a^2)}\frac{1}{R_i}$ in 10^{-20} N/kg	$1 + \cdots + \frac{3^3 \cdot 7^2 \cdot 11}{2^{17}}\left(\frac{a}{R_i}\right)^{10}$	$\frac{1}{m}G'(a)$ in 10^{-20} N/kg·m
Venus	14.41940343	17.96430487	1.13175575	15.00104842 + 20.33120533
Earth	2.46833671	7.00212803	1.06210568	2.51739142 + 7.43699995
Mars	0.02667098	0.19328299	1.02522783	0.02689154 + 0.19815910
Jupiter	0.15041401	13.51112955	1.00208302	0.15051830 + 13.53927350
Saturn	0.00215983	0.65436919	1.00061849	0.00216028 + 0.65477391
Uranus	0.00000998	0.01225242	1.00015264	0.00000998 + 0.01225429
Neptune	0.00000125	0.00375582	1.00006215	0.00000125 + 0.00375605

adding the numbers of the last column of Table 5.4 we get

$$\sum_{i=2}^{8} \frac{1}{m}G_i'(a) = \big((15.00104842 + 20.33120533) + (2.51739142 + 7.43699995) + (0.02689154 + 0.1981591)$$

$$+ (0.15051830 + 13.53927350) + (0.00216028 + 0.65477391) + (0.00000998 + 0.01225429)$$

$$+ (0.00000125 + 0.00375605)\big) \times 10^{-20}.$$

$$= 59.87444332 \times 10^{-20}.$$

Therefore,

$$a\frac{1}{m}\Phi'(a) = a\Big(-\frac{2GM_0}{a^3} - \sum_{i=2}^{8}\frac{1}{m}G_i'(a)\Big) = -\frac{2GM_0}{a^2} - a\sum_{i=2}^{8}\frac{1}{m}G_i'(a)$$

$$= -7.91491209 \times 10^{-2} - (57.909227 \times 10^9)(59.87444332 \times 10^{-20})$$

$$= -7.91491209 \times 10^{-2} - 3467.28272971651364 \times 10^{-11}$$

$$= -(7.91491209 + 0.000003467282729) \times 10^{-2}$$

$$= -7.91491556 \times 10^{-2}\,\mathrm{N/kg}.$$

By inserting the results of these computations and the fact (see Table 2.1) that the sidereal period of Mercury's orbit is $T = 0.2408467$ into the precession formula of section 5H, we get

$$\frac{\psi - 2\pi}{T} \approx \frac{2\pi\big((3 + a\frac{\Phi'(a)}{\Phi(a)})^{-\frac{1}{2}} - 1\big)}{T} \approx \frac{2\pi\big((3 + a\frac{\frac{1}{m}\Phi'(a)}{\frac{1}{m}\Phi(a)})^{-\frac{1}{2}} - 1\big)}{T}$$

$$\approx \frac{2\pi\big((3 + \frac{-7.91491556\times 10^{-2}}{3.95745372\times 10^{-2}})^{-\frac{1}{2}} - 1\big)}{0.2408467}$$

$$\approx \frac{2\pi\big((3 - 2.00000205)^{-\frac{1}{2}} - 1\big)}{0.2408467}$$

$$\approx \frac{2\pi(0.00000102)}{0.2408467} \approx 0.00002661.$$

So the perihelion of Mercury's orbit advances at a rate of approximately 0.00002661 radians per year. Converting this to degrees, minutes, and finally seconds of arc, we get that this precession is approximately

$$0.00002661 \times \tfrac{180}{\pi} \times 60 \times 60 \approx 5.49$$

arc seconds per year and hence 549 arc seconds per century. This is in satisfactory agreement with the 532 arc seconds per year that the literature reports.

We will see in the paragraph *The Perturbing Force of an Interior Planet* of the Problems and Discussions section of this chapter that the gravitational perturbing force on a planet P by a planet with orbit interior to the orbit of P is centripetal in the direction of the Sun S. The formula for its magnitude is analogous to the formula for $G(r)$ of section 5I. This formula (along with considerations similar to those above) makes it possible—after a slog of calculations—to compute for each of the planets the precession of the perihelion caused by the gravitational forces of the other planets.

5K. The Relativistic Component of Precession. Newton's theory of gravity celebrated a major triumph with the calculations that pointed to the existence of a planet beyond Uranus and the subsequent discovery of Neptune. Ironically, the same calculations also exposed a failure: its inability to account for the entire rotational precession of the perihelion of the orbit of Mercury. Observations had shown that Mercury's perihelion advances by about 575 arc seconds or 0.16 degrees per century. In the 1850s the French astronomer Urbain Le Verrier, who had earlier predicted the existence of Neptune, calculated that the collective gravitational forces of all the other planets could not account for this entire amount. By the 1880s, astronomers knew that Newton's theory could explain only 532 of the 575 arc seconds leaving the difference of 43 arc seconds per century unaccounted for. Perhaps surprisingly (given that 43 arc seconds per century is an extraordinarily small shift), scientists looked for explanations of the difference. They began with the assumption that there must be some as yet unknown body circling the Sun—within the orbit of Mercury—that causes the extra orbital rotation. When no such body was found, modifications to Newton's gravity were proposed—for instance, that the exponent of 2 in the inverse square law should be replaced by a number slightly larger than 2. But such assumptions also failed to clarify the discrepancy. Not until Einstein's theory of relativity appeared was there a satisfactory explanation of the gap between the observed and calculated precession of Mercury's orbit.

Einstein—always guided by his powerful intuition and probing thought experiments—had finalized his theory of special relativity by 1905. The theory asserts that the laws of physics are the same for any two non-accelerating observers, and that the speed of light c in a vacuum ($c = 299,792,458$ meters per second) never changes, even if the observer or the light source is moving. Special relativity unified space and time into a four-dimensional "space-time" geometric construct and laid out the relationship between energy E and mass m in the famous equation $E = mc^2$. General relativity added gravity to the theory of special relativity, explaining that moving matter causes depressions that ripple through space-time. The more massive the body, the deeper the flowing ripple. Lighter bodies move through the changing depression in space that the massive ones formed like a golf ball rolling on a green. Einstein's theory predicts that even light will curve as it moves around a massive, space-time bending object such as a galaxy cluster. This phenomenon has allowed astronomers to study very distant galaxies through the gravitational lenses molded by nearer ones.

For over 100 years, general relativity has withstood the test of time. A revolution when Einstein first proposed it in 1915, it is now accepted as the foundation on which our scientific understanding of the origin and evolution of the universe rests. Careful studies of the motion of matter and light throughout the universe have shown that ordinary matter (matter as we know it) cannot alone account for the way things move through space-time. In fact, observations suggest that only 5 percent of the universe is familiar matter and energy, while 25 percent is transparent, invisible "dark matter" that neither emits nor absorbs light and that reveals its existence only through its gravitational effects. The gravitational redshift of light from receding exploding stars, known as supernovas, is predicted by general relativity and tells us that the expansion of the universe is accelerating. The energy thought to be responsible for this expansion—and known to make up the remaining 70 percent of the universe—is a still mysterious "dark energy."

In our current context, general relativity also provides small corrections to the orbits of planets (as well as the spacecraft we send to all corners of our solar system). In particular, it explains the gap of 43 seconds of arc between the observed precession of Mercury's perihelion and the precession attributed to the gravitational pull of the other planets. We will have a look at this relativistic phenomenon from within the mathematical context that has already been developed.

Let P be any planet and S the Sun. Let m be the mass of the planet and M_0 the mass of the Sun. By concentrating their masses at their centers of mass, we'll assume that both S and P are point-masses. According to Newton's law of universal gravitation, the magnitude $F_S(r)$ of the Sun's attractive force on the planet is given by

$$\frac{GmM_0}{r^2}$$

where r is the variable distance between P and S.

General relativity corrects the magnitude of the gravitational force with which the Sun attracts the planet with the addition of the force

$$F(r) = \frac{3GM_0}{c^2} \frac{(mr^2 \cdot \frac{d\theta}{dt})^2}{mr^4},$$

where c is the speed of light and $mr^2 \cdot \frac{d\theta}{dt}$ is the angular momentum of the moving planet (namely the product of its moment of inertia mr^2 with its angular velocity $\frac{d\theta}{dt}$). Since the corrected force on P is centripetal, the discussion of Chapter 4D applies to tell us that $r^2 \frac{d\theta}{dt} = 2\kappa$ and hence that $mr^2 \cdot \frac{d\theta}{dt} = 2m\kappa$, where κ is Kepler's constant of the orbit. It follows that $F(r) = \frac{12GmM_0\kappa^2}{c^2 r^4}$ and hence that the magnitude of the corrected force that general relativity provides for the Sun's gravitational force on the planet is equal to

$$\frac{GmM_0}{r^2} + F(r) = \frac{GmM_0}{r^2} + \frac{12GmM_0\kappa^2}{c^2 r^4} = \frac{GmM_0}{r^2} + \frac{GmM_0}{r^2} \cdot \frac{12\kappa^2}{c^2 r^2}.$$

We'll now analyze the impact of this correction on the precession of the perihelion of the planet P by turning to the results of section 5H. Since

$$\Phi(r) = \frac{GmM_0}{r^2} + F(r) = \frac{GmM_0}{r^2} + \frac{GmM_0}{r^2} \cdot \frac{12\kappa^2}{c^2 r^2} = \frac{GmM_0}{r^2}\left[1 + \frac{12\kappa^2}{c^2 r^2}\right],$$

the product rule tells us that

$$\Phi'(r) = \frac{-2GmM_0}{r^3}\left[1 + \frac{12\kappa^2}{c^2 r^2}\right] + \frac{GmM_0}{r^2}\left[\frac{(-2)12\kappa^2}{c^2 r^3}\right].$$

After factoring out $\frac{2}{r}$ from the first factor of the first term and the second factor of the second term,

$$\Phi'(r) = \frac{-GmM_0}{r^2}\frac{2}{r}\left[1 + \frac{12\kappa^2}{c^2 r^2}\right] - \frac{GmM_0}{r^2}\frac{2}{r}\left[\frac{12\kappa^2}{c^2 r^2}\right] = -\frac{GmM_0}{r^2}\frac{2}{r}\left(1 + 2\cdot\frac{12\kappa^2}{c^2 r^2}\right)$$

and hence that

$$\frac{r\Phi'(r)}{\Phi(r)} = -2\frac{1 + 2\frac{12\kappa^2}{c^2 r^2}}{1 + \frac{12\kappa^2}{c^2 r^2}}.$$

We'll suppose that the relativistic force $F(r)$ disrupts the Newtonian orbit of P into a perturbed circle of radius s. An application of the conclusion of the discussion of section 5H, tells us that the force $\Phi(r)$ results in an advance or slippage of the perihelion of the orbit of the planet P of

$$\frac{2\pi\left(\left(3 + s\frac{\Phi'(s)}{\Phi(s)}\right)^{-\frac{1}{2}} - 1\right)}{T} = \frac{2\pi\left(\left(3 - 2\frac{1 + 2\frac{12\kappa^2}{c^2 s^2}}{1 + \frac{12\kappa^2}{c^2 s^2}}\right)^{-\frac{1}{2}} - 1\right)}{T}$$

radians per year, where T is the sidereal period in years of the orbit of the planet.

Let's return to the planet Mercury. By Table 5.1, the semimajor axis, eccentricity, and the perihelion period of its orbit are $a_1 = 57.909227 \times 10^6$ km, $\varepsilon_1 = 0.20563593$, and

$$T_1 = 0.2408489 \cdot 31{,}557{,}600 = 7600613.25 \,\text{sec}.$$

So the Kepler constant of Mercury's orbit is

$$\kappa_1 = \tfrac{a_1 b_1 \pi}{T_1} = \tfrac{\pi a_1^2 \sqrt{1-\varepsilon_1^2}}{T_1} = 13.56483955 \times 10^8 \,\text{km}^2/\text{sec}.$$

It follows that

$$\tfrac{12\kappa_1^2}{c^2 a_1^2} = \tfrac{12 \cdot 13.56483955^2}{299{,}792.458^2 \cdot 57.9092277^2} \times 10^4 = 0.000000073261132.$$

Notice therefore, that Einstein's relativistic correction is but a very small fraction of the value that Newton's law of universal gravitation provides for the Sun's attractive force on Mercury.

We now get that

$$\frac{a_1 \Phi'(a_1)}{\Phi(a_1)} = -2\,\frac{1 + 2\frac{12\kappa_1^2}{c^2 a_1^2}}{1 + \frac{12\kappa_1^2}{c^2 a_1^2}} = -2\,\frac{1 + 2 \cdot 0.000000073261132}{1 + 0.000000073261132} = -2.00000014652225.$$

Letting $T_1 = 0.2408467$ in years be the sidereal period of Mercury's orbit (from Table 2.1) we get

$$\frac{2\pi\left(\left(3 + a_1 \frac{\Phi'(a_1)}{\Phi(a_1)}\right)^{-\frac{1}{2}} - 1\right)}{T_1} = \frac{2\pi\left(\left(3 - 2.00000014652225\right)^{-\frac{1}{2}} - 1\right)}{0.2408467}$$

$$= 0.000019112125 \,\text{radians per year}.$$

Converting this to degrees, minutes, and finally seconds of arc, we get that the relativistic component of the precession of the perihelion of Mercury's orbit is approximately

$$0.000019112125 \times \tfrac{180}{\pi} \times 60 \times 60 \approx 0.394$$

arc seconds per year and hence 39.4 arc seconds per century. This is close to the commonly accepted value of 43 arc seconds per century for the relativistic component of Mercury's precession.

Let P be any of the other seven planets. Let a, ε and T be the semimajor axis, eccentricity and the perihelion period of the orbit of P. Since its Kepler constant is $\kappa = \frac{ab\pi}{T}$ where $b = a\sqrt{1-\varepsilon^2}$ is the semiminor axis of the orbit, $\frac{a}{T} = \frac{\kappa}{\pi b} = \frac{\kappa}{\pi a \sqrt{1-\varepsilon^2}}$. So $\frac{a^2}{T} = \frac{\kappa}{\pi \sqrt{1-\varepsilon^2}}$, and therefore $\frac{a^3}{T^2} = \frac{\kappa^2}{\pi^2 a(1-\varepsilon^2)}$. Kepler's third law tells us that $\frac{a^3}{T^2}$ is the same constant, say K, for all the planets. It follows that $\kappa^2 = K\pi^2 a(1-\varepsilon^2)$, and hence that $\frac{12\kappa^2}{c^2 a^2} = \frac{12 K \pi^2 a(1-\varepsilon^2)}{c^2 a^2} = \frac{12 K \pi^2}{c^2}\frac{1-\varepsilon^2}{a}$. Consider the constant $\frac{12\kappa_1^2}{c^2 a_1^2} = \frac{12 K \pi^2}{c^2}\frac{1-\varepsilon_1^2}{a_1}$ for Mercury, and notice that $\frac{12\kappa^2}{c^2 a^2}\big/\frac{12\kappa_1^2}{c^2 a_1^2} = \frac{1-\varepsilon^2}{1-\varepsilon_1^2} \cdot \frac{a_1}{a} = \frac{a_1}{a}\frac{1-\varepsilon^2}{1-\varepsilon_1^2}$. Therefore

$$\frac{12\kappa^2}{c^2 a^2} = \frac{a_1(1-\varepsilon^2)}{a(1-\varepsilon_1^2)}\frac{12\kappa_1^2}{c^2 a_1^2}.$$

Let's consider Venus, for example. Inserting information from Table 5.1, we get

$$\frac{12\kappa^2}{c^2 a^2} = \frac{(57.909227)(1-0.006776772^2)}{(108.209475)(1-0.20563593^2)}\,0.000000073261132 = 0.000000040935527.$$

So for Venus,

$$\frac{a\Phi'(a)}{\Phi(a)} = -2\frac{1 + 2\frac{12\kappa^2}{c^2a^2}}{1 + \frac{12\kappa^2}{c^2a^2}} = -2\frac{1 + 2 \cdot 0.000000040935527}{1 + 0.000000040935527} = -2.000000081871051.$$

Therefore with $T = 0.6151973$ years the sidereal period of Venus (taken from Table 2.1)

$$\frac{2\pi\left(\left(3 + a\frac{\Phi'(a)}{\Phi(a)}\right)^{-\frac{1}{2}} - 1\right)}{T} = \frac{2\pi\left(\left(3 - 2.000000081871051\right)\right)^{-\frac{1}{2}} - 1\right)}{0.6151973}$$

$$= 0.00000041808621 \text{ radians per year},$$

and hence 8.62 arc seconds per century. Unlike the case of Mercury for which the result derived above deviates a little from the accepted value, the value 8.62 for Venus is equal to the accepted value for the relativistic component of the orbital precession of its perihelion.

Example 5.13. Compute $\frac{12\kappa^2}{c^2a^2}$ for Earth, Mars, Jupiter, Saturn, Uranus, and Neptune. Then compute the relativistic correction to the precession for each. After checking entries three and four of the third row of the Table 5.5, complete the third row. Provide estimates for the entries

Table 5.5. About the precession of the perihelion of the planetary orbits.

orbital precession in arc seconds per century	Mercury	Venus	Earth	Mars	Jupiter	Saturn	Uranus	Neptune
as provided by observation	575	205	1145	1628	655	1950	3.34	0.36
due to the gravity of the other planets	532							
due to general relativity	43	8.6	3.8	1.35				

of the second row of the table under the assumption that the planetary and relativistic components add up to the precession as it is observed.

We have undertaken a study of the precession of the perihelion of the elliptical orbits of the planets. Over 90% of this phenomenon explained by the collective Newtonian gravitational pull of the other planets. The remaining few percentage points are the result of Einstein's relativistic correction of the gravitational pull of the Sun on the planets. Both components of these precessions are very small—a fraction of one degree per century. The components are small enough so that their combined effect can be obtained by adding them. The fact that the two components are small also means that their calculation is highly dependent on the accuracy of the planetary data as well as the methods with which they are computed.

Consider a planet other than Mercury. The computation of the planet's orbital precession requires—in addition to the facts and strategies developed in sections 5H, 5I, and 5J—the numerical study of the gravitational force exerted by a planet that moves *inside* the orbit of the given planet. This study is carried out in the Problems and Discussions section that concludes this chapter.

5L. Problems and Discussions. This problem set will explore issues that are related to the discussions of this chapter.

1. Moving Around the Ellipse. This segment considers matters that arise in section 5B.

Problem 5.1. Consider a semicircle of radius 5 and inscribe into it one half of the ellipse with semi-

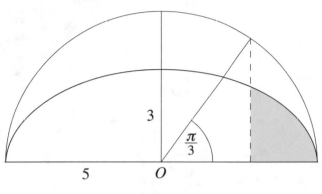

Figure 5.16

major axis 5 and semiminor axis 3 as shown in Figure 5.16. After reviewing the the discussion of Figure 5.4, compute the shaded area of the figure.

Problem 5.2. Figure 5.17 shows a coordinate system along with the upper half of a circle of radius 5 and the inscribed upper half of an ellipse with semimajor axis 5. The point $F = (2,0)$ is a focal point of the ellipse. The point P_0 is on the circle, the point P is on the ellipse, and the dashed

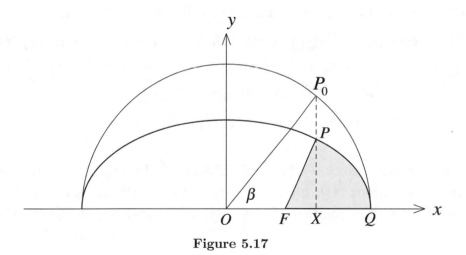

Figure 5.17

segment P_0PX is perpendicular to the x-axis.

i. Given that the y-coordinate of P_0 is 4, show that $X = (3,0)$. Use your calculator to find the angle β in radians.

ii. Compute the semiminor axis of the ellipse and write down an equation for the ellipse. Determine the y-coordinate of the point P.

iii. Study the discussion of Figure 5.4 and compute the area of the elliptical sector PFQ.

Problem 5.3. Figure 5.18 is a version of Figure 5.3 with P in the first and fourth quadrants,

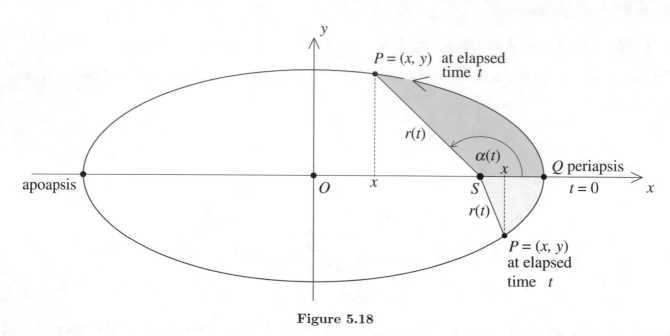

Figure 5.18

respectively. Go through the derivations of the results of section 5B in each of these two situations.

Problem 5.4. The function $r(t) = a\big(1 - \varepsilon \cos \beta(t)\big)$ measures the distance of the planet from the Sun. Consider it for one complete orbit, in other words, over the time interval $[0, T]$. If $\varepsilon = 0$, then $r(t) = a$ throughout, so assume that $\varepsilon > 0$. Over which part of the interval $[0, T]$ is $r(t)$ increasing and over which part is it decreasing? Answer first by studying Figure 5.18. Then check your answer by analyzing the derivative of $r(t)$ over the interval $0 \le t \le T$.

Problem 5.5. Solve $\cos \alpha(t) = \frac{\cos \beta(t) - \varepsilon}{1 - \varepsilon \cos \beta(t)}$ (established in section 5B) for $\cos \beta(t)$ to show that $\cos \beta(t) = \frac{\varepsilon + \cos \alpha(t)}{1 + \varepsilon \cos \alpha(t)}$. Insert this into the equation $r(t) = a(1 - \varepsilon \cos \beta(t))$ and conclude that $r(t) = \frac{a(1-\varepsilon^2)}{1+\varepsilon \cos \alpha(t)}$. It follows that r is the function $r(\alpha) = \frac{a(1-\varepsilon^2)}{1+\varepsilon \cos \alpha}$ of α.

2. Using Kepler's Equation. This segment explores some of the consequences of Kepler's equation and its solution.

Problem 5.6. Observe that when $\varepsilon = 0$ the orbit of P is a circle of radius $a = b$. Discuss Figures 5.3 and 5.4 as well as the functions $r(t)$, $\alpha(t)$, and $\beta(t)$ in this situation. What do Kepler's equation $\beta(t) - \varepsilon \sin \beta(t) = \frac{2\pi t}{T}$ and the definition of the angle $\beta(t)$ imply about the speed of the orbiting point-mass P?

Problem 5.7. Use Kepler's equation $\beta(t) - \varepsilon \sin \beta(t) = \frac{2\pi t}{T}$ with $\beta(t) = \pi$ to show that the point-mass P takes exactly as long, namely $\frac{T}{2}$, to go from periapsis to apoapsis as from apoapsis to periapsis.

Problem 5.8. A look at Figure 5.4 tells us that starting from periapsis the point-mass P will complete the first fourth of its orbit at the moment t for which $\beta(t) = \frac{\pi}{2}$. It will take P the time

$$t = \frac{T}{4} - \frac{T\varepsilon}{2\pi}$$

to complete this part of its orbit, where T is the period and ε the eccentricity.

i. Verify this formula in two different ways. First, by computing the area $A(t)$ in Figure 5.3 and using Kepler's second law, and then again by using Kepler's equation. How far is the point-mass from S at this time?

ii. Suppose that P is the center of mass of Earth or Halley's comet as each orbits the Sun. What is t equal to in days for the Earth? Show that $t = 7.3$ years for Halley. [Use the data provided by Table 5.1.]

iii. Show that the ratio of the time it takes for P to travel the first quarter of its orbit over the time it takes for the first half of its orbit is equal to $\frac{1}{2} - \frac{\varepsilon}{\pi}$. For ε close to 1, this is approximately equal to $\frac{1}{5}$.

Problem 5.9. Table 5.1 tells us that with its eccentricity of $\varepsilon \approx 0.0068$, the planet Venus is closer to being a circle than any of the other planets. Turn to Figure 5.4 and focus on the angular position α of Venus. How many days does it take for α to rotate from 0° to 60°? From 60° to 120°? And finally from 120° to 180°? [Hint: Combine the equations of Gauss and Kepler.]

Problem 5.10. Halley's orbital plane is separated by 17.7° from that of the Earth, but it orbits in the direction opposite to Earth's. In what follows we'll assume that the orbital planes of Halley and Earth are the same. By combining information in Table 5.1 with the fact that 1 au ≈ 149,598,000 km (see Chapter 1G), we'll take Halley's orbital data to be $a = 17.8341$ au, $\varepsilon = 0.9671$, and $T = 75.32$ years. It follows that the semiminor axis is $b = a\sqrt{1 - \varepsilon^2} = 4.5368$ au and the perihelion distance is $a(1 - \varepsilon) = 0.5867$ au. Place an xy-coordinate system so that the center of Halley's orbit is at the origin O. Use the scale 1 au = 1 centimeter to draw the right half of the ellipse $\frac{x^2}{a^2} + \frac{y^2}{b^2} = 1$ of

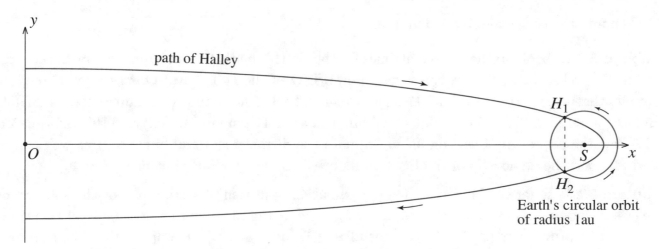

Figure 5.19

Halley's orbit as shown in Figure 5.19. The Sun S is at the point $(c, 0)$ where $c = a\varepsilon = 17.25$ cm. Finally draw Earth's orbit as a circle with center S and radius 1 cm (corresponding to 1 au). Assume that the comet moves counterclockwise around the Sun rather than clockwise. This puts the discussion into the framework of Chapter 5.

i. Determine an equation of Earth's circular orbit. Show that the x-coordinate of the points of intersection H_1 and H_2 of the two orbits is $x = \frac{a-1}{\varepsilon}$. Determine the numerical values of the x- and y-coordinates of H_1 and H_2. [Hints: In reference to the equation of the ellipse, it is better to work with the parameters a, b, c, etc., rather than their numerical values. Use the identity $c^2 = a^2 - b^2$. Since $\frac{a+1}{\varepsilon} > a$, note that $x = \frac{a+1}{\varepsilon}$ is not possible.]

ii. Consider section 5B and let P be Halley's comet. Let t be the time Halley requires to travel from perihelion to the point H_1. Make use of Figure 5.4 and the equations in section 5B together with the x- and y-coordinates computed in (i) to compute $r(t)$ and $\beta(t)$. Then use Kepler's Equation to determine t. For how many days will Halley be inside Earth's orbit?

iii. Consider the successive approximation method of section 5E for computing the angle $\beta(t)$ in the context of Halley's orbit. Give an estimate for the number of iterations necessary to do so with an accuracy of 0.0002. [Hint: Keep squaring $\varepsilon = 0.9672$ with a calculator.]

iv. Use the data from Table 5.1 in combination with the fact that $1\,\mathrm{au} \approx 149,758,000$ km to draw parts of the orbits of Mars, Jupiter, Saturn, Uranus, and Neptune into your copy of Figure 5.19. (To be able do so for Neptune and Uranus, your Figure 5.19 has to be extended to the left.)

Problem 5.11. Consider Mercury in its orbit after an elapsed time of t from its perihelion position.

i. Consider the successive approximation of the angle $\beta(t)$. How many steps will always be enough to insure that $\beta(t)$ has been computed with an accuracy of four decimal places? This is the number of steps required to achieve $\beta_{n+1} = \beta_n$ after both are rounded off to 4 decimal places.

ii. Carry out the computation of $\beta(t)$ with four decimal place accuracy when Mercury is in its orbit exactly $t = 20$ days after perihelion. Has Mercury travelled more than one quarter of its orbit by that time or less? [Use the equation that Problem 5.8 provides.]

iii. Compute the corresponding r and α.

Problem 5.12. Look up the dates and times for the Earth's perihelion and aphelion for the current year on the website https://www.usno.navy.mil/USNO of the U.S. Naval Observatory. (Note, for example, that the entry Jan 03 01:11 refers to January 3, 1 hour and 11 minutes after midnight. Similarly, July 03 23:10 refers to July 3, 10 minutes after 11 pm, and so on.) Use this information and follow the approach of section 5F to estimate the distance $r(t)$ and angle $\alpha(t)$ as well as the speed $v(t)$ and the angle $\gamma(t)$ of the Earth in its orbit around the Sun right now.

Problem 5.13. Refer to the successive approximation approach of section 5E for the solution of Kepler's equation $\beta(t) - \varepsilon \sin \beta(t) = \frac{2\pi t}{T}$ for $\beta(t)$. Review the mathematical principle of induction and use it to prove that $|\beta(t) - \beta_i| \le \varepsilon^i$ for all $i \ge 1$. Since $\varepsilon < 1$, this implies that the sequence $\beta_1, \beta_2, \ldots, \beta_i, \ldots$ converges to the solution $\beta(t)$ of the equation $\beta(t) - \varepsilon \sin \beta(t) = \frac{2\pi t}{T}$.

Problem 5.14. Review the Newton-Raphson method for finding the zeros of a differentiable function. Consider the function $f(x) = x - \varepsilon \sin x - \frac{2\pi t}{T}$, where t and T are constants. Kepler's equation tell us that $f(\beta(t)) = 0$. Let $\beta_1 = \frac{2\pi t}{T}$. Then let

$$\beta_{i+1} = \beta_i - \frac{f(\beta_i)}{f'(\beta_i)} = \beta_i - \frac{\beta_i - \varepsilon \sin \beta_i - \frac{2\pi t}{T}}{1 - \varepsilon \cos \beta_i}$$

and study the graph of $y = f(x)$ to explore conditions for $\beta_1, \beta_2, \beta_3, \ldots$ to converge to $\beta(t)$.

Problem 5.15. Turn to section 5F and the computation of $\beta(t)$ for Earth's summer solstice position in 2013. Check that the Newton-Raphson approximation with $\beta_1 = \frac{2\pi t}{T} = 2.924336$ radians, provides the correct result

$$\beta_2 = \beta_1 - \frac{f(\beta_1)}{f'(\beta_1)} = \beta_1 - \frac{\beta_1 - \varepsilon \sin \beta_1 - \frac{2\pi t}{T}}{1 - \varepsilon \cos \beta_1} = 2.927880.$$

after a single step. This suggests that the Newton-Raphson approximation converges more quickly in general than the successive approximation method of section 5E.

Problem 5.16. Section 5C derived the equality $\displaystyle\int_{x(t)}^{a} \sqrt{a^2 - x^2}\, dx = \frac{1}{2}a^2\beta(t) - \frac{1}{2}x(t)y_0(t)$ for P in the first half of its initial orbit. Review the topic of integration by trig substitution and then show that the substitution $x = a \sin\theta$ with $-\frac{\pi}{2} \le \theta \le \frac{\pi}{2}$ can also be used to derive this equality. [Hints: Use the formulas $\cos^2\theta = \frac{1}{2}(1 + \cos 2\theta)$, $\sin 2\theta = 2\sin\theta\cos\theta$, and $\cos\theta = \sqrt{1 - \sin^2\theta}$ to show that $\displaystyle\int \sqrt{a^2 - x^2}\, dx = \frac{1}{2}a^2 \sin^{-1}\frac{x}{a} + \frac{1}{2}x\sqrt{a^2 - x^2} + C$. A look at Figure 5.4 shows that $0 \le \beta(t) \le \pi$ and $\sqrt{a^2 - x(t)^2} = y_0(t)$, and basic trig formulas from Chapter 3A and another look at the figure show that $\sin(\frac{\pi}{2} - \beta(t)) = \frac{x(t)}{a}$. Use these facts to complete the derivation.]

3. About the Angles of Approach and Departure.
This segment relies on the discussion of section 5D and studies the angle $\gamma(t)$. The case of a circle in the two problems that follow is trivial, since $r(t) = a$ and $\gamma(t) = \frac{\pi}{2}$ are both constant. So suppose that the orbit is not a circle.

Problem 5.17. Let a point-mass P be in an elliptical orbit with focal point S, semimajor axis a, and eccentricity $\varepsilon < 1$. Consider P in its orbit at two different times t_1 and t_2. Suppose that the distances $r(t_1)$ and $r(t_2)$ from P to S satisfy $r(t_1) + r(t_2) = 2a$ and show that $\sin\gamma(t_1) = \sin\gamma(t_2)$. Show that if $r(t_1) + r(t_2) = 2a$ and P is moving from apoapsis to periapsis on both occasions or from periapsis to apoapsis on both occasions, then $\gamma(t_1) = \gamma(t_2)$. Show conversely, that if $\gamma(t_1) = \gamma(t_2)$, then either $r(t_1) = r(t_2)$ or $r(t_1) + r(t_2) = 2a$, and P is moving from apoapsis to periapsis on both occasions or from periapsis to apoapsis on both occasions.

Problem 5.18. Let P be a point-mass in an elliptical orbit with focal point S. Let a be its semimajor axis, $b = a\sqrt{1 - \varepsilon^2}$ its semiminor axis, and $\varepsilon < 1$ its eccentricity. Let $f(r)$ be the function defined by $f(r) = \frac{b}{\sqrt{r(2a - r)}}$, where r is the distance from P to S. Show that the domain of $f(r)$ as abstract function is $0 < r < 2a$ and that $f'(r) = -\frac{b(a - r)}{(r(2a - r))^{\frac{3}{2}}}$. Why is the domain of $f(r)$ equal to $a(1 - \varepsilon) \le r \le a(1 + \varepsilon)$ in the context of the orbital motion of P? Show that $f(r)$ is decreasing over $a(1 - \varepsilon) \le r \le a$ and increasing over $a \le r \le a(1 + \varepsilon)$ with maximum value $f(a(1 - \varepsilon)) = f(a(1 + \varepsilon)) = 1$ and minimum value $f(a) = \sqrt{1 - \varepsilon^2}$. Go to any standard calculus text, study the function $\sin^{-1} x$ and note that it is defined and increasing over the interval $[-1, 1]$. It follows that the function $\sin^{-1}\left(\frac{-b}{\sqrt{r(2a - r)}}\right)$ is decreasing over $a(1 - \varepsilon) \le r \le a$ and increasing over $a \le r \le a(1 + \varepsilon)$. Explore the implications of this for the angle $\gamma(t)$.

4. Using Definite Integrals. The next three problems consider definite integrals that are relevant in sections 5G and 5H.

Problem 5.19. Refer back to section 5G and consider the function $f(x) = \frac{b}{a}(a^2 - x^2)^{\frac{1}{2}}$ that has the upper half of the ellipse $\frac{x^2}{a^2} + \frac{y^2}{b^2} = 1$ as graph. Use the trig substitution $x = \sin\theta$ (and basic equalities involving a, b and the eccentricity ε of the ellipse) to show that circumference $4\int_0^a \sqrt{1 + f'(x)^2}\, dx$ of the ellipse is equal to $4a\int_0^{\frac{\pi}{2}} \sqrt{1 - \varepsilon^2 \sin^2\theta}\, d\theta$.

Problem 5.20. Consider a point-mass P in an elliptical orbit with focal point S and assume that P is at perihelion at time $t = 0$. Let the semimajor axis, eccentricity, and period of the orbit be a, ε, and T, respectively. For any $t \geq 0$, let $r(t)$ be the distance from P to S. Show that the average value of the function $r(t)^{-1}$ over the time interval $0 \leq t \leq T$ is a^{-1}. [Hint: Differentiate Kepler's equation $\beta(t) - \varepsilon\sin\beta(t) = \frac{2\pi t}{T}$ and use the equality $r(t) = a(1 - \varepsilon\cos\beta(t))$ to show that $\frac{T}{2a\pi}\beta'(t)r(t) = 1$. That $\frac{1}{T/2}\int_0^{T/2} r(t)^{-1}\, dt = a^{-1}$ is a consequence of this equality.]

Problem 5.21. Refer to section 5H for the discussion of the function $\frac{p(t)}{s} = \frac{q-s}{s}\cos\sqrt{C}t$. Show that the average value of $\frac{p(t)}{s}$ over the time $0 \leq t \leq \frac{2\pi}{\sqrt{C}}$ it takes for the planet P to move from one perihelion to the next is 0.

5. Taylor's Series and a Relativistic Term. The next problem considers the Taylor series for the function $g(x) = \frac{1+2x}{1+x}$ and the resulting approximation of the term $\frac{r\Phi'(r)}{\Phi(r)} \approx -2\frac{1 + 2\frac{12\kappa^2}{c^2 r^2}}{1 + \frac{12\kappa^2}{c^2 r^2}}$ involved in the computation of the relativistic component of the precession of a planet in section 5K.

Problem 5.22. Consider the function $g(x) = \frac{1+2x}{1+x}$. Check that $g'(x) = \frac{2(1+x)-(1+2x)}{(1+x)^2} = \frac{1}{(1+x)^2}$ and use mathematical induction to show that the higher derivatives of $g(x)$ are given $g^{(i)}(x) = (-1)^{i+1}i!(1+x)^{-(i+1)}$ for $i \geq 1$. It follows that the Taylor series of $g(x)$ centered at $x = 0$ is

$$\frac{1+2x}{1+x} = g(x) = g(0) + g'(0)x + \frac{g^{(2)}(0)}{2}x^2 + \frac{g^{(3)}(0)}{3!}x^3 + \frac{g^{(4)}(0)}{4!}x^4 + \dots$$

$$= 1 + x - x^2 + x^3 - x^4 + \dots .$$

Show that for all x with $|x| < 1$, this Taylor series converges to the function $g(x)$ that gives rise to it. To see that this is so, we'll start with the closely related power series

$$1 - x + x^2 - x^3 + \dots + (-1)^k x^k + \dots .$$

Let $S_n = 1 - x + x^2 - x^3 + \dots + (-1)^n x^n$ be the sum of the first $n+1$ terms of this series. Since $xS_n = x - x^2 + x^3 + \dots + (-1)^n x^{n+1}$, we get $S_n + xS_n = 1 + (-1)^n x^{n+1}$ and hence $S_n = \frac{1+(-1)^n x^{n+1}}{1+x}$. For $|x| < 1$, let n go to infinity to see that $1 - x + x^2 - x^3 + \dots + (-1)^k x^k + \dots$ converges to $\frac{1}{1+x}$. Conclude that the Taylor series of $g(x)$ centered at $x = 0$ converges to $2 - \frac{1}{1+x} = \frac{2(1+x)-1}{1+x} = \frac{1+2x}{1+x}$.

Turn to section 5K. We saw there that $x = \frac{12\kappa^2}{c^2 r^2}$ is very small. So for this x the higher terms of the Taylor series are negligible and hence

$$\frac{r\Phi'(r)}{\Phi(r)} = -2\,\frac{1 + 2\frac{12\kappa^2}{c^2 r^2}}{1 + \frac{12\kappa^2}{c^2 r^2}} \approx -2\Big(1 + \frac{12\kappa^2}{c^2 r^2}\Big) \qquad .$$

is a tight approximation of $\frac{r\Phi'(r)}{\Phi(r)}$. Check the approximation for the planet Mercury with $r = a_1$ the semimajor axis of its orbit.

6. The Perturbing Force of an Interior Planet. In this closing segment we'll study the gravitational force exerted on a planet P by a planet Q that orbits the Sun S inside the orbit of P. We'll let m be the mass of P and M the mass of the planet Q. Let r be the distance from P to S. As in section 5I we'll assume that the orbit of Q is a circle of radius R centered at S and that the mass of Q is spread uniformly throughout this circle. This time $R < r$. Under this "averaged" assumption, we will compute the magnitude of the force of Q on P and show that it is centripetal in the direction of S. With the masses of P and S regarded to be concentrated at their centers, both P and S are point-masses. The linear density of the thin, homogeneous circular ring of radius R with center S that represents the mass of Q is $\frac{M}{2\pi R}$.

Place P at the origin O of a polar coordinate system with angle coordinate ϕ. Refer to Figure 5.20. Let ϕ be any angle with the property that the ray it determines intersects the circular ring. The angle ϕ is at its maximum when the ray it determines is tangent to the circle. It follows that $\sin\phi_{\max} = \frac{R}{r}$ so that $\phi_{\max} = \sin^{-1}\frac{R}{r}$. The smallest ϕ is negative and by the symmetry of things, $\phi_{\min} = -\phi_{\max}$. Let Q_1 be a typical point on the circle, let ϕ be its angle, and let Q_2 be the second

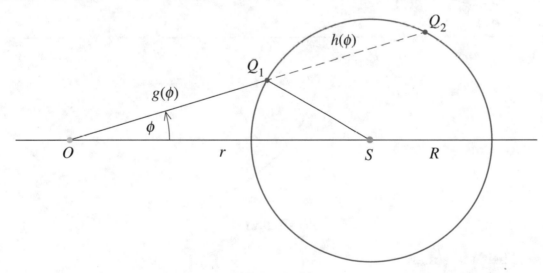

Figure 5.20

point on the circle that ϕ determines. Define the functions $r = g(\phi)$ and $r = h(\phi)$ by $g(\phi) = OQ_1$ and $h(\phi) = OQ_2$. Applying the law of cosines to the triangle $\triangle SOQ_1$ and then again to the triangle $\triangle SOQ_2$, we get that

$$R^2 = r^2 + g(\phi)^2 - 2rg(\phi)\cos\phi \quad \text{and} \quad R^2 = r^2 + h(\phi)^2 - 2rh(\phi)\cos\phi.$$

Solve for $g(\phi)$ and $h(\phi)$ with the quadratic formula to get

$$g(\phi) = \frac{2r\cos\phi \pm \sqrt{4r^2\cos^2\phi - 4(r^2 - R^2)}}{2} \quad \text{and} \quad h(\phi) = \frac{2r\cos\phi \pm \sqrt{4r^2\cos^2\phi - 4(r^2 - R^2)}}{2}.$$

Since $g(\theta) < h(\theta)$,

$$g(\phi) = r\cos\phi - \sqrt{R^2 - r^2\sin^2\phi} \quad \text{and} \quad h(\phi) = r\cos\phi + \sqrt{R^2 - r^2\sin^2\phi}.$$

We turn to compute the force that the ring exerts on P. In Figure 5.21, $d\phi$ is a sliver of an angle. The blue arcs that ϕ and $\phi + d\phi$ cut out are labeled 1 and 2 in the figure. Let ds_1 and ds_2 be their respective lengths. The geometry and the Pythagorean theorem provide the approximations

$$ds_1 \approx \sqrt{(g(\phi)d\phi)^2 + (g(\phi + d\phi) - g(\phi))^2} = \sqrt{g(\phi)^2 + \frac{(g(\phi+d\phi)-g(\phi))^2}{(d\phi)^2}}\, d\phi \approx \sqrt{g(\phi)^2 + g'(\phi)^2}\, d\phi, \text{ and}$$

$$ds_2 \approx \sqrt{(h(\phi)d\phi)^2 + (h(\phi + d\phi) - h(\phi))^2} = \sqrt{h(\phi)^2 + \frac{(h(\phi)-h(\phi+d\phi))^2}{(d\phi)^2}}\, d\phi \approx \sqrt{h(\phi)^2 + h'(\phi)^2}\, d\phi.$$

Since $d\phi$ is a small sliver of an angle and ds_1 and ds_2 are very short, regard each of the blue arcs as

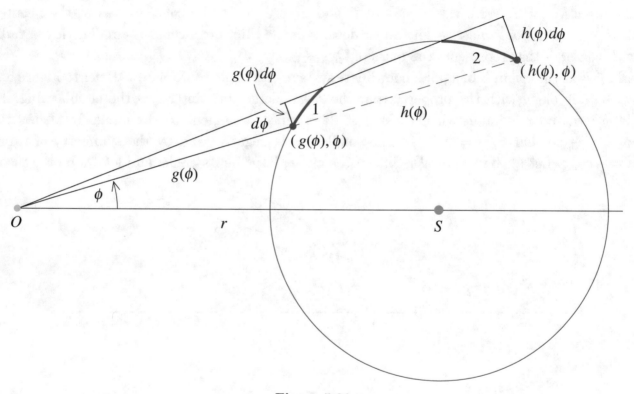

Figure 5.21

a point-mass. Their masses are

$$\frac{M}{2\pi R}ds_1 \approx \frac{M}{2\pi R}\sqrt{g(\phi)^2 + g'(\phi)^2}\, d\phi \quad \text{and} \quad \frac{M}{2\pi R}ds_2 \approx \frac{M}{2\pi R}\sqrt{h(\phi)^2 + h'(\phi)^2}\, d\phi,$$

respectively. By Newton's law of universal gravitation, the magnitudes of the forces with which the two arcs pull on P are

$$\frac{Gm}{g(\phi)^2}\left(\frac{Mds_1}{2\pi R}\right) \approx \frac{GmM}{2\pi R}\cdot\frac{\sqrt{g(\phi)^2+g'(\phi)^2}}{g(\phi)^2}\, d\phi \quad \text{and} \quad \frac{Gm}{h(\phi)^2}\left(\frac{Mds_2}{2\pi R}\right) \approx \frac{GmM}{2\pi R}\cdot\frac{\sqrt{h(\phi)^2+h'(\phi)^2}}{h(\phi)^2}\, d\phi,$$

So the magnitude of the combined pull of the two blue arcs on P is approximately

$$\frac{GmM}{2\pi R}\cdot\frac{\sqrt{g(\phi)^2+g'(\phi)^2}}{g(\phi)^2}\, d\phi + \frac{GmM}{2\pi R}\cdot\frac{\sqrt{h(\phi)^2+h'(\phi)^2}}{h(\phi)^2}\, d\phi = \frac{GmM}{2\pi R}\left(\frac{\sqrt{g(\phi)^2+g'(\phi)^2}}{g(\phi)^2} + \frac{\sqrt{h(\phi)^2+h'(\phi)^2}}{h(\phi)^2}\right)d\phi.$$

What follows next is the computation of the sum $\frac{\sqrt{g(\phi)^2+g'(\phi)^2}}{g(\phi)^2} + \frac{\sqrt{h(\phi)^2+h'(\phi)^2}}{h(\phi)^2}$. After some algebra,

$$\frac{\sqrt{g(\phi)^2+g'(\phi)^2}}{g(\phi)^2} + \frac{\sqrt{h(\phi)^2+h'(\phi)^2}}{h(\phi)^2} = \frac{h(\phi)^2\sqrt{g(\phi)^2+g'(\phi)^2} + g(\phi)^2\sqrt{h(\phi)^2+h'(\phi)^2}}{g(\phi)^2 h(\phi)^2}$$

$$= \frac{h(\phi)\sqrt{h(\phi)^2 g(\phi)^2 + h(\phi)^2 g'(\phi)^2} + g(\phi)\sqrt{g(\phi)^2 h(\phi)^2 + g(\phi)^2 h'(\phi)^2}}{g(\phi)^2 h(\phi)^2}.$$

Computations identical to those in the analogous step of section 5I(i) show that

$$\sqrt{h(\phi)^2 g(\phi)^2 + h(\phi)^2 g'(\phi)^2} \quad \text{and} \quad \sqrt{g(\phi)^2 h(\phi)^2 + g(\phi)^2 h'(\phi)^2}$$

are both equal to $\frac{R(r^2-R^2)}{\sqrt{R^2-r^2\sin^2\phi}}$, so that

$$\frac{\sqrt{g(\phi)^2+g'(\phi)^2}}{g(\phi)^2} + \frac{\sqrt{h(\phi)^2+h'(\phi)^2}}{h(\phi)^2} = \frac{h(\phi)+g(\phi)}{g(\phi)^2 h(\phi)^2}\frac{R(r^2-R^2)}{\sqrt{R^2-r^2\sin^2\phi}} = \frac{2r\cos\phi}{(r^2-R^2)^2}\frac{R(r^2-R^2)}{\sqrt{R^2-r^2\sin^2\phi}} = \frac{R}{r^2-R^2}\frac{2r\cos\phi}{\sqrt{R^2-r^2\sin^2\phi}}.$$

Conclude that the magnitude of the combined pull of the two blue arcs on the point-mass P is approximately equal to

$$\frac{GmM}{2\pi R}\left(\frac{R}{r^2-R^2}\frac{2r\cos\phi}{\sqrt{R^2-r^2\sin^2\phi}}\right)d\phi = \frac{GmMr}{\pi R(r^2-R^2)}\frac{\cos\phi}{\sqrt{1-\frac{r^2}{R^2}\sin^2\phi}}\,d\phi.$$

The smaller the sliver $d\phi$, the tighter this approximation (and all the approximations along the way). The magnitudes of the vertical and horizontal components of this pull (with regard to Figure 5.21) are

$$\frac{GmMr}{\pi R(r^2-R^2)}\frac{\sin\phi\cos\phi}{\sqrt{1-\frac{r^2}{R^2}\sin^2\phi}}\,d\phi \quad \text{and} \quad \frac{GmMr}{\pi R(r^2-R^2)}\frac{\cos^2\phi}{\sqrt{1-\frac{r^2}{R^2}\sin^2\phi}}\,d\phi,$$

respectively. By adding up these magnitudes over the entire circular ring of radius R using the strategy of integral calculus, we see that the magnitudes of the vertical and horizontal components of the gravitational force that this ring exerts on the point-mass P are equal "on the nose" to

$$\frac{GmMr}{\pi R(r^2-R^2)}\int_{-\phi_{max}}^{\phi_{max}}\frac{\sin\phi\cos\phi}{\sqrt{1-\frac{r^2}{R^2}\sin^2\phi}}\,d\phi \quad \text{and} \quad \frac{GmMr}{\pi R(r^2-R^2)}\int_{-\phi_{max}}^{\phi_{max}}\frac{\cos^2\phi}{\sqrt{1-\frac{r^2}{R^2}\sin^2\phi}}\,d\phi,$$

respectively. (Since ϕ_{max} satisfies $\sin\phi_{max} = \frac{R}{r}$, $\frac{r}{R}\sin\phi_{max} = 1$ and hence $\frac{r}{R}\sin(-\phi_{max}) = -1$. Notice therefore that these integrals are both improper integrals.) Surprisingly, these definite inte-

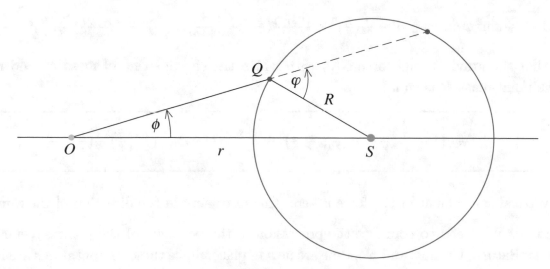

Figure 5.22

grals are—with the exception of the limits of integration—identical to those derived in section 5I(i).

A change of variables makes it possible to evaluate these integrals. Turn to Figure 5.22. By applying the law of sines to the triangle ΔOQS, we see that $\frac{\sin \phi}{R} = \frac{\sin(\pi - \varphi)}{r}$. Therefore by Figure 5.13, $\sin \phi = \frac{R}{r} \sin(\pi - \varphi) = \frac{R}{r} \sin \varphi$. Since $-\frac{\pi}{2} < \phi < \frac{\pi}{2}$, ϕ is the function $\phi = \sin^{-1}(\frac{R}{r} \sin \varphi)$ of φ. The fact that $\frac{r}{R} \sin \phi = \sin \varphi$, tells us that $1 - \frac{r^2}{R^2} \sin^2 \phi = 1 - \sin^2 \varphi = \cos^2 \varphi$, so that

$$\sqrt{1 - \tfrac{r^2}{R^2} \sin^2 \phi} = \cos \varphi.$$

In a similar way, $\cos^2 \phi = 1 - \sin^2 \phi = 1 - \frac{R^2}{r^2} \sin^2 \varphi$ and hence

$$\cos \phi = \sqrt{1 - \tfrac{R^2}{r^2} \sin^2 \varphi}.$$

Differentiate the equation $\sin \phi = \frac{R}{r} \sin \varphi$ implicitly with respect to φ, to get $\cos \phi \cdot \frac{d\phi}{d\varphi} = \frac{R}{r} \cos \varphi$ and hence

$$(\cos \phi) \, d\phi = \tfrac{R}{r} (\cos \varphi) \, d\varphi \, .$$

Notice from Figure 5.22 that as ϕ varies from $-\phi_{\max}$ to $\phi_{\max}$, the angle φ varies from $-\frac{\pi}{2}$ to $\frac{\pi}{2}$. By substituting the above equalities into the earlier force integrals, we can conclude that the magnitudes of the vertical and horizontal components of the gravitational force that the entire circular ring of radius R exerts on the point-mass P are equal to the definite integrals

$$\tfrac{GmMr}{\pi R(r^2 - R^2)} \int_{-\frac{\pi}{2}}^{\frac{\pi}{2}} \tfrac{R^2}{r^2} \sin \varphi \, d\varphi \quad \text{and} \quad \tfrac{GmMr}{\pi R(r^2 - R^2)} \int_{-\frac{\pi}{2}}^{\frac{\pi}{2}} \tfrac{R}{r} \sqrt{1 - \tfrac{R^2}{r^2} \sin^2 \varphi} \, d\varphi,$$

respectively, in the variable φ. Show that the first integral—representing the vertical force that the ring exerts on P—is equal to 0. So focus on the second. Note first that

$$\int_{-\frac{\pi}{2}}^{\frac{\pi}{2}} \sqrt{1 - \tfrac{R^2}{r^2} \sin^2 \varphi} \, d\varphi = 2 \int_0^{\frac{\pi}{2}} \sqrt{1 - \tfrac{R^2}{r^2} \sin^2 \varphi} \, d\varphi.$$

To solve this definite integral turn to the solution of $\int_0^{\frac{\pi}{2}} \sqrt{1 - \varepsilon^2 \sin^2 \theta} \, d\theta$ in section 5G. Use the formula derived there to show that

$$\int_0^{\frac{\pi}{2}} \sqrt{1 - \tfrac{R^2}{r^2} \sin^2 \varphi} \, d\varphi = \tfrac{\pi}{2} \left[1 - \tfrac{1}{2^2} (\tfrac{R}{r})^2 - \tfrac{(1 \cdot 3)^2}{(2 \cdot 4)^2} \tfrac{1}{3} (\tfrac{R}{r})^4 - \tfrac{(1 \cdot 3 \cdot 5)^2}{(2 \cdot 4 \cdot 6)^2} \tfrac{1}{5} (\tfrac{R}{r})^6 - \left(\tfrac{1 \cdot 3 \cdot 5 \cdot 7}{2 \cdot 4 \cdot 6 \cdot 8} \right)^2 \tfrac{1}{7} (\tfrac{R}{r})^8 - \cdots \right].$$

It follows that the gravitational force $G(r)$ with which the circular ring of mass M and radius R attracts the point-mass P is equal to

$$G(r) = \tfrac{GmM}{(r^2 - R^2)} \left[1 - \tfrac{1}{2^2} (\tfrac{R}{r})^2 - \tfrac{(1 \cdot 3)^2}{(2 \cdot 4)^2} \tfrac{1}{3} (\tfrac{R}{r})^4 - \tfrac{(1 \cdot 3 \cdot 5)^2}{(2 \cdot 4 \cdot 6)^2} \tfrac{1}{5} (\tfrac{R}{r})^6 - \left(\tfrac{1 \cdot 3 \cdot 5 \cdot 7}{2 \cdot 4 \cdot 6 \cdot 8} \right)^2 \tfrac{1}{7} (\tfrac{R}{r})^8 - \cdots \right]$$

Since the vertical component of the force is zero, this force acts in the direction of the Sun S.

All the tools you need to compute the precession of the perihelia of the planets Venus, Earth, Mars, Jupiter, Saturn, Uranus, and Neptune are now in place. Since these computations are laborious and similar in spirit to those already undertaken in section 5I, you'll get a pass.

Mathematics of Interplanetary Flight

6

The missions that have explored the inner and outermost reaches of our solar system over the past 50 years have been one of the most remarkable success stories of modern times. Dozens of spacecraft have journeyed far and wide to study the Sun, the planets, their moons, as well as some asteroids and comets. In flybys and in orbits around them, they have gathered volumes of information and captured thousands of images that could not have been imagined before. Since much of this has been described in Chapter 2, we will now turn to a discussion of the basic mathematics and physics that underly the design of the trajectories that send space probes to their targets. In so doing, this chapter will build on the description of rocket engines and the analysis of the flight path of NASA's *Juno* mission to Jupiter that concludes Chapter 2. As the mission of *Juno* already illustrated, the flight path of a spacecraft almost invariably consists of a carefully designed combination of elliptical segments together with hyperbolic flybys of planets that provide the craft with the additional velocity it needs. Accordingly, it is one of the primary concerns of this chapter to develop a mathematical study of hyperbolic trajectories that parallels what was done in Chapter 5 for elliptical orbits.

This chapter includes the study of the essentials about rocket engines, gravitational spheres of influence, Hohmann transfers, hyperbolic flybys, gravity assists, and orbit insertions. The *NEAR-Shoemaker*, *Voyager*, and *Cassini* missions will provide concrete illustrations of the particulars. The successes of the *Voyager* and *Cassini* missions have already been highlighted in Chapter 2. These flights are examples of NASA's large strategic missions, the most ambitious and costly (often in excess of $1 billion) of NASA's programs. The *NEAR-Shoemaker* mission was the first of NASA's program of smaller-scale projects that go from development to flight within an efficient three years (at a cost of no more than $150 million). It is this program's goal to explore the solar system with the latest technologies and to include the research of universities and the industrial sector. The *Near Earth Asteroid Rendezvous (NEAR)* craft was sent to investigate the asteroid 433 Eros. After spending one year in orbit around Eros, it concluded its mission in 2004 by making a soft landing on the asteroid. The *MESSENGER* mission to study Mercury and the *Dawn* mission to study the asteroids Ceres and Vesta were also carried out under this program.

© Alexander J. Hahn 2020
A. J. Hahn (ed.), *Basic Calculus of Planetary Orbits and Interplanetary Flight*,
https://doi.org/10.1007/978-3-030-24868-0_6

An installation that is critical for all the interplanetary missions that NASA and the Jet Propulsion Laboratory (JPL) operate is the *Deep Space Network* (DSN). This consists of an array of giant radio antennas that provides the indispensable communications link between all spacecraft and Earth. This largest and most sensitive scientific telecommunications system on our planet makes it possible to track and command these craft and to record the images and scientific information that they return from all parts of our solar system. The DSN operates the facilities at Goldstone, near Barstow, California; near Madrid, Spain; and near Canberra, Australia. These sites—spaced approximately 120 degrees apart in longitude around the 360° of Earth's circumference—permit constant communication with spacecraft as they move and our planet rotates. Before a distant spacecraft sinks below the horizon at one DSN site, another site picks up the signal and continues the contact.

An assumption that will be in effect whenever the focus is on a mathematical aspect of the trajectory of a spacecraft is the following. The main rocket engine or thruster that fires to increase or decrease the velocity of a craft during a trajectory correction maneuver can burn for as long as an hour or two on a few (rare) occasions during its mission. In the context of flights that almost always go on for several years, these are very short spans of time. In our study of the trajectory of a spacecraft, we will assume that the craft's engine is either not firing at all or that it fires in an instantaneous burst. When Newton's conclusions about gravity and trajectories—as these were developed in Chapter 4—are applied to the flight of a spacecraft, it will always be assumed that the craft's thrusters are not firing and that a single dominant gravitational force drives the craft in its flight. In this last regard we will assume that such a force on the spacecraft is a centripetal force directed to the center of mass of the attracting body and that Newton's law of universal gravitation applies to it. This means that what was set out in Chapter 5A applies in the current context and that the results of Chapters 5B to 5E apply in elliptical situations.

6A. *NEAR-Shoemaker* and Eros. The idea to send a spacecraft to explore an asteroid had been proposed since the early 1960s. It was the *NEAR* craft's mission to rendezvous with the asteroid 433 Eros, to go into orbit around it, and to study it for one year. The primary scientific objective of *NEAR* was to study properties of Eros "in-the-large", namely, its shape, composition, mineralogy, internal mass distribution, and magnetic field. While Eros is a Near Earth Asteroid (it has come to within 0.178 au or about 27 million km of Earth), it does not come close enough to pose a threat.

The *NEAR* spacecraft was launched with a Delta II rocket (see Figure 2.3) in February, 1996. This rocket has a total mass of 220,000 kg, a height of 38.20 m, a diameter of 2.44 m, and a thrust of 4 million N. (Recall that N is the abbreviation for newton). Its three stages fired in precisely timed succession to send *NEAR* off on a flawless launch. In spite of its power, the Delta II is

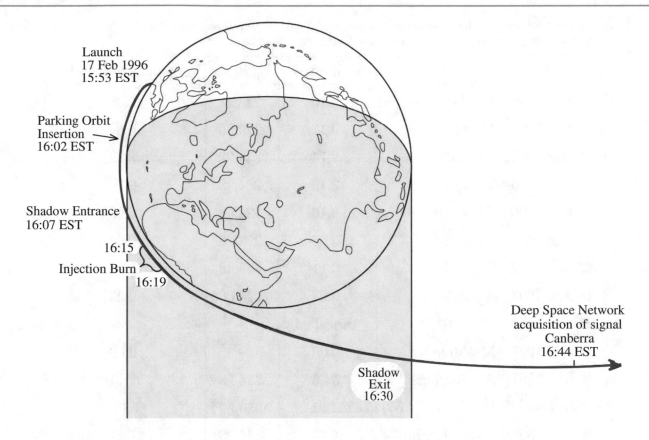

Figure 6.1. The flight of *NEAR* from launch to the acquisition of its signal in Canberra, Australia.

relatively small. The launch of *NEAR* was the first to use such a small rocket for an interplanetary flight. Taking advantage of the Earth's rotational speed, the Delta II moved into a low Earth parking orbit after about 10 minutes. After another 12 minutes the upper stage fired, to give *NEAR* its final boost. The craft separated from the upper stage 26 minutes after launch and was now on its way on an elliptical trajectory that had the Sun as its focal point. See Figure 6.1.

For its flight, *NEAR* had a propulsion system that consisted of a main engine capable of generating 470 N of thrust, four 21 N thrusters, and seven 3.5 N thrusters. The main thruster provided the large velocity adjustment—or LVA—maneuvers. A critical role of the smaller thrusters was to carefully position and orient the craft so that the thrusts that effected the velocity adjustments pointed in the intended direction with precision. The large thruster burned liquid hydrazine in combination with a liquid oxidizer. All the smaller thrusters burned hydrazine only. Of the total mass of 805 kg of the craft at launch, a full 40%, or 318 kg, was propellant (209 kg hydrazine and 109 kg oxidizer). In spite of the LVA maneuvers that the ample fuel supply made possible, JPL engineers chose a "Delta V−Earth Gravity Assist" trajectory for *NEAR*, meaning that the changes in velocity Δv included a velocity changing Earth flyby. (The flight path of the *Juno* spacecraft

Table 6.1. The table lists data for the initial two years of *NEAR*'s propulsion maneuvers. The craft's initial mass was 805.07 kg and it entered orbit around Eros with a mass of 502.13 kg.

date	event	Δv in m/sec	I_{sp} in sec	propellant mass in kg
17 Feb 1996	stabilization	0.16	228.48	0.06
24 Feb 1996	MCM-1	0.11	229.00	0.04
2 Mar 1996	TCM-1	9.74	234.98	3.39
13 Sep 1996	TCM-2A	2.13	234.98	0.74
13 Sep 1996	TCM-2B	0.16	228.48	0.06
6 Jan 1997	TCM-3	0.06	228.48	0.02
29 Jan 1997	TCM-4	0.11	228.48	0.04
18 Jun 1997	TCM-5	0.63	220.00	0.23
27 Jun 1997	TCM-6	cancelled	–	–
27 Jun 1997	Mathilde flyby	0.00	–	0.00
3 July 1997	settling burn	3.28	234.98	1.14
3 July 1997	DSM-1 (TCM-7)	261.01	313.55	65.05
3 July 1997	attitude/trim	4.23	234.98	1.47
23 July 1997	TCM-8	5.69	234.98	1.81
23 July 1997	attitude control	0.05	220.00	0.02
17 sep 1997	TCM 9	0.81	234.98	0.26
9 Jan 1998	TCM 10	0.08	220.00	0.03

to Jupiter was also a $\Delta VEGA$ trajectory. See Chapter 2H.)

Table 6.1 captures the initial sequence of *NEAR*'s propulsion maneuvers. The Δv column lists the change in velocity in meters per second that the maneuver achieved. A stabilization maneuver soon after launch was followed by a slight trajectory adjustment (the momentum control maneuver MCM-1) one week later and the first larger trajectory correction maneuver TCM-1 two weeks after launch. The specific impulse I_{sp} referred to in column four is directly related to the speed v_{ex} of the exhaust materials that the maneuver generates by the equation $v_{ex} = I_{sp} \cdot g_0$, where $g_0 = 9.80665$ m/sec^2 is the standard value for Earth's gravitational constant. Let $w(t)$ be the weight of the propellant at time t after the engine begins to fire. The connection between force and mass tells us that $w(t) = m(t)g_0$ where $m(t)$ is the mass of the propellant. It follows from the analysis of Chapter 2G that the momentum thrust of the engine is

$$F_{\text{mom}} = v_{\text{ex}}m'(t) = I_{\text{sp}}w'(t).$$

The trajectory correction maneuvers TCM-2 through TCM-5 from September 1996 to June 1997 fine-tuned *NEAR*'s path for its flyby of the asteroid 253 Mathilde. The first major correction of *NEAR*'s trajectory—the deep space maneuver DMS-1—occurred six days after the Mathilde flyby. On July 3rd, 1997, the 21 N thrusters fired to force the liquid oxidizer against its tank outlets to start the burn of the main engine. In a nearly perfect burn, *NEAR*'s LVA thruster fired for nearly 11 minutes to reduce the craft's velocity by 261 m/sec, putting the craft on course for the important January 1998 flyby of Earth. The maneuvers TCM 8, 9, and 10 fine-tuned the craft's approach to our planet (but TCM-11 was not needed and was cancelled). The flyby of Earth on January 23rd at an altitude of 540 km above the Earth's surface was critical. See Figure 6.2. It lowered the craft's

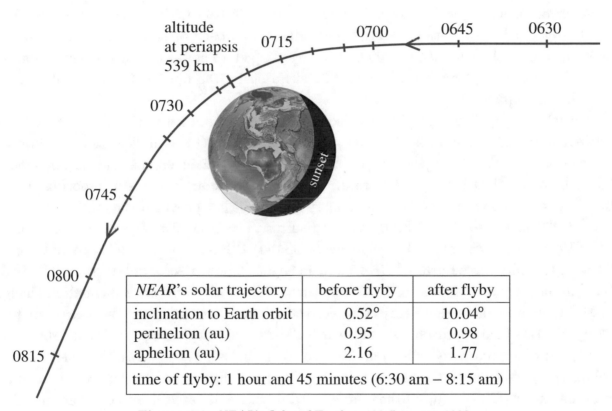

NEAR's solar trajectory	before flyby	after flyby
inclination to Earth orbit	0.52°	10.04°
perihelion (au)	0.95	0.98
aphelion (au)	2.16	1.77
time of flyby: 1 hour and 45 minutes (6:30 am − 8:15 am)		

Figure 6.2. *NEAR*'s flyby of Earth on 23 January, 1998.

speed so that it would be close to that of Eros at the time of the planned rendezvous. But the primary goal of the flyby was the alignment of *NEAR*'s orbital plane. The orbit planes of Eros and Earth around the Sun are separated by an angle of 11°. During the flyby, Earth's gravitational force pulled *NEAR* from its heliocentric orbit in Earth's orbital plane into a heliocentric orbit in the orbit plane of Eros. The flyby also decreased the aphelion distance of the craft's orbit and rotated its focal axis (the line joining perihelion and aphelion positions) to nearly match those of Eros.

Table 6.1 presents all the adjustment maneuvers of *NEAR*'s trajectory from launch to the Earth flyby. Other than the minor trajectory adjustments TCM-12 and TCM-15 of *NEAR*'s trajectory (TCM-13 was never scheduled and TCM-14 was cancelled), nothing further would be needed until the craft's approach of Eros at the end of 1998.

From late December 1998 into January 1999, *NEAR* was scheduled to perform four rendezvous maneuvers (RNDs). These were designed to slow *NEAR* by 949 m/sec, bring the craft to a velocity of 5 m/sec relative to Eros, and, ultimately, to settle it into orbit around the asteroid. The first and largest of these maneuvers called for a main thruster LVA burn to effect a velocity decrease of 650 m/sec. The burn began on schedule on Sunday, 20 December, at 22:00 hours (Greenwich Mean Time). The main engine fired, but suddenly—within seconds—the burn stopped. Soon thereafter the spacecraft's signal was lost and a spacecraft emergency was declared. "Black Sunday" had begun. After an agonizing wait of over 24 hours, communication with the craft was reestablished and *NEAR* returned to operational status. The original approach of Eros and the scheduled orbit insertion burn on 10 January 10th, 1999 had to be scrapped. (The reason for the engine shut down was later determined to have been a lateral acceleration by the craft that was greater than the limits set by the software.)

By December 23rd, 1998, *NEAR* had sailed past Eros by a distance of 3800 km. The information that *NEAR* obtained about Eros (its mass, volume, and density) as it flew past, facilitated the reconfiguration of *NEAR*'s trajectory that the misfire of its main engine had necessitated. The revised flight plan called for *NEAR* to return to Eros in February 2000. An important first step was the deep space maneuver DSM-2 on January 3rd, 1999 that provided the craft with a velocity decrease of 932 m/sec to slow *NEAR* relative to Eros. The DSM-2 maneuver was the last time *NEAR*'s LVA thruster was fired. The maneuvers DSM-1, DSM-2, and the failed rendezvous effort RND-1 were the only three burns of *NEAR*'s main engine. The trajectory corrections TCM-18 and TCM-19 (in January and August of 1999) fine tuned the velocity and orbit inclination changes of the DSM-2 maneuver. Four more trajectory correction maneuvers returned the spacecraft to Eros and set up *NEAR*'s final approach. One of them, TCM-22 in the beginning of February 2000 reduced *NEAR*'s speed of approach to Eros from 20 m/sec to 10 m/sec. When *NEAR* reached the desired orbit plane (perpendicular to the plane of Earth's orbit) at a distance of 330 km from Eros, an orbit insertion maneuver (OIM) reduced the craft's speed further and eased it into orbit around Eros on February 14th, 2000. Appropriately, it was Valentine's Day! After all, the asteroid had taken its name from the Greek god of love. The clear image of the large crater on Eros that *NEAR* sent back confirmed that it had achieved orbit.

Figure 6.3 presents a sketch of *NEAR*'s reconfigured trajectory that brought the craft from the aborted main engine burn to its first orbit around Eros. The plans for *NEAR*'s trajectory had

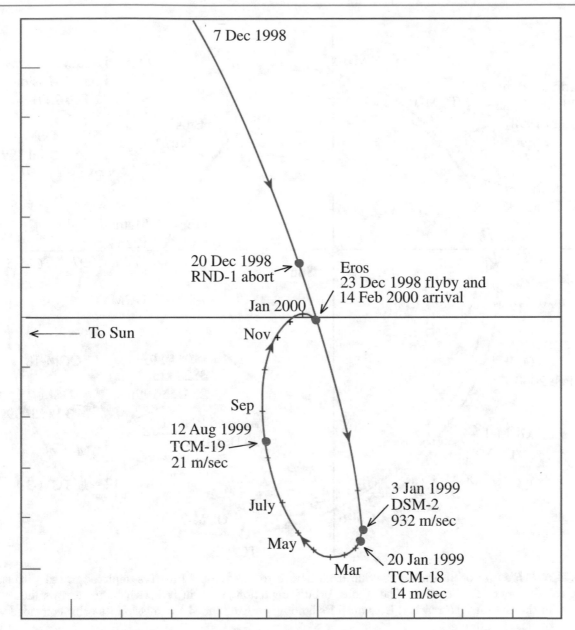

7 Dec 1998

20 Dec 1998
RND-1 abort

Eros
23 Dec 1998 flyby and
14 Feb 2000 arrival

Jan 2000

Nov

To Sun

Sep

12 Aug 1999
TCM-19
21 m/sec

July

3 Jan 1999
DSM-2
932 m/sec

May

20 Jan 1999
TCM-18
14 m/sec

Mar

Figure 6.3. A diagram of *NEAR*'s revised trajectory that brought the craft to its rendezvous with Eros.

included ample fuel margins and contingency options that made it possible for the craft to recover from the aborted first attempt at a rendezvous with Eros and to complete its mission to the asteroid. Figure 6.4 is a diagram of *NEAR*'s entire flight from launch to mission's end.

NEAR's first orbit around Eros had periapsis and apoapsis distances of 321 km and 366 km respectively. Because the mass of Eros had initially been overestimated by 9%, this first orbit was considerably smaller than originally planned. The gravitational field of Eros is weak, so that for *NEAR* to stay in orbit its orbital speed needed to be low. The speed in its first orbit was only about 1 m/sec or 3.6 km/hour. Figure 2.21 is an image taken by *NEAR* soon after its insertion into orbit. It tells us that Eros is a peanut-shaped rock covered with craters and that it has a large gouge in the center. Eros measures about 34 kilometers in length with a diameter of about 12 kilometers at its middle. The composite image of Figure 2.20 shows Eros dwarfed by Ceres and Vesta, two of the

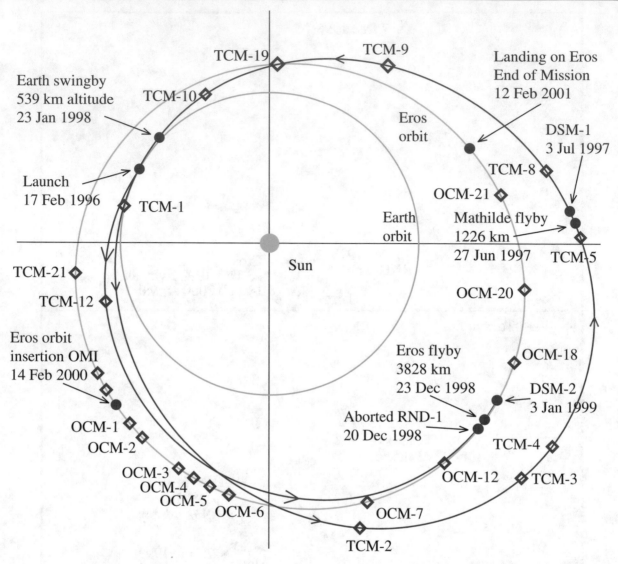

Figure 6.4. *NEAR*'s first orbit around the Sun from launch to the flyby of Earth is depicted in red with small red arrows indicated the craft's direction. The second orbit's depiction starts in red with larger arrows indicating the direction. From the time *NEAR* reached Eros until its landing on Eros ended the mission its orbit coincided with the orbit of Eros depicted in orange.

largest asteroids. A month after its insertion into orbit around Eros, the craft was renamed *NEAR-Shoemaker* in honor of the late Eugene Shoemaker, a pioneer in the study of asteroids and comets. (See Chapter 2E and Figure 2.26 for information about one of the comets he discovered.)

Table 6.2 is a record of *NEAR-Shoemaker*'s maneuvers around Eros. The first eight orbit correction maneuvers (OCMs) that followed orbit insertion decreased the size of the orbit and tilted the orbital plane. This was done in stages by first reducing the periapsis distance and then circularizing the orbit. By mid-July 2000 the orbit was close to a circle with a 35 km radius at right angles to the asteroid's equator. Thereafter, more correction maneuvers expanded and contracted *NEAR-Shoemaker*'s orbit several times. In the process, the orbital plane was aligned with the asteroid's equatorial plane. Maneuvers in mid and late October 2000 flattened the ellipse of the orbit and brought the craft to within 5.3 km of the surface of Eros. By mid-December 2000, *NEAR-Shoemaker* was back in a low circular orbit with a 35 km radius where it would remain until the end of its

Table 6.2. *NEAR's* changing orbits around Eros—from farther away, to closer—made it possible to study various aspects of the asteroid carefully. The axis of rotation of Eros is perpendicular to its long diameter. The asteroid's true equator (ATE) is determined by the plane perpendicular to its axis of rotation through its center of mass.

date	maneuver	periapsis × apoapsis in km of resulting orbit	orbit period in days	inclination to ATE in degs	time in the given orbit in days	Δv m/sec
14 Feb 2000	OIM	321 × 366	21.8	35	10	10.00
24 Feb 2000	OCM-1	204 × 365	16.5	34	8	0.13
3 Mar 2000	OCM-2	203 × 206	10.1	37	30	0.22
2 Apr 2000	OCM-3	100 × 209	6.7	55	9	0.50
11 Apr 2000	OCM-4	99 × 101	3.5	59	11	0.37
22 Apr 2000	OCM-5	50 × 101	2.2	64	8	0.45
30 Apr 2000	OCM-6	49 × 52	1.2	90	68	1.92
7 Jul 2000	OCM-7	35 × 51	1.0	90	7	0.32
14 Jul 2000	OCM-8	35 × 39	0.8	90	10	0.24
24 Jul 2000	OCM-9	36 × 56	1.1	90	7	0.34
31 Jul 2000	OCM-10	49 × 52	1.2	90	8	0.50
8 Aug 2000	OCM-11	50 × 52	1.2	105	18	1.01
26 Aug 2000	OCM-12	49 × 102	2.3	113	10	1.40
5 Sep 2000	OCM-13	100 × 103	3.5	115	38	0.96
13 Oct 2000	OCM-14	50 × 98	2.2	130	7	1.31
20 Oct 2000	OCM-15	50 × 52	1.2	133	5	0.58
25 Oct 2000	OCM-16	19 × 51	0.7	133	0.8	0.76
26 Oct 2000	OCM-17	64 × 203	5.4	145	8	1.66
3 Nov 2000	OCM-18	194 × 196	9.4	147	34	0.54
7 Dec 2000	OCM-19	34 × 193	4.2	179	6	0.96
13 Dec 2000	OCM-20	34 × 38	0.8	179	43	1.23
24 Jan 2001	OCM-21	22 × 35	0.6	179	4	0.54
28 Jan 2001	OCM-22	19 × 37	0.6	179	0.7	0.56
29 Jan 2001	OCM-23	35 × 36	0.8	179	5	0.68
2 Feb 2001	OCM-24	36 × 36	0.8	179	4	0.02
6 Feb 2001	OCM-25	36 × 36	0.8	179	6	0.01
12 Feb 2001	de-orbit	—	—	135	—	2.54

flight. This sequential configuration of *NEAR-Shoemaker*'s orbits around Eros had to be under-taken with great care, so that the craft would not be thrown out of orbit or pulled into a crash landing. Notice the delicate adjustments in the craft's orbital speed. To conclude the mission, the scientific team decided to attempt a controlled descent and landing on Eros, even though *NEAR-Shoemaker* had not been designed for such a maneuver. The landing on the surface of Eros took place on February 12th, 2001. A de-orbit burn of its smaller thrusters at an altitude of about 5 km followed by several breaking burns moved the craft to a slow descent that took several hours. *NEAR-Shoemaker*'s cameras obtained many high-resolution images along the way. The craft touched down in the "saddle" region of Eros with a speed of about 1.5 m/sec. The successful soft landing of the craft on the surface of Eros was the first by a spacecraft on an asteroid.

During the craft's orbital progression, its cameras, laser radar, and various instruments provided information about the physical characteristics of Eros with increasing accuracy. For the mass M of Eros, it was found that $GM = 4.4621 \times 10^{-4}$ km^3/sec^2 and explicitly that $M = 6.6904 \times 10^{15}$ kg. The average density of Eros is 2.67 grams/cm^3 is about half of Earth's 5.5 grams/cm^3. Eros does not spin around its long axis, instead it is the long axis of Eros that rotates with fixed point the asteroid's center of mass. The rotational speed is 4.5 revolutions per day. *NEAR-Shoemaker*'s instruments also mapped Eros's gravitational and magnetic fields. Its mission had met and exceeded all expectations.

Example 6.1. Consider *NEAR-Shoemaker*'s first orbit of Eros and let a and ε be its semimajor axis and eccentricity. The data in Table 6.2 tells us that $a(1 - \varepsilon) = 321$ km and $a(1 + \varepsilon) = 366$ km, respectively, so that $a = \frac{1}{2}(366 + 321) = 343.5$ km and $\varepsilon = \frac{1}{2}\frac{366-321}{343.5} \approx 0.066$. By Newton's version of Kepler's third law (refer to Chapter 1D), $\frac{a^3}{T^2} = \frac{GM}{4\pi^2}$, where M is the mass of Eros and T the period of the first orbit. It follows that

$$T^2 = \tfrac{4\pi^2 a^3}{GM} = \tfrac{4\pi^2 343.5^3}{4.4621 \times 10^{-4}} \approx 358,592,100 \text{ sec}^2$$

and hence that $T \approx 18,937$ seconds or 5.26 hours. An application of Example 5.1 tells us that the maximum and minimum velocities of *NEAR-Shoemaker* in this initial orbit were

$$\sqrt{\tfrac{GM(1+\varepsilon)}{a(1-\varepsilon)}} \approx \sqrt{\tfrac{4.4621(1.066) \times 10^{-4}}{321}} \approx 1.22 \text{ m/sec and } \sqrt{\tfrac{GM(1-\varepsilon)}{a(1+\varepsilon)}} \approx \sqrt{\tfrac{4.4621(0.934) \times 10^{-4}}{366}} \approx 1.07 \text{ m/sec},$$

respectively. The elongated shape of Eros and the way Eros rotates means that Newton's law of universal gravitation and therefore his version of Kepler's third law apply to *NEAR-Shoemaker*'s orbit around Eros only as rough approximations. (Refer to *Where the Law of Universal Gravitation Fails* in the Problems and Discussions section of Chapter 4.) So it is likely that these estimates for the period of *NEAR-Shoemaker*'s orbit and its orbital speeds are not very tight.

6B. Escape Velocity from Earth. The flight of *NEAR-Shoemaker* from its launch to the target of its mission points to a number of matters that call for further study. The same issues arise for most any spacecraft on a trajectory that takes it to a near or distant planet, one of its moons, an asteroid, or a comet. We'll regard a spacecraft as a point-mass. Since we'll assume that the bodies of the solar system that are relevant to its trajectory have their masses concentrated at their centers of mass, they too are regarded as point-masses.

Figure 6.1 shows the initial path of the *NEAR-Shoemaker* spacecraft. By the time it was first detected by the Deep Space Network facility in Canberra, it had left its parking orbit around Earth behind and was on its own elliptical near-Earth solar orbit. For such a trajectory transfer to be successful, a spacecraft needs to be provided with enough speed to allow it to escape from Earth's gravitational pull. Let's suppose that a spacecraft is in an elliptical orbit around Earth with semimajor axis a eccentricity ε. By Example 5.1, the maximum speed $v_{\max}$ of the craft in such an orbit is given by the formula

$$v_{\max} = \sqrt{\frac{GM_E(1+\varepsilon)}{a(1-\varepsilon)}} < \sqrt{\frac{2GM_E}{a(1-\varepsilon)}},$$

where M_E is the mass of the Earth. Since the term $a(1-\varepsilon)$ is the periapsis distance of the orbit, $\sqrt{a(1-\varepsilon)} > \sqrt{r_E}$, where r_E is the Earth's radius. It follows that

$$v_{\max} < \sqrt{\frac{2GM_E}{r_E}}.$$

Pure logic that tells us that if after its launch the spacecraft heads away from Earth with a speed of at least $\sqrt{\frac{2GM_E}{r_E}}$ then it cannot be in an elliptical orbit around the Earth, so that it will escape Earth's gravitational pull and be on its own near-Earth solar orbit. We'll refer to $v_{\mathrm{esc}} = \sqrt{\frac{2GM_E}{r_E}}$ as the *escape velocity* from Earth. Taking r_E to be equal to Earth's smaller polar radius $r_E = 6{,}357$ km and $GM_E = 398{,}600$ km^3/sec^2 from Table 2.3, we get that

$$v_{\mathrm{esc}} = \sqrt{\frac{2GM_E}{r_E}} = \sqrt{\frac{7.97200 \times 10^5}{6.357 \times 10^3}} \approx 11.2 \text{ km/sec}.$$

Suppose that the spacecraft has mass m and that it is moving with a velocity v. Then its kinetic energy is $\frac{1}{2}mv^2$. If the craft is launched into its trajectory with velocity $v > v_{\mathrm{esc}}$, then

$$\tfrac{1}{2}mv^2 - \tfrac{1}{2}mv_{\mathrm{esc}}^2 = \tfrac{1}{2}mv^2 - \frac{GmM_E}{r_E}$$

is the excess kinetic energy that the craft has available at the start of its initial trajectory around the Sun. Dividing this difference by m, we get the craft's *specific orbital energy* of $\frac{1}{2}v^2 - \frac{GM_E}{r_E}$. Twice this difference is the craft's *launch energy* or *characteristic energy*

$$C_3 = v^2 - \frac{2GM_E}{r_E}.$$

So the velocity with which the spacecraft proceeds into its initial orbit around the Sun is

$$v = \sqrt{C_3 + \frac{2GM_E}{r_E}}.$$

Since the Earth rotates from west to east, spacecraft are typically launched in the easterly direction, so that the launch can take advantage of the additional speed that the Earth's rotation adds. Since the Earth's equatorial radius is 6378 km, a point on Earth's equator moves with a speed of $\frac{2\pi(6378)}{24} \approx 1670$ km/hour, or 0.464 km/sec, due east. At the northern latitude of $28.5°$ at Cape Canaveral, Florida—the primary launch site for NASA spacecraft—such a point will travel at the slower speed of $\frac{2\pi(6378 \cdot \cos 28.5°)}{24} \approx 1474$ km/hour or 0.408 km/sec, again due east. So the additional speed that the Earth's rotation imparts to spacecraft at launch is marginal.

Example 6.2. The characteristic energy of *NEAR-Shoemaker* was $C_3 = 26.0$ km^2/sec^2. Since $\frac{2GM_E}{r_E} \approx 11.2^2 = 125.44$ km^2/sec^2, it follows that the craft's speed of

$$v = \sqrt{C_3 + \frac{2GM_E}{r_E}} \approx \sqrt{26.0 + 125.44} \approx 12.31 \text{ km/sec}$$

was enough to escape Earth's gravitational well and to propel it into its first solar orbit.

Taking advantage of the speed of the Earth in its orbit (an average of 29.78 km/sec according to Table 5.1), the upper stage of the Delta II rocket inserted *NEAR-Shoemaker* into its initial orbit around the Sun with a velocity of about 35 km/sec relative to the Sun. That this is not simply the sum of 12.31 km/sec and 29.78 km/sec is a consequence of the fact that the directions of the two component velocities differ.

A glance at Table 6.1 tells us that the first large velocity adjustment (LVA) of *NEAR-Shoemaker*'s flight path was the DSM-1 maneuver of July 3rd, 1997. This correction was the result of the firing of the craft's main engine. It was already established in Chapter 2G—as consequence of Tsiolkovsky's rocket equation—that any significant change in a spacecraft's velocity achieved by firing the craft's main engine requires significant quantities of fuel. We will see that this is also the case for *NEAR-Shoemaker*'s DSM-1 maneuver that decreased the spacecraft's velocity by 261 m/sec.

Example 6.3. The total mass of *NEAR-Shoemaker* at the time it began its flight was 805 kg. Table 6.1 provides the following information. The fuel the craft had consumed before the DSM-1 maneuver was 5.72 kilograms. So the craft's mass just before DSM-1 was $805.07 - 5.72 = 799.35$ kg. The fuel consumption for DCM-1 was 65.05 kilograms. So the post DSM-1 mass of the craft was $799.35 - 65.05 = 734.30$ kg. The speed of the expelled reaction mass v_{ex} is determined by the equality $v_{\text{ex}} = g_0 \cdot I_{\text{sp}}$, where $g_0 = 9.80665$ m/sec^2 and $I_{\text{sp}} = 313.55$ sec is the specific impulse. This provides a speed of $v_{\text{ex}} = 3074.88$ m/sec. Let t_1 be the instant that the DCM-1 maneuver began and t_2 the instant the maneuver was over. The rocket equation tells us that

$$v(t_2) - v(t_1) = \pm 3074.88 \left(\ln \tfrac{799.35}{734.30} \right) \approx \pm 261.00 \text{ m/sec}.$$

Since DSM-1 decreased the craft's speed, the $-$ option applies. This result is in agreement with the value 261.01 m/sec listed in Table 6.1 (where the minus sign is omitted).

The corrective maneuver DSM-2 necessitated by the failed *NEAR-Shoemaker* rendezvous with Eros produced a Δv of -932 m/sec. For this LVA, the primary thruster burned 54% of the craft's 209 kg total supply of hydrazine fuel.

6C. Gravitational Sphere of Influence. We now turn to a question that is of direct relevance to the story of *NEAR-Shoemaker*'s flight to Eros. For what distance from a massive body is it the case that a spacecraft or satellite moving within this distance is influenced almost exclusively by the gravitational force of this body without significant disruption from the gravitational pull of other, possibly more massive bodies farther away? Or put more concretely, what is the radius of a gravitational *Sphere of Influence* (SOI) around a planet (or moon, or asteroid) such that the force on a satellite or spacecraft in orbit or flyby within it, is dominated by the gravitational pull of the planet (or moon, or asteroid) without being measurably affected by the more distant or massive Sun? An early proposal for what such a radius should be came from the scientist Pierre-Simon Laplace (1749–1827). Laplace's five-volume *Mécanique Céleste*, a comprehensive mathematical account of the solar system, established him as the "Newton of France." Another version of such a radius was put forward by the American astronomer–mathematician George William Hill (1838–1914). Hill based his study on the work of Édouard Roche (1820–1883), so that his radius is also known as the Roche radius.

Suppose that a body P is in an orbit with semimajor axis a and period T around a much more massive body S. Suppose in turn that P is orbited by an object C that is much less massive than P. By taking the masses m, M, and m_0 of $P, S,$ and C to be concentrated at their respective centers of mass, P, S, and C are regarded as point-masses. Our concern is to gain a sense of the largest R such that if the orbit of C falls within a sphere with center P and radius R, then it is not greatly perturbed by the gravitational pull of S. We'll assume that C orbits at the outer limits of the gravitational influence of P. So we'll take the orbit of C to be a circle with center P and radius R, and explore the constraints that apply to R. Let τ be the period of the orbit of C. Figure 6.5

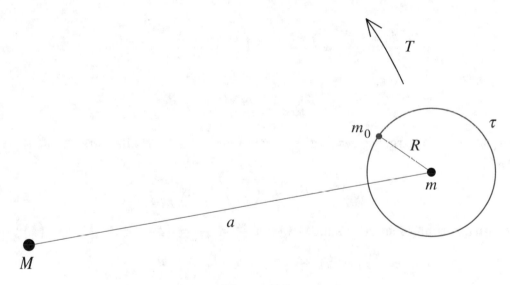

Figure 6.5

captures what has been described with a focus on the masses involved.

Laplace proposed the radius

$$R = R_{\mathrm{L}} = a\left(\frac{m}{M}\right)^{\frac{2}{5}}$$

and Hill proposed the alternative

$$R = R_{\mathrm{H}} = a\left(\frac{m}{3M}\right)^{\frac{1}{3}}.$$

Instead of going into the technical details, we'll provide considerations that support the values that Laplace and Hill put forward. Because $G\frac{m_0 M}{a^2}$ and $G\frac{m_0 m}{R^2}$ are the respective gravitational forces with which S and P pull on C, our assumptions tells us that

$$G\frac{m_0 M}{a^2} \approx G\frac{m_0 m}{R^2}.$$

So $\frac{M}{a^2} \approx \frac{m}{R^2}$, hence $R^2 \approx a^2 \frac{m}{M}$, and therefore,

$$R \approx a\left(\frac{m}{M}\right)^{\frac{1}{2}}.$$

The fact that m_0 moves has not as yet been considered. The example of our Moon tells us that this is relevant. (Were the Moon not to move relative to the Earth, gravity would cause the Moon and Earth to crash into each other.) The data in Table 5.1 tells us that the speed of a planet's (or Pluto's or the comet Halley's) revolution around the Sun reflects the impact of the Sun's gravitational force on it. The closer the body to the Sun, the greater the gravitational impact, the shorter the orbital period T, the greater the angular speed $\frac{2\pi}{T}$ of its motion around the Sun. We will now insert this consideration into our discussion. Since our assumptions imply that R is much smaller than a, we'll suppose that the angular speed $\frac{2\pi}{\tau}$ of C around P is greater than the average angular speed $\frac{2\pi}{T}$ of P around S. From Newton's version of Kepler's third law,

$$\frac{4\pi^2}{T^2} = \frac{GM}{a^3} \quad \text{and} \quad \frac{4\pi^2}{\tau^2} = \frac{Gm}{R^3},$$

so that

$$\frac{GM}{a^3} = \frac{4\pi^2}{T^2} = \left(\frac{2\pi}{T}\right)^2 < \left(\frac{2\pi}{\tau}\right)^2 = \frac{4\pi^2}{\tau^2} = \frac{Gm}{R^3}.$$

Therefore, $R^3 < a^3 \frac{m}{M}$ and hence,

$$R < a\left(\frac{m}{M}\right)^{\frac{1}{3}}.$$

Since $\frac{m}{M}$ is less than 1, a larger power of $\frac{m}{M}$ is smaller than a smaller power of $\frac{m}{M}$. So Laplace's radius R_L satisfies

$$a\left(\frac{m}{M}\right)^{\frac{1}{2}} < R_L = a\left(\frac{m}{M}\right)^{\frac{2}{5}} < a\left(\frac{m}{M}\right)^{\frac{1}{3}}.$$

From the assumption that $\frac{m}{M}$ is very small, we get $\frac{m}{M} < \frac{1}{243} = \left(\frac{1}{3}\right)^5$. So $\left(\frac{m}{M}\right)^{\frac{1}{15}} < \left(\frac{1}{3}\right)^{\frac{1}{3}}$ and

$$a\left(\frac{m}{M}\right)^{\frac{2}{5}} = a\left(\frac{m}{M}\right)^{\frac{1}{3}} \cdot \left(\frac{m}{M}\right)^{\frac{1}{15}} < a\left(\frac{1}{3}\right)^{\frac{1}{3}} \cdot \left(\frac{m}{M}\right)^{\frac{1}{3}} = a\left(\frac{m}{3M}\right)^{\frac{1}{3}}.$$

It follows that the radii of Laplace and Hill fit into the chain of inequalities:

$$a\left(\frac{m}{M}\right)^{\frac{1}{2}} < R_L = a\left(\frac{m}{M}\right)^{\frac{2}{5}} < R_H = a\left(\frac{m}{3M}\right)^{\frac{1}{3}} < a\left(\frac{m}{M}\right)^{\frac{1}{3}}.$$

Let's consider the Laplace and Hill radii for the Sun-Earth and the Sun-Eros systems.

Example 6.4. For S and P the Sun and Earth respectively, $GM = 1.32712 \times 10^{11}$ km³/sec², $Gm = 3.98600 \times 10^5$ km³/sec², and $a = 149{,}598{,}262$ km. (Refer to Chapter 1H and Tables 2.1 and 2.3 for these values.) So for the Sun-Earth system, the two radii are

$$R_{\mathrm{L}} = (1.49598 \times 10^8)\left(\tfrac{3.98600 \times 10^5}{1.32712 \times 10^{11}}\right)^{\frac{2}{5}} \approx 920{,}000 \text{ km and}$$

$$R_{\mathrm{H}} = (1.49598 \times 10^8)\left(\tfrac{3.98600 \times 10^5}{3(1.32712 \times 10^{11})}\right)^{\frac{1}{3}} \approx 1{,}497{,}000 \text{ km.}$$

Notice that the larger Hill radius R_H is very close to 1% of the semimajor axis of Earth's orbit. Since the aphelion distance of the Moon's orbit around Earth is $a(1 + \varepsilon) \approx 384{,}400(1.0554) \approx 410{,}000$ km, the Moon's orbit lies well within the radii of Laplace and Hill. This is consistent with what we have observed for centuries: a Moon that seems to be in no danger of drifting off into its own independent orbit around the Sun.

Example 6.5. For Eros, $Gm = 4.4621 \times 10^{-4}$ km^3/sec^2 and $a = 218{,}155{,}000$ km. So the radii for the Sun-Eros system are

$$R_{\mathrm{L}} = (2.18155 \times 10^8)\left(\tfrac{4.4621\times10^{-4}}{1.32712\times10^{11}}\right)^{\frac{2}{5}} \approx 342 \text{ km and}$$

$$R_{\mathrm{H}} = (2.18155 \times 10^8)\left(\tfrac{4.4621\times10^{-4}}{3(1.32712\times10^{11})}\right)^{\frac{1}{3}} \approx 2{,}270 \text{ km.}$$

Table 6.2 tells us that all orbits of *NEAR-Shoemaker* around Eros fit easily within the sphere that Hill's radius determines. The first two orbits are partly outside the sphere with Laplace's radius, but all the others fall inside it.

Since there are other factors at work, including gravitational fields of other planets as well as solar radiation pressure, an orbit is not always stable throughout the sphere that Hill's radius specifies. The tighter Laplace radius R_L is therefore a better measure of the extent of the gravitational *Sphere of Influence* (SOI) that a pair of massive bodies determines.

Let's continue by modifying the earlier discussion. We'll let S be the center of mass of the Sun, E Earth's center of mass, and C a point-mass. Their masses are M, m, and m_0, respectively. We'll let E and C be in circular orbits around the Sun centered at S. The orbit of E has radius a, and C orbits with a tighter radius than a. At a certain time $t = 0$, click a stopwatch. Suppose that at this time, C is on the segment SE at a distance R from E. Let time $t \geq 0$ flow, consider the revolving segment SE, and let $\theta = \theta(t)$ be the angle that it traces out. Figure 6.6 illustrates what has been described. As was already observed, the closer the orbit of an object to the Sun, the stronger the Sun's gravitational pull, the greater the angular speed of its revolution around the Sun. So the angular speed of C around S is greater than that of the segment SE, with the

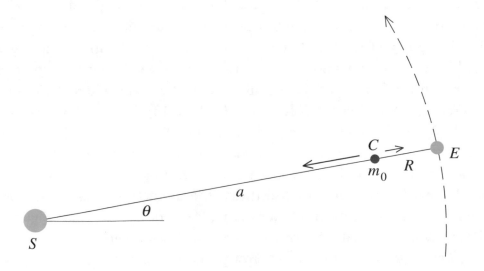

Figure 6.6

consequence that C will not maintain its position on the segment SE. However, if C is moved sufficiently close to E, then the Earth's gravitational force will in part counteract the Sun's pull on C. This in turn would lessen the angular speed of C. The question that arises is this: can the distance R be chosen in such a way that the position of C on the revolving segment SE is maintained? We'll suppose that this is so and investigate the implications for R.

The resultant of the two forces on C is a centripetal force in the direction of S of magnitude $F = \frac{GMm_0}{(a-R)^2} - \frac{Gmm_0}{R^2}$. An application of the centripetal force equation $F(t) = m\left[r(\frac{d\theta}{dt})^2 - \frac{d^2r}{dt^2}\right]$ of Chapter 4D with $r = r(t) = a - R$ tells us that

$$\frac{GMm_0}{(a-R)^2} - \frac{Gmm_0}{R^2} = m_0(a-R)\left(\frac{d\theta}{dt}\right)^2.$$

The approximation of Earth's orbit as a circle, tells us that its orbital speed is close to constant. So its angular speed is close to constant and hence $\frac{d\theta}{dt} \approx \frac{2\pi}{T}$, where T is the Earth's orbital period. By Newton's version of Kepler's third law, $\left(\frac{2\pi}{T}\right)^2 = \frac{4\pi^2}{T^2} = \frac{GM}{a^3}$, and therefore

$$\frac{GMm_0}{(a-R)^2} - \frac{Gmm_0}{R^2} \approx (a-R)\frac{GMm_0}{a^3}.$$

Hence $\frac{M}{(a-R)^2} - \frac{m}{R^2} \approx \frac{(a-R)M}{a^3}$. Since $(a-R) = a(1 - \frac{R}{a})$, we get $\frac{1}{a^2}(1 - \frac{R}{a})^{-2}M - \frac{m}{R^2} \approx \frac{(1-\frac{R}{a})M}{a^2}$, and hence $(1 - \frac{R}{a})^{-2}M - \frac{a^2m}{R^2} \approx (1 - \frac{R}{a})M$. Since $\frac{R}{a}$ is small, the binomial series (see Chapter 5G) with $k = -2$ and $x = -\frac{R}{a}$ tells us that $(1 - \frac{R}{a})^{-2} \approx 1 + 2\frac{R}{a}$. It follows that $(1 + 2\frac{R}{a})M - \frac{a^2}{R^2}m \approx (1 - \frac{R}{a})M$, and hence that $\frac{3R}{a}M \approx \frac{a^2}{R^2}m$. So $R^3 \approx \frac{a^3}{3}\frac{m}{M}$ and finally,

$$R \approx a\left(\frac{m}{3M}\right)^{\frac{1}{3}}.$$

So the equilibrium distance from C to E is approximated by the Hill radius $R_H \approx 1,497,000$ km of the Sun-Earth system.

The point C in Figure 6.6 with R equal to the Hill radius R_H is labeled L1 in Figure 6.7. The gravitational forces of the Sun S and Earth E on any mass placed at L1 are balanced in such a way that the mass moves in tandem with the Earth on the revolving segment SE. The point L2 moves in the same way outside Earth's orbit. The sphere with center E that the points L1 and L2 determine is the *Hill sphere* of the Sun-Earth system. The Moon orbits Earth well within it. The points L3, L4, and L5 are three more points that revolve with the same angular speed around S. This revolving frame around the Sun was first identified by the Swiss Leonhard Euler (1707–1783) and the Italian-born Frenchman Joseph-Louis Lagrange (1736–1813). Euler, who discovered equilibrium points L1, L2, and L3, was the most accomplished and prolific mathematician of the 18th century. He created a large body of mathematical tools that he applied to a broad set of problems, including the study of the solar system and in particular the computation of the orbits of comets. Lagrange added the points L4 and L5. He also established that the competing forces in any orbital situation similar to the Sun-Earth system give rise to a rotating frame of the sort that Figure 6.7 illustrates. The five equilibrium points of such a frame are called *Lagrange points* today.

Until recently, Lagrange points were of interest primarily for theoretical reasons. But they are now playing an increasingly important role in the exploration of space. The *Gaia* spacecraft—

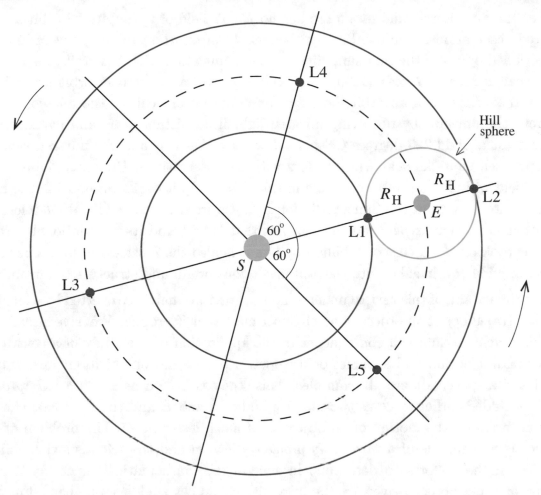

Figure 6.7. The assumption that Earth's orbit is a circle has simplified the development of this diagram. However, the reality is that the rotating scheme that it depicts is a scheme of ellipses and not of circles. As some of the other figures in this chapter, this diagram is not to scale.

its mission is described in Chapter 3H—is in an orbit centered at Lagrange point L2. *Gaia* curves in nearly closed loops around this point of equilibrium. Like the tip of a pendulum, it swings out a distance of about 150,000 km away from L2 in both directions with a period of about 180 days. Each loop is slightly different from the previous one so that the craft's path is a delicately shifting and revolving curve in three-dimensions. The craft passes successively above and below the Earth's orbital plane, at times slightly ahead of the Earth and at times slightly behind it. Once a month, one of *Gaia*'s small engines eases the craft into a small course-correcting maneuver that keeps its orbit around L2 within the targeted limits. The fact that the gravitational forces in the gravitational neighborhood of L2 are in balance means that these maneuvers require very little propellant. *Gaia*'s L2 orbit is advantageous for other reasons as well. It made it possible to equip the craft with a shield that faces both Sun and Earth and blocks the light and heat that both radiate. This is critical, as *Gaia*'s optics are so sensitive to changes of temperature that a variation of less than one thousandth of a degree over a few hours would disturb the alignment of the mirrors and degrade the quality of the images that it snaps. Solar panels on the outside surface of the shield power *Gaia*'s instruments. *Gaia*'s orbit was designed to keep it away from the shadow cast by the Earth. Were the craft to pass through it, its solar panels would not receive enough sunlight to generate sufficient power. Since the Sun, Earth, and Moon are behind *Gaia*'s field of vision, its L2 orbit also offers an unobstructed observational window. Finally, the fact that from the vantage point of Earth, *Gaia* is always positioned opposite the Sun simplifies the communication with the craft.

By the year 2021, the *James Webb Space Telescope* will also roam in orbit around the point L2. Aspects that make such an orbit ideal for *Gaia* are also critical for the telescope. The *Webb* will peer into the universe by observing infrared light from distant stars and galaxies. To do so, it will operate at a frigid 225 degrees Celsius below zero. Warmer, and the infrared radiation that the telescope itself emits as heat will interfere with its observations. Understandably, the *Webb*'s large heat shield will play an important role in this regard. As it orbits around L2, the heat shield will protect it not only from the Sun's radiation but also from Earth's. The *Webb* telescope is the most ambitious and complex astronomical project that NASA and its international partner space agencies ever took on. At a cost of 8 billion dollars it is also the most expensive. The telescope is expected to provide new insights into the still mysterious origin and structure of our universe.

It is the primary aim of this text to understand the gravitational force exerted by a single massive body and the trajectory of an object of much lesser mass that it propels. We have already seen that much of this understanding is a consequence of the application of the study of a centripetal force on a point-mass. The Sun and a planet or a comet, a planet and one of its moons, and a planet or asteroid and a spacecraft are all examples. This study is known as the *two body problem*. The discussion concluded earlier involves two massive bodies S and E and the combined effect of their gravitational forces on the motion of an object C of much lesser mass. The problem of analyzing its dynamics is an example of a *three body problem*. Newton and the scientists that followed him were stymied in their efforts to determine in general the mathematical functions that describe the motions of three bodies driven by the gravitational forces that act on them. Finally, in the 1950s, mathematicians demonstrated that the explicit determination of such functions is impossible. However, beginning in the 1960s, the use of high speed digital computers did make it possible to achieve better and better numerical approximations of these motions in the case of the *restricted three body problem*. The restriction is the assumption that two of the bodies are massive and that the third is so small that its gravitational pull on the other two is negligible. It also assumes that the two massive bodies are both in circular orbits about their common center of mass, and that the orbits of all three lie in the same plane. This restricted three body problem provides a useful model for the design of the trajectory of spacecraft in orbit around the Sun and in transit to a planet, or a craft that is making its way from Earth to the Moon. The discussion above about the Hill radius and the Lagrange points is a special case of the restricted three body problem.

6D. Modifying an Orbit. We have seen that the mission of *NEAR-Shoemaker* from launch to its landing on Eros took several years. With the exception of about four dozen trajectory corrections, most of them very minor, its flight path was determined by the gravitational attraction of a single massive body. It follows from Newton's theory of gravitation (see Chapter 4G) that between these corrective maneuvers the path of the spacecraft was a segment of a conic section—of an ellipse, a parabola, or a hyperbola—that had the center of mass of this massive body at a focal point.

After its launch, *NEAR-Shoemaker* was briefly in an elliptical parking orbit around Earth. The injection burn that followed expanded this orbit into a hyperbolic trajectory with Earth as focal point that transitioned the craft into an elliptical orbit around the Sun. *NEAR-Shoemaker*'s elliptical orbit around the Sun was punctuated and changed by several minor trajectory correction maneuvers until the major DSM-1 correction contracted the orbit to send the craft to its important flyby of Earth. This flyby was another hyperbolic segment that had Earth at a focal point. It corrected the inclination of the craft's orbit and contracted its aphelion distance, so that both the inclination and aphelion would match those of the orbit of Eros. The corrections that followed the aborted rendezvous maneuver paved the way for the injection of *NEAR-Shoemaker* into its 321×366 km elliptical orbit around Eros.

In the previous two sections we looked at basic mathematics that applied to the change of *NEAR-Shoemaker*'s trajectory from its parking orbit around Earth to its injection into its initial solar orbit. Thereafter, we studied concerns related to the transfer of the spacecraft from its third solar orbit into the gravitational sphere of influence of Eros. We turn next to the mathematics behind the correction maneuvers that sequentially adjusted *NEAR-Shoemaker*'s elliptical orbit around Eros. Table 6.2 informs us that during the year 2000 this orbit was repeatedly contracted until it achieved the nearly circular 35×39 km orbit of July 14th. For the remainder of its flight, *NEAR-Shoemaker*'s orbit around Eros was expanded and contracted over a dozen times with maneuvers that also changed the inclination of its orbital plane (relative to Eros's equatorial plane). From its final circular orbit of 36 km radius, the craft performed its slow descent to the surface of the asteroid. The study of Eros by the craft's instruments from so many different angles and distances provided accurate information about its surface features, shape, mass, density, and gravitational and magnetic fields.

Return to Table 6.2 and consider the initial sequence of *NEAR-Shoemaker*'s orbits around Eros. Notice that in consecutive, repeating steps, the orbit was made smaller and then circular (or nearly circular). The data exhibits the pattern: a maneuver that decreased the periapsis distance of an orbit while keeping the apoapsis distance the same was followed by a second maneuver that circularized the orbit by decreasing the apoapsis distance and keeping the periapsis distance the same. Each of this two-step tightening of the orbit was carried out with an engine burn that was directed against the direction of the craft's motion. For each, *NEAR-Shoemaker*'s smaller thrusters were deployed. What we have said also pertains to the orbit contractions that began on October 13th and December 7th. The expansions of *NEAR-Shoemaker*'s orbit—refer to the orbital data from July 14th to September 5th for instance—and the strategy for bringing them about also involved a two-step approach analogous to what has just been described. (This expansion will be taken up by two of the problems in the paragraph *More About NEAR-Shoemaker* of the Problems and Discussions of this chapter.)

While *NEAR-Shoemaker* and Eros is the example of interest, we will study the mathematics behind this orbital contraction process more generally for any spacecraft C in orbit within the gravitational sphere of influence of a single body of mass M and center of mass S. We'll assume that M is much greater than the mass of the craft. With their masses 6.69×10^{15} kg and a few hundred kilograms, respectively, Eros and *NEAR-Shoemaker* meet this requirement. When its thrusters are not firing, the craft is subject only to the gravitational force in the direction of S, so that its trajectory is a conic section with S at a focal point. During the time the engine of the craft fires, the craft is subject to the additional force of the engine's thrust. However, as soon as the burn stops, only the single gravitational force in the direction of S remains and the trajectory is once again a conic section with focal point S. Of course, this conic section will be different from the one before. Since we are assuming that what we have described occurs within the sphere of influence of the massive body, these conic sections are ellipses. It follows that the trajectory of the spacecraft C consists of a sequence of ellipses or segments of ellipses. It is an example of a *patched conic* trajectory.

Consider an elliptical orbit O_1 of the craft C with focal point S and let P and A be the periapsis and apoapsis of the orbit. Let a_1 be the semimajor axis and ε_1 the eccentricity of O_1, and let q_1 and d_1 be the periapsis and apoapsis distances, respectively (the distances between P and S and A and S). Refer to Figure 5.2 of Chapter 5B and observe that

$$a_1(1 - \varepsilon_1) = a_1 - a_1\varepsilon_1 = q_1 \quad \text{and} \quad a_1(1 + \varepsilon_1) = a_1 + a_1\varepsilon_1 = d_1.$$

Therefore $d_1 + q_1 = 2a_1$, $d_1 - q_1 = 2a_1\varepsilon_1$, and hence $a_1 = \frac{d_1+q_1}{2}$ and $\varepsilon_1 = \frac{d_1-q_1}{d_1+q_1}$. It follows that a_1 and ε_1 determine both q_1 and d_1, and conversely that q_1 and d_1 determine both a_1 and ε_1. By Example 5.1 of Chapter 5D, the velocities of C at P and A are

$$\sqrt{\frac{GM}{q_1}}\sqrt{1 + \varepsilon_1} \qquad \text{and} \qquad \sqrt{\frac{GM}{d_1}}\sqrt{1 - \varepsilon_1},$$

respectively. They are the maximum and minimum velocities of C in orbit O_1. We'll use analogous notation for subsequent elliptical orbits O_2 and O_3 of the craft C (with focal point S). So the semimajor axis, eccentricity, periapsis distance, and apoapsis distance of O_2 are denoted by a_2, ε_2, q_2, and d_2, respectively, and similarly for O_3. What was asserted about the velocities of the craft in orbit O_1 also holds for orbits O_2 and O_3.

Suppose that the spacecraft C is in orbit O_1. At the instant it reaches its apoapsis A, the craft's engine fires. The vector representing the force that is generated is tangential to the orbit and hence lies within the plane of the orbit. The force acts against the direction of the craft's motion, decreases the speed of the craft at A, and alters the craft's orbit. After the engine shut-off, the craft's new orbit O_2 starting at A is again a conic section. What can be said about this new trajectory?

Turn to Figure 6.8a. The craft C travels in its orbit O_1, depicted in blue, in a counterclockwise direction. The velocity of C on arrival at apoapsis A is $v_1 = \sqrt{\frac{GM}{d_1}}\sqrt{1 - \varepsilon_1}$. The craft's engine fires with a single burst at the instant the craft arrives at A with its thrust acting tangentially against the motion of C. Let v_2 with $0 < v_2 < v_1$ be the velocity with which the craft begins its new orbit O_2 at A. At the instant the new orbit begins, its distance from S is d_1. Since $v_2 < v_1$, the new orbit O_2 is a contraction of O_1, so that d_1 continues to be the maximum distance from C to S. So the orbit O_2 is an ellipse with apoapsis A and $d_2 = d_1$. Observe that the velocity of the craft at the beginning of its new orbit at A is

$$v_2 = \sqrt{\frac{GM}{d_1}}\sqrt{1 - \varepsilon_2}.$$

The ratio $\frac{v_2}{v_1}$ is equal to

$$\frac{v_2}{v_1} = \frac{\sqrt{GM}}{\sqrt{d_1}}\sqrt{1 - \varepsilon_2} \cdot \frac{\sqrt{d_1}}{\sqrt{GM}} \frac{1}{\sqrt{1 - \varepsilon_1}} = \frac{\sqrt{1 - \varepsilon_2}}{\sqrt{1 - \varepsilon_1}}.$$

Given that O_1 and hence d_1, ε_1, and v_1 are understood, it follows that v_2 determines the eccentricity ε_2 of the new orbit O_2. Since the semimajor axis a_2 of O_2 satisfies $a_2(1 + \varepsilon_2) = d_1$, v_2 also determines

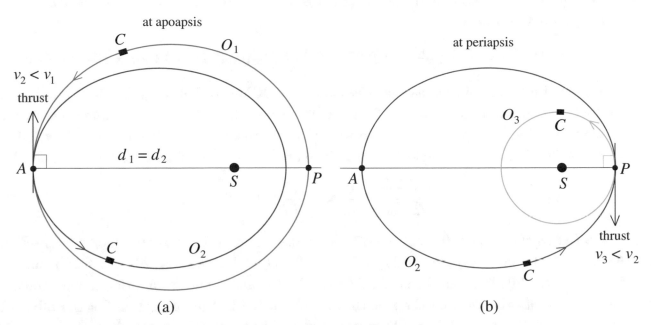

Figure 6.8. In each of the two situations, the craft C orbits counterclockwise. Its thruster fires tangentially to the orbit against the motion of the craft C. In case (a), it fires at apoapsis A thereby decreasing the periapsis distance. In case (b) it fires at periapsis and decreases the apoapsis distance.

a_2 and hence the orbit O_2. Since $v_2 < v_1$, it follows that $1 - \varepsilon_2 < 1 - \varepsilon_1$, so that $\varepsilon_2 > \varepsilon_1$. Since $a_1(1 + \varepsilon_1) = d_1 = d_2 = a_2(1 + \varepsilon_2)$, we get $a_2 < a_1$. Finally, from $d_1 + q_1 = 2a_1$ and $d_2 + q_2 = 2a_2$, it follows that $q_2 < q_1$. So the maneuver has decreased the periapsis distance of the orbit while keeping the apoapsis distance the same.

A question that remains is this. Let O be any ellipse with focus S and apoapsis A, hence apoapsis distance d_1, and periapsis distance $q < q_1$. Can a velocity v_2 be chosen so that the new orbit O_2 of C obtained by reducing the craft's velocity at A in orbit O_1 to v_2 is equal to the preassigned orbit O? We know that the semimajor axis and the eccentricity of O are $a = \frac{d_1 + q}{2}$ and $\varepsilon = \frac{d_1 - q}{d_1 + q}$, respectively. The fact that $q < q_1$, tells us that $d_1 - q_1 < d_1 - q$ and $d_1 + q_1 > d_1 + q$, so that

$$\varepsilon_1 = \tfrac{d_1 - q_1}{d_1 + q_1} < \tfrac{d_1 - q}{d_1 + q} = \varepsilon \text{ and } \sqrt{\tfrac{GM}{d_1}}\sqrt{1 - \varepsilon} < \sqrt{\tfrac{GM}{d_1}}\sqrt{1 - \varepsilon_1} = v_1.$$

Let O_2 be the orbit obtained by an engine burn that decreases the velocity of C at A to $v_2 = \sqrt{\frac{GM}{d_1}}\sqrt{1 - \varepsilon}$. The equality $\frac{v_2}{v_1} = \frac{\sqrt{1 - \varepsilon_2}}{\sqrt{1 - \varepsilon_1}}$ tells us that the eccentricity ε_2 of the orbit O_2 is equal to the eccentricity ε of O. Since the apoapsis distances of O_2 and O are also the same, it follows that the two orbits are the same.

The following has been established. Suppose that the craft C is in an elliptical orbit O_1 with focal point S, apoapsis A, and apoapsis and periapsis distances d_1 and q_1. Let q_2 satisfy $0 < q_2 < q_1$. By firing the craft's engine tangentially to the orbit O_1 at A against its motion and reducing its velocity at A to $v_2 = \sqrt{\frac{GM}{d_1}}\sqrt{1 - \varepsilon}$ where $\varepsilon = \frac{d_1 - q_2}{d_1 + q_2}$, the craft is brought into the elliptical orbit O_2 with focal point S, apoapsis A, and periapsis distance q_2.

Example 6.6. Turn to Table 6.2 and let O_1 be the initial orbit of *NEAR-Shoemaker* around Eros on February 14, 2000. This orbit had apoapsis distance $d_1 = 366$ km and periapsis distance $q_1 = 321$ km. Since $\varepsilon_1 = \frac{d_1-q_1}{d_1+q_1} = \frac{45}{687} = 0.0655$ and GM for Eros is 4.4621×10^{-4} km^3/sec^2, *NEAR-Shoemaker*'s velocity at apoapsis in orbit O_1 was

$$v_1 = \sqrt{\frac{GM}{d_1}}\sqrt{1-\varepsilon_1} = \frac{\sqrt{4.4621\times 10^{-4}}}{\sqrt{366}}\cdot\sqrt{0.9345} \approx \frac{2.112\times 10^{-2}}{19.131}(0.9667) \approx 0.1067 \times 10^{-2} \text{ km/sec}$$

or 1.067 m/sec. Table 6.2 informs us that the maneuver OCM-1 of February 24, 2000 reduced the craft's velocity at apoapsis to $v_2 = v_1 - 0.13 = 0.937$ m/sec. Let O_2 be *NEAR-Shoemaker*'s post OCM-1 orbit. To align its apoapsis distance with that of orbit O_1, we will take $d_2 = d_1 = 366$ km (instead of the 365 km that the table provides). The analysis above informs us that for orbit O_2,

$$\frac{\sqrt{1-\varepsilon_2}}{\sqrt{1-\varepsilon_1}} = \frac{v_2}{v_1} \approx \frac{0.937}{1.067} \approx 0.878.$$

So $1 - \varepsilon_2 \approx (0.878)^2(1-\varepsilon_1) \approx (0.771)(0.9345) \approx 0.720$, and hence $\varepsilon_2 \approx 0.280$. Since $\varepsilon_2 = \frac{d_1-q_2}{d_1+q_2}$, we get $366 - q_2 \approx 0.280(366 + q_2)$. So $1.280q_2 \approx 366(1 - 0.280)$ and hence $q_2 \approx 205.875 \approx 206$ km. This value is close to the 204 km that Table 6.2 lists for this distance. The assumption made about the apoapsis distance of orbit O_2 and the fact that the Δv listed in the last column of the table was actually closer to 0.135 m/sec (than 0.13 m/sec) explain the discrepancy with the table.

We'll continue to explore the differences between the orbits O_1 and O_2. For O_2, we'll now use the data of Table 6.2 and take $q_2 = 204$ and $d_2 = 365$ km. So $\varepsilon_2 = \frac{d_2-q_2}{d_2+q_2} \approx \frac{161}{569} = 0.2830$. The velocity of the craft at apoapsis of orbit O_2 was

$$v_2 = \sqrt{\frac{GM}{d_2}}\sqrt{1-\varepsilon_2} \approx \frac{\sqrt{4.4621\times 10^{-4}}}{\sqrt{365}}\cdot\sqrt{0.7170} \approx 0.0936 \times 10^{-2} \text{ km/sec}$$

or 0.936 m/sec. The maximum velocity reached by *NEAR-Shoemaker* at periapsis of its first orbit O_1 was

$$\sqrt{\frac{GM}{q_1}}\sqrt{1+\varepsilon_1} \approx \frac{\sqrt{4.4621\times 10^{-4}}}{\sqrt{321}}\sqrt{1.0655} \approx 0.1217 \times 10^{-2} \text{ km/sec}$$

or 1.217 m/sec, and the maximum velocity that the craft reached in its second orbit O_2 was

$$\sqrt{\frac{GM}{q_2}}\sqrt{1+\varepsilon_2} \approx \frac{\sqrt{4.4621\times 10^{-4}}}{\sqrt{204}}\sqrt{1.2830} \approx 0.1675 \times 10^{-2} \text{ km/sec}$$

or 1.675 m/sec. Observe that the minimum speed of *NEAR-Shoemaker* was greater in orbit O_1, but that its maximum speed was greater in orbit O_2. The orbit O_2 is tighter than orbit O_1. So the planetary data of Table 5.1 suggests that *NEAR-Shoemaker*'s average velocity around O_2 should be greater than its average velocity around O_1. This is in fact the case. To verify this, note first that the semimajor axes a_1 and a_2 of the two orbits are equal to $a_1 = \frac{d_1+q_1}{2} = \frac{687}{2} = 343.5$ km and $a_2 = \frac{d_2+q_2}{2} = \frac{569}{2} = 284.5$ km, respectively. The formula $T^2 = \frac{4\pi^2 a^3}{GM}$ tells us that the periods of the two orbits around Eros are $T_1 = 1,893,653$ and $T_2 = 1,427,360$ seconds, respectively. The formula

$$C = 2\pi a\left(1 - \tfrac{1}{4}\varepsilon^2 - \tfrac{3}{64}\varepsilon^4 - \tfrac{45}{2304}\varepsilon^6 - \tfrac{1575}{147456}\varepsilon^8 - \cdots\right)$$

for the circumference of an ellipse with semimajor axis a and eccentricity ε (see Chapter 5G) tells us that the circumferences of *NEAR-Shoemaker*'s two orbits are $C_1 \approx 2155.96$ km and $C_2 \approx 1751.22$ km, respectively. As anticipated, the average velocity $\frac{C_2}{T_2} \approx 1.23$ m/sec is greater than the average velocity $\frac{C_1}{T_1} \approx 1.14$ m/sec.

Let's return to the craft C and its orbit O_2. What changes to O_2 are achieved by firing its engine tangentially to this orbit against the direction of its motion, this time at its periapsis? Turn to Figure 6.8b. By an earlier formula, the velocity of C at periapsis P in orbit O_2 is $v_2 = \sqrt{\frac{GM}{q_2}}\sqrt{1+\varepsilon_2}$ (with q_2 the periapsis distance of O_2). The engine burn slows the craft at P, reducing its velocity to $v_3 < v_2$ at the beginning of its new elliptical orbit O_3 (depicted in green in the figure). Because the angle of the velocity vector at P makes an angle of $90°$ with the focal axis of both orbits, we know by Example 5.2 of Chapter 5D, that the point P is either the periapsis or apoapsis for orbit O_3.

i.) Suppose that $v_3 \geq \frac{\sqrt{GM}}{\sqrt{q_2}}$. Then P is the periapsis for O_3. If not, it is the apoapsis of O_3. So $d_3 = q_2$ and $v_3 = \sqrt{\frac{GM}{d_3}}\sqrt{1-\varepsilon_3} = \sqrt{\frac{GM}{q_2}}\sqrt{1-\varepsilon_3}$. This contradicts $v_3 \geq \frac{\sqrt{GM}}{\sqrt{q_2}}$. Hence P is the periapsis of O_3. Therefore $q_3 = q_2$ and $v_3 = \sqrt{\frac{GM}{q_3}}\sqrt{1+\varepsilon_3}$, where ε_3 is the eccentricity of O_3. So

$$\frac{v_3}{v_2} = \frac{\sqrt{GM}}{\sqrt{q_3}}\sqrt{1+\varepsilon_3} \cdot \frac{\sqrt{q_2}}{\sqrt{GM}}\frac{1}{\sqrt{1+\varepsilon_2}} = \frac{\sqrt{1+\varepsilon_3}}{\sqrt{1+\varepsilon_2}}.$$

Since $v_3 < v_2$, it follows that $\varepsilon_3 < \varepsilon_2$. Since $a_3(1-\varepsilon_3) = q_3 = q_2 = a_2(1-\varepsilon_2)$, we get $a_3 < a_2$. From $d_2 + q_2 = 2a_2$ and $d_3 + q_3 = 2a_3$, it follows that $d_3 < d_2$. So the maneuver has decreased the apoapsis distance of the orbit while keeping the periapsis distance the same.

ii.) Suppose that $0 < v_3 < \frac{\sqrt{GM}}{\sqrt{q_2}}$. If P were the periapsis of orbit O_3, then $q_3 = q_2$ and $v_3 = \sqrt{\frac{GM}{q_2}}\sqrt{1+\varepsilon_3}$. Since this cannot be so, P is the apoapsis of orbit O_3. So $q_3 < d_3 = q_2 < d_2$. In this case, the maneuver has decreased both the apoapsis and periapsis distances of the orbit.

Let O be any ellipse with focus S, periapsis P, and hence periapsis distance q_2. Suppose that its apoapsis distance d satisfies $q_2 \leq d \leq d_2$ (where d_2 is the apoapsis distance of the orbit O_2). Is there a velocity v_3 such that the orbit O_3 depicted in Figure 6.8b is equal to O? This can be shown to be the case by an argument similar to the one that established the analogous statement in the earlier situation of periapsis decrease. See Figure 6.8a.

Example 6.7. We'll now consider the maneuver OCM-2 of *NEAR-Shoemaker* and the transition from its February 24, 2000 orbit to its March 3, 2000 orbit. This maneuver kept the periapsis distance (essentially) the same and reduced the apoapsis distance in such a way that the resulting orbit was nearly circular. The maneuver OCM-2 is therefore an example of what is depicted in Figure 6.8b and described above. Let O_2 be the orbit of *NEAR-Shoemaker* before OCM-2. We'll take the apoapsis and periapsis distances of O_2 to be $d_2 = 365$ km and $q_2 = 204$ km, respectively, as listed in Table 6.2. The eccentricity of O_2 was equal to $\varepsilon_2 = \frac{d_2-q_2}{d_2+q_2} = \frac{161}{569} = 0.2830$ and *NEAR-Shoemaker*'s periapsis velocity in orbit O_2 was

$$v_2 = \sqrt{\frac{GM}{q_2}}\sqrt{1+\varepsilon_2} \approx 0.1675 \times 10^{-2} \text{ km/sec}$$

or $v_2 \approx 1.675$ m/sec. Let O_3 be the post OCM-2 orbit of March 3, 2000. Let its apoapsis distance be $d_3 = 206$ km and its periapsis distance $q_3 = 204$ km (instead of the 203 km of the table to conform to the earlier discussion). The eccentricity of O_3 was $\varepsilon_3 = \frac{d_3-q_3}{d_3+q_3} = \frac{2}{410} \approx 0.0049$. By an earlier formula, *NEAR-Shoemaker*'s periapsis velocity v_3 for orbit O_3 satisfies

$$\frac{v_3}{v_2} = \frac{\sqrt{1+\varepsilon_3}}{\sqrt{1+\varepsilon_2}} \approx \frac{\sqrt{1.0049}}{\sqrt{1.2830}} \approx 0.8850.$$

So $v_3 \approx (0.8850)v_2 \approx 1.482$ m/sec. Therefore, $\Delta v = v_2 - v_3 \approx 1.675 - 1.482 \approx 0.193$ m/sec. This is in reasonable agreement with the entry 0.22 m/sec listed for OCM-2 in the last column of Table 6.2. The difference $0.22 - 0.193 = 0.0207$ m/sec was a factor in changing the inclination of the orbit (relative to the equator of Eros) from 34° to 37°.

The discussion about the velocity changes of a spacecraft at either periapsis or apoapsis of its orbit assumed that these were achieved by instantaneous bursts of its thruster. The engine burns of most trajectory correction maneuvers during the mission of a spacecraft last only a few seconds (but some can take up an hour or two) so that this assumption is appropriate. The correction maneuvers of *NEAR-Shoemaker*'s orbits around Eros achieved its small changes in the velocity of the craft with engine burns of short duration, so that this assumption did lead to good approximations of what actually happened (as Examples 6.6 and 6.7 confirm). Our discussion of these maneuvers only touched on the fact that in addition to modifying an orbit's size and shape, it was their role to change the inclination of the plane of the orbit and/or the orientation of its focal axis. For example, *NEAR-Shoemaker*'s OCM-11 on August 8, 2000 increased the periapsis distance of *NEAR-Shoemaker*'s orbit around Eros by only 1 km (from 49 km to 50 km) and kept the apoapsis fixed (at 52 km), but changed the inclination of its orbital plane by 15 degrees. Evidently, this was a primary goal of the Δv of 1.01 m/sec of OCM-11.

The flight of *NEAR-Shoemaker* relied on all the important maneuvers that spacecraft undergo during their missions. This includes the orbit expansion that occurred soon after its launch. The injection burn that took place about 25 minutes after launch expanded the craft's brief, near-Earth orbit around the Sun (essentially a small section of the blue orbit depicted in Figure 6.4) into its initial elliptical solar orbit (depicted in red in Figure 6.4). This expansion of *NEAR-Shoemaker*'s trajectory, driven by maneuver TCM-1 (with later adjustments by TCM-2 and TCM-3), is an example of a *Hohmann Transfer Orbit* discussed in Chapter 2G. The initial solar orbit of *NEAR-Shoemaker* had a perihelion distance 0.99 au, aphelion distance 2.19 au, and an angle of inclination 0.78° with the plane of Earth's solar orbit. The subsequent orbit contraction maneuver DSM-1 decreased this to 0.95 au, 2.16 au, and 0.52°, respectively, and lined up the craft for its flyby around Earth that changed these parameters to 0.98 au, 1.77 au, and 10.04°, respectively (see Figure 6.2), and brought *NEAR-Shoemaker* into an orbit close to that of Eros. This flyby was the crucial maneuver that sent *NEAR-Shoemaker* to its rendezvous with the asteroid. The fact that the Laplace radius for Earth is 920,000 km (refer to Example 6.4) tells us that *NEAR-Shoemaker* was well within the sphere of Earth's gravitational influence during this flyby. It follows that the trajectory of *NEAR-Shoemaker* around Earth was (part of) a conic section with Earth's center of mass at a focal point. But what conic section was it? Figure 6.2 tells us that it was hyperbolic. (In Figure 6.4, the radius of the blue circle (the orbit of Earth) corresponds to 1 au. Since Laplace's radius is about 0.006 au, the hyperbolic segment of *NEAR-Shoemaker*'s flyby around the Earth was too fine a detail to be captured by Figure 6.4).

The hyperbolic flyby of *Juno* around Earth (described in Chapter 2H) was also a critical aspect of its flight to Jupiter. The fact is that hyperbolic flybys have been of central importance to the navigation of all spacecraft with mission to explore the solar system. We will soon see this to be the case for the two *Voyager* missions to the outer planets as well as the *Cassini* mission to Saturn. It was noted in Chapter 2E that some comets approach the Sun, turn, and leave the solar system along hyperbolic paths. Hyperbolic trajectories are therefore important within the aims of this text. The next several sections of this chapter take up their study.

We start with the development of the mathematical tools that this study requires. First among them are the functions that play exactly the same role for the analysis of the hyperbola that the trigonometric functions play for the ellipse.

6E. Hyperbolic Functions. Let's begin with a review of the basics about hyperbolic curves as these were discussed in Chapters 1C and 3D. Turn to Figure 6.9 and observe that a hyperbola is a curve that has two branches. The corresponding focal points are denoted by S and F in the figure. The points S and F determine the focal axis of the hyperbola and the midpoint O of the segment

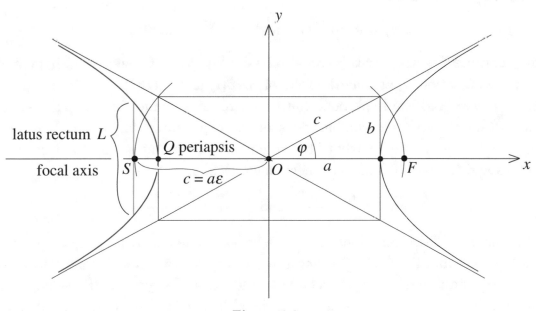

Figure 6.9

SF is the center of the hyperbola. The distance from O to either one of the points of intersection of the hyperbola with the focal axis is the semimajor axis a of the hyperbola. Let c be the length of the segments SO and OF. The semiminor axis of the hyperbola is $b = \sqrt{c^2 - a^2}$ and its eccentricity is $\varepsilon = \frac{c}{a}$. Notice that $\varepsilon > 1$ and that $c = a\varepsilon$. The rectangle with sides $2a$ and $2b$ determines the two intersecting asymptotes of the hyperbola. The angle φ that the segment OF makes with the asymptote satisfies $\tan \varphi = \frac{b}{a}$, so that $\varphi = \tan^{-1} \frac{b}{a}$. The length of the segment (in green in the figure) that the hyperbola cuts from the perpendicular through either of the focal points F or S is the *latus rectum* L of the hyperbola. We saw in Chapter 3D that $L = \frac{2b^2}{a}$.

By adding an xy-coordinate system to the figure as shown, we see that the center O is the origin, the focal axis is the x-axis, that $S = (-c, 0)$, $F = (c, 0)$, and that the points of intersection of the hyperbola with the x-axis are $(-a, 0)$ and $(a, 0)$. The asymptotes are the two lines $y = \frac{b}{a}x$ and $y = -\frac{b}{a}x$ and the equation of the hyperbola is

$$\frac{x^2}{a^2} - \frac{y^2}{b^2} = 1.$$

When it comes to the description of the hyperbolic trajectory of a spacecraft (or comet) only one branch of the hyperbola is needed. Figure 6.9 singles one of them out in red. The intersection of this branch with the x-axis is the periapsis for this branch and the distance $a\varepsilon - a = a(\varepsilon - 1)$ is the periapsis distance. It is the minimum distance between S and this branch of the hyperbola. Incidentally, much of the literature about aerodynamics takes the semimajor axis of the hyperbola to be the negative number $-a$. (With this convention, the periapsis distance has the expression $a(1 - \varepsilon) = (-a)(\varepsilon - 1)$ for both the ellipse and the hyperbola.)

Suppose that it is our task to analyze the trajectory of a spacecraft (possibly also a comet) that is in a hyperbolic flyby around a body of much larger mass. A look back to Chapters 5A to 5D tells us that the analysis of elliptical orbits made important use of the fact that the x- and y-coordinates of a point-mass moving in such an orbit can be written as functions of time t in terms of the trigonometry involved, as

$$x(t) = a \cos \beta(t) \quad \text{and} \quad y(t) = b \sin \beta(t)$$

The question that arises is whether this can also be done as part of a successful hyperbolic theory? If so, what are the functions that play the roles of $\cos \beta(t)$ and $\sin \beta(t)$?

The answer is yes! And the hyperbolic functions are tailor-made for this purpose. They arise, somewhat surprisingly, as simple combinations of the exponential functions e^x and e^{-x}. Again, surprisingly, they behave very similarly to the trigonometric functions $\sin x$ and $\cos x$ in terms of their derivatives and the formulas that connect them. Define

$$\sinh x = \frac{e^x - e^{-x}}{2} \quad \text{and} \quad \cosh x = \frac{e^x + e^{-x}}{2}$$

for any real number x. These functions are the *hyperbolic sine* and *hyperbolic cosine*, respectively. They get these names and the "suffix" h because they are related to the hyperbola $x^2 - y^2 = 1$ in the same way that $\sin x$ and $\cos x$ are related to the circle $x^2 + y^2 = 1$. (See Chapter 3A in this regard.) The graphs of $\sinh x$ and $\cosh x$ are depicted in Figure 6.10, both for small values of x. For larger x, they continue their rapid rise (or fall). The most basic relationship between $\sinh x$ and $\cosh x$ is

$$\cosh^2 x - \sinh^2 x = 1.$$

As in trigonometric situations, $\sinh^2 x$ means $(\sinh x)^2$, $\cosh^2 x$ means $(\cosh x)^2$ and similarly for the other hyperbolic functions to follow. The verification of this identity is easy,

$$\cosh^2 x - \sinh^2 x = \tfrac{1}{4}\left[(e^x)^2 + 2e^x e^{-x} + (e^{-x})^2)\right] - \tfrac{1}{4}\left[(e^x)^2 - 2e^x e^{-x} + (e^{-x})^2)\right] = e^x e^{-x} = 1.$$

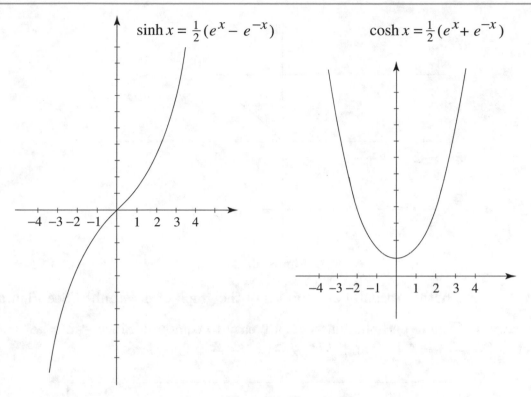

$$\sinh x = \tfrac{1}{2}(e^x - e^{-x}) \qquad \cosh x = \tfrac{1}{2}(e^x + e^{-x})$$

Figure 6.10

Example 6.8. Show that the addition formulas

    i. $\sinh(x + y) = (\sinh x)(\cosh y) + (\cosh x)(\sinh y)$

    ii. $\cosh(x + y) = (\cosh x)(\cosh y) + (\sinh x)(\sinh y)$

are direct consequences of properties of the exponential function.

The striking similarities between the trigonometric sine and cosine and the hyperbolic sine and cosine suggest that additional *hyperbolic functions* analogous to their trigonometric counterparts, should be singled out and considered. We'll only need the hyperbolic tangent and the hyperbolic secant. Predictably, they are defined by

$$\tanh x \; = \; \frac{\sinh x}{\cosh x} \quad \text{and} \quad \operatorname{sech} x = \frac{1}{\cosh x}.$$

Let's turn to the study of the graph of $y = \tanh x$. Observe that $\cosh x \geq 1$ for all x and that $\sinh x \geq 0$ for $x \geq 0$ and $\sinh x < 0$ for $x < 0$. Observe also that

$$\cosh x - \sinh x = e^{-x} > 0 \text{ and } \cosh x - (-\sinh x) = \cosh x + \sinh x = e^x > 0$$

for all x. In particular, $\cosh x > \sinh x$ and $\cosh x > -\sinh x$ for all x. Because $|\cosh x| = \cosh x$ and $|\sinh x| = \pm \sinh x$, it follows that $|\cosh x| > |\sinh x|$. Therefore

$$|\tanh x| < 1 \text{ for all } x.$$

So the graph of $y = \tanh x$ lies between the lines $y = 1$ and $y = -1$. Since $\lim\limits_{x\to+\infty} e^{-x} = \lim\limits_{x\to+\infty} \frac{1}{e^x} = 0$, it follows that $\lim\limits_{x\to+\infty} \tanh x = \lim\limits_{x\to+\infty} \frac{e^x - e^{-x}}{e^x + e^{-x}} = 1$. Similarly, $\lim\limits_{x\to-\infty} \tanh x = -1$. Therefore the lines

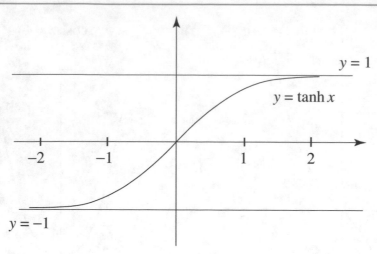

Figure 6.11

$y = 1$ and $y = -1$ are both horizontal asymptotes of the graph of $y = \tanh x$. See Figure 6.11.

The derivatives of the hyperbolic functions are easy to compute. Since $\frac{d}{dx} e^x = e^x$ and $\frac{d}{dx} e^{-x} = -e^{-x}$, we get $\frac{d}{dx}\left(\frac{e^x - e^{-x}}{2}\right) = \frac{e^x + e^{-x}}{2}$ and $\frac{d}{dx}\left(\frac{e^x + e^{-x}}{2}\right) = \frac{e^x - e^{-x}}{2}$. Therefore

$$\frac{d}{dx} \sinh x = \cosh x \quad \text{and} \quad \frac{d}{dx} \cosh x = \sinh x$$

Example 6.9. Use the quotient and chain rules to show that

$$\frac{d}{dx} \tanh x = \operatorname{sech}^2 x \quad \text{and} \quad \frac{d}{dx} \operatorname{sech} x = -(\operatorname{sech} x)(\tanh x).$$

The fact that $\frac{d}{dx} \tanh x$ is always positive tells us that $y = \tanh x$ is an increasing function. Check that $\frac{d^2}{dx^2} \tanh x = \frac{d}{dx} \operatorname{sech}^2 x = -2(\tanh x)(\operatorname{sech}^2 x)$. This confirms another feature of the graph of $y = \tanh x$. It is concave up for $x < 0$ and concave down for $x > 0$.

Example 6.10. Since the graph of the function $y = \sinh x$ is increasing, it has an inverse function $y = \sinh^{-1} x$. Apply the chain rule and the identity $\cosh x = \sqrt{\sinh^2 x + 1}$ to the equality $\sinh(\sinh^{-1} x) = x$ to show that $\frac{d}{dx} \sinh^{-1} x = \frac{1}{\sqrt{x^2+1}}$.

6F. Moving along the Hyperbola. We'll now consider a spacecraft (or comet) in a hyperbolic flyby of a much more massive body. What is set out in Chapter 5A applies to our discussion. But instead of applying it to the point-mass P, we'll apply it to a spacecraft (or comet) C that we regard as a point-mass. Consider Figure 5.1 and assume that the trajectory depicted there is given by the left branch of the hyperbola $\frac{x^2}{a^2} - \frac{y^2}{b^2} = 1$ of Figure 6.9. The point $Q = (-a, 0)$ is the periapsis. The time t is assigned to a position of the craft as follows. When C is at Q, the time is $t = 0$. When C is on approach to periapsis, t is negative. After C passes periapsis, t is positive. The craft's time of travel to or from periapsis is $|t|$ in either case. The position of C on the left branch of the hyperbola depends on t and we'll let its x- and y-coordinates be given by the functions $x(t)$ and $y(t)$ of t. The focal point S of the hyperbola is $(-c, 0)$, where $c = \sqrt{a^2 + b^2}$. The distance from C to S is given by the function $r(t)$, and $\alpha(t)$ is the angle in radians between the segments SC and SQ. The angle $\alpha(t)$ is measured counterclockwise for $t \geq 0$ and clockwise for $t < 0$. So $\alpha(t) \geq 0$ for $t \geq 0$ and $\alpha(t) < 0$ for $t < 0$.

A look at the graph of the function $f(x) = \sinh x$ in Figure 6.10 tells us that no matter what the value of $\frac{y(t)}{b}$ is, there is some number that we'll denote by β, such that $\sinh \beta = \frac{y(t)}{b}$. Since $f(x) = \sinh x$ is an increasing function, there is only one such number β for a given $\frac{y(t)}{b}$. Since $\frac{y(t)}{b}$ depends on t, the number β does also, so that $\beta = \beta(t)$ is a function of t. It follows for any time t, that $\sinh \beta(t) = \frac{y(t)}{b}$, and hence that $y(t) = b \sinh \beta(t)$. Because $\frac{x(t)^2}{a^2} - \frac{y(t)^2}{b^2} = 1$,

$$\frac{x(t)^2}{a^2} = \frac{y(t)^2}{b^2} + 1 = \sinh^2 \beta(t) + 1 = \cosh^2 \beta(t),$$

and we see that $x(t)^2 = (a \cosh \beta(t))^2$. Since $x(t) < 0$ and $\cosh \beta(t) > 0$, we get $x(t) = -a \cosh \beta(t)$. From the graph of $f(x) = \sinh x$ we see that $\sinh x > 0$ when $x > 0$ and $\sinh x < 0$ when $x < 0$. Since $y(t) < 0$ for $t < 0$ and $y(t) > 0$ for $t > 0$, it follows that $\beta(t) < 0$ for $t < 0$ and $\beta(t) > 0$ for $t > 0$. At time $t = 0$, C is at periapsis, so that $x(0) = -a, y(0) = 0$, and $\beta(0) = 0$.

In contrast to the situation of the ellipse, where $\beta(t)$ is an angle with a geometric meaning, this hyperbolic $\beta(t)$ is an abstractly defined function. Other than that, the functions $\beta(t), x(t) = -a \cosh \beta(t)$, and $y(t) = b \sinh \beta(t)$ play the same role in the hyperbolic context that the functions $\beta(t), x(t) = a \cos \beta(t)$, and $y(t) = b \sin \beta(t)$ played in the analysis of elliptical orbits in Chapter 5.

Figure 6.12 captures the geometry of the hyperbolic trajectory with the craft in a typical position at $(x(t), y(t)) = (-a \cosh \beta(t), b \sinh \beta(t))$. The figure also sets out the relevant notation. Notice

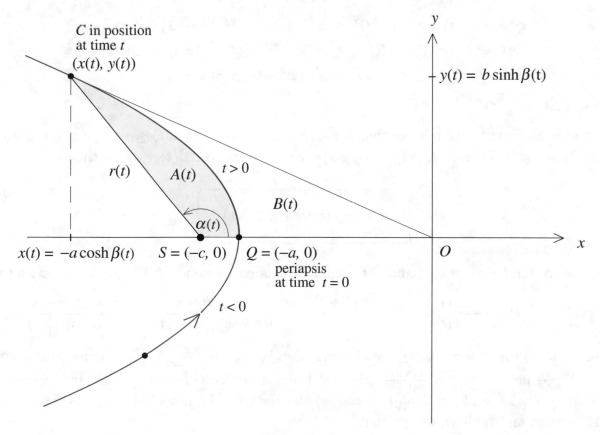

Figure 6.12

that $x(t) \leq -a$ for any t. For $t \geq 0$, let $A(t)$ be the area that the segment SC sweeps out from time $t = 0$ to time t. Recall from Chapter 5A that

$$A(t) = \kappa t,$$

where κ is Kepler's constant of the trajectory. For a negative t, this equality holds also with the understanding that $A(t)$ is the negative of the area that the segment SC sweeps out from time t to time $t = 0$. In the same way, $B(t)$ is the area (or the negative of the area) of the curving triangle that the points Q, C, and the origin O determine. Check that $A(t) + B(t) = \frac{1}{2}cy(t)$ for any t. As t moves through $-\infty < t < \infty$ the craft C, positioned at the point $(x(t), y(t))$ with

$$x(t) = -a \cosh \beta(t) \quad \text{and} \quad y(t) = b \sinh \beta(t),$$

traces out the entire branch of the hyperbola of Figure 6.12 from the bottom to the top. We will assume that the motion of the spacecraft is completely smooth. This means that $x = x(t)$, $y = y(t)$, and $\beta = \beta(t) = \sinh^{-1} \frac{y(t)}{b}$ as well as $r(t)$ and $\alpha(t)$ are all differentiable functions of t.

Let any elapsed time t be given. The position of C in its trajectory at time t is pinpointed by the values $r(t)$ and $\alpha(t)$. The problem of determining these values for the given t is solved in the same way as in the elliptical case of Chapter 5. The first step expresses $r(t)$ and $\alpha(t)$ in terms of $\beta(t)$. A second step develops a hyperbolic version of Kepler's equation that links $\beta(t)$ and t. With t given, a final step solves this equation for $\beta(t)$. After this $\beta(t)$ is inserted into the equations for r and α, the position of C at time t has been determined and the solution is complete.

We'll show first that $r(t) = a(\varepsilon \cosh \beta(t) + 1)$. Refer to Figure 6.12 to see that

$$\begin{aligned}
r(t)^2 &= (-x(t) - c)^2 + y(t)^2 = (c + x(t))^2 + b^2\left(\frac{x(t)^2}{a^2} - 1\right) \\
&= c^2 + 2cx(t) + x(t)^2 + \frac{b^2}{a^2}x(t)^2 - b^2 = \left(\frac{a^2 + b^2}{a^2}\right)x(t)^2 + 2cx(t) + a^2 \\
&= \frac{c^2}{a^2}x(t)^2 + 2cx(t) + a^2 = \varepsilon^2 x(t)^2 + 2a\varepsilon x(t) + a^2 \\
&= (\varepsilon x(t) + a)^2.
\end{aligned}$$

Another look at Figure 6.12 tells us that $x(t)$ is negative and that $x(t) \leq -a$. So $\varepsilon x(t) < -a$, and hence $\varepsilon x(t) + a < 0$. Therefore, $r(t) = -(\varepsilon x(t) + a) = -(-\varepsilon a \cosh \beta(t) + a)$, so that

$$\boxed{r(t) = a(\varepsilon \cosh \beta(t) - 1)}$$

The connection between $\alpha(t)$ and $\beta(t)$ begins with a consequence of Figure 6.12. Notice that

$$\cos(\pi - \alpha(t)) = \frac{-x(t) - c}{r(t)} = \frac{-(x(t) + c)}{r(t)} = \frac{-(-a \cosh \beta(t) + a\varepsilon)}{a(\varepsilon \cosh \beta(t) - 1)} = \frac{\cosh \beta(t) - \varepsilon}{\varepsilon \cosh \beta(t) - 1}.$$

Because $\cos \alpha(t) = -\cos(\pi - \alpha(t))$, it follows that $\cos \alpha(t) = \frac{\varepsilon - \cosh \beta(t)}{\varepsilon \cosh \beta(t) - 1}$. Using the identities $\tan^2 \frac{\alpha}{2} = \frac{1 - \cos \alpha}{1 + \cos \alpha}$ and $\tanh^2 \frac{\beta}{2} = \frac{\cosh \beta - 1}{\cosh \beta + 1}$ (the first is a consequence of the two standard trig identities $\cos^2 \theta = \frac{1}{2}(1 + \cos 2\theta)$ and $\sin^2 \theta = \frac{1}{2}(1 - \cos 2\theta)$ and the second follows quickly from the definitions of $\sinh x, \cosh x$, and $\tanh x$), we get that

$$\tan^2 \tfrac{\alpha(t)}{2} = \frac{1-\cos\alpha(t)}{1+\cos\alpha(t)} = \frac{1 - \frac{\varepsilon - \cosh\beta(t)}{\varepsilon\cosh\beta(t)-1}}{1 + \frac{\varepsilon-\cosh\beta(t)}{\varepsilon\cosh\beta(t)-1}} = \frac{\frac{\varepsilon\cosh\beta(t)-1-\varepsilon+\cosh\beta(t)}{\varepsilon\cosh\beta(t)-1}}{\frac{\varepsilon\cosh\beta(t)-1+\varepsilon-\cosh\beta(t)}{\varepsilon\cosh\beta(t)-1}}$$

$$= \frac{(\varepsilon+1)\cosh\beta(t) - (\varepsilon+1)}{(\varepsilon-1)\cosh\beta(t) + (\varepsilon-1)} = \frac{(\varepsilon+1)(\cosh\beta(t)-1)}{(\varepsilon-1)(\cosh\beta(t)+1)}$$

$$= \left(\tfrac{\varepsilon+1}{\varepsilon-1}\right)\tanh^2\tfrac{\beta(t)}{2} \ .$$

Note that $-\pi < \alpha(t) < \pi$ and that $\alpha(t)$ and $\beta(t)$ are either both positive together or negative together. Since $-\frac{\pi}{2} < \frac{\alpha(t)}{2} < \frac{\pi}{2}$, it follows from a comparison of the graphs of the tangent and hyperbolic tangent functions (see Figure 6.11) that $\tan\frac{\alpha(t)}{2}$ and $\tanh\frac{\beta(t)}{2}$ have the same sign for any t. Therefore $\tan\frac{\alpha(t)}{2} = \sqrt{\frac{\varepsilon+1}{\varepsilon-1}}\tanh\frac{\beta(t)}{2}$. This last equation is the hyperbolic version of Gauss's equation. Since $-\frac{\pi}{2} < \frac{\alpha(t)}{2} < \frac{\pi}{2}$, we can conclude that

$$\boxed{\alpha(t) = 2\tan^{-1}\left(\sqrt{\tfrac{\varepsilon+1}{\varepsilon-1}}\,\tanh\tfrac{\beta(t)}{2}\right)}$$

Turn to the angle $\varphi = \tan^{-1}\frac{b}{a}$ of Figure 6.9. A review of the discussion about the hyperbola in Chapter 3D and a comparison of Figures 3.19 and 6.12 tells us that when $\alpha(t) \geq 0$ then $\alpha(t) < \pi - \varphi$, and that when $\alpha(t) < 0$ then $\alpha(t) > -(\pi - \varphi)$. So $-(\pi - \tan^{-1}\frac{b}{a}) < \alpha(t) < \pi - \tan^{-1}\frac{b}{a}$. It is a consequence of this comparison that as t varies over $-\infty < t < \infty$, the angle $\alpha(t)$ assumes all values within these bounds.

6G. The Hyperbolic Kepler Equation. This step links t and $\beta(t)$ with an equation that involves both. Our computations will concentrate on the case $t \geq 0$. The argument as well as the conclusion is the same for $t < 0$ (if minus signs are carefully attended to). The two areas $A(t)$ and $B(t)$ of Figure 6.12 play the key role. The area $B(t)$ of the triangular region is bounded by the hyperbola and the segments OC and OQ. Solving $\frac{x^2}{a^2} - \frac{y^2}{b^2} = 1$ for y tells us that the upper half of the hyperbola is the graph of the function $y = \frac{b}{a}\sqrt{x^2 - a^2}$. Since $B(t)$ is the difference between the area of the triangle with base $-x(t)$ and height $y(t)$ and the area under the hyperbola $y = \frac{b}{a}\sqrt{x^2 - a^2}$ over the interval $x(t) \leq x \leq -a$, it follows that

$$B(t) = \tfrac{1}{2}(a\cosh\beta(t))(b\sinh\beta(t)) - \int_{-a\cosh\beta(t)}^{-a} \frac{b}{a}\sqrt{x^2 - a^2}\,dx\ .$$

We'll start the computation of $B(t)$ by computing $B'(t)$. Let $D(x)$ be an antiderivative of the function $\frac{b}{a}\sqrt{x^2 - a^2}$. By the fundamental theorem of calculus,

$$\int_{-a\cosh\beta(t)}^{-a} \frac{b}{a}\sqrt{x^2 - a^2}\,dx = D(-a) - D(-a\cosh\beta(t)).$$

By the chain rule, the derivative of the function of t on the right is

$$0 - D'(-a\cosh\beta(t))\cdot(-a\sinh\beta(t))\cdot\beta'(t) = -\tfrac{b}{a}\sqrt{(-a\cosh\beta)^2 - a^2}\cdot(-a\sinh\beta(t))\cdot\beta'(t)$$

$$= \tfrac{b}{a}\sqrt{a^2(\cosh^2\beta(t) - 1)}\cdot(a\sinh\beta(t))\beta'(t)$$

$$= ab\sqrt{\sinh^2\beta(t)}\cdot\sinh\beta(t)\cdot\beta'(t)$$

$$= ab\sinh^2\beta(t)\cdot\beta'(t).$$

By making use of this equality and by applying the product and chain rules, we see that

$$B'(t) = \tfrac{1}{2}\big[a\sinh\beta(t)\cdot\beta'(t)\cdot b\sinh\beta(t) + a\cosh\beta(t)\cdot b\cosh\beta(t)\cdot\beta'(t)\big] - ab\sinh^2\beta(t)\cdot\beta'(t)$$

$$= \tfrac{1}{2}ab\big[\sinh^2\beta(t) + \cosh^2\beta(t)\big]\beta'(t) - ab\sinh^2\beta(t)\cdot\beta'(t)$$

$$= \tfrac{1}{2}ab\big[\cosh^2\beta(t) - \sinh^2\beta(t)\big]\beta'(t) = \tfrac{1}{2}ab\beta'(t).$$

Since $B(0) = 0$ and $\beta(0) = 0$, we can conclude that

$$B(t) = \tfrac{1}{2}ab\beta(t).$$

This equality provides the geometric interpretation $\beta(t) = \frac{2B(t)}{ab}$ of the function $\beta(t)$. It also says that the area $B(t)$ is analogous to the area $B(t)$ of Figure 5.4 in the elliptical case of Chapter 5B.

Since the area $A(t) + B(t)$ of Figure 6.12 is equal to that of a triangle with base c and height $y(t)$, we get $A(t) + B(t) = \tfrac{1}{2}cb\sinh\beta(t)$. Since $c = a\varepsilon$, this implies that

$$A(t) = \tfrac{1}{2}cb\sinh\beta(t) - \tfrac{1}{2}ab\beta(t) = \tfrac{1}{2}ab(\varepsilon\sinh\beta(t) - \beta(t)).$$

Since the area $A(t)$ is also equal to $A(t) = \kappa t$, where κ is Kepler's constant for the trajectory, it follows that

$$\varepsilon\sinh\beta(t) - \beta(t) = \tfrac{2\kappa t}{ab}.$$

We'll now recall some basic facts from Chapter 5A. Let m and M be the masses of the moving craft C and the attracting body respectively, and let $F(t)$ be the magnitude of the gravitational force on C. Newton's inverse square law and his law of universal gravitation assert that

$$F(t) = \frac{8\kappa^2 m}{L}\cdot\frac{1}{r(t)^2} \quad\text{and}\quad F(t) = \frac{GmM}{r(t)^2},$$

where κ and L are Kepler's constant and the *latus rectum* of the hyperbolic trajectory of C, and G is the universal gravitational constant. By setting Newton's two force equations equal to each other, we get $\frac{8\kappa^2}{L} = GM$. So $\kappa = \sqrt{\frac{GML}{8}}$. Refer to part **C** of Chapter 3D for the fact that $L = \frac{2b^2}{a}$ for the hyperbola $\frac{x^2}{a^2} - \frac{y^2}{b^2} = 1$. So $\kappa = \sqrt{\frac{GLM}{8}} = \sqrt{\frac{GM}{a}\frac{b^2}{4}} = \frac{b}{2}\sqrt{\frac{GM}{a}}$, so that $\frac{2\kappa}{ab} = \sqrt{\frac{GM}{a^3}}$.

This completes the derivation of the hyperbolic Kepler equation

$$\boxed{\varepsilon\sinh\beta(t) - \beta(t) = \tfrac{2\kappa t}{ab} = \sqrt{\frac{GM}{a^3}}\,t}$$

As in the elliptical case, the quantities $\frac{2\kappa t}{ab} = \sqrt{\frac{GM}{a^3}}\,t$, $\beta(t)$, and $\alpha(t)$ are referred to by the historical terms *mean anomaly, eccentric anomaly,* and *true anomaly*, respectively.

As might be suspected, the solution of the hyperbolic Kepler equation for $\beta(t)$ with t given, is similar to the solution of the elliptical version. Before we present it, we'll use the equation to derive formulas for the velocity of the craft.

For any time t, let $v(t)$ be the speed of C in its hyperbolic trajectory and let $\gamma(t)$ be the angle between the segment SC and the tangent to the trajectory at C. See Figure 6.13. As in the elliptical case, $\gamma(t)$ is measured counterclockwise from the tangent to the segment SC, so that $\gamma(t) > 0$ for any t. Notice that $\gamma < \frac{\pi}{2}$ during the craft's approach to periapsis, that $\gamma = \frac{\pi}{2}$ at periapsis, and that $\gamma > \frac{\pi}{2}$ after the craft's departure from periapsis. Observe also that $\gamma(t)$ is an increasing function of t.

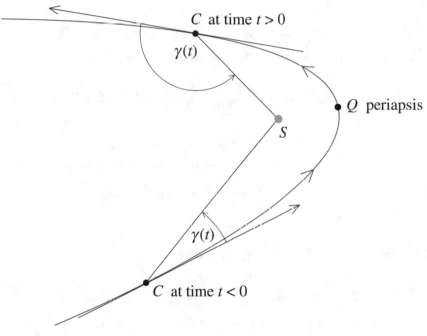

Figure 6.13

Given all the parallels between the hyperbolic and elliptical situations, it should come as no surprise that the formulas for $v(t)$ and $\gamma(t)$ as well as their derivations are analogous to those in Chapter 5D in the elliptical case. Both depend only on the distance $r(t)$ and constants of the trajectory.

Using the fact that the derivatives of $x(t) = -a\cosh\beta(t)$ and $y(t) = b\sinh\beta(t)$ are $x'(t) = -a\sinh\beta(t)\cdot\beta'(t)$ and $y'(t) = b\cosh\beta(t)\cdot\beta'(t)$, we get

$$
\begin{aligned}
v(t)^2 = x'(t)^2 + y'(t)^2 &= \left[a^2\sinh^2\beta(t) + b^2\cosh^2\beta(t)\right]\beta'(t)^2 \\
&= \left[a^2\sinh^2\beta(t) + (c^2 - a^2)\cosh^2\beta(t)\right]\beta'(t)^2 \\
&= \left[a^2\varepsilon^2\cosh^2\beta(t) - a^2(\cosh^2\beta(t) - \sinh^2\beta(t))\right]\beta'(t)^2 \\
&= \left[a^2\varepsilon^2\cosh^2\beta(t) - a^2\right]\beta'(t)^2 = a^2\left[\varepsilon^2\cosh^2\beta(t) - 1\right]\beta'(t)^2 \\
&= a^2\left[(\varepsilon\cosh\beta(t) - 1)(\varepsilon\cosh\beta(t) + 1)\right]\beta'(t)^2
\end{aligned}
$$

By differentiating the hyperbolic Kepler equation

$$\varepsilon \sinh \beta(t) - \beta(t) = \sqrt{\tfrac{GM}{a^3}}\, t$$

we get $\sqrt{\tfrac{GM}{a^3}} = \varepsilon \cosh \beta(t) \cdot \beta'(t) - \beta'(t) = (\varepsilon \cosh \beta(t) - 1)\beta'(t)$, so that

$$\beta'(t) = \sqrt{\tfrac{GM}{a^3}} \cdot \frac{1}{\varepsilon \cosh \beta(t) - 1}.$$

It follows that

$$v(t)^2 = a^2[(\varepsilon \cosh \beta(t) - 1)(\varepsilon \cosh \beta(t) + 1)] \cdot \frac{GM}{a^3} \cdot \frac{1}{(\varepsilon \cosh \beta(t) - 1)^2}$$

$$= \frac{GM}{a} \cdot \frac{\varepsilon \cosh \beta(t) + 1}{\varepsilon \cosh \beta(t) - 1}.$$

From the formula $r(t) = a(\varepsilon \cosh \beta(t) - 1)$ of section 6F, we get $\varepsilon \cosh \beta(t) - 1 = \frac{r(t)}{a}$ and therefore $\varepsilon \cosh \beta(t) = \frac{r(t)}{a} + 1$. So by simple substitutions and a little algebra,

$$v(t)^2 = \frac{GM}{a} \cdot \frac{\frac{r(t)}{a} + 2}{\frac{r(t)}{a}} = \frac{GM}{a} \cdot \frac{\frac{r(t)+2a}{a}}{\frac{r(t)}{a}} = \frac{GM}{a} \frac{r(t)+2a}{r(t)} = \frac{GM}{a}\left(\frac{2a}{r(t)} + 1\right).$$

We have completed the derivation of the hyperbolic speed formula

$$\boxed{\; v(t) = \sqrt{\frac{GM}{a}} \sqrt{\frac{2a}{r(t)} + 1} = \sqrt{GM}\sqrt{\frac{2}{r(t)} + \frac{1}{a}} \;}$$

Example 6.11. The craft attains its maximum speed

$$v_{\max} = \sqrt{\frac{GM}{a}} \sqrt{\frac{2a}{a(\varepsilon - 1)} + 1} = \sqrt{\frac{GM}{a}} \sqrt{\frac{\varepsilon + 1}{\varepsilon - 1}} = \sqrt{\frac{GM}{a(\varepsilon - 1)}} \sqrt{\varepsilon + 1}$$

at periapsis when $r(t)$ is at its minimum $a(\varepsilon - 1)$. Observe that the craft's speed is always greater than its limiting speed $v_\infty = \sqrt{\frac{GM}{a}}$. Notice that this is the limit on the speed of the craft as the distance $r(t)$ gets larger and larger.

The derivation of the formula for $\gamma(t)$ carried out in Chapter 5D in the elliptical case applies step by step to the hyperbolic situation (at the very end the hyperbolic speed formula and the hyperbolic equalities $2\kappa = b\sqrt{\frac{GM}{a}}$ and $b = a\sqrt{\varepsilon^2 - 1}$ have to be inserted) with the result that

$$\boxed{\; \sin \gamma(t) = \frac{a\sqrt{\varepsilon^2 - 1}}{\sqrt{r(t)(2a + r(t))}} \;}$$

By applying $\sin^{-1}$ to both sides of this equation and recalling that the inverse sine of any number between -1 and 1 needs to lie between $-\frac{\pi}{2}$ and $\frac{\pi}{2}$, we get the formulas

$$\boxed{\; \gamma(t) = \sin^{-1}\left(\frac{a\sqrt{\varepsilon^2 - 1}}{\sqrt{r(t)(2a + r(t))}}\right) \quad \text{and} \quad \gamma(t) = \pi - \sin^{-1}\left(\frac{a\sqrt{\varepsilon^2 - 1}}{\sqrt{r(t)(2a + r(t))}}\right) \;}$$

where the first equality applies when the craft is on approach to periapsis and the second after its departure from periapsis. Check that $\gamma = \frac{\pi}{2}$ at periapsis. Observe that $\lim_{t \to \pm\infty} \frac{a\sqrt{\varepsilon^2 - 1}}{\sqrt{r(t)(2a + r(t))}} = 0$ and conclude that 0 and π are optimal lower and upper bounds for the angle $\gamma(t)$.

6H. Solving the Hyperbolic Kepler Equation. The task is to solve, for a given t, the hyperbolic Kepler's equation

$$\varepsilon \sinh \beta(t) - \beta(t) = \sqrt{\frac{GM}{a^3}}\, t$$

for $\beta(t)$. Consider the function $f(x) = \varepsilon \sinh x - x$. Since $f'(x) = \varepsilon \cosh x - 1 > \cosh x - 1 \geq 0$, we know that $f(x)$ is an increasing function. The very rapid rise of the graph of the function implies that $\lim_{x \to -\infty} (\sinh x - x) = -\infty$ and $\lim_{x \to \infty} (\sinh x - x) = \infty$. Since $f(x) = \varepsilon \sinh x - x$ is continuous, it follows from this that the hyperbolic Kepler equation has a unique solution $\beta(t)$ for any given $\sqrt{\frac{GM}{a^3}}\, t$. This is the $\beta(t)$ that we need to find. Given that both $r(t)$ and $\alpha(t)$ have already been expressed in terms of $\beta(t)$, this $\beta(t)$ "closes the loop" in the sense that it completes, for a given t, the determination of the corresponding $r(t)$ and $\alpha(t)$.

The successive approximation approach that solved Kepler's equation for $\beta(t)$ in the elliptical situation in Chapter 5E works here too, but not "as is." The approximation step in the elliptical case relied on the inequality $|\sin x_1 - \sin x_2| \leq |x_1 - x_2|$. This inequality in turn depended on the fact that both $x - \sin x$ and $x + \sin x$ are increasing functions of x. But in the current hyperbolic situation the function $x - \sinh x$ is decreasing, so that this approach does not get off the ground. However, the inverse hyperbolic sine saves the day. The fact that the derivative of $\sinh^{-1} x$ is $\frac{1}{\sqrt{x^2 + 1}}$ (see Example 6.10) has the consequence that the derivatives of the functions $g(x) = x - \sinh^{-1} x$ and $h(x) = x + \sinh^{-1} x$ are both positive for all x with the single exception $g'(0) = 0$. This implies that $y = g(x)$ and $y = h(x)$ are both increasing functions of x. It follows, as in the analogous step of the elliptical case, that $|\sinh^{-1} x_1 - \sinh^{-1} x_2| \leq |x_1 - x_2|$ for any x_1 and x_2.

From the graph of the hyperbolic sine in Figure 6.10 we know that if small positive and negative x are excluded, then $|\sinh x|$ is much greater than $|x|$. Thus with the exception of small positive and negative $\beta(t)$, the term $\varepsilon \sinh \beta(t)$ dominates the left side of the hyperbolic Kepler equation, so that $\varepsilon \sinh \beta(t) \approx \sqrt{\frac{GM}{a^3}}\, t$. By applying $\sinh^{-1}$ to both sides of $\sinh \beta(t) \approx \frac{1}{\varepsilon}\left(\sqrt{\frac{GM}{a^3}}\, t\right)$, we get $\beta(t) \approx \sinh^{-1} \frac{1}{\varepsilon}\left(\sqrt{\frac{GM}{a^3}}\, t\right)$. It therefore makes sense, for a given elapsed time t, to take

$$\beta_1 = \sinh^{-1} \frac{1}{\varepsilon}\left(\sqrt{\frac{GM}{a^3}}\, t\right)$$

as the first approximation for the solution $\beta(t)$ of Kepler's equation.

By rearranging things algebraically and then taking $\sinh^{-1}$ of both sides, the hyperbolic Kepler equation can be rewritten as

$$\beta(t) = \sinh^{-1} \frac{1}{\varepsilon}\left(\sqrt{\frac{GM}{a^3}}\, t + \beta(t)\right).$$

The inequality $|\sinh^{-1} x_1 - \sinh^{-1} x_2| \leq |x_1 - x_2|$ implies that

$$|\beta(t) - \beta_1| = \left| \sinh^{-1} \frac{1}{\varepsilon}\left(\sqrt{\frac{GM}{a^3}}\, t + \beta(t)\right) - \sinh^{-1} \frac{1}{\varepsilon}\left(\sqrt{\frac{GM}{a^3}}\, t\right) \right| \leq \left| \frac{1}{\varepsilon}\left(\sqrt{\frac{GM}{a^3}}\, t + \beta(t)\right) - \frac{1}{\varepsilon}\sqrt{\frac{GM}{a^3}}\, t \right| = \frac{1}{\varepsilon}|\beta(t)|.$$

Therefore $|\beta(t) - \beta_1| \le \frac{1}{\varepsilon}|\beta(t)|$. For any $i \ge 1$, define β_{i+1} inductively by

$$\beta_{i+1} = \sinh^{-1}\frac{1}{\varepsilon}\left(\sqrt{\frac{GM}{a^3}}\,t + \beta_i\right).$$

Let's use mathematical induction to show that $|\beta(t) - \beta_i| \le \frac{1}{\varepsilon^i}|\beta(t)|$ for all $i \ge 1$. The case $i = 1$ was just done. It remains to assume that $|\beta(t) - \beta_i| \le \frac{1}{\varepsilon^i}|\beta(t)|$ and to show that $|\beta(t) - \beta_{i+1}| \le \frac{1}{\varepsilon^{i+1}}|\beta(t)|$. Because $\beta(t) = \sinh^{-1}\frac{1}{\varepsilon}\left(\sqrt{\frac{GM}{a^3}}\,t + \beta(t)\right)$ and $\beta_{i+1} = \sinh^{-1}\frac{1}{\varepsilon}\left(\sqrt{\frac{GM}{a^3}}\,t + \beta_i\right)$, we get

$$\begin{aligned}
|\beta(t) - \beta_{i+1}| &= \left|\sinh^{-1}\frac{1}{\varepsilon}\left(\sqrt{\frac{GM}{a^3}}\,t + \beta(t)\right) - \sinh^{-1}\frac{1}{\varepsilon}\left(\sqrt{\frac{GM}{a^3}}\,t + \beta_i\right)\right| \\
&\le \left|\frac{1}{\varepsilon}\left(\sqrt{\frac{GM}{a^3}}\,t + \beta(t)\right) - \frac{1}{\varepsilon}\left(\sqrt{\frac{GM}{a^3}}\,t + \beta_i\right)\right| = \left|\frac{1}{\varepsilon}\beta(t) - \frac{1}{\varepsilon}\beta_i\right| \le \frac{1}{\varepsilon^{i+1}}|\beta(t)|
\end{aligned}$$

so that our verification of the inequality $|\beta(t) - \beta_i| \le \frac{1}{\varepsilon^i}|\beta(t)|$ for all $i \ge 1$ is complete. Since $\varepsilon > 1$, it follows that the sequence $\beta_1, \ldots, \beta_i, \ldots,$ converges to the solution $\beta(t)$ of the hyperbolic Kepler equation $\varepsilon \sinh\beta(t) - \beta(t) = \sqrt{\frac{GM}{a^3}}\,t$. The speed of the convergence depends on the rate at which $\frac{1}{\varepsilon^i}|\beta(t)|$ goes to zero for increasing i. This depends on both ε and the magnitude of $\beta(t)$. The larger the $|\beta(t)|$, the longer it takes. On the other hand, it follows from a point made earlier that if $\beta(t)$ is large, then β_1 is already a good approximation of $\beta(t)$.

Let's summarize what has been accomplished. For a given elapsed time t, the position of the craft or comet C with respect to the center of mass S of the Sun, a planet, moon, or asteroid can be determined by solving Kepler's equation for $\beta(t)$ and inserting this value into the equations

$$r(t) = a(\varepsilon\cosh\beta(t) - 1) \quad\text{and}\quad \alpha(t) = 2\tan^{-1}\left(\sqrt{\frac{\varepsilon+1}{\varepsilon-1}}\,\tanh\frac{\beta(t)}{2}\right).$$

Refer back to Figure 6.12. By substituting this $r(t)$ in turn into the formulas for the speed $v(t)$ and the angle $\gamma(t)$, the velocity of C at time t can be determined as well.

Let T be the time it takes for C to move through the part of its hyperbolic orbit that is cut by the segment through S that defines the latus rectum. See Figure 6.9. Refer to Figure 6.12 and notice that $y(\frac{T}{2}) = \frac{L}{2}$. It follows that $b\sinh\beta(\frac{T}{2}) = \frac{b^2}{a}$. Therefore

$$\sinh\beta\left(\tfrac{T}{2}\right) = \frac{b}{a} = \frac{\sqrt{c^2 - a^2}}{a} = \frac{\sqrt{a^2(\varepsilon^2 - 1)}}{a} = \sqrt{\varepsilon^2 - 1}\,.$$

The standard formula $\sinh^{-1} x = \ln|x + \sqrt{x^2 + 1}|$ implies that

$$\beta\left(\tfrac{T}{2}\right) = \sinh^{-1}\left(\sqrt{\varepsilon^2 - 1}\right) = \ln\left(\sqrt{\varepsilon^2 - 1} + \varepsilon\right).$$

By inserting $t = \frac{T}{2}$ and the equalities above into the hyperbolic Kepler equation and solving for T, we get

$$T = 2\sqrt{\frac{a^3}{GM}}\left[\varepsilon\left(\sqrt{\varepsilon^2 - 1}\right) - \ln\left(\sqrt{\varepsilon^2 - 1} + \varepsilon\right)\right].$$

This equation is of important consequence for our understanding of the solar system. If T, a, and ε are known for a hyperbolic flyby by a spacecraft of a planet, a moon, or asteroid of mass M, then GM and hence M can be estimated.

We'll illustrate the discussion above by applying it to *NEAR-Shoemaker*'s gravity assist flyby of Earth. The relevant parameters of the craft's hyperbolic trajectory were $a = 8.500675 \times 10^6$ m, $\varepsilon = 1.813524$ and with M the Earth's mass, $GM = 3.986004 \times 10^{14}$ m^3/sec^2. Rotate the hyperbolic arc of Figure 6.2 and move it into the position of the one in Figure 6.12 with S representing the Earth's center of mass.

Example 6.12. By inserting the data of *NEAR-Shoemaker*'s flyby into the formula for T,

$$T = 2\sqrt{\tfrac{a^3}{GM}}\left[\varepsilon(\sqrt{\varepsilon^2 - 1}) - \ln(\sqrt{\varepsilon^2 - 1} + \varepsilon)\right]$$

$$= \tfrac{2(8.500675\times10^6)^{\frac{3}{2}}}{\sqrt{3.986004\times10^{14}}}\left[1.813524(\sqrt{1.813524^2 - 1}) - \ln(\sqrt{1.813524^2 - 1} + 1.813524)\right] \approx 3828 \text{ sec}$$

or about 64 minutes. This is consistent with the timeline of the flyby that Figure 6.2 displays.

Example 6.13. How long after it passed periapsis did *NEAR-Shoemaker* reach the boundary of the Sun-Earth Laplace radius of 920,000 km? We'll set $r(t) = 9.200000 \times 10^8$ m and determine t in seconds. Consider the equation $r(t) = a(\varepsilon \cosh \beta(t) - 1)$ and check that $\frac{r(t)+a}{a\varepsilon} = \cosh \beta(t)$. Since $\frac{r(t)+a}{a\varepsilon} \geq 1$, we get

$$\beta(t) = \cosh^{-1}(\varepsilon^{-1}(a^{-1}r(t) + 1))$$

$$= \cosh^{-1}\left(1.813524^{-1}((8.500675 \times 10^6)^{-1}(9.200000 \times 10^8) + 1)\right) \approx 4.791232.$$

Solving Kepler's equation for t, we see that $t = a\sqrt{\tfrac{a}{GM}}\left(\varepsilon \sinh \beta(t) - \beta(t)\right)$. Feeding in the values for the various terms, gives us

$$t = (8.500675 \times 10^6)\sqrt{\tfrac{8.500675\times10^6}{3.986004\times10^{14}}}\left(1.813524 \sinh(4.791232) - 4.791232\right) \approx 129{,}627 \text{ sec,}$$

or almost exactly 36 hours. The corresponding angle $\alpha(t)$ is

$$\alpha(t) = 2\tan^{-1}\left(\sqrt{\tfrac{\varepsilon+1}{\varepsilon-1}} \tanh \tfrac{\beta(t)}{2}\right) = 2\tan^{-1}\left(\sqrt{\tfrac{1.813524+1}{1.813524-1}} \tanh \tfrac{4.791232}{2}\right) \approx 2.119367 \text{ radians}$$

or 121.43°. The craft's speed relative to Earth at the time was

$$v(t) = \sqrt{GM}\sqrt{\tfrac{2}{r(t)} + \tfrac{1}{a}}$$

$$= \sqrt{3.986004 \times 10^{14}}\sqrt{\tfrac{2}{9.200000\times10^8} + \tfrac{1}{8.500675\times10^6}} \approx 6911 \text{ m/sec,}$$

or 6.911 km/sec. It is not surprising, given the distances involved, that this is close to the craft's limiting speed

$$v_\infty = \sqrt{\tfrac{GM}{a}} = \sqrt{\tfrac{3.986004\times10^{14}}{8.500675\times10^6}} \approx 6848 \text{ m/sec} = 6.848 \text{ km/sec.}$$

Example 6.14. How far from Earth was *NEAR-Shoemaker* 12 hours after it passed the periapsis of its flyby of Earth? The first step is to take $t = 12 \cdot 3600 = 43{,}200$ seconds and to solve $\varepsilon \sinh \beta(t) - \beta(t) = \frac{2\kappa t}{ab} = \sqrt{\tfrac{GM}{a^3}}\, t$ for $\beta(t)$. Taking

$$\sqrt{\tfrac{GM}{a^3}} = \sqrt{\tfrac{3.986004\times10^{14}}{(8.500675\times10^6)^3}} = 0.080554 \times 10^{-2} = 8.0554 \times 10^{-4}$$

and applying the method of successive approximations, we get

$$\beta_1 = \sinh^{-1}\tfrac{1}{\varepsilon}\left(\sqrt{\tfrac{GM}{a^3}}\,t\right) = \sinh^{-1}\tfrac{1}{1.813524}\left((8.0554\times10^{-4})\,(4.3200\times10^4)\right) = 3.648152$$

$$\beta_2 = \sinh^{-1}\tfrac{1}{\varepsilon}\left(\sqrt{\tfrac{GM}{a^3}}\,t+\beta_1\right) = \sinh^{-1}\tfrac{1}{1.813524}\left((8.0554\times10^{-4})\,(4.3200\times10^4)+3.648152\right)$$
$$= 3.822006$$

$$\beta_3 = \sinh^{-1}\tfrac{1}{\varepsilon}\left(\sqrt{\tfrac{GM}{a^3}}\,t+\beta_2\right) = \sinh^{-1}\tfrac{1}{1.813524}\left((8.0554\times10^{-4})\,(4.3200\times10^4)+3.822006\right)$$
$$= 3.829582$$

$$\beta_4 = \sinh^{-1}\tfrac{1}{\varepsilon}\left(\sqrt{\tfrac{GM}{a^3}}\,t+\beta_3\right) = \sinh^{-1}\tfrac{1}{1.813524}\left((8.0554\times10^{-4})\,(4.3200\times10^4)+3.829582\right)$$
$$= 3.829911$$

$$\beta_5 = \sinh^{-1}\tfrac{1}{\varepsilon}\left(\sqrt{\tfrac{GM}{a^3}}\,t+\beta_4\right) = \sinh^{-1}\tfrac{1}{1.813524}\left((8.0554\times10^{-4})\,(4.3200\times10^4)+3.829911\right)$$
$$= 3.829926$$

$$\beta_6 = \sinh^{-1}\tfrac{1}{\varepsilon}\left(\sqrt{\tfrac{GM}{a^3}}\,t+\beta_5\right) = \sinh^{-1}\tfrac{1}{1.813524}\left((8.0554\times10^{-4})\,(4.3200\times10^4)+3.829926\right)$$
$$= 3.829926,$$

so that with an accuracy of 6 decimal places, $\beta(t) = 3.829926$ radians. Inserting this value into the equation $r(t) = a(\varepsilon\cosh\beta(t)-1)$, tells us that *NEAR-Shoemaker* was

$$r(43200) = (8.500675\times10^6)(1.813524\cosh(3.829926)-1) \approx 346.694551\times10^6 \text{ m },$$

or 346,695 km from Earth's center. The corresponding angle $\alpha(t)$ was

$$\alpha(t) = 2\tan^{-1}\left(\sqrt{\tfrac{\varepsilon+1}{\varepsilon-1}}\tanh\tfrac{\beta(t)}{2}\right) = 2\tan^{-1}\left(\sqrt{\tfrac{1.813524+1}{1.813524-1}}\tanh\tfrac{3.829926}{2}\right) \approx 2.118194 \text{ radians,}$$

or 121.36°. A comparison of this value for $\alpha(t)$ with the one computed in Example 6.13 tells us that the convergence of $\alpha(t)$ to its upper bound $\pi - \tan^{-1}\tfrac{b}{a} = 123.46°$ is slow.

The speed with which *NEAR-Shoemaker* moved away from Earth 12 hours after periapsis was

$$v(t) = \sqrt{GM}\sqrt{\tfrac{2}{r(t)}+\tfrac{1}{a}} = \sqrt{3.986004\times10^{14}}\sqrt{\tfrac{2}{346.694551\times10^6}+\tfrac{1}{8.500675\times10^6}} \approx 7014 \text{ m/sec,}$$

or 7.014 km/sec, while the craft's maximum speed, reached at periapsis, was

$$v_{\max} = \sqrt{\tfrac{GM}{a}}\sqrt{\tfrac{\varepsilon+1}{\varepsilon-1}} = \sqrt{\tfrac{3.986004\times10^{14}}{8.500675\times10^6}}\sqrt{\tfrac{1.813524+1}{1.813524-1}} \approx 1.273451\times10^4 \text{ m/sec,}$$

or 12.735 km/sec. And 12 hours after periapsis, the craft moved away from Earth at the angle

$$\gamma(t) = \pi - \sin^{-1}\left(\tfrac{a\sqrt{\varepsilon^2-1}}{\sqrt{r(t)(2a+r(t))}}\right) = \pi - \sin^{-1}\left(\tfrac{(8.500675\times10^6)\sqrt{1.813524^2-1}}{\sqrt{(9.200000\times10^8)(2(8.500675\times10^6)+9.200000\times10^8)}}\right) \approx 3.131753$$

radians, or 179.44°.

We'll now return to the analysis of the strategic problem of sending a spacecraft from its initial parking orbit around Earth on to its first solar orbit, then on a transfer to an expanded solar orbit, and, after the trajectory changes typically brought about by one or more hyperbolic flybys of nearer planets, onward to its programmed rendezvous with one or more of the solar system's more distant objects.

6I. Hohmann Transfer Orbits. The most critical aspects of the flights of both the *Juno* space-craft to Jupiter and the *NEAR-Shoemaker* craft to the asteroid Eros were the following: the first was the transfer from the craft's brief near-Earth orbit around the Sun to an expanded elliptical solar orbit that was carefully designed to move the craft into position for a course-changing hyper-bolic flyby of a planet that then sent the craft to the target of its mission. In the case of both *Juno* and *NEAR-Shoemaker*, the craft was on the return leg of its expanded elliptical orbit and the flyby planet was Earth. The missions that have sent spacecraft to the outer planets and beyond, to Jupiter and Saturn, to some asteroids and comets, to Pluto and a few Kuiper belt objects, have all deployed such a strategy. Limited rocket thrust and constraints on budgets have required that one or several hyperbolic flyby maneuvers needed to be programmed into the design of their trajectories.

Let C be a spacecraft flying in its own near-Earth solar orbit. In order to expand this initial orbit, the craft's main engine is fired in a burst tangential to the craft's motion to increase the craft's speed. At the completion of this maneuver, the engine shuts down, and the gravitational pull of the Sun is the only force on the craft. The craft is now in a wider transfer orbit that is

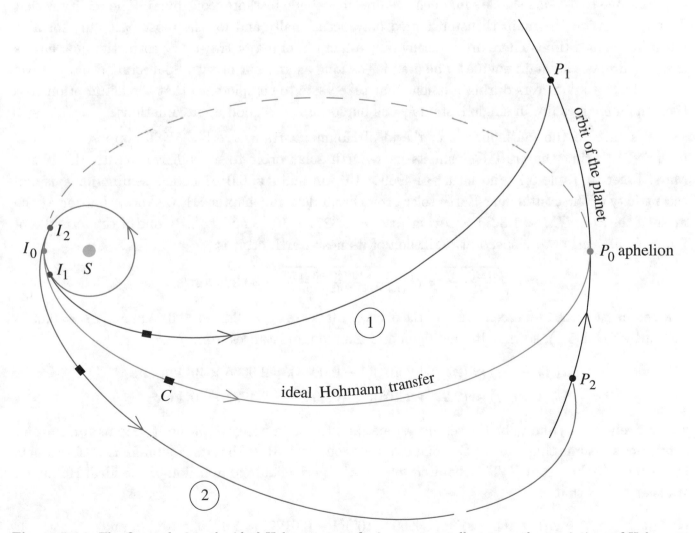

Figure 6.14. The figure depicts the ideal Hohmann transfer in green as well as two other variations of Hohmann orbits in red and purple. All three are possible trajectories of a spacecraft for a flight from an initial near-Earth solar orbit to a rendezvous with a distant planet.

designed to bring it to the vicinity of a more distant planet. This new trajectory is elliptical and has the center of mass of the Sun as focal point. Figure 6.14 depicts the Sun with its center of mass S, the Earth's orbit in blue, and in red, green, and purple, three different possible transfer trajectories for the craft. The points I_0, I_1, and I_2 designate the points on the craft's near-Earth orbit at which the main engine fires and the insertion into the transfer trajectory occurs. Their colors in red, green, and purple, correspond to those of the three trajectories. The points P_0, P_1, and P_2 mark the targeted points of rendezvous with the distant planet. The ideal Hohmann transfer orbit is depicted in green. The point of insertion I_0 is the perihelion of the near-Earth orbit. Since the transfer orbit expands the near-Earth orbit, the point I_0 is also the perihelion of the transfer orbit. The ideal Hohmann transfer has the point of rendezvous P_0 at its aphelion. In variation 1 of a Hohmann transfer, the point of insertion I_1 occurs after the perihelion of the craft's near-Earth orbit, and in version 2, the point of insertion I_2 occurs before perihelion. Note also that the points of rendezvous P_1 and P_2 can be anywhere along these trajectories.

Our mathematical analysis of the trajectory correction maneuver of the flight of a spacecraft will continue to assume that its main engine fires in a single instantaneous burst. The reality is that during a mission the main thruster is fired only occasionally and for the most part only for at a few minutes at a time. There are typically only a handful of major trajectory correction maneuvers that require for the main engine to be fired for as long as an hour or two. Spacecraft usually travel for millions of kilometers during missions that take years to complete, so that the assumption that their main engines fire in single instantaneous bursts provides good approximations.

Let's turn to the mathematics of the ideal Hohmann transfer orbit. We'll suppose that it is intended to bring the craft C from its near-Earth solar orbit to a rendezvous with the planet Mars. Refer to Table 5.1, and let $a \approx 1.4960 \times 10^8$ km and $\varepsilon \approx 0.0167$ be the semimajor axis and eccentricity of the craft's near-Earth solar orbit. From data in Chapter 1H, we know that for M the mass of the Sun, $GM \approx 1.3271 \times 10^{20}$ m^3/sec$^2 = 1.3271 \times 10^{11}$ km^3/sec^2. By one of the formulas of Example 5.1, the craft's speed at perihelion of its near-Earth orbit is

$$v = \sqrt{\tfrac{GM(1+\varepsilon)}{a(1-\varepsilon)}} \approx \sqrt{\tfrac{(1.3271\times10^{11})(1+0.0167)}{(1.4960\times10^8)(1-0.0167)}} \approx 30.286 \text{ km/sec.}$$

The semimajor axis and eccentricity of the orbit of Mars are $a_M \approx 2.2794 \times 10^8$ km and $\varepsilon_M \approx 0.0934$ (again by Table 5.1), so that its minimum and maximum distances from S are

$$a_M(1 - \varepsilon_M) \approx (2.2794 \times 10^8)(1 - 0.0934) \approx 2.0665 \times 10^8 \text{ km and}$$
$$a_M(1 + \varepsilon_M) \approx (2.2794 \times 10^8)(1 + 0.0934) \approx 2.4923 \times 10^8 \text{ km,}$$

respectively. Let a_0 and ε_0 be the semimajor axis and the eccentricity of the craft's Hohmann transfer orbit. Let's assume that the craft's point of rendezvous with Mars that its Hohmann transfer targets is 2.0700×10^8 km from S. The requirements for the perihelion and aphelion of the ideal Hohmann transfer tell us that

$$a_0(1 - \varepsilon_0) = a(1 - \varepsilon) \approx (1.4960 \times 10^8)(1 - 0.0167) \approx 1.4710 \times 10^8 \text{ km and}$$
$$a_0(1 + \varepsilon_0) \approx 2.0700 \times 10^8 \text{ km.}$$

Since $2a_0 \approx (1.4710 + 2.0700) \times 10^8 \approx 3.5410 \times 10^8\,\text{km}$ and $2a_0\varepsilon_0 \approx (2.0700 - 1.4710) \times 10^8 \approx 0.599 \times 10^8\,\text{km}$, we get

$$a_0 \approx 1.7705 \times 10^8\,\text{km} \quad \text{and} \quad \varepsilon_0 \approx \tfrac{0.599}{3.541} \approx 0.1692.$$

Applying the formula of Example 5.1 for the speed at perihelion again, we find that for the successful placement of the craft into this Hohmann transfer orbit, it needs to be provided with a speed of

$$v_0 = \sqrt{\frac{GM(1+\varepsilon_0)}{a_0(1-\varepsilon_0)}} \approx \sqrt{\frac{(1.3271\times10^{11})(1+0.1692)}{(1.7705\times10^8)(1-0.1692)}} \approx 31.986\ \text{km/sec}$$

at the point of insertion into the orbit. It follows that the insertion of the craft into this Hohmann transfer to Mars requires an increase of approximately $31.986 - 30.286 = 1.70\,\text{km/sec}$ or $1700\,\text{m/sec}$ in the speed of the craft when it arrives at the perihelion of its near-Earth solar orbit.

In order to time a rendezvous of the craft with Mars, the duration of the transfer is an important concern. In the case of the ideal Hohmann transfer this is $\frac{T_0}{2}$, where T_0 is the full period of the expanded orbit. The transfer orbit of the craft satisfies Newton's version of Kepler's second law, $\frac{a_0^3}{T_0^2} = \frac{GM}{4\pi^2}$, so that $\frac{T_0^2}{4\pi^2} = \frac{a_0^3}{GM}$ and (in MKS) that

$$\frac{T_0}{2} = \pi\sqrt{\frac{a_0^3}{GM}} \approx \frac{\pi\sqrt{(1.7705\times10^{11})^3}}{\sqrt{1.3271\times10^{20}}} \approx 20{,}316{,}175\,\text{sec}.$$

Since one day has $86{,}400$ seconds (see Chapter 1G), it follows that this ideal Hohmann transfer to Mars would take about 235 days. It turns out that a craft that is inserted into its transfer orbit at perihelion of its near-Earth orbit—as is the case with the ideal Hohmann transfer—takes the longest time to get to its destination, but the required speed increase that gets it there is least.

A look back at Example 2.2 and the discussion of the flight of *Juno* to Jupiter in Chapter 2H makes it clear that the main engine of a smaller spacecraft (such as *Juno* with its 645 newton

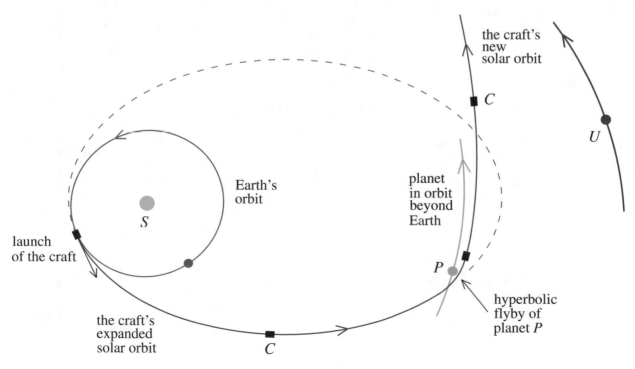

Figure 6.15

thruster) would not be capable of providing it with the speed increase of 1700 m/sec that the successful injection into this Hohmann transfer orbit requires. The fact is that a flight to Mars is possible only with either a speed increasing hyperbolic flyby of an interior planet (such as the one that propelled *Juno* towards Jupiter), or a powerful engine (of the sort that sent the *Voyagers* on their way to Jupiter. See the upcoming section 6L).

6J. Gravity Assist Flybys. Let's assume that the continuing mission of a spacecraft C calls for the exploration of an outer planet. To get there, the craft will need another boost in its velocity relative to the Sun in order to break out of an earlier tighter, elliptical solar orbit. Refer Figure 6.15. If things are timed and calibrated carefully, the craft can be brought into the gravitational neighborhood of a planet P with the result that the gravitational pull of P will redirect the craft, increase its speed, and bring it on a course to the targeted outer planet U.

We turn our attention to the craft's hyperbolic flyby and the changes in the Sun-relative motion of the craft that it brings about. Suppose that the craft has entered the gravitational sphere

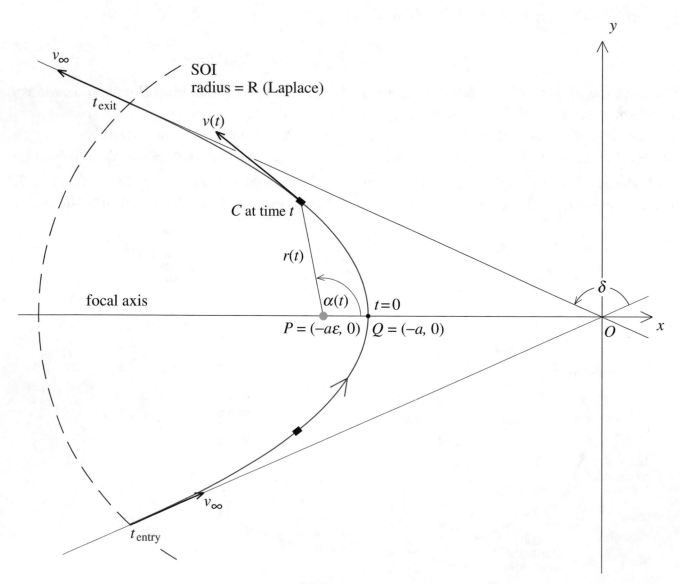

Figure 6.16

of influence (SOI) of the planet P that the Laplace radius of the Sun-planet system determines. Suppose also that the craft's speed is great enough so that it is drawn into a hyperbolic flyby—but not into orbit—around P. Figure 6.16 shows the part of the hyperbolic trajectory that falls inside the planet-centered SOI indicated by the dashed circular arc. The coordinates of the point P (understood to be the center of mass of the planet) and Q (the periapsis of the flyby) are determined by the semimajor axis a of the hyperbola and its eccentricity ε. Given that the Laplace radius is relatively large, the two slanting lines emanating from the origin O and tangent to the trajectory are close approximations of the hyperbola's asymptotes. A time t is assigned to the position of the craft during its flyby by taking $t < 0$ on the craft's approach to periapsis, $t = 0$ at periapsis, and $t > 0$ thereafter, in such a way that $|t|$ is the craft's time of flight to or from its position to periapsis (as in section 6F and Figure 6.12). The instants at which the craft enters and departs the SOI are denoted by t_{entry} and t_{exit}, respectively. The distance of C from P at any time t is denoted by $r(t)$.

We'll let M_P be the mass of P and make use of the results of section 6G. The velocity of the craft relative to P at time t in its flyby is given by the speed formula

$$v(t) = \sqrt{\frac{GM_P}{a}}\sqrt{\frac{2a}{r(t)} + 1} = \sqrt{GM_P}\sqrt{\frac{2}{r(t)} + \frac{1}{a}}$$

together with the angle $\gamma(t)$ of Figure 6.13 and the formulas for $\gamma(t)$ that conclude section 6G. At the time t_{entry} the craft enters the SOI and at the time t_{exit} that it departs from it, its distance $r(t)$ from P is large relative to a. So $\frac{2a}{r(t)}$ is small and

$$v(t_{\text{entry}}) \approx v_\infty = \sqrt{\frac{GM_P}{a}} \approx v(t_{\text{exit}}).$$

We'll now study the velocity vectors of the flyby both at the time of entry into and departure

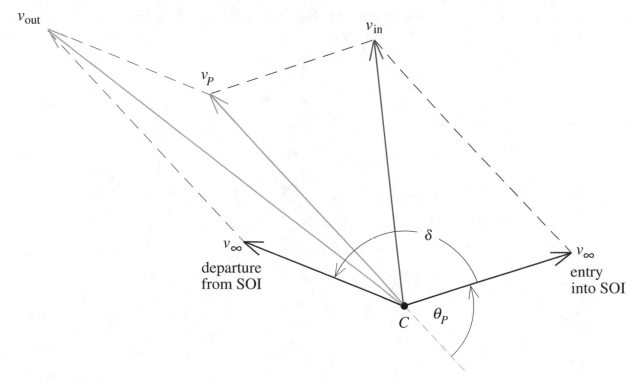

Figure 6.17

from the SOI and the effect of the flyby on the motion of the spacecraft C relative to the Sun. Our study relies on the fact that P in its orbit around the Sun drags its SOI as well the hyperbolic flyby of C with it.

Turn to Figure 6.17. The velocity of the planet P relative to S is represented by the vector v_P that has length the speed of P and direction given by the tangent to its orbital path. The figure depicts the vector v_P and the angle θ_P between it and the vector v_∞ of the craft's entry into the planet's SOI. Table 5.1 tells us that the eccentricities of the outer planets are all small. It follows from a comparison of the formulas for $v_{\max}$ and $v_{\min}$ of Example 5.1 that the speed of the planet does not vary much over time. Given that the planet P takes years to orbit the Sun but the craft's flyby only a few hours (see Example 6.13 for instance), we can assume that the velocity vector v_P of P relative to S is constant. Figure 6.17 also depicts the velocity vector v_∞ at the point of the craft's departure from the planet's SOI. The respective resultants of v_P and the two vectors v_∞, each determined by the parallelogram law, are drawn into the figure as v_{in} and v_{out} in blue and green, respectively. The vector v_{in} represents the velocity of the craft C relative to the Sun at the point of entry into the SOI and the vector v_{out} the velocity of the craft relative to the Sun at the point of departure from the SOI. The fact that v_{out} is longer than v_{in} tells us (in the particular situation being considered) that the hyperbolic cruise around P has increased the speed of the craft relative to the Sun. The angle δ is the *angle of deflection*. A careful look at Figure 6.9 tells us that $\delta = \pi - 2\varphi$. So $\frac{\delta}{2} = \frac{\pi}{2} - \varphi$ and $\sin\frac{\delta}{2} = \frac{a}{c} = \frac{a}{a\varepsilon} = \frac{1}{\varepsilon}$. Therefore $\frac{\delta}{2} = \sin^{-1}\left(\frac{1}{\varepsilon}\right)$ and

$$\delta = 2\sin^{-1}\left(\tfrac{1}{\varepsilon}\right).$$

The magnitudes of v_{in} and v_{out} as well as the angle between them can be computed by making use of basic trigonometry. The computation relies on the Figure 6.18 extracted from Figure 6.17. By the law of cosines applied to the triangle that the vectors v_P and v_{in} determine, we get (now also using $v_\infty, v_P, v_{\text{in}}$, and v_{out} for the magnitudes of these vectors) that $v_{\text{in}}^2 = v_\infty^2 + v_P^2 - 2v_\infty v_P \cos\theta_P$

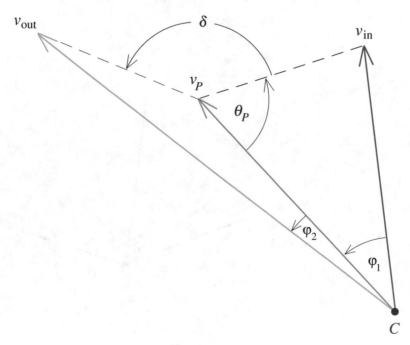

Figure 6.18

and hence that

$$v_{\text{in}} = \sqrt{v_\infty^2 + v_P^2 - 2v_\infty v_P \cos\theta_P}\,.$$

By the law of cosines applied to the triangle formed by the vectors v_P and v_{out}, we get $v_{\text{out}}^2 = v_\infty^2 + v_P^2 - 2v_\infty v_P \cos(2\pi - (\theta_P + \delta)) = v_\infty^2 + v_P^2 - 2v_\infty v_P \cos(-(\theta_P + \delta))$, so that,

$$v_{\text{out}} = \sqrt{v_\infty^2 + v_P^2 - 2v_\infty v_P \cos(\theta_P + \delta)}\,.$$

The angle $\varphi_1 + \varphi_2$ measures the change in the direction of the motion of the craft C relative to the Sun that it undergoes as a result of the hyperbolic flyby. The angles φ_1 and φ_2 can be found by the law of sines. Two applications of this law tell us that

$$\frac{\sin\varphi_1}{v_\infty} = \frac{\sin\theta_P}{v_{\text{in}}} \quad \text{and} \quad \frac{\sin\varphi_2}{v_\infty} = \frac{\sin(2\pi - (\theta_P + \delta))}{v_{\text{out}}} = -\frac{\sin(\theta_P + \delta)}{v_{\text{out}}}.$$

Therefore the change in the direction in the craft's trajectory relative to the Sun that the flyby around P brings about is given by the angle

$$\varphi = \varphi_1 + \varphi_2 = \sin^{-1}\left(\frac{v_\infty}{v_{\text{in}}}\sin\theta_P\right) - \sin^{-1}\left(\frac{v_\infty}{v_{\text{out}}}\sin(\theta_P + \delta)\right),$$

where $v_\infty = \sqrt{\frac{GM_P}{a}}$ and the magnitudes v_{in} and v_{out} are provided by the formulas above.

The derivations of the formulas derived above have relied on Figure 6.17 and the assumption that the vector v_P is positioned between the two vectors v_∞ at the entry and exit of the SOI of the

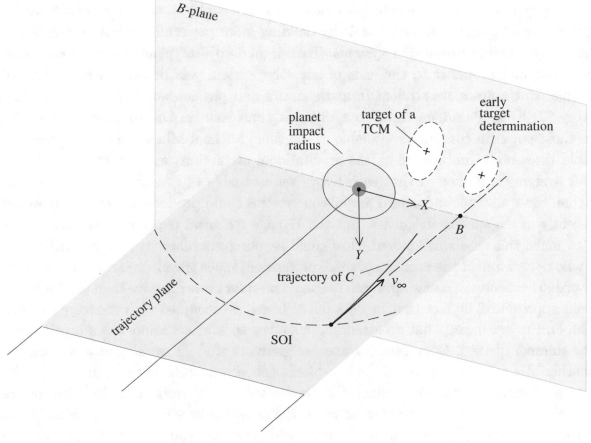

Figure 6.19

flyby. The same formulas (with similar derivations) also hold for the other locations of the vector v_P relative to the vectors v_{in} and v_{out} of the flyby. Refer to the paragraph *Returning to the Voyagers* in the Problems and Discussions section of this chapter.

It is obvious that if a craft's gravity assist flyby of planet P is to achieve its navigational goals, then it needs to be tightly timed and carefully configured. An important tool in the execution of such a maneuver is the construct of a target plane known as the B-plane (B for Body). Turn to Figure 6.19. At the time of the craft's arrival in the SOI of the planet, consider the asymptote of its hyperbolic trajectory (this is essentially the line in Figure 6.16 determined by the vector v_∞ at time t_{in}). The B-plane is the plane perpendicular to this asymptote and through the center of mass of the planet P (the planet in brown, its center of mass in black). This is a plane in which the craft can be tracked. The *aim point* is the point B of intersection of the asymptote with the B-plane. The impact radius of P is the circular region (elliptical in perspective) in the B-plane that the craft needs to steer clear of so as not to risk a collision with the planet. The vector X in the B-plane can be the specified in different ways to lie, for example, in the orbital plane of P or the equatorial plane of P. The vector Y in the B-plane is perpendicular to X. The pair X and Y provide a coordinate system for the B-plane. The points designated by a $+$ indicate projected positions of the craft or targets of trajectory correction maneuvers and the white ellipses that surround them in the B-plane give indication of the possible errors involved.

6K. The *Voyagers* and *Cassini*. In view of the considerable weight that the required quantities of propellant add to any spacecraft, it had been thought—even by the expert scientists and engineers who were engaged in the post-Sputnik space race—that the exploration of the more distant reaches of the solar system would only be possible by building more powerful conventional rocket engines or entirely new nuclear propulsion systems. That tight flybys of planets could boost the speed of a spacecraft and propel it to the ends of our solar system was beyond their concept of what was possible. But a graduate student in mathematics and physics working at the Jet Propulsion Laboratory (JPL) in Pasadena came to a different conclusion in the summer of 1961. Working with the most powerful computers available at the time, Michael Minovitch studied the seemingly intractable three-body problem (e.g., the gravitational attractions and motions of a Sun-planet-spacecraft system) and showed that such flybys were capable of generating vehicle velocities far greater than those the most advanced propulsion systems could produce and that they would do so independently of the spacecraft's mass. But the 47 page technical paper that Minovitch presented to JPL to make this case was ignored. How could a young graduate student in mathematics and physics who never studied the problem of space propulsion, space travel, or astrodynamics before the summer of 1961 conceive of a new approach to space travel far more effective than the "only possible" traditional approach? The fact that in 1962 the JPL was preoccupied with the support of NASA's Apollo Moon Project meant that no attention was given to Michael Minovitch's breakthrough.

In the summer of 1965, Gary Flandro, another graduate student working at JPL, began to consider whether Minovitch's solution of the Sun-planet-spacecraft problem could inform the challenge of sending a spacecraft to the outer planets. He began to make a study of the locations of the outer planets in the solar system for the coming years. His investigations revealed, much to his surprise, that in the late 1970s, Jupiter, Saturn, Uranus, and Neptune would all be aligned on the same side of the Sun. His computations confirmed that Minovitch's ideas implied that a single mission launched from Earth in 1977, could sling a spacecraft past all four of these planets within 12 years. Such an opportunity would not present itself again for another 176 years. Eventually, NASA and JPL embraced both Minovitch's gravity assist propulsion concept and Flandro's idea for a "grand tour" of the planets.

In the fall of 1977, two *Voyager* spacecraft were launched from Cape Canaveral, Florida on exploratory cruises through the solar system. For over 40 years these craft gathered information about the planets and their moons, sent back the data they collected, and captured thousands of astonishing images. *Voyager 1* was launched on September 5th, 1977 aboard a Titan III-E/Centaur rocket. The spacecraft was attached to an Injection-Propulsion Unit (IPU) that was driven by a powerful 76,500 N solid-propellant rocket engine and four smaller liquid fuel powered thrusters. The combined mass of the craft and the IPU was 2,060 kg. The IUP had a mass of 1,235 kg, most of it the 1120 kg solid fuel rocket engine. Three days after *Voyager 1*'s launch, the IPU's rocket engine fired for 43 seconds, burning its 1,040 kg of propellant in the process. During the burn, the smaller thrusters pulsated on and off to stabilize the vehicle's direction. After the fuel was spent and the required speed increase achieved, the IPU was jettisoned and *Voyager 1* was off on its own on a Sun-focused trajectory that took it toward Jupiter. The identical craft *Voyager 2* had been launched in the same way two weeks before *Voyager 1*. But *Voyager 1*, placed on a more direct trajectory through the solar system, soon caught up to its twin and overtook it.

Each of the *Voyagers* started its cruise to Jupiter with a mass of 825 kg that included 103 kg of liquid propellant for its set of small thrusters. The thrusters provided each craft with the capability to stabilize it in flight, to adjust its orientation, and to execute small trajectory correction maneuvers. A burn of one of them generated a thrust of about 0.834 newtons. This is a very small push. A single penny has a mass of 2.50 grams or 0.0025 kg and therefore a weight of $0.0025 \times 9.80665 = 0.0245$ newtons. It follows that 34 pennies in the palm of your hand exert the same push as one of the thrusters. Consequently, only a few minor trajectory correction maneuvers were planned for the *Voyagers*. The mission of *Voyager 2*, for example, called for only eight trajectory correction maneuvers, the first two early in the cruise to Jupiter, three more prior to the encounter with Jupiter, and another four between Jupiter and Saturn.

Example 6.15. Use the fact that the specific impulse of one of *Voyager*'s thrusters is $I_{sp} \approx 200$ sec and the equation $v_{ex} = I_{sp} \cdot g_0$ from section 6A, to show that the exhaust gases that it generates have a velocity of approximately 2 km/sec. Use the rocket equation of Chapter 2G, to conclude that the total change in velocity Δv that the thrusters could have generated by burning all available fuel, would have been $\Delta v = v_{ex} \ln \frac{M_1}{M_2} \approx 2 \ln \frac{825}{722} \approx 0.27$ km/sec.

Even if the *Voyagers*' thrusters would have been deployed exclusively for increasing their speeds—rather than to also stabilize or reorient them—these thrusters would have had essentially no impact on the ability of these spacecraft to escape their initial orbits. Only the gravity assist flybys of the outer planets made their exploration of the outer solar system possible.

The *Voyager* missions succeeded magnificently. They discovered that Jupiter has a complicated atmosphere and that its great red storm rotates once every six days pulling in smaller eddies as it does. They detected Jupiter's rings and discovered that the moon Io has active volcanoes.

They provided a detailed sense of the complexity of Saturn's ring system. These feature braids, kinks, and spokes. The *Voyagers* discovered new 'shepherd' moons that keep Saturn's rings stable. Their instruments detected the smoggy, mostly nitrogen-containing atmosphere of Saturn's moon Titan. *Voyager 2*'s investigations of Uranus and Neptune revealed that both had large and unusual magnetic fields. It determined the chemical compositions of their atmospheres. Its cameras detected ten previously unknown moons of Uranus, studied the planet's ring system, discovered two new rings and the fine detail of those previously known. *Voyager 2* also discovered active, geyser-like features on Neptune's largest moon Triton.

Both *Voyagers* left the realm of the planets long ago. *Voyager 1* is currently about 150 au from the Sun moving away from it at a speed of about 3.6 au per year, and *Voyager 2* is about 120 au from the Sun speeding along at about 3.3 au per year. *Voyager 1* left the plane of the orbits of the planets after its flyby of Jupiter at an angle of about 35° with this plane and *Voyager 2* left the plane of the planets after its flyby of Neptune at an angle of 48°. Now exiting the solar system, the two craft continue to send data as they probe for the limits of the influence of the solar wind—the stream of charged particles that is released from the upper atmosphere of the Sun.

The *Cassini* spacecraft was launched with a Titan IV-Centaur rocket in October 15th, 1997, on a mission to investigate Saturn, its moons, and its ring system. With its equatorial radius of 60,268 km and mass of 5.6851×10^{26} kg, Saturn is the solar system's second largest planet (after Jupiter). The spacecraft was named after Dominique Cassini, the French-Italian astronomer of the 17th century whose parallax studies informed us about the true size of the solar system, who discovered four of Saturn's major moons, and who was first to observed gaps in Saturn's ring system. At the time of its launch the craft had a total mass to 5,630 kg, including 3,130 kg of liquid propellant for its rocket engines. The challenge to provide the massive craft with the velocity necessary to bring it to its gravity-assist flyby of Jupiter and then to Saturn was met very differently for *Cassini* than it had been for the *Voyagers*. Instead of relying on a powerful injection-propulsion unit, the design of *Cassini*'s flight plan called for several gravity assist maneuvers. *Cassini*'s path took it from Earth to two flybys of Venus on to another flyby of Earth that directed it to a rendezvous with Jupiter. The flyby of Jupiter finally gave *Cassini* the velocity needed to bring it to Saturn. Had the resulting velocity increases been left to the craft's main thruster, the amount of propellant required would have exceeded in mass many times the total mass that the craft had at launch.

After taking 7 years to journey to Saturn, *Cassini* circled around the planet, its moons, and its rings for 13 years. During one of its early orbits around Saturn, *Cassini* ejected the probe *Huygens* toward the moon Titan. *Huygens* studied Titan's atmosphere on its descent, and after landing softly on its surface, took images of the flat, sandy, pebble-strewn plain nearby. After an orbit correction maneuver in August of 2004, *Cassini* had used up much of its propellant. To chase down Saturn's moons and to analyze its rings, the spacecraft had to rely on the velocity changes that close flybys of Titan provided. *Cassini* looped around Saturn almost 300 times in orbits that varied widely in size, orientation, and inclination. With the information that its instruments gathered during the years of its mission, initially scheduled to run until 2008 but extended until 2017, the scientific story about Saturn, its moons, and its ring system was completely rewritten.

After 20 years in space, *Cassini*'s fuel supply was exhausted. In its final bow on the stage that Saturn provided, it was sent on two dozen daring, risky final orbits. In these final loops, the spacecraft dove through the gap between Saturn and its icy, inner rings and skimmed through the outer edges of the planet's atmosphere. Along the way, *Cassini*'s particle detectors sampled icy ring particles and observed how Saturn's magnetic field funneled them into its atmosphere, while its cameras took amazing, closeup images of the planet's rings and clouds. On September 15, 2017, *Cassini* was on its final approach to the giant planet. Sending data for as long as its small thrusters could keep its antenna pointed at Earth, it burst through Saturn's atmosphere and burned up like a disintegrating meteor. This fiery end ensured that the craft would not collide with Enceladus and Titan and contaminate environments that have given evidence of harboring some of the essential ingredients of life.

We have seen that spacecraft perform a variety of tasks during a mission. They receive instructions from mission scientists and engineers and respond to them. This includes navigational aspects, such as trajectory correction maneuvers where smaller thrusters orient the craft, and precisely directed and quantified burns of the main engine that adjust its velocity. Spacecraft need to respond to the commands involved with high degrees of accuracy. They send back to mission control what their sensors and instruments collect, the data their computers record, and the images that their cameras capture. The electronic and mechanical systems of spacecraft must be heated to be fully operational, and backup systems must be on standby. All this requires a significant amount of electrical power. Standard sources of power do not measure up to the demands. Conventional batteries have limited lifetimes in the frigid conditions of space and cannot supply power for missions of years in duration. Solar panels are effective for missions to the inner planets and for missions to Jupiter. (See Figure 2.4 for instance.) But at distances beyond the orbit of Jupiter, solar radiation is too weak—for example, Saturn is about 10 times farther from the Sun than Earth, so that the sunlight it receives has only 1% of the intensity of the sunlight that strikes Earth—and current solar panels not advanced enough to produce sufficient power. The technology that has been used to provide power both for *Cassini* and the *Voyagers* is the radioisotope thermoelectric generator (RTG). This highly reliable nuclear battery produces heat through slow radioactive decay, of plutonium-238 for instance, which is in turn converted to electricity. These batteries have powered all spacecraft on missions to the outer solar system throughout their long journeys.

The websites https://voyager.jpl.nasa.gov/ and https://saturn.jpl.nasa.gov/ provide much of the information that the *Voyager* and *Cassini* missions have collected. Some of what they transmitted was discussed in Chapter 2. We will now turn to illustrate much of the analysis of this chapter by studying some of the navigational aspects of the trajectories of the two *Voyagers* and *Cassini*. Throughout this study, we'll let S be the center of mass of the Sun.

6L. The Cruise of *Voyager 1*. The essential information about the segments of *Voyager 1*'s trip through the solar system is provided by NASA and collected in Table 6.3. The first row of the table lists basic data about the craft's elliptical, Sun-focused trajectory to Jupiter. The other rows pertain to the hyperbolic flybys of Jupiter and Saturn, and the Sun-focused hyperbolic cruises to Saturn and to the outer reaches of the solar system beyond Saturn. (Note that in each case the eccentricity is greater than one.) The dates and times for the interplanetary flights are those of the start in the

Table 6.3. *Voyager 1*'s cruise through the solar system. The segments Earth−Jupiter, Jupiter−Saturn, and post Saturn are Sun-focused trajectories. The two flybys are planet-focused. The times listed refer to Eastern Standard Time. The data are taken from https://voyager.jpl.nasa.gov/mission/science/hyperbolic-orbital-elements/.

Voyager 1 trajectory	date, time	semimajor axis (km)	eccen-tricity	inclination (deg)	mean ano-maly (deg)
Earth−Jupiter	9/8/77, 9:08:17	745,761,000	0.797783	1.032182	0.304932
flyby Jupiter	3/5/79, 12:05:26	1,092,356	1.318976	3.979134	
Jupiter−Saturn	4/24/79, 7:33:03	593,237,000	2.302740	2.481580	19.156329
flyby Saturn	11/12/80, 23:46:30	166,152	2.107561	65.893904	
post Saturn	1/1/91, 00:00	480,926,000	3.724716	35.762854	688.967795

given trajectory. Those for the hyperbolic flybys reference the closest approaches at periapsis. The information about the inclination of the planes of *Voyager 1*'s trajectory together with information in Table 5.1 tells us that the craft's Sun-focused approaches to both Jupiter and Saturn were tightly aligned with the orbital planes of these planets. The flyby of Saturn lifted the plane of *Voyager 1*'s final trajectory away from the planets' orbital plane by 36°. Given the precision of the data of the table, we'll be computing with an accuracy of six significant figures.

We begin our study with the Earth−Jupiter leg of *Voyager 1*'s journey. The craft traversed only a part of the elliptical orbit that the injection-propulsion unit (IPU) had put it on (because after it arrived at Jupiter, it was redirected toward Saturn). Had it completed this solar orbit, Kepler's second law $\frac{a^3}{T^2} = \frac{GM}{4\pi^2}$ and the fact that $GM = 1.3271244 \times 10^{11}$ km^3/sec^2 for the Sun would have specified its period T to have been

$$T = \sqrt{\frac{4\pi^2 a^3}{GM}} = \sqrt{\frac{4\pi^2 (754,761,000)^3}{1.3271244 \times 10^{11}}} = 357,633,529 \text{ sec} \approx 11.3327 \text{ years.}$$

Let's consider the injection of *Voyager 1* into its orbit. Let t_1 be the elapsed time of the craft's orbit injection from perihelion (assuming that it had come from there on a previous orbit). Since the mean anomaly at time t_1 is 0.304932° and hence $0.304932 \cdot \frac{\pi}{180} = 0.0053144$ radians, we get that $\frac{2\pi t_1}{T} = 0.0053144$ (by a concluding remark of Chapter 5C). It follows that *Voyager 1*'s orbit insertion occurred $t_1 = \frac{0.0053144}{2\pi} 11.3327 = 0.009585$ years or $0.009585 \cdot 365.25 = 3.50$ days after perihelion. Turn next to Chapter 5E and the solution of Kepler's equation $\beta(t) - \varepsilon \sin \beta(t) = \sqrt{\frac{GM}{a^3}} t$ in the current situation of $t = t_1$. The first approximation of the solution $\beta(t_1)$ is given by the mean anomaly $\beta_1 = \sqrt{\frac{GM}{a^3}} t_1 = \frac{2\pi t_1}{T} = 0.0053144$. By applying the approximation step $\beta_{i+1} = \beta_1 + \varepsilon \sin \beta_i$ a total of 52 times (this is boring and laborious)[1], we get

$$\beta_1 = 0.0053144, \beta_2 = 0.0095541, \beta_3 = 0.0129364, \ldots, \beta_{10} = 0.0235315, \ldots,$$

$$\beta_{20} = 0.0259838, \ldots, \beta_{30} = 0.0262315, \ldots, \beta_{40} = 0.0262649, \ldots, \beta_{50} = 0.0262684,$$

$$\beta_{51} = 0.0262685, \beta_{52} = 0.0262686, \beta_{53} = 0.0262686.$$

---

[1]but manageable when carried out with a calculator such as the one provided by https://web2.0calc.com/.

Since the sequence has stabilized, it follows that $\beta(t_1) = 0.0262686$. (There is little doubt that the Newton-Raphson method—see the paragraph *Using Kepler's Equation* of the Problems and Discussion section of Chapter 5—would have converged to this solution more quickly.)

By inserting the value $\beta(t_1) = 0.0262686$ into the equations for $r(t)$ and $\alpha(t)$ of Chapter 5B, we find that the distance of *Voyager 1* from S at the time of the insertion into its transfer orbit to Jupiter was

$$r(t_1) = a(1 - \varepsilon \cos \beta(t_1)) = 754{,}761{,}000(1 - 0.797783 \cos(0.0262686)) = 152{,}833{,}242 \text{ km}$$

with corresponding angle $\alpha(t_1)$ (as in Figure 5.3) equal to

$$\alpha(t_1) = 2\tan^{-1}\left(\sqrt{\tfrac{1+\varepsilon}{1-\varepsilon}} \tan \tfrac{\beta(t_1)}{2}\right) = 2\tan^{-1}\left(\sqrt{\tfrac{1+0.797783}{1-0.797783}} \tan(\tfrac{0.0262686}{2})\right)$$
$$= 0.078324 \text{ radians} \approx 4.49°.$$

The equations for $v(t)$ and $\gamma(t)$ of Chapter 5D inform us about the velocity of *Voyager 1* relative to the Sun at the instant of the insertion into its transfer orbit. It had a speed of

$$v(t_1) = \sqrt{GM}\sqrt{\tfrac{2}{r(t_1)} - \tfrac{1}{a}} = \sqrt{1.3271244 \times 10^{11}}\sqrt{\tfrac{2}{152{,}833{,}242} - \tfrac{1}{754{,}761{,}000}} \approx 39.51 \text{ km/sec}$$

and was moving at the angle

$$\gamma(t_1) = \pi - \sin^{-1}\left(\tfrac{a\sqrt{1-\varepsilon^2}}{\sqrt{r(t_1)(2a-r(t_1))}}\right) = \pi - \sin^{-1}\left(\tfrac{754{,}761{,}000\sqrt{1-0.797783^2}}{\sqrt{152{,}833{,}242(2\cdot754{,}761{,}000-152{,}833{,}242)}}\right)$$
$$= 1.605536 \text{ radians} \approx 91.99°$$

relative to the Sun.

Having studied the insertion of *Voyager 1* into its transfer orbit to Jupiter on September 8th, 1977, we turn to the spacecraft's rendezvous with Jupiter on March 5th, 1979. The date and time data of Table 6.3 tells us that this occurred 542 days and 3 hours or $542.125/365.25 = 1.4843$ years after the craft's orbit insertion. So the craft's travel time from periapsis to its Jupiter rendezvous was $t_2 = 1.4843 + t_1 = 1.4843 + 0.0096 = 1.4939$ years. To determine the distance $r(t_2)$ of *Voyager 1* from the Sun and the corresponding angle $\alpha(t_2)$ at the time of the rendezvous, we need to compute $\beta(t_2)$. Returning to Chapter 5E and taking $\beta_1 = \tfrac{2\pi t_2}{T} = \tfrac{2\pi 1.4939}{11.3327} = 0.828263$, we get (much more quickly than the last time)

$$\beta_2 = 1.416036, \ \beta_3 = 1.616511, \ \beta_4 = 1.625213, \ \beta_5 = 1.624865, \ \beta_6 = 1.624880, \ \beta_7 = 1.624880,$$

so that $\beta(t_2) = 1.624880$. It follows that at the time of the rendezvous, Jupiter was

$$r(t_2) = a(1 - \varepsilon \cos \beta(t_2)) = 754{,}761{,}000(1 - 0.797783 \cos(1.624880)) = 787{,}310{,}826 \text{ km},$$

or about $787{,}000{,}000$ km from the Sun. The angle $\alpha(t_2)$ was

$$\alpha(t_2) = 2\tan^{-1}\left(\sqrt{\tfrac{1+\varepsilon}{1-\varepsilon}} \tan \tfrac{\beta(t_2)}{2}\right) = 2\tan^{-1}\left(\sqrt{\tfrac{1+0.797783}{1-0.797783}} \tan(\tfrac{1.624880}{2})\right)$$
$$= 2.526331 \text{ radians} \approx 144.75°.$$

In terms of the velocity relative to the Sun on its arrival at Jupiter, *Voyager 1* had a speed of

$$v(t_2) = \sqrt{GM}\sqrt{\tfrac{2}{r(t_2)} - \tfrac{1}{a}} = \sqrt{1.3271244 \times 10^{11}}\sqrt{\tfrac{2}{787,310,826} - \tfrac{1}{754,761,000}} \approx 12.70 \text{ km/sec},$$

and was traveling away from the Sun at an angle of

$$\gamma(t_2) = \pi - \sin^{-1}\left(\frac{a\sqrt{1-\varepsilon^2}}{\sqrt{r(t_2)(2a - r(t_2))}}\right) = \pi - \sin^{-1}\left(\frac{754,761,000\sqrt{1 - 0.797783^2}}{\sqrt{787,310,826(2 \cdot 754,761,000 - 787,310,826)}}\right)$$

$$= 2.493702 \text{ radians} \approx 142.88°.$$

Figure 6.20 captures the information that has been developed about *Voyager 1*'s elliptical cruise

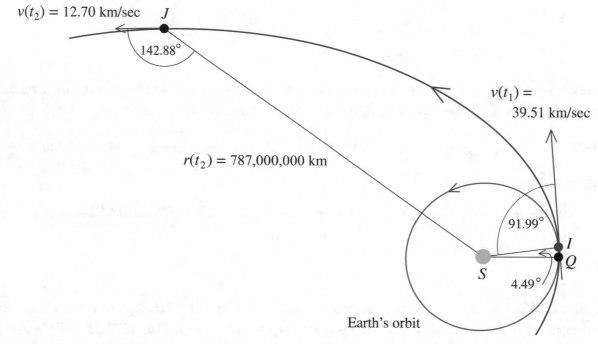

Figure 6.20. The figure lists important numerical data of *Voyager 1*'s elliptical Hohmann transfer trajectory from Earth orbit to Jupiter. The point of insertion into the trajectory is denoted by I and Q is the periapsis of the ellipse. The location of Jupiter is denoted by J. The flyby at Jupiter is not (yet) taken into account.

from its beginning near Earth to its approach to Jupiter.

We turn next to *Voyager 1*'s flyby of Jupiter. We'll let P be the planet Jupiter and refer to the discussion of section 6J and Figure 6.17 in particular. The gravitational constant $GM_P = 1.26712765 \times 10^8$ km^3/sec^2 with M_P the mass of Jupiter is taken from Table 6.4. The values $v_P = 12.83$ km/sec for the Sun-relative speed of Jupiter at the time of the flyby (Table 5.1 tells us that Jupiter's average orbital speed is 13.06 km/sec) and $\theta_P = 63.8°$ for the angle between Jupiter's

Table 6.4 Data taken from Cesarone, A Gravity Assist Primer, of the References for Chapter 6.

Voyager 1 hyperbolic flyby of	GM_P in km^3/sec^2	v_P in km/sec	θ_P in degrees	inclination relative to Earth's orbit
Jupiter	126,712,765	12.83	63.8	1.03° → 3.98° → 2.48°

velocity and the direction of *Voyager 1*'s approach to Jupiter come from Table 6.4. Using data in Table 6.3, we get that

$$v_\infty = \sqrt{\frac{GM_P}{a}} = \sqrt{\frac{126{,}712{,}765}{1{,}092{,}356}} \approx 10.77 \text{ km/sec}$$

for *Voyager 1*'s flyby and $a(\varepsilon - 1) = 1{,}092{,}356(1.318976 - 1) = 348435$ km for the distance of the craft from Jupiter's center of mass at periapsis of the flyby. Since $\varepsilon = 1.318976$, we get $\delta = 2\sin^{-1}(\frac{1}{\varepsilon}) = 2\sin^{-1}(\frac{1}{1.318976}) = 98.61°$ for the angle of deflection. It follows that

$$v_{\text{in}} = \sqrt{v_\infty^2 + v_P^2 - 2v_\infty v_P \cos\theta_P}$$
$$\approx \sqrt{10.77^2 + 12.83^2 - 2(10.77)(12.83)\cos 63.8}$$
$$\approx 12.59 \text{ km/sec (close to the 12.70 km/sec calculated earlier), and}$$
$$v_{\text{out}} = \sqrt{v_\infty^2 + v_P^2 - 2v_\infty v_P \cos(\delta + \theta_P)}$$
$$\approx \sqrt{10.77^2 + 12.83^2 - 2(10.77)(12.83)\cos(98.61 + 63.8)}$$
$$\approx 23.32 \text{ km/sec.}$$

The flyby resulted in a change of direction of

$$\varphi = \varphi_1 + \varphi_2 = \sin^{-1}\left(\frac{v_\infty}{v_{\text{in}}}\sin\theta_P\right) - \sin^{-1}\left(\frac{v_\infty}{v_{\text{out}}}\sin(\delta + \theta_P)\right)$$
$$\approx 50.13° - 8.03° = 42.10°$$

in the craft's trajectory. With its direction changed and with $23.32 - 12.59 = 10.73$ km/sec added to its speed relative to the Sun, *Voyager 1* was now on its way to the exploration of Saturn and its ring system. A comparison of *Voyager*'s initial elliptical orbit with Saturn's orbit tells that without the boost that Jupiter gave it, *Voyager 1* would have turned back in the direction of Earth, many millions of kilometers short of a rendezvous with Saturn. (In case you're wondering, the main engine of the *Voyagers* was not nearly powerful enough and its fuel supply not nearly large enough to have provided such a boost. Refer back to Example 6.15.)

Our final computations of *Voyager 1*'s trajectory deal with the segment from Jupiter to Saturn after the flyby of Jupiter. Over a month and a half after *Voyager 1* passed the periapsis of this flyby, the gravitational impact of Jupiter had become negligible. Table 6.3 informs us of the date (the 24th of April, 1979) and the exact time at which the craft had started on a new Sun-focused trajectory. Since its eccentricity—listed as 2.302740—was greater than 1, this new trajectory was hyperbolic. The table tells us that its semimajor axis was $a = 593{,}237{,}000$ km and that when *Voyager 1* started into this trajectory, the mean anomaly was 19.156329 degrees or $19.156329\frac{\pi}{180} = 0.334341$ radians. Therefore (see the definition of mean anomaly in section 6G) $\sqrt{\frac{GM}{a^3}}\, t_1 = 0.334341$, where M is the mass of the Sun and t_1 is the time of the craft's travel from the perihelion of its new hyperbolic trajectory to its starting point in it. To clarify, t_1 is the time it would have taken the craft to travel from the perihelion to the starting point had it been inserted into this hyperbolic trajectory before perihelion. As for the Earth–Jupiter segment of *Voyager 1*'s trip, the key to the analysis of the craft's motion at the start of its new trajectory is the value of $\beta(t_1)$, but this time in the hyperbolic context.

To compute $\beta(t_1)$, turn to section 6H. Since $\beta_1 = \sinh^{-1}\frac{1}{\varepsilon}\left(\sqrt{\frac{GM}{a^3}}\,t_1\right) = \sinh^{-1}\frac{1}{2.302740}(0.334341)$, we get $\beta_1 = 0.144673$ by using a calculator. The iteration of the approximation step

$$\beta_{i+1} = \sinh^{-1}\frac{1}{\varepsilon}\left(\sqrt{\frac{GM}{a^3}}\,t_1 + \beta_i\right) = \sinh^{-1}\frac{1}{2.302740}(0.334341 + \beta_i)$$

and the continued use of the calculator, provide the sequence

$$\beta_2 = 0.206547, \beta_3 = 0.232781, \ldots, \beta_{14} = 0.251918, \beta_{15} = 0.251919, \beta_{16} = 0.251919,$$

so that $\beta(t_1) = 0.251919$. The substitution of $\beta(t_1)$ into the equations for $r(t)$ and $\alpha(t)$ of section 6F informs us that at the beginning of *Voyager 1*'s hyperbolic trajectory from Jupiter to Saturn, the craft's distance from the Sun was

$$r(t_1) = a(\varepsilon \cosh \beta(t_1) - 1) = (593{,}237{,}000)\big(2.302740 \cosh(0.251919) - 1\big) = 816{,}410{,}896 \text{ km}$$

and the corresponding angle $\alpha(t_1)$ was

$$\alpha(t_1) = 2\tan^{-1}\left(\sqrt{\frac{\varepsilon+1}{\varepsilon-1}}\,\tanh\frac{\beta(t_1)}{2}\right) = 2\tan^{-1}\left(\sqrt{\frac{2.302740+1}{2.302740-1}}\,\tanh\frac{0.251919}{2}\right)$$
$$= 0.393837 \text{ radians} \approx 22.57°.$$

After substituting $r(t_1)$ into the formulas for $v(t)$ and $\gamma(t)$ of section 6G, we get that the craft began its new trajectory with a Sun-relative speed of

$$v(t_1) = \sqrt{GM}\sqrt{\frac{2}{r(t_1)} + \frac{1}{a}} = \sqrt{1.3271244 \times 10^{11}}\sqrt{\frac{2}{816{,}410{,}896} + \frac{1}{593{,}237{,}000}} \approx 23.43 \text{ km/sec},$$

traveling at an angle of

$$\gamma(t_1) = \pi - \sin^{-1}\left(\frac{a\sqrt{\varepsilon^2-1}}{\sqrt{r(t_1)(2a+r(t_1))}}\right) = \pi - \sin^{-1}\left(\frac{593{,}237{,}000\sqrt{2.302740^2-1}}{\sqrt{816{,}410{,}896(2(593{,}237{,}000)+816{,}410{,}896))}}\right)$$
$$= 1.846245 \text{ radians} \approx 105.78°$$

away from the Sun.

Table 6.3 tells us that the cruise from Jupiter to Saturn took close to 568 days and 16 hours, or 49133088 sec. Adding t_1 to this time, let $t_2 = 49133088 + t_1$. So t_2 is the time from perihelion of the post-Jupiter hyperbolic trajectory to the rendezvous with Saturn. Since

$$t_1 = (0.334341)\sqrt{\frac{a^3}{GM}} = (0.334341)\sqrt{\frac{593{,}237{,}000^3}{1.3271244\times10^{11}}} = 13260997 \text{ sec},$$

$t_2 = 49133088 + 13260997 = 62394085$ seconds. As in similar situations before, our understanding of *Voyager 1*'s motion on its approach to Saturn at time t_2 depends on the value $\beta(t_2)$. Turning to section 6G again, we get

$$\beta_1 = \sinh^{-1}\frac{1}{\varepsilon}\left(\sqrt{\frac{GM}{a^3}}\,t_2\right) = \sinh^{-1}\frac{1}{2.302740}\left(\sqrt{\frac{1.3271244\times10^{11}}{593{,}237{,}000^3}}\,62394085\right) = 0.638803.$$

Using the approximation step

$$\beta_{i+1} = \sinh^{-1} \tfrac{1}{\varepsilon}\left(\sqrt{\tfrac{GM}{a^3}}\, t_2 + \beta_i\right) = \sinh^{-1} \tfrac{1}{2.302740}\left(\sqrt{\tfrac{1.3271244\times10^{11}}{593{,}237{,}000^3}}\, 62394085 + \beta_i\right)$$

$$= \sinh^{-1} \tfrac{1}{2.302740}\left(1.573102 + \beta_i\right),$$

we obtain

$$\beta_2 = 0.853204, \beta_3 = 0.918813, \dots, \beta_{10} = 0.946386, \beta_{11} = 0.946390,$$

$$\beta_{12} = 0.946391, \beta_{13} = 0.946391, \beta_{14} = 0.946391,$$

so that therefore, $\beta(t_2) = 0.946391$. Applying earlier formulas once more, we get the following information about *Voyager 1*'s approach to Saturn. Its distance from the Sun was

$$r(t_2) = a(\varepsilon \cosh \beta(t_2) - 1) = (593{,}237{,}000)\big(2.302740 \cosh(0.946391) - 1\big) = 1{,}431{,}644{,}288 \text{ km}$$

and the corresponding angle $\alpha(t_2)$ was

$$\alpha(t_2) = 2\tan^{-1}\left(\sqrt{\tfrac{\varepsilon+1}{\varepsilon-1}}\, \tanh \tfrac{\beta(t_2)}{2}\right) = 2\tan^{-1}\left(\sqrt{\tfrac{2.302740+1}{2.302740-1}}\, \tanh \tfrac{0.946391}{2}\right)$$

$$= 1.223897 \text{ radians} \approx 70.12°.$$

The craft's Sun-relative speed and its angle of departure from the Sun were

$$v(t_2) = \sqrt{GM}\sqrt{\tfrac{2}{r(t_2)} + \tfrac{1}{a}} = \sqrt{1.3271244 \times 10^{11}}\sqrt{\tfrac{2}{1{,}431{,}644{,}288} + \tfrac{1}{593{,}237{,}000}} \approx 20.23 \text{ km/sec}$$

and

$$\gamma(t_2) = \pi - \sin^{-1}\left(\tfrac{a\sqrt{\varepsilon^2-1}}{\sqrt{r(t_2)(2a+r(t_2))}}\right) = \pi - \sin^{-1}\left(\tfrac{593{,}237{,}000\sqrt{2.302740^2-1}}{\sqrt{1{,}431{,}644{,}288(2(593{,}237{,}000)+1{,}431{,}644{,}288))}}\right)$$

$$= 2.452811 \text{ radians} \approx 140.54°.$$

The data of Tables 6.3 and 6.4 in combination with the mathematical tools developed in this and the previous chapter have allowed us to reconstruct basic relevant numerical information about the cruise of *Voyager 1* from Earth to Jupiter to Saturn.

It was an underlying assumption of our study of gravitational assists undertaken in section 6J that the plane of the flyby trajectory of the spacecraft and the orbital plane of the planet around which the flyby maneuver occurs are nearly the same. The entry $1.03° \rightarrow 3.98° \rightarrow 2.48°$ of Table 6.4 together with the information about the inclinations of the orbit of Jupiter in Table 5.1 tells us that the planes of *Voyager 1*'s trajectory before and during the flyby of Jupiter were closely aligned with the orbital plane of the planet. The information about the inclination of Saturn's orbital plane tells us that this was also the case for *Voyager 1*'s approach of Saturn. But this was no longer so for *Voyager 1*'s flyby of Saturn. The plane of the flyby was at an angle of $65.89°$ with the plane of Jupiter's orbit. After this maneuver *Voyager 1* sped away from the orbital plane of the planets. on a hyperbolic trajectory at an angle of $35.76°$ with their orbital plane. The mathematics of the changes in *Voyager 1*'s velocity that the flyby of Saturn brought about is much more complex. The three-dimensional vector calculus that is required is beyond the intention and scope of this text.

6M. The Cruise of *Voyager 2*. The mission of *Voyager 2* took advantage of a rare geometric alignment of the outer planets (such an alignment occurs once every 175 years) and threaded four successive needles to slide past Jupiter, Saturn, Uranus and, finally, Neptune. *Voyager 2*'s trajectory depended exclusively on the increases in the craft's velocity that each of the four flybys provided. *Voyager 2*'s small thrusters oriented the craft and facilitated minor trajectory corrections, but played

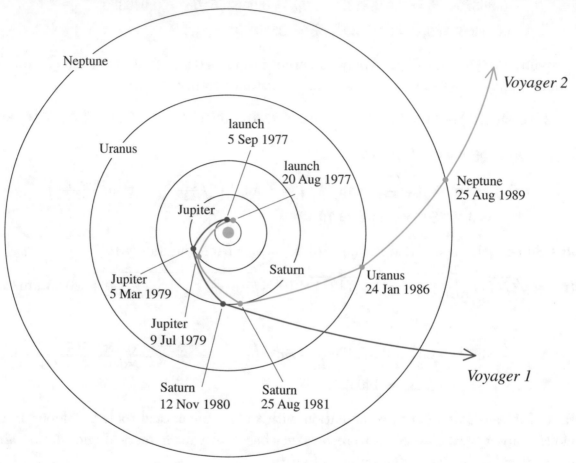

Figure 6.21. This diagram of the trajectories of the *Voyagers* and the planetary orbits is not to scale. Diagram adapted from https://commons.wikimedia.org/wiki/File:Voyager_Path.svg.

no role in boosting the craft's velocity. Figure 6.21 shows the delicately executed flight path of *Voyager 2* and the simpler course of *Voyager 1* side by side.

Table 6.5 provides essential details about the trajectory of *Voyager 2*. The dates and times for the interplanetary segments are those of the insertion into the trajectory. Their eccentricities tell us that with the exception of the flight segment to Jupiter, they are all hyperbolic. All have the center of mass S of the Sun at a focal point. The dates and times for the hyperbolic flybys reference the craft's closest approaches at periapsis. The flight of *Voyager 2* began on August 23, 1977, when the Injection-Propulsion Unit injected it from its near-Earth orbit into a Hohmann transfer toward Jupiter. The fact that its eccentricity was $\varepsilon = 0.724429$ tells us that this was an elliptical trajectory. Its semimajor axis was $a = 544{,}470{,}000$ km. Since $GM = 1.3271244 \times 10^{11}$ km³/sec² for the Sun, the period of this orbit would have been (had *Voyager 2* completed it)

$$T = \sqrt{\frac{4\pi^2 a^3}{GM}} = \sqrt{\frac{4\pi^2 (544{,}470{,}000)^3}{1.3271244 \times 10^{11}}} = 219{,}121{,}496 \text{ sec} \approx 6.9435 \text{ years.}$$

We see from the table that the mean anomaly at the point of the craft's injection into the transfer orbit was $-0.888403°$ or $-0.888403 \cdot \frac{\pi}{180} = -0.015506$ radians. The expression $\frac{2\pi t_1}{T} = \sqrt{\frac{GM}{a^3}}\, t_1$ for the

Table 6.5. *Voyager 2*'s cruise through the solar system. The segments from one planet to the next are all Sun-focused trajectories. The four flybys are planet-focused. The times listed refer to Eastern Standard Time. The data are taken from https://voyager.jpl.nasa.gov/mission/science/hyperbolic-orbital-elements/.

Voyager 2 trajectory	date, time	semimajor axis in km	eccen-tricity	inclination (deg)	mean ano-maly (deg)
Earth$-$Jupiter	8/23/77, 11:29	544,470,000	0.724429	4.825717	-0.888403
flyby Jupiter	7/9/79, 22:30	2,184,140	1.330279	6.913454	
Jupiter$-$Saturn	9/15/79, 11:07	2,220,315,000	1.338264	2.582320	4.798319
flyby Saturn	8/26/81, 03:25	332,965	1.482601	3.900931	
Saturn$-$Uranus	10/17/81, 18:44	579,048,000	3.480231	2.665128	10.350850
flyby Uranus	1/24/86, 18:00	26,694	5.014153	11.263200	
Uranus$-$Neptune	6/9/87, 00:00	448,160,000	5.806828	2.496223	315.018680
flyby Neptune	8/25/89, 03:57	24,480	2.194523	115.956093	
post Neptune	1/1/91, 00:00	601,124,000	6.284578	78.810177	342.970736

mean anomaly (see Chapter 5C) tells us that $t_1 = -(0.015506)\frac{219,121,496}{2\pi} = -540760$ seconds. The fact that this is negative means that *Voyager 2* was injected into this trajectory $t_1 = 540760$ seconds (or about 6 days and 6 hours) before it reached its perihelion. To understand the specifics about the insertion into this orbit, we need to compute $\beta(t_1)$.

Refer to the solution of Kepler's equation in Chapter 5E and notice that the successive approximation sequence for $\beta(t_1)$ begins with $\beta_1 = -0.015506$. The subsequent steps are given by $\beta_{i+1} = \frac{2\pi t_1}{T} + \varepsilon \sin \beta_i = -0.015506 + (0.724429) \sin \beta_i$. The repetitive use of this formula and a slog with a calculator, provides the values (that a reader might check),

$$\beta_2 = -0.026739, \beta_3 = -0.034874, \ldots, \beta_{10} = -0.053979, \ldots, \beta_{20} = -0.056105, \ldots,$$

$$\beta_{30} = -0.056187, \beta_{31} = -0.056188, \beta_{32} = -0.056189, \beta_{33} = -0.056190, \beta_{34} = -0.056190,$$

so that $\beta(t_1) = -0.056190$ radians.

Inserting this value into the equations for $r(t)$ and $\alpha(t)$ of Chapter 5B, we find that the distance of *Voyager 2* from S at the time of the insertion into its transfer orbit to Jupiter was

$$r(t_1) = a(1 - \varepsilon \cos \beta(t_1)) = 544,470,000(1 - 0.724429 \cos(-0.056190)) = 150,662,648 \text{ km}$$

and that the corresponding angle $\alpha(t_1)$ was

$$\alpha(t_1) = 2 \tan^{-1}\left(\sqrt{\tfrac{1+\varepsilon}{1-\varepsilon}} \tan \tfrac{\beta(t_1)}{2}\right) = 2 \tan^{-1}\left(\sqrt{\tfrac{1+0.724429}{1-0.724429}} \tan(\tfrac{-0.056190}{2})\right)$$
$$= -0.266248 \text{ radians} \approx -15.25°.$$

By applying the equations for $v(t)$ and $\gamma(t)$ from Chapter 5D, we get that the speed of *Voyager 2* at the instant of its insertion into its transfer orbit was

$$v(t_1) = \sqrt{GM}\sqrt{\tfrac{2}{r(t_1)} - \tfrac{1}{a}} = \sqrt{1.3271244 \times 10^{11}}\sqrt{\tfrac{2}{150,662,648} - \tfrac{1}{544,470,000}} \approx 38.96 \text{ km/sec}$$

and (since the craft was on approach to periapsis) that the direction of its velocity relative to the Sun was

$$\gamma(t_1) = \sin^{-1}\left(\frac{a\sqrt{1-\varepsilon^2}}{\sqrt{r(t_1)(2a-r(t_1))}}\right) = \sin^{-1}\left(\frac{544,470,000\sqrt{1-0.724429^2}}{\sqrt{150,662,648(2\cdot544,470,000-150,662,648)}}\right)$$
$$= 1.511846 \text{ radians} \approx 86.62°.$$

We'll leave the study of *Voyager 2*'s arrival at Jupiter for the Problems and Discussions section of this chapter and turn instead to its flyby of Jupiter. Inserting data from Tables 6.5 and 6.6 tells

Table 6.6. The data of the second column is taken from Table 2.6 and that of the last column from Table 6.5. The data from the middle columns comes from Cesarone, A Gravity Assist Primer, of the References for Chapter 6.

Voyager 2 hyperbolic flyby of	GM_P in km^3/sec^2	v_P in km/sec	θ_P in degs	inclination relative to Earth's orbit
Jupiter	126,712,765	12.69	48.3	4.83° → 6.91° → 2.58°
Saturn	37,940,585	9.59	98.2	2.58° → 3.90° → 2.67°
Uranus	5,794,549	6.71	106.0	2.67° → 11.26° → 2.50°

us that when *Voyager 2* entered Jupiter's sphere of gravitational influence (SOI) as determined by the Laplace radius, it had an speed of

$$v_\infty = \sqrt{\frac{GM_P}{a}} = \sqrt{\frac{1.26712765\times10^8}{2.184140\times10^6}} \approx 7.616754 \text{ km/sec}$$

relative to Jupiter. Substituting the data of Table 6.6 into the formulas of section 6J tells us that the speed of the craft relative to the Sun was

$$v_{\text{in}} = \sqrt{v_\infty^2 + v_P^2 - 2v_\infty v_P \cos\theta_P}$$
$$\approx \sqrt{7.6168^2 + 12.69^2 - 2(7.6168)(12.69)\cos(48.3)}$$
$$\approx \sqrt{219.0517 - 193.3144 \cdot 0.6652} = \sqrt{219.0517 - 128.5927} \approx 9.51 \text{ km/sec}$$

at the beginning of the flyby and

$$v_{\text{out}} = \sqrt{v_\infty^2 + v_P^2 - 2v_\infty v_P \cos(\theta_P + \delta)}$$
$$\approx \sqrt{7.6168^2 + 12.69^2 - 2(7.6168)(12.69)\cos(48.3 + 97.479)}$$
$$\approx \sqrt{219.0517 - 193.3144 \cdot (-0.8269)} \approx \sqrt{219.0517 + 159.8517} \approx 19.47 \text{ km/sec}$$

at its conclusion. So the hyperbolic flyby of Jupiter increased the speed of *Voyager 2* relative to the Sun by about $19.47 - 9.51 = 9.96$ km/sec. After inserting the data into the angle formulas of section 6J we get

$$\varphi_1 \approx \sin^{-1}\left(\tfrac{v_\infty}{v_{\text{in}}}\sin\theta_P\right) \approx \sin^{-1}\left(\tfrac{7.6168}{9.51}\sin 48.3\right) \approx 36.73° \text{ and}$$
$$\varphi_2 \approx -\sin^{-1}\left(\tfrac{v_\infty}{v_{\text{out}}}\sin(\theta_P + \delta)\right) \approx -\sin^{-1}\left(\tfrac{7.6168}{19.47}\sin(48.3 + 97.479)\right) \approx -12.72°.$$

Therefore the flyby bent the direction of the craft's flight by $36.73° - 12.72° \approx 24°$.

Voyager 2 was now on its way to Saturn, arriving there about 9 months after its faster twin. Velocity boosting gravitational assist maneuvers around Saturn and Uranus followed. The data of Tables 6.5 and 6.6 and a repetition of the calculations above show that the Saturn flyby increased the spacecraft's speed by $20.25 - 15.33 = 4.92$ km/sec and changed its direction by $45.28°$ and that the Uranus flyby increased the spacecraft's speed by $19.66 - 17.79 = 1.87$ km/sec and changed its direction by $17.31°$. Figure 6.22 depicts the graphs of the speeds of the two *Voyagers*. The jumps in the graphs record the speed changes that the hyperbolic flybys generated. The last column of

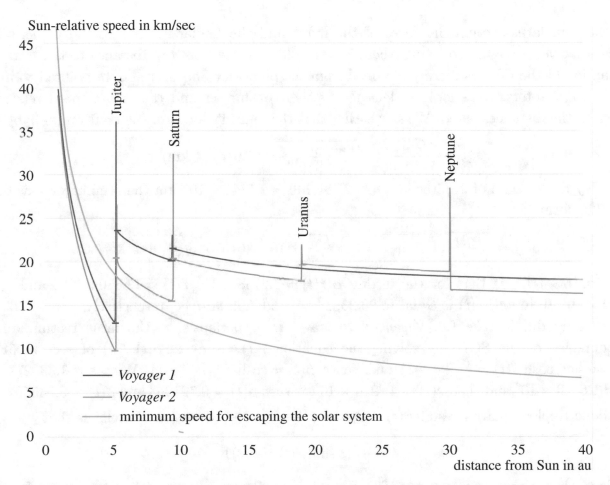

Figure 6.22. The figure graphs the Sun-relative speeds of the *Voyagers* in terms of their distances from the Sun. The jumps in the graph measure the changes in the speeds that the flybys of the planets provided. Notice that each flyby increased the spacecrafts' speed (except for *Voyager 2*'s flyby of Neptune which resulted in a decrease). Recall that 1 au ≈ 150 million km. Adapted from https://commons.wikimedia.org/wiki/File:Voyager_2_Heliocentric_ Velocity.png.

Table 6.6 confirms that, as required, the plane of *Voyager*'s trajectory was close to those of the orbits of the three planets that pulled the craft along. The last flyby around Neptune tilted the plane of *Voyager*'s trajectory away from the plane of the planets by almost $80°$. As Figure 6.22 shows, this maneuver simultaneously decreased the speed of the craft by about 2 km/sec.

We'll conclude our discussion of the mission of *Voyager 2* with a quantitative look at the distances, speeds, and forces related to its flybys with a focus on that of Jupiter. The data of Table 6.7 gives a sense of the dimensions involved by comparing the equatorial radius of the planet, the

Table 6.7. Distances are to the planet's center of mass. For L the latus rectum, "distance at latus rectum" is $\frac{L}{2}$.

Voyager 2 hyperbolic flyby of	the planet's radius in km	distance at periapsis in km	distance at latus rectum in km	Laplace radius of Sun-planet system
Jupiter	6.9911×10^4	7.2138×10^5	1.6810×10^6	4.8210×10^7
Saturn	5.8232×10^4	1.6069×10^5	0.3990×10^6	5.4551×10^7
Uranus	2.5362×10^4	1.0715×10^5	0.6445×10^6	5.1764×10^7

periapsis and latus rectum distances of the flyby, and the Laplace radius of the planet's SOI. Notice, that very roughly speaking, there is a tenfold increase from one distance to the next. The information of the table is readily obtained from the planetary and orbital data pointed to in this and earlier chapters. Let's look at *Voyager 2*'s flyby of Jupiter and the craft's speed relative to Jupiter at the distances listed. We saw earlier that the limiting speed of the craft on its flyby was

$$v_\infty = \sqrt{\frac{GM_J}{a}} = \sqrt{\frac{1.26712765 \times 10^8}{2.184140 \times 10^6}} \approx 7.616754 \text{ km/sec},$$

where M_J is the mass of Jupiter and $a = 2,184,140 = 2.1841 \times 10^6$ km the semimajor axis of the flyby. The formula

$$v(t) = \sqrt{\frac{GM_J}{a}} \sqrt{\frac{2a}{r(t)} + 1} = v_\infty \sqrt{\frac{2a}{r(t)} + 1} = 7.616754 \sqrt{\frac{2a}{r(t)} + 1} \text{ km/sec}$$

in section 6G tells us that for the distances $r(t) = 7.2138 \times 10^5$, $r(t) = 1.6810 \times 10^6$, and $r(t) = 4.8210 \times 10^7$ all in km, $v(t)$ is equal to 20.23, 14.45, and 7.95 km/sec, respectively.

How long did it take for *Voyager 2* to travel from periapsis to the latus rectum and to the boundary of the SOI? By solving the equation $r(t) = a(\varepsilon \cosh \beta(t) - 1)$ of section 6F for $\beta(t)$, we get $\cosh \beta(t) = \frac{1}{\varepsilon}\left(\frac{r(t)}{a} + 1\right)$ and hence $\beta(t) = \cosh^{-1} \frac{1}{\varepsilon}\left(\frac{r(t)}{a} + 1\right)$. With $\varepsilon = 1.330279$, and $r(t) = 1.6810 \times 10^6$ and then $r(t) = 4.8210 \times 10^7$, we get $\beta(t) = 0.791903$ and $\beta(t) = 3.545597$. The hyperbolic Kepler equation $\varepsilon \sinh \beta(t) - \beta(t) = \sqrt{\frac{GM}{a^3}}\, t = \frac{v_\infty}{a} t$ of section 6G tells us that

$$t = \frac{a}{v_\infty}\left(\varepsilon \sinh \beta(t) - \beta(t)\right).$$

It follows that *Voyager 2* took $t = 107578$ seconds or about 1 day and 6 hours to get from the periapsis of its flyby to the latus rectum point and $t = 5588618$ seconds or about 64 days and 16 hours to get from periapsis to the boundary of the SOI.

A repetition of the earlier analysis of *Voyager 1*'s approach to Jupiter shows that *Voyager 2* was about 791,483,000 km from the Sun at the time of its rendezvous with the planet. With M the mass of the Sun and $m = 800$ kg the mass of *Voyager 2* at that time, and computing in MKS, we get that the Sun's pull on the craft was $\frac{GMm}{r^2} \approx \frac{(1.3271244 \times 10^{20})(800)}{(7.91483 \times 10^{11})^2} \approx 0.17$ newtons. When the spacecraft entered Jupiter's SOI, it was about 48,210,000 km from the planet, so that the Jupiter's pull on it was $\frac{GM_J m}{r^2} \approx \frac{(1.26712765 \times 10^{17})(800)}{(4.8210 \times 10^{10})^2} \approx 0.044$ newtons. Even though *Voyager 2* was in Jupiter's SOI, the Sun's force on the craft was 4 times greater than that of the planet. But as the craft sped towards Jupiter, the planet's force began to dominate. At the distances of 20 million km, 10 million km,

5 million km, and 1 million km, Jupiter's gravitational pull increased successively to 0.25, 1.01, 4.05, and 101 newtons.

6N. The *Cassini-Huygens* Mission. As Figure 6.23 illustrates and as was already observed, *Cassini*'s flightpath took it on successive flybys of Venus, Earth, and Venus again, the last of which sent the craft on its way to Jupiter. The final flyby around Jupiter propelled *Cassini* to Saturn, the

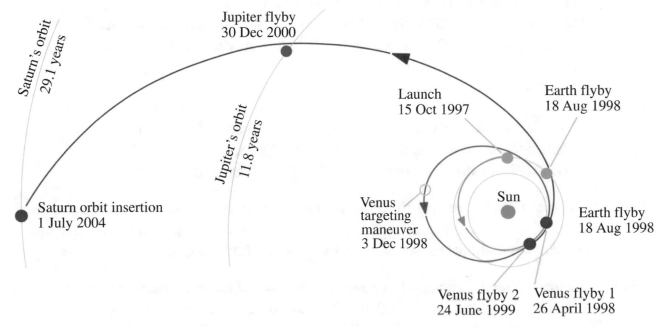

Figure 6.23. *Cassini*'s gravity assist cruise to Saturn. The diagram is adapted from https://solarsystem.nasa.gov/news/13070/scenic-route-to-saturn/

target of its mission.

The "*Cassini* Cruise Event Summary" that follows informs us that by the middle of May, 2004, *Cassini* was on a hyperbolic approach to Saturn, that it was injected into orbit around Saturn on July 1, 2004, and that it released the probe *Huygens* toward a soft landing on Saturn's moon Titan on December 24, 2004. The probe was named after Christiaan Huygens, the 17th century Dutch scientist who first discovered Saturn's rings and its largest moon Titan. Notice that the hyperbolic orbital parameters of *Cassini*'s path to Saturn varied slightly throughout its approach. The focal axis of the hyperbola changed position as as well. These changes were due not only to the programmed Trajectory Correction Maneuvers TCMs, but also (to a lesser extent) to the perturbing effects on the trajectory caused by the gravitational pull of the Sun (and Jupiter). The adjustments to *Cassini*'s initial orbit around Saturn were achieved by Orbit Trim Maneuvers OTMs. The abbreviation UTC refers to Coordinated Universal Time.[2] See Figure 6.24 for a diagrammatic overview of *Cassini*'s flight from its hyperbolic approach of Saturn in June, 2004, to the descent of *Huygens* to the moon Titan in January, 2005. The discussion about *Cassini* that follows uses information about Saturn

---

[2]Coordinated Universal Time is the primary time standard by which the world regulates clocks and the flow of time. Adopted in 1970, its based on the time as measured by atomic clocks. The *leap second* that it makes use of is occasionally subtracted to compensate for the fact that the Earth's rotation is slowing and to keep UTC aligned to the more familiar Greenwich Mean Time (GMT).

from the sections that studied the *Voyagers*. Statements about its distance from Saturn refer to its distance from Saturn's center of mass. Any reference to its speed is speed relative to Saturn.

May 14. *Cassini*'s orbital data were $a = 1.387719 \times 10^6$ km and $\varepsilon = 1.045856$.

May 27. TCM-20 occurred. This correction took place when *Cassini* was about 1.86×10^7 km from Saturn. It reduced the speed of the craft by 0.034 km/sec. After this maneuver the orbital data were $a = 1.401380 \times 10^6$ km and $\varepsilon = 1.055423$. What we know about Saturn's Laplace radius tells us that TCM-20 occurred within Saturn's gravitational sphere of influence.

June 11. *Cassini* flyby of Saturn's moon Phoebe.

June 16. TCM-21 occurred. This correction took place when *Cassini* was about 8.77×10^6 km from Saturn. It reduced *Cassini*'s speed by 0.004 km/sec. Before TCM-21 the orbital data were $a = 1.400853 \times 10^6$ km and $\varepsilon = 1.056965$. Thereafter they were $a = 1.401824 \times 10^6$ km and $\varepsilon = 1.058198$.

A look at Figure 6.24 informs us that the diagram of *Cassini*'s flight that is depicted begins right after the trajectory correction TCM-21 and that the critical insertion maneuver that put the spacecraft into orbit around Saturn was next on the agenda of its mission.

July 1. *Cassini* crossed through a large gap between two of Saturn's rings at a distance of 158,500 km from Saturn's center of mass. Its orbital parameters at the time were $a = 1.388176 \times 10^6$ km and $\varepsilon = 1.058353$. At 1:12 UTC the main engine burn began and generated 445 newtons of thrust in the direction opposite to the motion of the craft. During this Saturn Orbit Injection maneuver, *Cassini* reached its closest approach of 80,234 km from Saturn's center. The injection burn was terminated after 96 minutes at 2:48 UTC. It reduced *Cassini*'s speed by 0.626 km/sec. With the craft at a distance of 81,211 km from Saturn's center, *Cassini* was now in an elliptical orbit around Saturn with parameters $a = 4.600545 \times 10^6$ km and $\varepsilon = 0.982487$. After another 1 hour and 44 minutes, *Cassini* descended back through the gap in the rings near the latus rectum position of its elliptical orbit.

Example 6.16. Use data from the May 14th entry of the cruise event summary to show that the speed v_∞ of *Cassini*'s hyperbolic approach to Saturn was approximately 5.229 km/sec. Show that *Cassini*'s speed and angle of approach to Saturn after the trajectory correction of May 27th were $v \approx 5.582$ km/sec and $\gamma \approx 1.36°$, respectively. (See Figure 6.13.)

Example 6.17. Show that on its hyperbolic trajectory on July 1, *Cassini* zipped through Saturn's ring plane at a speed of 22.496 km/sec. Show that after the termination of the main engine burn, the craft moved in its elliptical orbit with a speed of 30.432 km/sec. Conclude that the maximum speed that *Cassini* reached just before the injection burn was 31.058 km/sec. Assume that at the

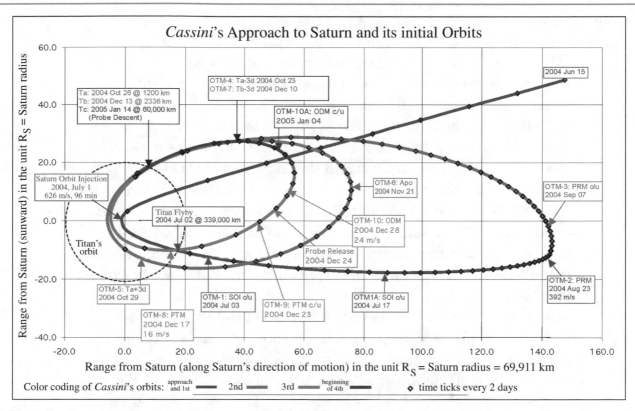

Figure 6.24. *Cassini*'s approach to Saturn and its orbits through the descent of the *Huygens* probe. Image provided by Jeremy Jones, leader of the navigation team of the *Cassini* mission to Saturn, NASA/JPL.

end of the injection burn with *Cassini* 81,211 km from Saturn's center of mass, the craft's speed had been reduced by 0.500 km/sec (rather than 0.626 km/sec). Would this have been sufficient to bring *Cassini* into an elliptical orbit around Saturn? If so, what would the semimajor axis of this orbit have been? Discuss what would have happened with a speed reduction of 0.480 km/sec. [Hint: look at both the elliptical and hyperbolic speed formulas. Then compare $v(t)$ and $\sqrt{GM}\sqrt{\frac{2}{r(t)}}$.]

Example 6.18. Compute the periapsis distance of *Cassini*'s initial elliptical orbit around Saturn. Check your result against the information from the cruise event summary for July 1. Show that without subsequent trajectory corrections *Cassini*'s initial elliptical orbit would have had a period of about $116\frac{1}{2}$ days and would have taken it as far as 9.12 million km away from Saturn.

The immediate goal after the orbit injection was to maneuver *Cassini* into position for the release of the probe *Huygens* towards the moon Titan. To achieve this, *Cassini*'s initial elliptical orbit needed to be trimmed and opened. This is what OTM-1 to OTM-10 and the flybys of Titan were designed to do.[3]

> July 2. *Cassini*'s first flyby of Titan. Since the distance of closest approach was 339,000 km, it had little impact on the orbit.

---

[3]For the most part, the data for *Cassini*'s trajectory were generated by JPL's HORIZONS system. Refer to the site https://ssd.jpl.nasa.gov/horizons.cgi?s_body=1#top. Refer also to the upcoming last section of this chapter. The author thanks Jeremy Jones, the chief of *Cassini*'s navigation team for most of its mission, for supplying him with data for *Cassini*'s trajectory early on. The author is grateful to Duane Roth, who succeeded Jeremy Jones, for providing data that cleared up issues related to OTM-2 and the October 26, 2004 flyby of Titan.

July 3.   The maneuver OTM-1 resulted in minor changes to *Cassini*'s orbit.

July 17.   OTM-1A made more small corrections to the spacecraft's orbit.

Aug 23.   OTM-2 was a major orbit correction maneuver. Just before it, *Cassini* was heading away from Saturn in a very flat elliptical orbit with parameters $a = 4.585367 \times 10^6$ km and $\varepsilon = 0.982347$. OTM-2 took place 9,075,000 km from Saturn's center of mass. The telemetry data showed that OTM-2 was a burn of about 51 minutes. It increased *Cassini*'s speed by 0.392 km/sec and widened the ellipse. After OTM-2 the craft was in an elliptical orbit with parameters $a = 4.789736 \times 10^6$ km and $\varepsilon = 0.895845$. OTM-2 was the third longest engine burn of the craft's main engine. The Saturn injection burn had been 96 minutes long and long and the targeting maneuver on 3 December 1998 (see Figure 6.23) that aligned *Cassini* for its second gravity assist flyby of Venus had been 88 minutes long.

Sep 4.   The maneuver OTM-3 made corrections to OTM-2. Both OTM-2 and and OTM-3 increased the periapsis distances of *Cassini*'s orbit. The designation PRM in Figure 6.24 is short for periapsis raise maneuver.

Oct 26.   In this flyby of Titan, *Cassini* passed 1200 km from the moon. The impact on *Cassini*'s orbit was considerable. The parameters changed from $a = 4.828958 \times 10^6$ km and $\varepsilon = 0.896228$ to $a = 2.554039 \times 10^6$ km and $\varepsilon = 0.854052$. This maneuver both opened and trimmed the ellipse further. Two days later, *Cassini* reached periapsis (of the orbit depicted in green in Figure 6.24) at a distance of 372,200 km from Saturn's center.

Dec 13.   Another close flyby of Titan opened the orbit slightly by decreasing the eccentricity from 0.853929 to 0.852098, but it trimmed the orbit significantly by decreasing the semimajor axis from 2.550182×10^6 km to 1.944093×10^6 km.

Dec 25.   After several more minor OTMs, the *Huygens Probe* was ready for release. It separated from the orbiting *Cassini*, entered Titan's atmosphere and descended to land on its surface on January 14, 2005. It was the first landing undertaken in the outer solar system. *Huygens* returned data and images of Titan's surface successfully to Earth using *Cassini* as a relay.

Example 6.19. Let S be the center of mass of Saturn and consider *Cassini* at a point C in its orbit just before the orbit trim maneuver OTM-2. Show that *Cassini* had a speed of about 0.295 km/sec and that it moved toward the apoapsis of its orbit at an angle of about $\gamma = 1.97$ radians $\approx 113°$ with the segment CS. Show that right after OTM-2, *Cassini* moved with a speed of 0.664 km/sec at an angle of about $\gamma = 1.67$ radians $\approx 96°$.

Example 6.20. Use the data of Table 2.5 to show that the distance of the moon Titan from Saturn is approximately 1,220,000 km. Verify that this information along with the data from *Cassini*'s

October 26 flyby of Titan implies that the spacecraft's speed relative to Saturn decreased from about 7.372 km/sec before the flyby to about 6.881 km/sec thereafter. Show that it also implies that the flyby tightened *Cassini*'s angle of approach to Saturn from about 41.9° to about 37.6°.

We have described the first eight months of *Cassini*'s orbital dance around Saturn. As was already discussed earlier, *Cassini*'s extraordinary *pas de deux* with Saturn would continue for over 12 more years.

6O. Orbits and their Ephemerides. Our discussions of the orbits of the planets and their moons, comets and asteroids, and the spacecraft that explore them have considered their shapes and sizes, and have shown that each orbit is a conic section determined by its semimajor axis a and eccentricity ε. We have also analyzed exactly how they trace their orbits out by deriving formulas for the distance $r(t)$, velocity $v(t)$, and angles $\alpha(t)$ and $\gamma(t)$ as functions of the elapsed time t (from periapsis). These functions tell us in quantitative terms how the motion along an orbit unfolds. But this text has said next to nothing about the location of the plane of an orbit and how the orbit's position within it is specified. We'll discuss the essentials of these matters now.

We'll designate the moving planet, moon, comet, asteroid, or spacecraft by P and—as before— refer to the path that P travels as its orbit or trajectory. Figure 6.25 presents the orbit as an ellipse, but allows for the possibility that it is a parabola or a hyperbola. We'll refer to the center of mass of the body that is being orbited, or (depending on the context) the barycenter (center of mass) of the surrounding system, as the *central body*. The central body—we'll denote it by S—is at a focal point of the orbit. The first basic aspect is the *reference plane* (in gray in the figure) through S. In the case the central body is the Sun, the reference plane is typically the orbital plane of the Earth—known as *ecliptic* since ancient times—and in the case of a planet and an orbiting moon, it is usually the plane determined by the equator of the planet. The *reference direction* in the reference plane points from S typically to a star or clusters of stars (or powerful sources of radiation) so distant that they are essentially fixed relative to the orbital motion.

The plane of the orbit (in yellow in the figure) contains S and intersects the reference plane in a line called the *nodal line*. The point on the nodal line at which the moving P rises through the reference plane from south to north (north as determined for the ecliptic, for example, by Earth's north pole) is the *ascending node* and the *descending node* is the point of the descent of P through the reference plane. The nodal line in the reference plane is determined by S and the positive angle between the reference direction and the line to the ascending node. This angle is denoted by Ω and referred to as the *longitude of the ascending node*. (In the case of Earth's orbit, the reference plane and Earth's orbital plane coincide. So there is no nodal line for the Earth in its orbit as specified in the figure. The line that takes its place is the line that the two equinox positions determine. (See Chapter 1A.) The *inclination* is the angle i between the reference plane and the plane of the orbit. For Earth's orbit, $i = 0°$. The example of Halley's comet (See Table 5.1) informs us of the following convention. The angles 17.7° and 162.3° both provide the same amount of separation between a given orbital plane and the reference plane. So they determine the same orbital plane. The larger angle $i = 162.3°$ listed for Halley means that Halley circles the Sun in a direction opposite that of Earth. (The alternative $i = 17.7°$ would have meant that it moves in the same direction.) Observe

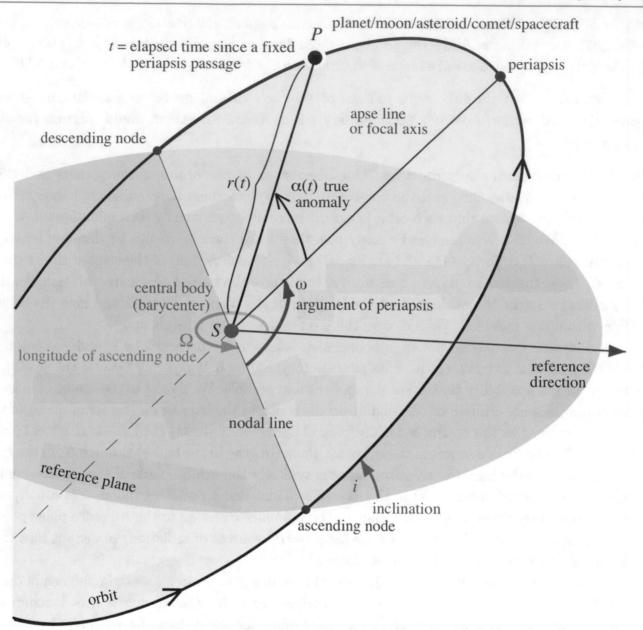

Figure 6.25. Adapted from https://sco.wikipedia.org/wiki/File:Orbit1.svg, an image attributed to Lasunncty at the English Wikipedia under the *GNU Free Documentation License*, Version 1.2.

that the point S, the reference plane, the reference direction, and the angle Ω together determine the nodal line. In turn, the nodal line and the angle of inclination i determine the plane in which the orbit of P lies.

The angle ω between the ascending side of the nodal line and the line from S to the periapsis of the orbit is the *argument of periapsis*. Since S is a focal point of the orbit of P, the angle ω determines the position of the periapsis, and hence the orientation of the orbit within its plane. The orbital elements $a, \varepsilon, \Omega, i$, and ω are called *ephemerides* of the orbit. They determine the shape, size, location, and orientation of the orbit of P (and hence that of any planet, moon, asteroid, comet, and spacecraft). Two more ephemerides tell us where in the orbit P starts and where it is at any time thereafter. Choose a unit of time, assign time $t = 0$ to the periapsis position, and let time $t > 0$ flow (take $t < 0$ if the concern is the past). The *mean anomaly* is an angle that provides the time—positive if after the periapsis, and negative if before—of the first observation of P in its orbit. The *true anomaly* is the angle $\alpha(t)$ that provides the precise position of P in its orbit at any time.

A given listing of orbital ephemerides is a snapshot—at a particular time—of the orbit under consideration. Since these orbital elements change (often by very small amounts, but sometimes significantly), the time at which a snapshot is taken needs to be stated and the flow of time needs to be carefully measured. These matters are technical. Briefly put, a starting time is specified as an *epoch*. For example, "J2000 with respect to the mean ecliptic and equinox" (or 12 pm on January 1, 2000) refers to a standard epoch. Einstein's theory of general relativity correctly asserts that gravity warps space-time, and in particular that clocks that are far from massive bodies run more quickly, and clocks close to massive bodies run more slowly. These effects are extremely small. For example, over the time-span of the 4.6 billion years of the Earth's existence, a clock set at the peak of Mount Everest (farther from Earth's center of mass) would have run about 39 hours ahead of a clock set at sea level (closer to Earth's center of mass). However, the accuracy of both atomic clocks and modern observational methods are so exact that these small differences need to be considered when orbits and astronomical ephemerides of planets, moon, asteroids, comets, and interplanetary spacecraft are calculated. *Barycentric Dynamical Time* (TDB, from the French *Temps Dynamique Barycentrique*) is a current time scale that takes these relativistic effects into account.

Consider a moving spacecraft at a fixed a moment in time. Analyze the gravitational forces on the craft due to the Sun, planets, their moons, and major asteroids. When the craft is near a planet, include the effects of atmospheric drag and mass distribution (such as mountain ranges). Consider all other relevant forces acting on the spacecraft, both large and small, such as thruster burns, attitude control jet firings, solar radiation pressure on its surfaces, heat radiation, and relativistic bending of space-time due to motion and mass. Lots of very complex mathematics (such as second order differential equations, vector calculus, spherical harmonic functions, probability theory, and statistical methods) along with hundreds and even thousands of pages of computer code are required to construct a dynamic model of all relevant forces operative on the craft at the given time. This is done relative to a preferred fixed reference point such as the center of mass of the Sun, Jupiter, Saturn, or the barycenter of the solar system. Once the position and the velocity at the given time (and relative to the reference point) are inserted, this dynamical force model is used to predict what future measurements should be. When these future measurements are obtained (via the Deep Space Network), they are used to correct the parameters in the dynamical model. In this way, the dynamical force model is iteratively improved and brought into tight agreement with the actual measured reality. With the resultant of the forces that this dynamical force model delivers regarded to have been rotated in the direction of the preferred reference point, this resultant is—at the given instant—a centripetal force acting on the spacecraft. Therefore, by Newton's theory (as summarized in Chapter 1D or Chapter 4G), the position of the spacecraft at this instant lies on a conic section that has the fixed reference point as a focal point. The shape of this conic section, the plane it lies on, its orientation on this plane, and the craft's position on it, determine and are determined

Table 6.8. The table lists *Voyager 1*'s ephemerides chronologically on relevant dates from 9/9/1977 to 4/16/1979 (each at time 00:00:00 in TDB). The lower block of four ephemeris categories and its thirteen rows of data are understood to be a continuation of the upper block with the same dates and times. The data was generated by JPL's HORIZONS at https://ssd.jpl.nasa.gov/horizons.cgi.

	Voyager 1 ephemerides date/time	semimajor axis a in km	eccentricity ε	inclination i in degrees
1	9/9/1977	$7.448213729378 \times 10^8$	0.795276258678564	1.038338998067602
2	9/15/1977	$7.454362080855 \times 10^8$	0.7977038078142186	1.036233172112535
3	10/29/1977	$7.442743552332 \times 10^8$	0.7973864163644588	1.035944011210695
4	10/30/1977	$7.442740595266 \times 10^8$	0.7973862771492531	1.035942937212381
5	1/29/1979	$7.622177526760 \times 10^8$	0.8082043145984138	1.094399155248978
6	1/30/1979	$7.626888952444 \times 10^8$	0.8085380133431345	1.096458434622106
7	2/20/1979	$7.881927352477 \times 10^8$	0.8251284265944755	1.205102454375513
8	2/21/1979	$7.915697046984 \times 10^8$	0.8270808927668779	1.218596759296712
9	3/4/1979	$13.866068504325 \times 10^8$	0.9307976746986110	2.016299387050204
10	3/5/1979	$17.195185203440 \times 10^8$	1.057138936447251	1.838178207476771
11	3/19/1979	$5.410635313032 \times 10^8$	2.427264625174953	2.515232928416584
12	4/2/1979	$5.804692046893 \times 10^8$	2.331231389384568	2.494371943539552
13	4/16/1979	$5.880232277174 \times 10^8$	2.314183772973725	2.481095839236705
	longitude of ascending node Ω in degrees	argument of periapsis ω in degrees	mean anomaly in degrees	true anomaly $\alpha(t)$ in degrees
1	343.2092914732131	359.1562745993133	0.3604648640488028	5.297885334011201
2	343.1903064295933	359.1467166156721	0.8935101230088243	13.06542275194955
3	343.1824902296153	359.1534947967443	4.803757691248907	59.40756121732325
4	343.1823915312585	359.1536240329527	4.892575671980855	60.20081395065512
5	340.6092100828861	1.585149171019804	43.18339423970573	143.6183896482964
6	340.5345312912726	1.652774014051279	43.20693954289691	143.6765600182676
7	337.1167715395085	4.735653644448552	41.75602327003497	145.0471112205921
8	336.7482613708431	5.068132710538456	41.44049037146785	145.1302962281412
9	324.5898208446674	17.54826931022829	15.15029758770040	145.2951903107177
10	326.3063676308474	26.41193215557000	9.134612249193491	134.7549304002062
11	113.8116714980738	358.2961314890333	16.48175506308393	17.49928016638131
12	113.6687634184156	358.4905081971434	16.82392951475677	19.45220561937466
13	113.5636297070193	358.5830522533213	18.32638293453528	21.42159673607886

by—as Figure 6.25 illustrates—the ephemerides of the spacecraft at that time. The ephemerides are a numerical snapshot of the position and motion of the spacecraft at the chosen instant.

We have described the process with which JPL's HORIZONS program specifies the ephemerides of the planets, moons, asteroids, comets, and spacecraft at any instant of time. By specifying a sequence of instances of time, say noon for a sequence of days, and determining the ephemerides for each instant, we get, frame by frame, a moving flow of snapshots of the spacecraft's motion. If the forces on the spacecraft change slowly over time, the conic sections that model the trajectory and the corresponding ephemerides will change slowly over time. If the forces change more quickly, these changes will be more rapid and more pronounced. In either case, the evolving sets of ephemerides give a quantitative sense of the changing trajectory of the craft.

Let's illustrate the discussion above by considering the cruise of *Voyager 1* to Jupiter and the thirteen rows of ephemerides data of Table 6.8. We'll focus on the five ephemerides that determine the shape and location of the orbit, namely the semimajor axis, the eccentricity, the inclination, the longitude of the ascending node, and the argument of periapsis. (The mean and true anomaly specify the location of the craft in the orbit.) The two rows of the table labeled 1 list the craft's ephemerides on September 9th, 1977, a day after it was inserted into its elliptical trajectory. On September 11th and 13th, 1977, *Voyager 1* underwent its first trajectory correction maneuver in two parts. The fact that the ephemerides set out for September 15th in the two rows labeled 2 are close to those of September 9th, 1977, tells us that these early trajectory corrections changed the orbit only very slightly. The next correction on October 29th, 1977, cleaned up small flight path inaccuracies. The third trajectory correction took place on January 29th, 1979, with a 22-minute 36-second thruster burn that adjusted *Voyager 1*'s flight path and changed its speed by about 4 m/sec. Comparisons of the ephemerides in rows 3 and 4, and then again in rows 5 and 6, tell us that these two maneuvers had little impact on Voyager 1's trajectory. We see, however, that smaller forces (most likely the pull of planets) acting in the intervening $16\frac{1}{2}$ months did effect the ephemerides in a modest way. One final pre-Jupiter burn on February 20th, 1979, fine-tuned the spacecraft's approach. Its thrusters fired for a little over 2 minutes, changing the velocity and direction to deliver the spacecraft to Jupiter's doorstep. A comparison of rows 7 and rows 8 shows the impact of the maneuver on the trajectory.

By January 17th, 1979, *Voyager 1* had approached Jupiter to within the 4.8×10^7 km of the planet's Laplace radius with its cameras already capturing images of details of Jupiter's moons and of the turbulent atmosphere surrounding the Great Red Spot, that had never been seen before (with Earth-based telescopes). On February 10th, the craft crossed the orbit of Sinope, one of Jupiter's smaller, outermost satellites at a distance of 2.3×10^7 km from Jupiter. All the while—as the changes in the craft's ephemerides from those in rows 6 of January 30th to those in rows 7 of February 20th confirm—Jupiter's gravitational pull began to have an impact *Voyager 1*'s trajectory. The changes in the data from February 21st (rows 8), to March 4th (rows 9), to the time of the craft's closest approach to Jupiter on March 5th (rows 10), provide more dramatic evidence. Note that from March 4th to March 5th, the trajectory changed from elliptical to hyperbolic. The table shows that it took another 40 days for the gravitational force of Jupiter on *Voyager 1* to dissipate and for the hyperbolic trajectory that would take it to Saturn to become stable.

The study in section 6L of *Voyager 1*'s elliptical flight path from its near-Earth orbit to Jupiter— its conclusions are summarized by Figure 6.20—was based on the craft's September 8th, 1977 ephemerides. The fact that the ephemeris data is changing throughout its trajectory implies that the numerical specifics that were derived in this study can only be approximations. Notice, however, that the ephemeris data changed did not change significantly over the first $16\frac{1}{2}$ months (from September 9th, 1977 to January 30th, 1979) of the 18 months of the craft's flight, so that these approximations are relatively accurate. (The mean anomaly and the true anomaly did change significantly, but this reflects the obvious fact that *Voyager 1* was on the move.)

Our study has focused on the changing ephemerides of the trajectory of the *Voyager* spacecraft. The ephemerides of the orbits of the planets and their moons experience changes as well. The stability of these orbits tells us that these changes are considerably more gradual than those of a spacecraft. But just a few examples show that they are not insignificant. The discovery of Neptune was the consequence of the distortions of its elliptical orbit by the slight gravitational pull of Uranus. The precession of the perihelion of the orbit of one planet, caused primarily by the gravitational pull of all the others, is a more general example. The fact that five orbital periods of Jupiter (adding to 59.31 years) are nearly equal to two orbital periods of Saturn (adding to 58.91 years) brings about alignments that causes larger perturbations of both of their orbits in a cycle of some 900 years. The orbit of the planet Venus currently has the smallest eccentricity and is therefore the most circular. But in 25,000 years, Earth's eccentricity will be smaller than that of Venus. We saw in Chapter 2E that the gravitational forces of the Sun and Jupiter have caused severe orbital perturbations of the orbits of a number of comets that have led to the fiery end of some of them.

6P. Problems and Discussions. This problem set explores many of the conceptual and computational issues that arise in this chapter. They also add to the scope of the discussions.

1. NEAR–Shoemaker's Orbits of Eros. These problems deal with the elliptical orbits of *NEAR-Shoemaker* around Eros and the analysis in section 6D of the orbital changes that the various OCMs brought about. Their solutions tell us that the predictions of this analysis are not always in line with the data of Table 6.2. To understand the reasons for the discrepancy, let's consider the velocity-changing firing of the craft's engine at periapsis. It is the underlying assumption of the analysis of section 6D that the force vector of the thrust is tangent to the orbit at periapsis and that it therefore lies in the plane of the orbit. If this force vector does not lie in the plane of the orbit, then it has a component perpendicular to the orbital plane. As a consequence, the craft moves outside its original orbital plane so that the position of the orbital plane changes. If the force vector lies in the plane of the orbit, but is not tangential to the orbit, then the position of the periapsis changes and, along with it, the orientation of the ellipse (namely the direction of its focal axis). Similar things apply to velocity changes at the apoapsis. The analysis of section 6D computed the change in the speed (at periapsis or apoapsis) that was required to bring about the change in the shape of the orbit attributed to a given OCM. The additional change in the speed necessary to bring about the changes in the position of the orbital plane—both its inclination and the direction of the nodal line (see Figure 6.25)—or the orientation of the orbit was not considered. The Δv in the table does take the required additional change in the speed into account. This appears to be primary reason why the speeds resulting from our analysis are usually somewhat less than those listed in Table 6.2. The results of the problems below reflect the observations just made.

Problem 6.1. Consider the post OCM-4 orbit of *NEAR-Shoemaker* on April 11th, 2000. It had periapsis distance $q_1 = 99$ km and apoapsis distance $d_1 = 101$ km. Determine the eccentricity ε_1 of this orbit and compute the speed v_1 of the spacecraft at apoapsis. Use the analysis of section 6D to show that the burn OCM-5 at apoapsis that kept the apoapsis distance of *NEAR-Shoemaker*'s orbit at $d_2 = 101$ km while tightening the periapsis distance to $q_2 = 50$ km, decreased the craft's speed at apoapsis from 2.09 m/sec to 1.71 m/sec. The difference of 0.38 m/sec is less than the $\Delta v = 0.45$ m/sec of Table 6.2. Note that OCM-5 also changed the inclination of the orbital plane by 5°.

Problem 6.2. Assume that the maneuver OCM-6 brought *NEAR-Shoemaker* into a 50×52 km orbit, rather than the 49×52 km orbit listed in Table 6.2. Show that a burn that decreased the craft's speed at periapsis from 3.46 m/sec to 3.02 m/sec could have resulted in the orbital change that this maneuver effected. One reason why the difference of 0.44 m/sec is far off the $\Delta v = 1.92$ m/sec that Table 6.2 lists, is the fact that OCM-6 changed the inclination of the orbital plane by a substantial 26°.

Problem 6.3. Use a calculation to show that the burn OCM-8 at periapsis that kept the periapsis distance of *NEAR-Shoemaker*'s orbit at 35 km while decreasing its apoapsis distance from 51 km to 39 km, decreased the craft's speed at periapsis from 3.89 m/sec to 3.67 m/sec. Note that the difference of 0.22 m/sec is in close agreement with the $\Delta v = 0.24$ m/sec that Table 6.2 provides. Note that this OCM did not change the inclination of the orbital plane.

Problem 6.4. The maneuver OCM-20 kept the periapsis distance of *NEAR-Shoemaker*'s orbit at 34 km, while decreasing its apoapsis distance from 193 km to 38 km. Show that the analysis of section 6D predicts that the engine burn at periapsis that resulted in this change in the apoapsis distance, decreased the craft's speed at periapsis from 4.72 m/sec to 3.72 m/sec. The difference of 1.00 m/sec is reasonably close to the $\Delta v = 1.23$ m/sec that Table 6.2 lists. But note that this OCM did not change the inclination of the orbital plane.

For the most part, we have studied the contraction of *NEAR-Shoemaker*'s early larger orbits of Eros to later tighter ones. As the data of Table 6.2 shows, there were several occasions during its orbital cruise around Eros in which the orbits were expanded. We'll now turn to the study of the expansion of the orbit of a spacecraft in the context of Hohmann transfers.

2. Hohmann Transfers to Mars. Figure 6.26 depicts the orbit of Mars and the near-Earth elliptical orbit of a spacecraft. The point I is the perihelion of this orbit. It is also the point of insertion into three possible elliptical Hohmann transfer orbits of the craft. They are depicted in three shades of green in the figure. Each of the three Hohmann transfers is designed bring the craft to a rendezvous with Mars. The near-Earth orbit has semimajor axis $a = 1.4960 \times 10^8$ km and eccentricity $\varepsilon = 0.0167$. The three targeted points of rendezvous with Mars are labeled P_0, P_1, and P_2. The trajectory from I to P_0 is the ideal Hohmann transfer orbit. Turn to section 6I and review the discussion of this transfer. Recall that at the perihelion of its near-Earth orbit the craft has an orbital speed of 30.286 km/sec. The increase of this speed at perihelion to 31.986 km/sec inserted the craft into the ideal Hohmann transfer that brought it to its rendezvous with Mars in 235 days.

A look at Figure 6.26 tells us that the paths from the insertion point I to the rendezvous points P_0, P_1 and P_2 become successively shorter. So the times of travel are successively shorter as well. But there is a tradeoff. The ellipses involved expand successively from one transfer to the next as the paths they provide to the rendezvous points get shorter. This means that the speed that is imparted to the craft at the point of the insertion is greater for a shorter transfer than a longer one.

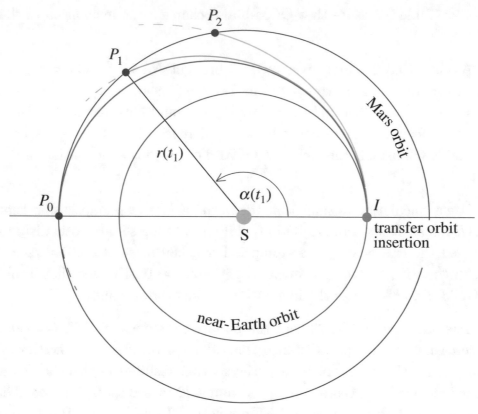

Figure 6.26

This in turn means that more propellant is required for a shorter transfer than a longer one. This increases the mass of the craft for a shorter transfer and hence the cost at launch. We'll now confirm these conclusions with a quantitative study.

Consider the elliptical transfer orbit from the point of insertion I to the point P_1. Let t_1 be the time it takes for the craft to make this trip. The angle $\alpha(t_1)$ and the distance $r(t_1)$ from S to P_1 are known to be $130° = \frac{13}{18}\pi$ radians and 2.0700×10^8 km, respectively. The questions of interest are these. What speed is required to insert the craft into this transfer orbit at its perihelion I and what is t_1 equal to? The answers rely on information developed in Chapters 5B, 5C, and 5D. Start by computing the semimajor axis a_1 and the eccentricity ε_1 of the transfer orbit. As in the situation of the ideal Hohmann transfer, $a_1(1 - \varepsilon_1) = 1.4710 \times 10^8$. Combine this with the formula $r(t_1) = \frac{a_1(1-\varepsilon_1^2)}{1+\varepsilon_1 \cos \alpha(t_1)}$ from Chapter 5B to show that $\varepsilon_1 \approx 0.2138$ and $a_1 \approx 1.8710 \times 10^8$ km. The formula of Example 5.1 tells us that the speed that the craft needs to be given at its insertion point I is about 33.092 km/sec. Next use Gauss's formula to show that $\beta(t_1) \approx 2.0913$. Finally, solving Kepler's equation for t_1 and converting the result to days, tells us that $t_1 \approx 155$ days.

Problem 6.5. Repeat these computations for the Hohmann transfer from I to P_2. Let t_2 be the time required and suppose that $\alpha(t_2) = 100°$ and $r(t_2) = 2.0700 \times 10^8$ km.

The *Mariner* missions focused on the exploration of Mars. *Mariner 4* was the first spacecraft to go to Mars. Departing in July 1965, it arrived in its Hohmann transfer orbit 228 days later. *Mariner 6* and *Mariner 7* took 155 days and 128 days, respectively, to get to Mars for their flybys in 1969. The craft *Mariner 9*, sent to Mars in 1971, took 168 days in its Hohmann transfer to get there. It was the first spacecraft to orbit Mars. The spacecraft *Viking 1* and *Viking 2* both landed on Mars in 1975. Their Hohmann transfers took 304 days and 333 days, respectively. The various Mars missions since have taken from 200 days to 308 days to get to the planet.

Since the missions of a spacecraft to Mars involve either an orbit insertion around Mars or a velocity changing flyby of it, mission engineers have an interest in the Laplace and Hill radii of the Sun-Mars system.

Problem 6.6. Let M be the mass of the Sun and m that of Mars. Recall that $GM = 1.32712 \times 10^{11}$ km^3/sec^2 and, from Table 2.3, that $Gm = 4.28284 \times 10^4$ km^3/sec^2. A look at Table 2.1 tells us that the semimajor axis of the orbit of Mars is $a = 227,944,000$ km. Use this information to show that the Laplace radius R_L and the Hill radius R_H for the Sun-Mars system are $R_L \approx 577,000$ km and $R_H \approx 1,084,000$ km, respectively.

Venus orbits the Sun inside Earth's orbit. Therefore a Hohmann transfer of a spacecraft from a near-Earth solar orbit to Venus involves a contraction of this solar orbit that can be achieved by a speed reduction at its perihelion. Such maneuvers have already been discussed extensively in the context of *NEAR-Shoemaker*'s orbits of the asteroid Eros.

Problem 6.7. Consider the mission of a spacecraft to Venus and draw a diagram analogous to that of Figure 6.26. Include three different Hohmann transfer orbits that bring the craft to a rendezvous with the planet.

3. With MESSENGER to Mercury. The *Voyagers* and *Cassini* are examples of missions to the outer planets that would not have been possible without the velocity boosting hyperbolic flybys of Jupiter (and other outer planets for the *Voyagers*) to counteract the persistent gravitational pull of the Sun. When it comes to a spacecraft on a mission to a planet with an orbit inside that of Earth, counteracting the gravitational pull of the Sun is not an issue since the target lies in the general direction of the Sun's pull on the craft. Getting to Venus is relatively unproblematic as it can be accomplished by carefully trimming a craft's near-Earth solar orbit. Mercury, however, presents difficulties. Table 5.1 informs us that a craft in a near-Earth solar orbit moves at about 30 km/sec, but that Mercury's average orbital speed is 47.4 km/sec. A successful Hohmann transfer from a near-Earth orbit to Mercury needs to bring a spacecraft to its rendezvous with Mercury with a speed that matches that of the planet. This requires a high velocity at the point of the insertion of the craft into its Hohmann trajectory. Unless this flight is very carefully attended to, the Sun's gravity will accelerate the craft past Mercury to a fiery impact on the Sun. The strategy that made missions to Mercury possible was the same as the one that enabled the *Voyagers* and *Cassini* to reach the outer planets: the gravitational assist flyby.

The first craft that negotiated its journey to Mercury successfully was *Mariner 10*. Launched in November of 1973, it was sent to study Venus and Mercury and was the first ever mission for which the strategy of gravity assist was deployed. A successfully timed flyby of Venus aligned the craft's trajectory with the orbit of Mercury. On its three passes of Mercury, the last bringing it to within 327 km of its north pole, *Mariner 10*'s cameras surveyed Mercury's surface by snapping thousands of images.

NASA's second foray to Mercury was the *MESSENGER* mission launched in August of 2004. The spacecraft was designed and built by the Applied Physics Laboratory (APL) of Johns Hopkins University. The name *MESSENGER* is an acronym derived from a contraction of MErcury Surface, Space ENvironment, GEochemistry, and Ranging, that is intended to suggest the scientific goals of the mission. The mission to put *MESSENGER* into orbit around Mercury was complex. A combi-

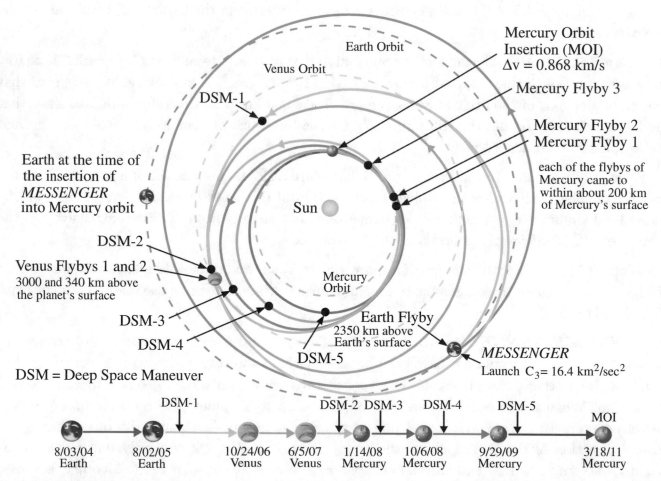

Figure 6.27. This diagram of *MESSENGER*'s flight path to Mercury with its multiple gravity assist flybys is adapted from https://commons.wikimedia.org/wiki/File:MESSENGER_trajectory.svg. Image Credit: NASA, JHU APL.

nation of multiple gravity assist flybys and many targeted trajectory corrections did the trick. Figure 6.27 illustrates what was involved. *MESSENGER*'s journey to Mercury included one flyby of Earth, two flybys of Venus, and three flybys of Mercury. A total of 35 minor Trajectory Correction Maneuvers (TCMs) and 5 major Deep Space Maneuvers (DSMs) during its $5\frac{1}{2}$ year flight fine-tuned the spacecraft's approaches to the flybys, so that they could be executed with the required precision.

Table 6.9 lists ephemerides data for *MESSENGER*'s solar orbits before and after each of its six planetary flyby maneuvers and related TCMs (as measured at the midpoint of each orbit segment). As before, a, ε, i, and ω, denote the semimajor axis, eccentricity, inclination of the orbit plane relative

Table 6.9. Data from http://messenger.jhuapl.edu/About/mission-design/details-propulsion-activity/MOE.html, except for Mercury's orbit in the last row that were provided by JPL's HORIZONS. Many thanks to James McAdams, *MESSENGER*'s Mission Design Lead Engineer, for generously and unfailingly responding to many issues, both small and large, involving the *MESSENGER* mission.

Initial Event Final Event date/time	a in au	ε	i in degrees	ω in degrees
MESSENGER's initial orbit on Aug 3 and 4, 2004	1.00124769	0.076000037	6.37291	255.64018
TCM-6 Earth flyby 21 July 2005 2 Aug 2005	1.00073173	0.077559813	6.40539	255.29239
Earth flyby TCM-9 2 Aug 2005 12 Dec 2005	0.80890445	0.254969749	2.53035	2.08998
TCM-12 Venus flyby 1 5 Oct 2006 24 Oct 2006	0.82948991	0.273427405	2.57977	1.78382
Venus flyby 1 TCM-13 24 Oct 2006 2 Dec 2006	0.72262040	0.244658760	8.17246	57.84991
TCM-16 Venus flyby 2 25 May 2007 5 June 2007	0.72340047	0.244271182	8.16396	58.13654
Venus flyby 2 TCM-18 5 June 2007 17 Oct 2007	0.53857757	0.383581489	6.77134	356.38575
TCM-19 Merc flyby 1 19 Dec 2007 14 Jan 2008	0.53442413	0.391714040	6.79802	356.64574
Merc flyby 1 TCM-23 14 Jan 2008 19 Mar 2008	0.50663630	0.381599938	6.92133	6.49836
TCM-23 Merc flyby 2 19 Mar 2008 6 Oct 2008	0.50775313	0.378597781	6.88791	6.56324
Merc flyby 2 TCM-29 6 Oct 2008 6 Dec 2008	0.46626923	0.351628516	7.00307	19.44225
TCM-29 Merc flyby 3 6 Dec 2008 29 Sep 2009	0.46998082	0.341312317	6.99426	19.86922
Merc flyby 3 TCM-35 29 Sep 2009 24 Nov 2009	0.43502378	0.303835103	7.00838	32.48547
TCM-35 MOI 24 Nov 2009 18 Mar 2011	0.43753386	0.296341326	7.02855	32.44229
Mercury's orbit on 18 Mar 2011	0.38709827	0.205624606	7.00432	29.15565

to Earth's orbit plane, and the argument of periapsis, respectively. The data of the table is listed in paired rows with each row of pre-flyby orbital data followed by a row of post-flyby orbital data. A comparison of the data of each of these rows with the data of the preceding row confirms that the impact of each flyby was substantial. By contracting and reorienting *MESSENGER*'s orbit in successive steps, they brought the spacecraft from its initial orbit close to Mercury. In March 2011, after a 7.9 billion km journey, *MESSENGER* arrived at at Mercury's doorstep with a velocity low enough and fuel supply ample enough for its insertion into orbit around the planet.

4. Hyperbolas and Hyperbolic Flybys. The first problem gives a sense of how the shape of a hyperbola relates to its numerical parameters. The paragraph then turns to some matters about hyperbolic flybys and trajectory calculations for *Voyager 2*.

Problem 6.8. Figure 6.28 depicts four hyperbolas. They are labeled ①,②,③, and ④. All have the x-axis as the focal axis and all have the same semimajor axis a. Their semiminor axes, b_1, b_2, b_3, and b_4 satisfy $b_1 < b_2 < b_3 < b_4$ with $b_3 = a$. For each i, the coordinates of the focal point are $(c_i, 0)$, where $c_i = \sqrt{a^2 + b_i^2}$. The focal points are color coded to correspond to the color of the hyperbola.

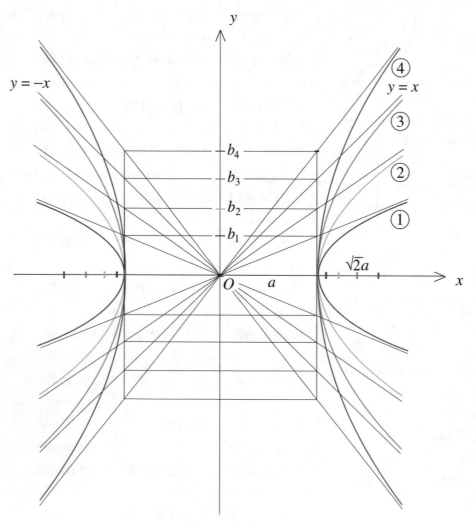

Figure 6.28

Show that their eccentricities are given by $\varepsilon_i = \sqrt{1 + \frac{b_i^2}{a^2}}$. Notice that $1 < \varepsilon_1 < \varepsilon_2 < \varepsilon_3 < \varepsilon_4$ with $\varepsilon_3 = \sqrt{2}$. The equations of the hyperbolas are $\frac{x^2}{a^2} - \frac{y^2}{b_i^2} = 1$. Notice that the closer ε_i is to 1, the closer the focal point is to the line $x = a$, the tighter the hyperbola is. Why is this observation consistent with the interpretation of the hyperbola as the trajectory of a moving object that is subject to the gravitational pull of a massive body located at the focal point?

Suppose that $a = 5$ and $b_1 = 2, b_2 = 3, b_3 = 5$ and $b_4 = 6$. Show that $\varepsilon_i \approx 1.077, 1.166, 1.281,$ 1.414, and 1.562, respectively. For each hyperbola compute the coordinates of the two focal points and the latus rectum. Write down an equation for each of the hyperbolas, and for each hyperbola write down equations for its two asymptotes.

Let's return to section 6J and the derivations of the equations for $v_{\text{in}}, v_{\text{out}}$, and $\varphi = \varphi_1 + \varphi_2$ that govern hyperbolic gravity assist flybys.

Problem 6.9. Verify the formulas for v_{in} and v_{out} in the situation of Figure 6.29.

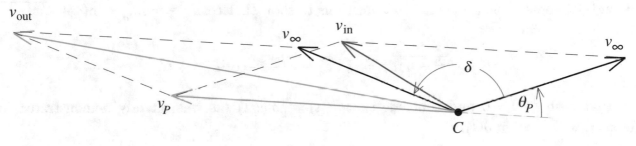

Figure 6.29

Problem 6.10. Verify the formula for $\varphi = \varphi_1 + \varphi_2$ in the situation of Figure 6.30.

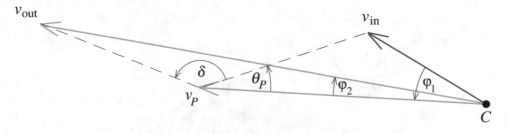

Figure 6.30

Problem 6.11. Follow the study in section 6L of the hyperbolic flyby of Jupiter by *Voyager 1* to analyze the hyperbolic flybys of Saturn as well as Uranus by *Voyager 2*. Show that the flyby of Saturn increased the speed of *Voyager 2* relative to the Sun from 15.33 to 20.25 km/sec and changed its direction by 45.28°, and that the flyby of Uranus increased its speed from 17.79 to 19.66 km/sec and changed its direction by 17.31°.

Problem 6.12. The analysis in section 6L of the elliptical trajectory of *Voyager 1* from Earth to Jupiter and its depiction in Figure 6.20 used the Earth–Jupiter data of Table 6.3. Use the same strategy to show that *Voyager 2* was 791,483,000 km from the Sun when it rendezvoused with Jupiter and that it moved toward it with a Sun-relative speed of 9.57 km/sec.

5. Dealing with the Hyperbolic Kepler Equation. This segment explores a number of issues having to do with the hyperbolic Kepler equation and the discussion in section 6G.

Problem 6.13. Show that $f(x) = \frac{b}{a}\sqrt{x^2 - a^2}$ is a function that has the upper part of the hyperbola $\frac{x^2}{a^2} - \frac{y^2}{b^2} = 1$ as its graph. Go to a standard calculus text or an online source for the formula

$$\int \sqrt{x^2 - a^2}\, dx = \frac{x}{2}\sqrt{x^2 - a^2} - \frac{a^2}{2}\ln(x + \sqrt{x^2 - a^2}) + C, \text{where } x \ge a.$$

Turn to Figure 6.12 and focus on the area $B(t)$. As in the figure, consider the situation with $t \ge 0$ and hence $y(t) \ge 0$. Use the symmetry of the hyperbola about the y-axis to explain that

$$B(t) = -\tfrac{1}{2}x(t)y(t) - \int_a^{a\cosh\beta(t)} \frac{b}{a}\sqrt{x^2 - a^2}\, dx.$$

The formula $\cosh^{-1} u = \ln(u + \sqrt{u^2 - 1})$ for $u \ge 1$ expresses a connection between the hyperbolic cosine and the natural log. It can be found in basic calculus texts. Let $u = \frac{x}{a}$ with $x \ge a$ in the formula and use some basic properties of logarithms to show that $\cosh^{-1}\frac{x}{a} = \ln\frac{1}{a} + \ln(x + \sqrt{x^2 - a^2})$. Notice that therefore

$$\int \sqrt{x^2 - a^2}\, dx = \frac{x}{2}\sqrt{x^2 - a^2} - \frac{a^2}{2}\cosh^{-1}\frac{x}{a} + C.$$

Show that $\cosh^{-1}(1) = 0$ and conclude that $B(t) = \frac{1}{2}ab\beta(t)$ (as was already demonstrated in a different way in section 6G).

Problem 6.14. Turn to section 6G and repeat the derivation of the hyperbolic Kepler equation in the case of a negative t (for a craft on approach to periapsis) and negative $B(t)$ and $A(t)$.

Problem 6.15. Turn to Figure 6.31. Let P be a point on the circle $x^2 + y^2 = a^2$ and let β be the

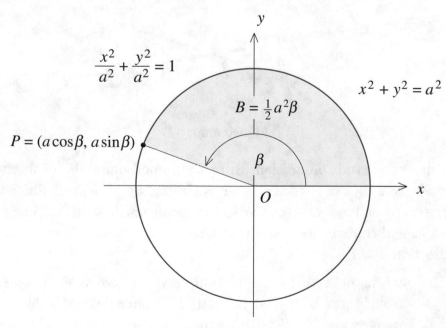

Figure 6.31

indicated angle. Show that the point P has coordinates $P = (a\cos\beta, a\sin\beta)$ for some positive real number β. (Review the trigonometry in Chapter 3C if needed.) Go polar with the circle to show that the area B of the highlighted sector is equal to $\frac{1}{2}a^2\beta$.

Problem 6.16. We'll consider a hyperbolic version of the conclusion of the previous problem. In Figure 6.32, $P = (x, y)$ is any point on the right branch of the hyperbola $x^2 - y^2 = a^2$. Consider the graphs of the hyperbolic functions sinh and cosh and show that $x = a\cosh\beta$ and $y = a\sinh\beta$ for some real number β. Let B be the area of the highlighted hyperbolic wedge. Use facts and strategies

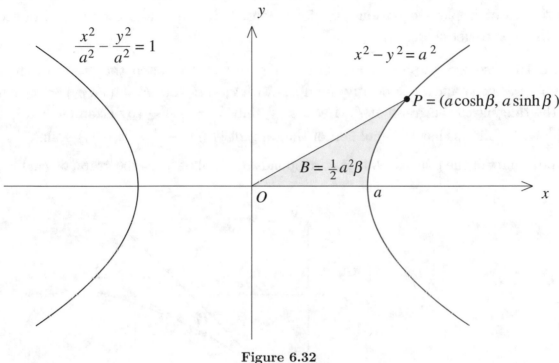

Figure 6.32

deployed in the solution of Problem 6.13 to verify that $B = \frac{1}{2}a^2\beta$.

Problem 6.17. Let $y = f(x)$ be a differentiable function with $f(0) = 0$. Suppose that $f'(x) > 0$ for $x \neq 0$ and that the graph of $y = f(x)$ is concave up for $x \geq 0$ and concave down for $x \leq 0$. Now consider $y = f(x) - c$ with c a constant and suppose that $f(x_0) - c = 0$. Turn to the case $c \geq 0$. Since $f(x_0) = c \geq 0$, it follows that $x_0 \geq 0$. Review the Newton-Raphson method for finding the zeros of a differentiable function. Show by making use of a generic graph of $y = f(x) - c$, that if x_1 with $x_1 > x_0$ is a stab at an approximation of x_0, then $x_2 = x_1 - \frac{f(x_1) - c}{f'(x_1)}$ is a better approximation of x_0, and that the sequence $x_1, x_2, x_3, \dots x_i, \dots$ given by $x_{i+1} = x_i - \frac{f(x_i) - c}{f'(x_i)}$ converges to x_0. Consider the hyperbolic Kepler equation $\varepsilon \sinh x - x = \sqrt{\frac{GM}{a^3}}\, t$, where $t \geq 0$ is a constant. Show that what was just set out applies with $y = f(x) = \varepsilon \sinh x - x$, $c = \sqrt{\frac{GM}{a^3}}\, t$, and $x_0 = \beta(t)$. Check that with $\beta_1 = \sinh^{-1}\frac{1}{\varepsilon}\left(\sqrt{\frac{GM}{a^3}}\, t\right)$ as a first stab at $\beta(t)$, the sequence $\beta_{i+1} = \beta_i - \frac{(\varepsilon \sinh\beta_i - \beta_i) - \sqrt{\frac{GM}{a^3}}\, t}{\varepsilon \cosh\beta_i - 1}$ converges to the solution $\beta(t)$ of the hyperbolic Kepler equation.

Problem 6.18. Return to the discussion of the hyperbolic trajectory of *Voyager 1*'s flight from Jupiter to Saturn in section 6L. The eccentricity of the trajectory was $\varepsilon = 2.302740$ and $\sqrt{\frac{GM}{a^3}}\,t_1 = 0.334341$, where M is the mass of the Sun, a the semimajor axis of the hyperbola, and t_1 the time between perihelion and the craft's insertion into its hyperbolic trajectory. The first approximation β_1 of the solution $\beta(t_1)$ of the hyperbolic Kepler equation $\varepsilon \sinh x - x - \sqrt{\frac{GM}{a^3}}\,t_1 = 0$ is $\beta_1 = \sinh^{-1}\frac{1}{\varepsilon}\left(\sqrt{\frac{GM}{a^3}}\,t_1\right) = 0.144687$ (with six decimal place accuracy). Show that the sequence that the method of Newton-Raphson provides for the solution of $\beta(t_1)$ is $\beta_2 = 0.253730$, $\beta_3 = 0.251920$, $\beta_4 = 0.251919$, and $\beta_5 = 0.251919$. So the Newton-Raphson method arrives at the solution $\beta(t_1) = 0.251919$ of Kepler's equation after four steps. It took the simpler method of section 6H a total of 15 steps to get there.

Problem 6.19. Use the graph of $y = \sinh x$ of Figure 6.10 to sketch the graph of the function $f(x) = \sinh^{-1} x$. Let x_1 and x_2 be positive numbers with $x_1 < x_2$. Let $P_1 = (x_1, y_1) = (x_1, \sinh^{-1} x_1)$. Let L be the tangent to the graph at P_1. Because $\frac{d}{dx}\sinh^{-1} x = \frac{1}{\sqrt{x^2+1}}$ (by Example 6.10), the slope of L is $\frac{1}{\sqrt{x_1^2+1}}$. It follows that the equation of the tangent is $y = \frac{1}{\sqrt{x_1^2+1}}(x - x_1) + \sinh^{-1} x_1$. Let y_2 be the y-coordinate of the point on L that corresponds to x_2. Show that the graph of $f(x) = \sinh^{-1} x$

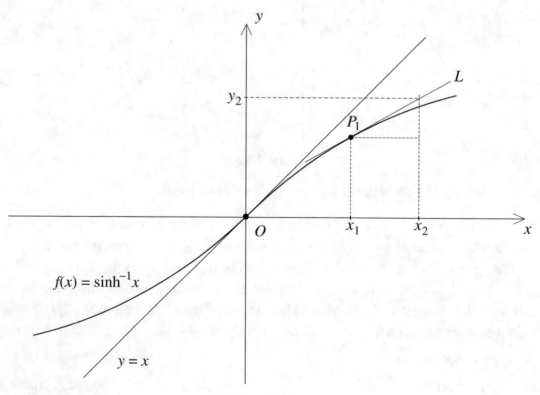

Figure 6.33

is concave down for $x \geq 0$ and conclude from Figure 6.33 that

$$\left|\sinh^{-1} x_2 - \sinh^{-1} x_1\right| < |y_2 - y_1| = \left|\frac{1}{\sqrt{x_1^2+1}}(x_2 - x_1) + \sinh^{-1} x_1 - \sinh^{-1} x_1\right| = \frac{1}{\sqrt{x_1^2+1}}|x_2 - x_1|.$$

Refer to the solution of the hyperbolic Kepler equation of section 6H and note that the inequality above is stronger than the inequality $\left|\sinh^{-1} x_2 - \sinh^{-1} x_1\right| < |x_2 - x_1|$ (when $x_1 \neq 0$). This means that the sequence $\beta_1, \beta_2, \beta_3, \ldots$ converges to the solution $\beta(t)$ of Kepler's equation more quickly than the inequality $|\beta(t) - \beta_i| \leq \frac{1}{\varepsilon^i}|\beta(t)|$ suggests.

6. Cassini and HORIZONS. Figure 6.24 tells us that three of the critical maneuvers that placed *Cassini* into orbit around Saturn were the Saturn Orbit Insertion (SOI) on July 1st, 2004, the Orbit Trim Maneuver OTM-2 of August 23rd, 2004, and the Titan flyby of October 26th, 2004.

The three problems that follow are invitations to explore these three maneuvers with the HORIZONS system and to compare the numerical data that is generated with the geometric information of Figure 6.24.

Go to the website https://ssd.jpl.nasa.gov/horizons.cgi and consider **Current Settings**. Under Ephemeris Type[change], click on change, Select Orbital Elements, and click on Use Selection Above. Under Target Body[change], click on change and type Cassini into the box Lookup the specified body, then Search, and Select MB: Cassini (spacecraft), and click on Select Indicated Body. Under Center[change] click on change and type @Saturn into the box Specify Center, then Search, Select Saturn (body center), and click Use Selected Location. Relevant Time Span settings follow in the problems below. For Table Settings and Display/Output use the *defaults*.

Problem 6.20. To study the SOI, return to **Current Settings**. Under Time Span[change], click on change. Under Start Time insert 2004-July-01 01:12, under Stop Time insert 2004-July-01 02:48, and under Step Size insert 1 and minutes. Then click Use Specified Time. Finally click Generate Ephemeris. The data that HORIZONS generates provides a minute by minute picture of *Cassini*'s changing trajectory during the SOI.

Problem 6.21. To study the OTM-2, insert 2004-August-23 15:56 under Start Time and 2004-August-23 16:45 under Stop Time and use 1 and minutes for the Step Size. The ephemerides that HORIZONS sets out give a minute by minute snapshot of *Cassini*'s changing elliptical orbit.

Problem 6.22. For the Titan flyby of October 26, 2004, insert 2004-October-26 14:30 under Start Time and 2004-October-26 16:00 under Stop Time and use 1 minute for the Step Size. As a consequence of the flyby, the orbit's eccentricity increased, but the periapsis distance and the semimajor axis both became smaller. So the orbit became less circular, but smaller.

7. Parabolic Trajectories. While there is no such thing as a perfect circular orbit, circles do serve as close approximations to the orbits of many planets and their moons. This is confirmed by the data in Tables 2.1, 2.2, 2.4, and 2.5. They tell us that the eccentricities of many planets and moons are very close to zero. Parabolas play a similar role for the comets. While there is no such thing as a perfect parabolic trajectory, a number of comets have trajectories that are close to parabolic. This is confirmed by the eccentricity data of Table 2.8. It tells us that a number of comets have eccentricities close to one. This segment turns to a discussion of parabolic trajectories. Basic facts about parabolas are discussed in Chapter 1C. More basic facts are developed in the problems that follow.

Problem 6.23. Figure 6.34 depicts a parabola in an xy-coordinate plane with focal axis the x-axis, focal point $(d, 0)$ where $d > 0$ and directrix the line $x = -d$. Use the definition of a parabola as the set of points in the plane whose distance from the focal point is equal to its (perpendicular) distance from the directrix to show that $y^2 = 4dx$ is an equation of the parabola. Let $P = (x, y)$ be any point on the parabola with $y \geq 0$ and let B be the area of the region shown in the figure. Use integral

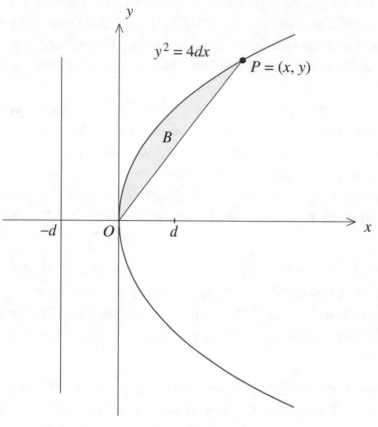

Figure 6.34

calculus to verify that

$$B = \tfrac{4}{3}\sqrt{d}\,x^{\frac{3}{2}} - \tfrac{1}{2}xy = \tfrac{4}{3}\sqrt{d}\,x^{\frac{3}{2}} - \sqrt{d}\,x^{\frac{3}{2}} = \tfrac{1}{3}\sqrt{d}\,x^{\frac{3}{2}}.$$

Now let C be a point-mass representing a comet or a spacecraft in parabolic flyby of the Sun or a planet. Let S be the center of mass of the Sun or planet. Assume that the gravitational force of S is the only force on C and that the trajectory of C is a parabola that has S at its focal point. Place a coordinate system into the plane of the trajectory in such a way that the focal axis of the parabola lies on the x-axis, the parabola crosses the x-axis at $x = 0$, and the parabola opens in the direction of the negative x-axis. With q the periapsis distance of the trajectory of C, the focal point is positioned at $(0, -q)$. See Figure 6.35. The directrix is the line $x = q$ and (referring to Problem 6.23) the equation of the parabola is

$$y^2 = -4qx.$$

Check that $y = \pm 2\sqrt{q}(-x)^{\frac{1}{2}}$ with $+$ in effect for the upper half of the parabola, and $-$ for the lower half. With $x = -q$, we get $L = 4q^{\frac{1}{2}}q^{\frac{1}{2}} = 4q$ for the latus rectum.

We'll continue in a way that is analogous to the hyperbolic discussion of section 6F and Figure 6.12. So we suppose that the craft approaches periapsis along the lower part of the parabola and that it departs along the upper part. A time t is assigned to a position C of the craft as follows. At periapsis, $t = 0$. For C on approach to periapsis t is negative, and for C on departure t is positive. The craft's time of travel from C to O on approach and from O to C on departure is $|t|$. Again turn to Figure 6.35. Let $\alpha(t)$ be the angle between the segment SO and the segment SC. We take $\alpha(t) \geq 0$ when $t \geq 0$, and $\alpha(t) < 0$ when $t < 0$. As t progresses through $-\infty < t < \infty$, the angle $\alpha(t)$ satisfies $-\pi < \alpha(t) < \pi$. For $t \geq 0$, let $A(t)$ be the area swept out by the segment SC from

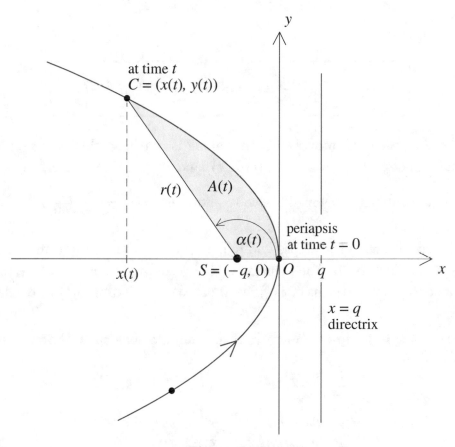

Figure 6.35

$t = 0$ to t. For a negative t, $A(t)$ is minus the area swept out from t to $t = 0$.

Let the coordinates of the position C of the craft at time t be $x(t)$ and $y(t)$ so that $C = (x(t), y(t))$. Let $r(t)$ be the distance from S to C. Since this distance is equal to the distance from C to the directrix,

$$r(t) = -x(t) + q.$$

Using the trig identity $\cos \alpha(t) = -\cos(\pi - \alpha(t))$, we get $\cos \alpha(t) = -\left(\frac{-x(t)-q}{r(t)}\right) = \frac{x(t)+q}{r(t)}$. So

$$1 - \cos \alpha(t) = 1 - \tfrac{x(t)+q}{r(t)} = \tfrac{r(t)-x(t)-q}{r(t)} = \tfrac{-2x(t)}{r(t)} \text{ and } 1 + \cos \alpha(t) = 1 + \tfrac{x(t)+q}{r(t)} = \tfrac{r(t)+x(t)+q}{r(t)} = \tfrac{2q}{r(t)}.$$

It follows from the trig formula $\tan^2 \frac{\alpha(t)}{2} = \frac{1-\cos \alpha(t)}{1+\cos \alpha(t)}$ that $\tan^2 \frac{\alpha(t)}{2} = \frac{-x(t)}{q}$. Therefore,

$$\tan \tfrac{\alpha(t)}{2} = \pm \tfrac{(-x(t))^{\frac{1}{2}}}{\sqrt{q}},$$

with $+$ for $t \geq 0$ and $-$ for $t < 0$. Now define $\beta(t)$ by

$$\boxed{\beta(t) = \tan \tfrac{\alpha(t)}{2} = \pm \tfrac{(-x(t))^{\frac{1}{2}}}{\sqrt{q}}}$$

Because $-x(t) = q\beta(t)^2$ and $y(t) = \pm 2\sqrt{q}\,(-x(t))^{\frac{1}{2}}$, it follows that

$$x(t) = -q\beta(t)^2 \quad \text{and} \quad y(t) = 2q\beta(t).$$

Expressed in terms of $\beta(t)$,

$$\boxed{r(t) = q(\beta(t)^2 + 1) \quad \text{and} \quad \alpha(t) = 2\tan^{-1}\beta(t)}$$

We will assume that the craft moves smoothly along its path, so that $x = x(t)$ and $y = y(t)$ are differentiable functions of t. It follows that $r(t), \alpha(t)$, and $\beta(t)$ are differentiable as well.

Problem 6.24. Use formulas developed above to show that $r(t) = \frac{2q}{1+\cos\alpha(t)}$.

As in the earlier situations of the ellipse and hyperbola, we are interested in solving the following problem: For any time t determine the position of C by computing $r(t)$ and $\alpha(t)$. Since both $r(t)$ and $\alpha(t)$ have been expressed in terms of $\beta(t)$, we'll proceed by finding $\beta(t)$ in terms of t. As before, area is the key.

Suppose first that $t \geq 0$. Figure 6.35 and routine computations show that

$$
\begin{aligned}
A(t) &= \int_{x(t)}^{0} 2\sqrt{q}(-x)^{\frac{1}{2}}\,dx - \tfrac{1}{2}(-x(t) - q)y(t) \\
&= \tfrac{4}{3}\sqrt{q}(-x(t))^{\frac{3}{2}} - (-x(t) - q)\sqrt{q}\,(-x(t))^{\frac{1}{2}} \\
&= \tfrac{1}{3}\sqrt{q}(-x(t))^{\frac{3}{2}} + q\sqrt{q}\,(-x(t))^{\frac{1}{2}}.
\end{aligned}
$$

For $t \leq 0$ and $y(t) \leq 0$, and this computation shows that $A(t) = -\left(\tfrac{1}{3}\sqrt{q}(-x(t))^{\frac{3}{2}} + q\sqrt{q}\,(-x(t))^{\frac{1}{2}}\right)$.

By results of Chapter 5A, we know that $A(t) = \kappa t$, where the Kepler constant of the orbit of C. We also know that $\kappa = \sqrt{\frac{GLM}{8}} = \sqrt{\frac{GMq}{2}}$, where M is the mass of the attracting body S. After dividing the equation

$$\pm\left(q\sqrt{q}(-x(t))^{\frac{1}{2}} + \tfrac{1}{3}\sqrt{q}(-x(t))^{\frac{3}{2}}\right) = A(t) = \kappa t$$

through by q^2 and substituting $\beta(t)$ for $\pm\frac{(-x(t))^{\frac{1}{2}}}{\sqrt{q}}$, we get the parabolic version of Kepler's equation

$$\boxed{\beta(t) + \tfrac{1}{3}\beta(t)^3 = \frac{\kappa t}{q^2} = \sqrt{\frac{GM}{2q^3}}\,t}$$

In contrast to the elliptical situation (in Chapter 5E) and the hyperbolic situation (in section 6H), where successive approximation methods were used to solve Kepler's equation for $\beta(t)$ in terms of a given t, the parabolic Kepler equation can be solved for $\beta(t)$ in terms of t in an explicit form.

After multiplying Kepler's equation through by $3q^2$, we get $q\beta(t)\big(3q + q\beta(t)^2\big) = 3\kappa t$. After squaring both sides, $q^2\beta(t)^2\big(3q + q\beta(t)^2\big)^2 = \frac{9}{2}GMqt^2$, so that

$$q\beta(t)^2\big(3q + q\beta(t)^2\big)^2 = \tfrac{9}{2}GMt^2.$$

We'll solve this equation for $q\beta(t)^2$.

Problem 6.25. Consider the function $y = f(x) = x(x + 3q)^2 - \frac{9}{2}GMt^2$. Show that

$$f'(x) = 3(x + 3q)(x + q) = 3(x^2 + 4qx + 3q^2) \text{ and } f''(x) = 6(x + 2q).$$

Use the second derivative test to conclude that $y = f(x)$ has a local maximum at $x = -3q$, a local minimum at $x = -q$, and a point of inflection for $x = -2q$. Check that $f(-3q) = -\frac{9}{2}GMt^2$ and $f(-q) = -4q^3 - \frac{9}{2}GMt^2$ and that $y = f(x)$ is increasing and concave up for $x > -q$. Since $f(0) = -\frac{9}{2}GMt^2$, it follows that $y = f(x)$ has a unique positive real root.

Since $q\beta(t)^2$ is a positive real root of $f(x) = x(x + 3q)^2 - \frac{9}{2}GMt^2$, it is the unique positive real root. As first step in the solution for $q\beta(t)^2$, we use the substitution $x = X - 2q$ to transform $x(x + 3q)^2 - \frac{9}{2}GMt^2$ into the cubic polynomial

$$X^3 - 3q^2X - (2q^3 + \tfrac{9}{2}GMt^2).$$

The study of polynomial equations of the form $X^3 - uX - v = 0$ has a rich history. Two mathematicians of the Italian Renaissance, Scipio del Ferro of Bologna and Gerolamo Cardano of Pavia are credited with the solution. Del Ferro discovered it and Cardano, giving due recognition to Del Ferro, published it in his *Artis Magnae* (*The Great Art*) in 1545. The proof—a fairly involved but elementary algebra exercise—can be found in the literature. The fact we use asserts that if u and v are positive real numbers with $(\frac{v}{2})^2 \geq (\frac{u}{3})^3$, then $(\frac{v}{2} + z)^{\frac{1}{3}} + (\frac{v}{2} - z)^{\frac{1}{3}}$ where $z = \sqrt{(\frac{v}{2})^2 - (\frac{u}{3})^3}$, is a real root of $X^3 - uX - v = 0$.

Let $u = 3q^2$ and $v = (2q^3 + \frac{9}{2}GMt^2)$. Since $(\frac{v}{2})^2 = \big(q^3 + \frac{9}{4}GMt^2\big)^2 \geq q^6 = (q^2)^3 = (\frac{u}{3})^3$, this result applies to tell us that $\big(\frac{9}{4}GMt^2 + q^3 + z(t)\big)^{\frac{1}{3}} + \big(\frac{9}{4}GMt^2 + q^3 - z(t)\big)^{\frac{1}{3}}$, where $z(t) = \sqrt{(\frac{9}{4}GMt^2 + q^3)^2 - q^6}$, is a real root of the transformed cubic. It follows that

$$\big(\tfrac{9}{4}GMt^2 + q^3 + z(t)\big)^{\frac{1}{3}} + \big(\tfrac{9}{4}GMt^2 + q^3 - z(t)\big)^{\frac{1}{3}} - 2q,$$

is the unique positive real root of $x(x + 3q)^2 - \frac{9}{2}GMt^2$. It is therefore equal to $q\beta(t)^2$. Solving for $\beta(t)$, we get

$$\beta(t) = \pm\sqrt{\tfrac{1}{q}\big(\tfrac{9}{4}GMt^2 + z(t) + q^3\big)^{\frac{1}{3}} + \tfrac{1}{q}\big(\tfrac{9}{4}GMt^2 - z(t) + q^3\big)^{\frac{1}{3}} - 2}$$

where the $+$ applies if $t \geq 0$, and the $-$ if $t < 0$. By substituting this formula for $\beta(t)$ into earlier equations, we get both $r(t)$ and $\alpha(t)$ expressed as functions of t.

The formulas for the speed and direction of the motion of C follow next. After differentiating $x(t) = -q\beta(t)^2$ and $y(t) = 2q\beta(t)$, we get

$$v(t)^2 = x'(t)^2 + y'(t)^2 = 4q^2\beta(t)^2 \cdot \beta'(t)^2 + 4q^2\beta'(t)^2 = 4q^2\big(\beta(t)^2 + 1\big)\beta'(t)^2.$$

By differentiating Kepler's equation, $\beta'(t) + \beta(t)^2\beta'(t) = \frac{\kappa}{q^2}$, and hence $\beta'(t) = \frac{\kappa}{q^2} \cdot \frac{1}{\beta(t)^2+1}$. Since $r(t) = q(\beta(t)^2 + 1)$, we get $\beta'(t) = \frac{\kappa}{q}\frac{1}{r(t)}$, and therefore

$$v(t)^2 = 4qr(t)\big(\tfrac{\kappa}{q} \cdot \tfrac{1}{r(t)}\big)^2 = \tfrac{4\kappa^2}{q}\tfrac{1}{r(t)}.$$

Finally, since $4\kappa^2 = 2GMq$,

$$\boxed{v(t) = \sqrt{GM}\sqrt{\tfrac{2}{r(t)}}}$$

This formula extends the speed formulas $v(t) = \sqrt{GM}\sqrt{\frac{2}{r(t)} - \frac{1}{a}}$ and $v(t) = \sqrt{GM}\sqrt{\frac{2}{r(t)} + \frac{1}{a}}$ of the elliptical and hyperbolic cases (where a is the semimajor axis in both) to the parabolic situation.

As before, we'll take the direction of the motion of the craft to be defined by the angle $\gamma(t)$ of Figure 6.13. The derivation of the formula $\sin\gamma(t) = \frac{2\kappa}{r(t)v(t)}$ undertaken in Chapter 5D applies in the elliptical, the hyperbolic, as well as the parabolic situation. It is more useful in rewritten form. We know from Chapter 5A, that $\kappa = \sqrt{\frac{L}{8} \cdot GM}$, where M is the mass of the attracting body S. Paragraphs A and C of Chapter 3D, tell us both for the ellipse and the hyperbola that the latus rectum L is equal to $L = \frac{2b^2}{a}$, where a and b are the semimajor and semiminor axes. A look at Figures 5.2 and 6.9 tells us that for the ellipse, $L = \frac{2b^2}{a} = \frac{2(a^2-a^2\varepsilon^2)}{a} = \frac{2(a-a\varepsilon)(a+a\varepsilon)}{a} = 2q(1+\varepsilon)$, and for the hyperbola, $L = \frac{2b^2}{a} = \frac{2(a^2\varepsilon^2-a^2)}{a} = \frac{2(a\varepsilon-a)(a\varepsilon+a)}{a} = 2q(1+\varepsilon)$, where q is the periapsis distance in both cases. For the parabola, $\varepsilon = 1$ so that $L = 2q(1+\varepsilon)$ also holds in the parabolic case. It follows that $\kappa = \sqrt{\frac{L}{8} \cdot GM} = \frac{1}{2}\sqrt{GMq(1+\varepsilon)}$ and therefore that

$$\sin\gamma(t) = \frac{\sqrt{GMq(1+\varepsilon)}}{r(t)v(t)}$$

in all three situations. The fact that the inverse sine of any number between -1 and 1 needs to lie between $-\frac{\pi}{2}$ and $\frac{\pi}{2}$ and a look back at Figure 6.13 tells us that on approach to periapsis and on departure from periapsis

$$\gamma(t) = \sin^{-1}\frac{\sqrt{GMq(1+\varepsilon)}}{r(t)v(t)} \ \text{ and } \ \gamma(t) = \pi - \sin^{-1}\frac{\sqrt{GMq(1+\varepsilon)}}{r(t)v(t)},$$

respectively.

The paragraph *Sungrazing Comets and their Speeds* of the Problems and Discussion section of Chapter 2 briefly described the Sun-grazing comet C/2011 W3 as one of the fastest to ever speed around the Sun. More definitive data about its most recent passage through the solar system (than listed earlier) specifies its orbital eccentricity to have been $\varepsilon = 0.99992942$ with a perihelion distance of 0.00555381 au. Since ε is very close to 1, the comet's trajectory is nearly parabolic, so that the discussion above can be applied to it. Since 1 au $=$ 149597870.7 km, we'll take $q = 8.3083815 \times 10^5$ km.

Problem 6.26. The parabolic speed formula tells us that the maximum speed of a comet on a parabolic or nearly parabolic trajectory around the Sun occurs at perihelion and is given by $v_{\max} \approx \sqrt{GM}\sqrt{\frac{2}{q}}$ with $GM = 1.3271244 \times 10^{11}$ km^3/sec^2 (for the Sun). Use this formula to show that the maximum speed of comet C/2011 W3 was 565.21 km/sec. The formula of the Problems and Discussion section of Chapter 2 provided a maximum speed of 568 km/sec. (See Problem 2.15.) Discuss the basic difference between the two calculations. Why is it that in the current situation, the parabolic formula probably provides the more accurate result?

Problem 6.27. The last perihelion for C/2011 W3 occurred on December 16th, 2011 (with the previous one on January 4th of the year 1329 in the Julian calendar). Show that exactly one week after perihelion, the comet was $r(t) = 59{,}359{,}828$ km from the Sun, that it had slowed to $v(t) = 66.87$ km/sec, and that it was moving away from the Sun at an angle $\gamma(t) = 3.023$ radians or 173.21°. Incredibly, one short week after perihelion the comet was about as far away from the Sun as Mercury. (Table 5.1 tells us that the semimajor axis of Mercury's orbit is 57,909,227 km.) [Suggestions: use the fact that 1 day has 86,000 seconds and take $t = 7(86{,}000) = 6.04800 \times 10^5$ sec. Show that $\beta(t) = 8.39319475$. Along the way check that $z(t) = 10.92244830 \times 10^{22}$, $\left(\frac{9}{4}GMt^2 + z(t) + q^3\right)^{\frac{1}{3}} = 6.02259045 \times 10^7$, and that $\left(\frac{9}{4}GMt^2 - z(t) + q^3\right)^{\frac{1}{3}} = -0.00352381 \times 10^7$].

On August 30th, 2019, the amateur astronomer Gennady Borisov discovered a comet from an observatory on the Crimean peninsula. Initially, there was nothing about the comet, now designated C/2019 Q4, that seemed remarkable. But when it was observed again a few days later on September 10th and 12th, it was estimated to have been about 420 million kilometers from the Sun traveling toward the inner solar system with a speed of about 150 thousand kilometers per hour. This information caused a stir. But why?

Problem 6.28. Turn to the parabolic speed formula and compare the left side $v(t)$ against the right side $\sqrt{GM}\sqrt{\frac{2}{r(t)}}$ at the time of the observation. The two values are $v(t) \approx \frac{150 \times 10^3}{3.60 \times 10^3} \approx 41.67$ km/sec and $\sqrt{GM}\sqrt{\frac{2}{r(t)}} \approx \sqrt{(1.327 \times 10^{11})\frac{2}{4.20 \times 10^8}} \approx 25.14$ km/sec. Why does a comparison of these two numbers tells us that the comet's trajectory is a hyperbola? Use the hyperbolic speed formula to derive the estimate of 120,000,000 km for the semimajor axis of its hyperbola. The comet was at perihelion on December 8th, 2019, with a perihelion distance of about 300 million kilometers. Let ε be the eccentricity of the hyperbola and show that $\varepsilon \approx 3.50$. (The Minor Planet Center of the International Astronomical Union lists the comet's eccentricity as 3.3565551.)

The wide, open hyperbola of its trajectory (a look at Problem 6.8 gives a sense of the flatness of this hyperbola) is strong indication that Borisov's comet is an interstellar comet, one that originated from outside the solar system. Since C/2019 Q4 is traveling in the inner solar system, it will be possible to observe and study the comet for several months. The analysis of the light that it reflects will provide information about its chemical composition. It might provide insights into the evolution and composition of other star systems and exoplanets in them. One scientist put it this way: "We've got an object out there that's throwing out material that formed around another star in another part of our galaxy. So this will be our first real chance to do a detailed analysis of those molecules and those compounds, compare it with what we see in our solar system, compare it with what we see in interstellar space and hopefully ... start coming up with an overall picture of how the environments where planets and - potentially - life form vary throughout the galaxy."

8. Orbit Insertion. The thought process that identified the orbit of comet Borisov as hyperbolic, has other applications as well. Let's consider a spacecraft in flight within the gravitational sphere of influence of a planet. Let M be the mass of the planet and let P be its center of mass. Tables 2.3 and 2.6 provide accurate values of the gravitational term GM. We'll assume that the gravitational pull of P is the dominant force on the craft and hence that its trajectory is a conic section that has P at a focal point. Any position C of the craft has a time t attached to it in the same way as before. At periapsis $t = 0$, at any position after periapsis t is the time for the craft to get there from periapsis, and for any position of the craft before periapsis $t < 0$ where $|t| = -t$ is the time it takes for the craft to get to periapsis. For any position C and corresponding time t, we'll let $r(t)$ be the distance from C to P and $v(t)$ the speed of C relative to P. Conclusions of the previous segment inform us that for any t,

$$v(t) = \sqrt{GM}\sqrt{\frac{2}{r(t)} - \frac{1}{a}}, \ v(t) = \sqrt{GM}\sqrt{\frac{2}{r(t)}}, \ \text{or} \ v(t) = \sqrt{GM}\sqrt{\frac{2}{r(t)} + \frac{1}{a}}$$

in the elliptical, parabolic, and hyperbolic cases, respectively (where $a > 0$ is the semimajor axis in the elliptical and hyperbolic cases).

Fix any instant instant t_0 during the craft's approach and assume that telemetry readings provide the values of $r(t_0)$ and $v(t_0)$. Clearly, $v(t_0)$ is either less than, equal to, or greater than $\sqrt{GM}\sqrt{\frac{2}{r(t_0)}}$. The equations above tell us that the craft's trajectory is an ellipse in the first case, a parabola in the second, and a hyperbola in the third. It follows that the data point $(r(t_0), v(t_0))$ tells us whether the craft is in an elliptical, parabolic, or hyperbolic orbit. In fact, more can be deduced from this single data point.

Problem 6.29. Use Figure 6.36 to show the following. If $v(t_0) < \sqrt{GM}\sqrt{\frac{2}{r(t_0)}}$, then $v(t_0) = \sqrt{GM}\sqrt{\frac{2}{r(t_0)} - D}$ and if $v(t_0) > \sqrt{GM}\sqrt{\frac{2}{r(t_0)}}$, then $v(t_0) = \sqrt{GM}\sqrt{\frac{2}{r(t_0)} + D}$, where in each case, D is some positive constant. What facts about the function $f(x) = \sqrt{x}$ are relevant?

Put $D = \frac{1}{a}$ for some a in the elliptical and hyperbolic cases. From the speed equations above we know that a is the semimajor axis of the craft's orbit in either case. Since $r(t_0)$ and $v(t_0)$ together determine D, they determine the semimajor axis.

The direction of the motion of the craft is given by the angle $\gamma(t)$ as described in Figure 6.13. We know that

$$\sin \gamma(t) = \frac{\sqrt{GMq(1+\varepsilon)}}{r(t)v(t)},$$

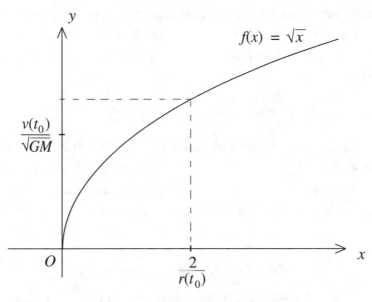

Figure 6.36

where q is the periapsis distance and ε the eccentricity of the craft's trajectory. If the craft's telemetry also provides the angle $\gamma(t_0)$, then it provides the value of the term $\frac{\sqrt{GMq(1+\varepsilon)}}{r(t_0)v(t_0)}$ and therefore, the value of $q(1+\varepsilon)$. In the parabolic case, $\varepsilon = 1$, so that the periapsis distance q is determined and hence also the shape of the parabola. (See Figure 6.34.) In the elliptical case $q(1+\varepsilon) = a(1-\varepsilon)(1+\varepsilon)$. Since a is already determined so is $1 - \varepsilon^2$. Letting $1 - \varepsilon^2 = \delta$, we get $\varepsilon^2 = 1 - \delta$, and hence $\varepsilon = \sqrt{1-\delta}$. It follows that the data point $(r(t_0), v(t_0)), \gamma(t_0)$ determines both a and ε and hence the shape of the elliptical trajectory. A similar computation shows that this is also true in the hyperbolic case.

Since t_0 is given, the appropriate Kepler equation (elliptical, parabolic, or hyperbolic) can be solved for $\beta(t_0)$ so that $\alpha(t_0)$ can be found as well. This in turn specifies the focal axis and hence the orientation of the conic section relative to the location of the focal point P. Therefore the single measurement $(r(t_0), v(t_0)), \gamma(t_0)$ determines both the shape and the position of the trajectory of the craft within its orbital plane relative to the location of P.

Suppose now that the spacecraft is on a hyperbolic approach to the planet or on hyperbolic departure from it. The mission calls for the craft to be inserted into an elliptical orbit around the planet. It is clear that the craft's thrusters need to be fired to change its trajectory. Intuitively, the craft needs to be slowed down and redirected. The smaller thrusters orient the craft and the main thruster will decrease its speed. We will now explore in basic terms how such a trajectory correction maneuver can be executed. The important fact to keep in mind is this: At any time that the thrusters are firing, the resultant thrust is a second force on the craft. So the fundamental assumption—that the spacecraft is subject to a single centripetal force—is no longer operative. However, at any time the thrusters have stopped firing, this assumption is back in force and the craft's trajectory will once again be a conic section with P at a focal point. Of course, this conic section will invariably be different from the one the craft was on before.

To begin, let t_0 be the instant at which the craft's main thruster begins its burn. Note that t_0 is negative, zero, or positive depending on whether the craft is on approach to periapsis, at periapsis, or on departure from it. At the initial moment t_0 the thrust is still zero. Since the hyperbolic version of the velocity formula applies to the craft,

$$v(t_0) = \sqrt{GM}\sqrt{\tfrac{2}{r(t_0)} + \tfrac{1}{a_0}},$$

where a_0 is the semimajor axis of the hyperbola. Let τ be the time that elapses from t_0 onward and monitor both the distance $r(t_0 + \tau)$ of the craft from P and the velocity $v(t_0 + \tau)$ of the craft relative to P during the time the thruster is fired. If a_1 is the semimajor axis of the desired elliptical orbit, fire the thrusters (with the main thruster generally directed opposite to the craft's motion) to slow the craft until

$$v(t_0 + \tau) = \sqrt{GM}\sqrt{\tfrac{2}{r(t_0+\tau)} - \tfrac{1}{a_1}}.$$

At this instant, with $\tau = \tau_0$, the thruster is shut off. At time $t_1 = t_0 + \tau_0$ from the time of the original periapsis (with t_0 negative or positive) relative to its new orbit, $v(t_1) = \sqrt{GM}\sqrt{\tfrac{2}{r(t_1)} - \tfrac{1}{a_1}}$ so that the new orbit is elliptical with a_1 its semimajor axis. Our earlier discussion tells us that the data point $(r(t_1), v(t_1)), \gamma(t_1))$ determines both the shape and the orientation of this elliptical orbit.

What has been described are the principles behind the insertion of the *Cassini* spacecraft into its initial elliptical orbit around Saturn. Refer back to section 6N, especially to Figure 6.24 and the July 1st entry of the *Cassini* Cruise Event Summary.

9. The Three Body Problem and Low Δv Trajectories. The patched conic strategy relies on the solution of the *two body problem,* and, in particular, the fact that the trajectory a craft that is subject to the gravitational attraction of a single celestial body is a conic section with the center of mass of the body at a focal point. When a third object or body is taken into consideration, then, with three gravitational forces involved, the problem of analyzing each of the motions is known as the *three body problem.* If a fourth body is considered, the problem is the *four body problem.* Newton and the powerful mathematicians in the centuries that followed were stymied in their efforts to determine the mathematical functions that describe the motions of the three (or four) bodies for any given set of initial conditions (the positions and velocities of the bodies). Finally, in the 1950s, mathematicians demonstrated that the explicit determination of such functions was impossible. However, beginning in the 1960s the use of high speed digital computers did make it possible (as already described in section 6K) to achieve better and better numerical approximations of the motions in the case of the *restricted three-body problem.* The restriction consists of the assumptions that two of the bodies are massive and that the third is so small that its gravitational pull on the other two is negligible. It also assumes that the two massive bodies are both in circular orbits about their common center of mass, and that the orbits of all three lie in the same plane. This restricted three body problem provides a useful model for the design of the trajectories of spacecraft in orbit around the Sun and in transit to a planet, or a craft that is making its way from Earth to the Moon. To a less accurate extent, the restricted three body problem also models the Earth-Moon-Sun system. (The *restricted four body problem* plays a role as well. This models the motion of a craft and two massive bodies—Earth and Moon, for instance—that are both moving in circular orbits around their common center of mass with this center of mass in turn in a circular orbit around a third massive object such as the Sun.)

The figure eight, depicted in Figure 6.37, illustrates what is involved. If the craft C is given the correct position and velocity on departure from a parking orbit around Earth, it can achieve a stable orbit around the Earth-Moon system of the sort shown in the figure. Such figure eight orbits

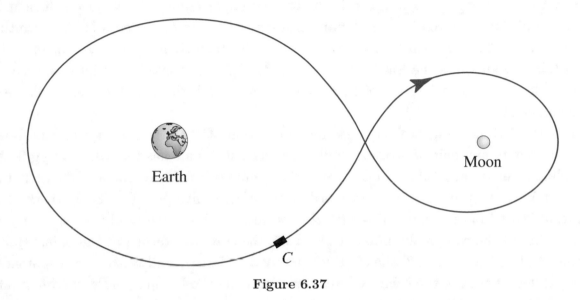

Figure 6.37

were deployed in some of the missions that brought *Apollo* spacecraft to the Moon. Notice that each of the two loops is almost a closed orbit around each of the two bodies. So if the goal is to put the craft into orbit around the Moon, the additional maneuvers needed are minor. While a Hohmann transfer from an Earth orbit to the Moon blasts a spacecraft through Earth's gravitational pull, the figure eight trajectory relies on the more subtle, interactive gravitational currents of the Earth-Moon system.

In the 1980s, Edward Belbruno, then a mathematician at the JPL, began research on trajectories that would bring a spacecraft from a parking orbit around Earth to the Moon with minimal amounts of propellant. He discovered that it was possible to insert a craft into orbit around the Moon with minimal reliance on the thrust from its engines. The delicate interplay between the gravitational tugs of Earth and Moon on a spacecraft and a subtle balance between the speed and the direction of the motion of the craft can draw it into a region called *weak stability boundary* within which such a *ballistic capture* can occur. The restricted three-body problem described in section 6C applies not only to the Sun and Earth, but also to the Earth and Moon. A rotating frame exactly like the one in Figure 6.7 with the Sun at the origin, Earth as fixed point, also exists with the Earth-Moon barycenter at the origin and the Moon as fixed point. The weak stability boundary is a complex shape in the Earth-Moon rotating frame, but Belbruno was able to provide a mathematical description of its geometry. The second difficult problem he faced was the determination of the trajectory that would transfer the craft from its parking orbit around Earth to the weak stability boundary at the desired distance from the Moon with the required velocity. Here too, the subtle interaction between the gravitational forces of Earth and Moon sets out a path that uses minimal amounts of propellant.

The path follows a natural channel or tube that exists as a mathematical construct called a manifold within position-velocity space.

The Japanese spacecraft *Hiten* (Japanese for "celestial maiden"), while in Earth orbit, had ejected a probe that was designed to go into orbit around the Moon. Unfortunately, the communications system failed and the probe was lost. In order to be able to declare its first mission to the Moon a success, the Japanese wanted to place the much larger *Hiten* into a lunar orbit instead. However, since this was not the original intention, the larger craft did not have sufficient fuel to reach the Moon with a conventional Hohmann transfer. Belbruno proposed a ballistic capture trajectory for *Hiten*. The route that *Hiten* traveled from Earth took it 1 million kilometers beyond the Moon (about four times the Earth-Moon distance), before it floated back into ballistic capture around the Moon. It was a journey of three months, but the little fuel that was available to the craft was sufficient.

The restricted three-body problem described in section 6C applies not only to Sun-Earth and Earth-Moon, but to any pair of orbiting bodies and each such pair has its own rotating frame and corresponding channels and tubes. In each case, the Lagrange points L1, L2, and L3 are not stable, but L4 and L5 are. A craft precisely positioned at any of these five points will stay there. But even the slightest thrust from its engine firing for the shortest possible time (and then shut off) will set the craft adrift. In the case of the points L4 and L5, the craft will return to the point. But in the case of the points L1, L2, and L3, the craft will gently drift farther and farther away, potentially to an entirely different part of the solar system. Understanding the kind of gentle gravitational drifts that a craft experiences near an unstable Lagrange point of the Earth-Moon system lies at the heart of Belbruno's weak stability trajectories. The various tubes spiral in the solar system towards and away from the unstable Lagrange points. The remarkable fact is that many of these tubes link to form a network that crisscrosses the solar system in ever changing patters. A craft floating in one of these tubes is swept along by gravitational tugs to or away from the Lagrange points. With a tiny thrust near a Lagrange point, a spacecraft can switch to a tube that spins off in a completely different destination. In this way, a craft can travel vast distances on minimal amounts of propellant.

The network of tubular manifolds just described has been referred to as the "interplanetary superhighway." This label seems somewhat of a misnomer given that the vehicles that use it travel at a snail's pace. The interplanetary superhighway is not static and the interchanges keep shifting. Each set of Lagrange points and tubes is stationary only within its rotating frame. Computers are used to chart these moving highways and their rotating interchanges. The drawbacks of interplanetary trajectories that use the system of tubes are related to the long transfer time and their sensitivity to subtle gravitational changes. A long transfer time increases the risk of failure as well as the operational costs. Finally, these trajectories are very sensitive to perturbations. A trajectory of consecutive tubes requires robust algorithms to correct for interfering perturbations. A patched conic trajectory, on the other hand, requires only occasional corrections. Belbruno's trajectories are very sensitive to the initial conditions. Think of two leaves that fall side by side into a lively descending brook. They land very close to each other. Their initial conditions are almost identical as they are carried by the current side by side. The brook widens, gains volume, and becomes a stream. It widens more, absorbs tributaries, its waters slow, and it becomes a mighty river. Would you expect the two leaves to float along in the river side by side? Hardly. This means that two spacecraft in trajectories of the sort just described, can start out in nearly the same place with nearly the same velocities but spiral off into completely different regions of space. It means that Belbruno's trajectories have to be computed with much greater accuracy than conventional trajectories and trajectory correction maneuvers needed to be very carefully attended to. New numerical methods using ideas from chaos theory and dynamical systems theory, and computer-executed numerical algorithms needed to be developed.

A technological development that was described in the paragraph *Dawn's Ion Propulsion Engine* of the Problems and Discussions section of Chapter 2 goes hand in glove with what has just been described. This is the ion-engine that generates a very small thrust that it sustains over periods of months or even years. *SMART*-1 was the inaugural mission of the European Space Agency's (ESA) Small Missions for Advanced Research in Technology (SMART) program. With its washing machine size and its mass of 367 kg, *SMART*-1 was indeed small. Launched late in 2003, the primary goal of the mission was to test its ion engine. The ESA's first Moon probe and the second spacecraft

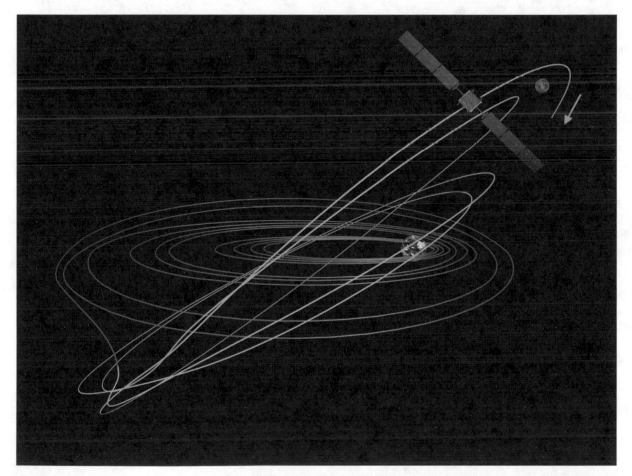

Figure 6.38. A *SMART-1* was a spacecraft of the European Space Agency of Swedish design. Its ion-engine accelerated the craft by a mere 0.7 m/sec per hour. Unlike conventional thrusters that operate in short bursts, the ion engine of *SMART-1* fired for one third to one half of every orbit to drive the craft on the spiraling trajectory that took it from the Earth to the Moon. Image credit: ESA 2002, CC BY-SA 3.0 IGO https://creativecommons.org/licenses/by-sa/3.0/igo/.

to use ion propulsion, *SMART*-1 was launched into a low Earth orbit. Its small engine generated a thrust of 0.068 newtons. (The fact that one penny has a weight of 0.0245 newtons, means that the thrust that the engine generated was less than the weight of three pennies.) Month after month, first imperceptibly slowly and then faster, with its engine firing about half of the time, the craft spiraled away from Earth in ever expanding loops. This part of the craft's trip to the Moon did not follow a Belbruno trajectory. See Figure 6.38. However, over 14 months after its launch, *SMART*-1 drifted through a position 310,000 km from the Earth and 90,000 km from the Moon in and was captured by the Moon's gravitational field. It's entry into orbit around the Moon relied on Belbruno's ballistic capture strategy. The route that *SMART*-1 had taken was lengthy. Most of the craft's journey was a winding spiral of over 80 million kilometers in length. But for the entire, extraordinary trip, it used only 82 kg of the Xenon gas (22% of its total mass) that fueled its engine.

The many websites cited in the References provide visually enriching details for many of the topics in the text. If a particular website is no longer active, it should be a routine matter to use information from the URL address to find an equivalent or updated version.

Chapter 1

Chapter 1 presents a historical overview of astronomy from the early thoughts of the Greeks, to the advances by Copernicus, Galileo, Kepler and Newton during the Scientific Revolution, to the discovery of Uranus and Neptune, and finally to Hubble's realization that there are a multitude of clusters of myriads of stars—galaxies—at vast distances from us. The references below consist of some of the important original sources, some references that play a specific role, and some that expand the conversation into related areas.

1. Olaf Pedersen, *Early Physics and Astronomy: A Historical Introduction,* revised edition, Cambridge University Press, 1993.

2. Nicolaus Copernicus, *On the Revolutions of the Heavenly Spheres,* trans. A. M. Duncan, Barnes & Noble, New York, 1976,

3. Johannes Kepler, *New Astronomy*, trans. William H. Donahue, Cambridge University Press, Cambridge, 1992.

4. Owen Gingerich, "The Great Martian Catastrophe and How Kepler Fixed It," *Physics Today* 64 (2011), http://sites.apam.columbia.edu/courses/ap1601y/PhysToday-2011-Kepler.pdf.

5. Isaac Newton, *Mathematical Principles of Natural Philosophy*, trans. Robert Thorp, W. Strahan and T. Cadell, London, 1777. Reprinted by Dawsons of Pall Mall, London, 1969.

6. I. Bernard Cohen, Newton's Discovery of Gravity, *Scientific American*, Vol. 244, No. 3 (March 1981), pp. 166–181, http://www.jstor.org/stable/24964334

7. D. W. Hughes, The 'Principia' and Comets, Notes and Records of the Royal Society of London, Vol. 42, No. 1, Newton's 'Principia' and Its Legacy (Jan., 1988), pp. 53–74, Stable URL: http://www.jstor.org/stable/531369

8. D. W. Hughes, Measuring the Moon's Mass, *The Observatory, NASA Astrophysics Data System*, Vol. 122, No. 1167, 2002, 61–70.

© Alexander J. Hahn 2020
A. J. Hahn (ed.), *Basic Calculus of Planetary Orbits and Interplanetary Flight*,
https://doi.org/10.1007/978-3-030-24868-0

9. Parker Moreland, *The JJMO Mars Parallax Project* and *The 2003 JJMO Mars Parallax Project*, http://www.jjmo.org/astro/astmenu2.htm

10. G. Foderà Serio, A. Manara, and P. Sicoli, Guiseppe Piazzi and the Discovery of Ceres, https://www.lpi.usra.edu/books/AsteroidsIII/pdf/3027.pdf

11. NASA, *Images of Uranus*, https://www.jpl.nasa.gov/education/images/pdf/ss-uranus.pdf

12. NASA, *Images of Neptune*, https://www.jpl.nasa.gov/education/images/pdf/ss-neptune.pdf

13. ENCYCLOPEDIA BRITANNICA, *The Metric System of Measurement*, https://www.britannica.com/science/measurement-system/The-metric-system-of-measurement

14. D. W. Hughes, Six stages in the history of the astronomical unit, *Journal of Astronomical History and Heritage*, 4(1), 2001, 15–28.

15. National Institute of Standards and Technology, *Solving the mystery of the big G controversy*, November 17, 2016, https://phys.org/news/2016-11-mystery-big-controversy.html

16. Big History Project, *Edwin Hubble, Evidence for an Expanding Universe*, https://school.bighistoryproject.com/media/khan/articles/U2_Edwin_Hubble_2014_850L.pdf

17. NASA, *The Sun*, https://www.nasa.gov/sun

18. E. V. Pitjeva, Determination of the Value of the Heliocentric Gravitational Constant ($GM_\odot$) from Modern Observations of Planets and Spacecraft, *Journal of Physical and Chemical Reference Data* 44, 031210, 2015, https://doi.org/10.1063/1.4921980

19. The Infrared Processing & Analysis Center (IPAC) at the California Institute of Technology, *Galaxies*, http://coolcosmos.ipac.caltech.edu/page/galaxies

20. Martin Gutzwiller, Moon-Earth-Sun: The oldest three-body problem, *Rev. Mod. Phys.*, Vol. 70, No. 2, April 1998, 589–639.
http://sites.apam.columbia.edu/courses/ap1601y/Moon-Earth-Sin%20RMP.70.589.pdf

21. Alexander Hahn, *Basic Calculus: From Archimedes to Newton to its Role in Science*, Springer-Verlag, New York, 1998. https://www3.nd.edu/~hahn/

Chapter 2

Chapter 2 is a story about the dozens of spacecraft that the American, Russian, and more recently also the European, Japanese, and Chinese space programs have sent into orbit and flybys around the planets, moons, asteroids, and comets of our solar system, with an emphasis on the incredible images and vast numerical information that they have sent back to us. In the case of Mars, this has included the deployment of several robotic craft that have explored the planet's terrain. In the case of our Moon, it has involved highly complex and astonishing American Apollo missions that landed a dozen men on its surface. Some of them went on excursions in exotic vehicles to collect lunar soil and rocks that they brought back to Earth.

1. The Exploration of Space in $5\frac{1}{2}$ minutes, https://www.youtube.com/watch?time_continue=299&v=gK1bAWbbrh4

2. Gene Kranz, *Failure is not an Option: Mission Control from Mercury to Apollo 13 and Beyond*, Simon and Schuster Paperbacks, New York, London, 2000.

3. National Aeronautics and Space Administration (NASA), *Hubble Space Telescope*, https://www.nasa.gov/mission_pages/hubble/main/index.html

4. NASA/JPL-Caltech *Solar System Dynamics, Planets and Pluto: Physical Characteristics*, https://ssd.jpl.nasa.gov/?planet_phys_par

5. W. M. Folkner, J. G. Williams, D. H. Boggs, R. S. Park, and P. Kuchynka, The Planetary and Lunar Ephemerides DE430 and DE431, *Interplanetary Network Progress Report*, 42–196, February 15, 2014. https://ipnpr.jpl.nasa.gov/progress_report/42-196/196C.pdf

6. NASA/JPL-Caltech *Solar System Dynamics, Planetary Satellite Physical Parameters*, https://ssd.jpl.nasa.gov/?sat_phys_par

7. NASA/JPL-Caltech *Solar System Dynamics, JPL Small-Body Database Browser*, https://ssd.jpl.nasa.gov/sbdb.cgi

8. NASA, *Solar System Exploration*, http://www.jpl.nasa.gov/solar-system

9. NASA, *Planets*, https://solarsystem.nasa.gov/planets/

10. NASA, *Blue Marble–Image of the Earth from Apollo 17*, https://www.nasa.gov/content/blue-marble-image-of-the-earth-from-apollo-17

11. The video https://www.nasa.gov/sites/default/files/thumbnails/image/dscovrepicmoontransitfull.gif taken by the *DISCOVR* satellite illustrates how the Moon behaves in its orbit.

12. NASA, *Mars Today: Robotic Exploration*, https://www.nasa.gov/mars

13. NASA, *Venus*, https://www.nasa.gov/venus

14. NASA, *Project Mercury*, https://www.nasa.gov/mercury

15. NASA, *Jupiter*, https://www.nasa.gov/jupiter

16. NASA, *Saturn*, https://www.nasa.gov/saturn

17. NASA, *Uranus*, https://www.nasa.gov/uranus

18. NASA, *Neptune*, https://solarsystem.nasa.gov/planets/neptune/overview/

19. NOVA, Public Broadcasting Service (PBS), *THE PLANETS*, The life of our solar system told in five dramatic stories spanning billions of years. https://www.pbs.org/wgbh/nova/series/planets/

20. NASA, *Solar System Exploration, Asteroids*, https://solarsystem.nasa.gov/small-bodies/asteroids

21. NASA's spacecraft *OSIRIS-REx* and the asteroid *Bennu* https://solarsystem.nasa.gov/missions/osiris-rex/in-depth/

22. NASA, *Solar System Exploration, Comets,*
https://solarsystem.nasa.gov/small-bodies/comets

23. International Astronomical Union, *Pluto and the Developing Landscape of Our Solar System,*
https://www.iau.org/public/themes/pluto/

24. NASA, *Juno,* https://www.nasa.gov/mission_pages/juno/main/index.html

25. G. H. Born, E. J. Christensen, A. J. Ferrari, J. F. Jordan, and S. L. Reinbold, The determination of the satellite orbit of Mariner 9, *Celestial Mechanics,* May 1974, Volume 9, Issue 3, 395–414.

26. The fateful mission of Apollo 13, https://www.youtube.com/watch?v=1e4fYb-zwdE

27. Astronomy Picture of the Day Archive http://antwrp.gsfc.nasa.gov/apod/archivepix.html

Chapter 3

Chapter 3 is a self-contained treatment of calculus with focus on polar coordinates and polar functions. It does assume that the reader has a sense for the kinds of limits that arise in the definitions of the derivative and the definite integral. The first reference offers a review of standard calculus as well as practice problems that deal with calculus, but also with algebra, Cartesian coordinate geometry, and trigonometry. The second reference sets out basic calculus comprehensively, and presents historical aspects and applications.

1. S. K. Chung, *Understanding Basic Calculus.* The author makes this text available on line at
http://www.math.nagoya-u.ac.jp/~richard/teaching/f2015/BasicCalculus.pdf.
With its emphasis on explanations of concepts and solutions to examples, it is a very nice basic treatment of calculus as it had been taught to students at the University of Hong Kong in recent years. The author tells us that his text is "dedicated to all the people who have helped me in my life."

2. Alexander J. Hahn, *Calculus in Context: Background, Basics, and Applications,* Johns Hopkins University Press, Baltimore, MD, 2017.
https://jhupbooks.press.jhu.edu/content/calculus-context

3. The Infrared Processing & Analysis Center (IPAC) at the California Institute of Technology, *The Galactic Center,* http://coolcosmos.ipac.caltech.edu/page/galactic_center

4. B. L. Davis, J. C. Berrier, et al, Measurement of galactic logarithmic spiral arm pitch angles using two-dimensional fast Fourier transform decomposition, *The Astrophysical Journal Supplement Series, Volume 199, Number 2, 2012.*
https://iopscience.iop.org/article/10.1088/0067-0049/199/2/33

5. Dorota M. Skowron, A three-dimensional map of the Milky Way using classical Cepheid variable stars, *Science,* 365, August 2019, 478–482. The Milky Way galaxy is warped and twisted, not flat, https://www.bbc.com/news/world-us-canada-49182184

Chapter 4

Chapter 4 applies the calculus of polar coordinates of the previous chapter in a self-contained way to derive Kepler's three laws of planetary orbits from Newton's fundamental laws of motion and his law of universal gravitation. Following Newton's original theory in the *Principia*, this is done in the abstract context of a centripetal force acting on a point-mass. The assumption that the Sun and the planets are nearly spherical with the Sun much more massive than the planets, allows Newton to show that his theory applies to the orbits of the planets of the solar system. This chapter confines itself strictly to the mathematics, in fact only to its most essential "first order" aspects. The more subtle "second order" mathematical matters are not considered, nor are related philosophical issues. In this regard, see the article

1. George E. Smith, Closing the Loop: Testing Newtonian Gravity, Then and Now, in *Newton and Empiricism*, ed. Zvi Beiner and Eric Schliesser, Oxford University Press, 2014, 262–351. http://strangebeautiful.com/other-texts/smith-closing-the-loop.pdf and http://web.stanford.edu/dept/cisst/visitors.html.

The discussion of Chapter 4 relies on very basic laws of physics, but it does not explore the underlying principles of physics (such as the conservation of momentum and conservation of energy) in any detail. For such an exploration we refer to *The Feynman Lectures on Physics*. Richard Feynman was one of the most famous physicists of the second half of the 20th century. He was not only superb theoretician whose research was awarded the Nobel Prize in 1965, he was also a legendary teacher. In the early 1960s, he responded to a request to "spruce up" the teaching of undergraduates at the California Institute of Technology. After devoting three years to the task, he produced a series of lectures that would become The Feynman Lectures on Physics, perhaps the most popular physics book ever written. A 2013 review described the book as having "simplicity, beauty, unity ... presented with enthusiasm and insight". More than 1.5 million English-language copies have been sold, and many more in a dozen foreign-language editions. In 2013, Caltech made the book available in the new online edition

2. *The Feynman Lectures on Physics*, New Millennium Edition, http://www.feynmanlectures.caltech.edu/info/

In November 1964, Feynman gave the *Messenger Lectures* on "The Character of Physical Law" at Cornell University. Offering an overview of selected physical laws, he drew on his Caltech lectures to gather the common features of the laws of physics into one broad principle of invariance. In January 2016, Bill Gates (the founder of Microsoft Corporation and the world's most generous philanthropist) referred to Feynman's talents as a teacher as the inspiration for his acquisition of the rights to the original videos of Feynman's Cornell lectures and for making them available online. They are a bit grainy, but they also show "The Great Explainer" in action.

3. Richard Feynman, *Messenger Lectures* (1964), http://www.cornell.edu/video/playlist/richard-feynman-messenger-lectures

Chapter 5

One of the key considerations of Chapter 5 is the Kepler Equation for elliptical orbits. Its solution, the angle $\beta(t)$, provides the angle $\alpha(t)$ and the distance $r(t)$ as functions of the elapsed time t that pinpoint the position of the point-mass P in its orbit around the center of force S. In turn, the distance $r(t)$ together with the orbital constants determines the functions $v(t)$ and $\gamma(t)$ that specify the velocity of the point-mass in its orbit around S.

1. Peter Colwell, *Solving Kepler's Equation over Three Centuries*, Willmann-Bell Inc., Richmond, Virginia, 1993.

2. Richard Fitzpatrick, *Introduction to Celestial Mechanics,*
   https://farside.ph.utexas.edu/teaching/celestial/Celestial/Celestialhtml.html.
   What is relevant from this comprehensive work to the discussion of Chapter 5 are the three chapters Keplerian Orbits, Perihelion Precession of Mercury, and Orbits in Central Force Fields.

3. M. P. Price and W. F. Rush, Nonrelativistic contribution to Mercury's perihelion precession, *American Journal of Physics* 47(6), 1979, 531–534. https://doi.org/10.1119/1.11779.

4. M. G. Stewart, Precession of the perihelion of Mercury's orbit, *American Journal of Physics* 73(8), 2005, 730–734. http://dx.doi.org/10.1119/1.1949625

5. Kin-Ho Lo, K. Young, and B. Y. P. Lee, Advance of perihelion, *American Journal of Physics,* 81(9), 2103, 695–702. http://dx.doi.org/10.1119/1.4813067

The discussion of the orbital precession of Mercury in this chapter follows the approach of Price and Rush [3]. The details differ however. As Lo, Young, and Lee [5] correctly observe, the tightly accurate value for Mercury's precession that Price and Rush achieve is the result of the cancelation of inaccuracies introduced by approximations that are too loose. The presentation in this chapter—in Chapter 5I—corrects this flaw. The definite integral expresses the magnitude of the force of the planetary ring on the point-mass m precisely, and the infinite series that solves the integral is summed up accurately.

Chapter 6

The references listed below provide additional information about the various topics that Chapter 6 studies. These range from the *NEAR-Shoemaker* mission, to the gravitational Sphere of Influence and its Hill and Laplace radii, to Hohmann transfer orbits, to texts about spaceflight in general, to information about the *Voyager, Cassini,* and *MESSENGER* missions.

1. D. W. Dunham, J. V. McAdams, and R. W. Farquhar, NEAR Mission Design, *Johns Hopkins APL Technical Digest,* Vol 23, No 1, 2002, 18–33.
   https://www.jhuapl.edu/techdigest

2. J. V. McAdams, D. W. Dunham, J. V. McAdams, L. E. Mosher, J. C. Ray, P. G. Andreasian, D. E. Helfrich, J. K. Miller, Maneuver History for the NEAR Mission: Launch through Eros Orbit Insertion, *American Institute of Aeronautics and Astronautics* 2000, 244–261.

3. R. Farquhar, J. Kawaguchi, C. Russell, G. Schwehm, J. Veverka, and D. Yeomans, *Spacecraft Exploration of Asteroids: The 2001 Perspective,* in Asteroids III, 2002, 367–376.

4. D. K. Yeomans, J.-P. Barriot, et al, Estimating the Mass of Asteroid 253 Mathilde from Tracking Data During the NEAR Flyby, *Science* 278, January 1998, 2106–2109.

5. NASA Science, Solar System Exploration, Missions, https://solarsystem.nasa.gov/missions/near-shoemaker/in-depth/

6. R. A. N. Araujo, O. C. Winter, A. F. B. A. Prado, and R. Vieira Martins, Sphere of influence and gravitational capture radius: a dynamical approach, *Monthly Notices of the Royal Astronomical Society* 391, 2008, 675–684.

7. William I. McLaughlin, Walter Hohmann's Roads In Space, *Journal of Space Mission Architecture,* Issue 2, 2000, pp. 114. http://vesta.astro.amu.edu.pl/~breiter/lectures/astrody/Hohmann_renamed.pdf

8. NASA Science, Solar System Exploration, https://solarsystem.nasa.gov

9. NASA Science, Solar System Exploration, Basics of Space Flight, http://www2.jpl.nasa.gov/basics

10. Robert A. Braeunig, ORBITAL MECHANICS, http://www.bracunig.us/space/orbmech.htm

11. R. R. Bate, D. D. Mueller, W. W. Saylor, and J. E. White, *Fundamentals of Astrodynamics,* Dover Books on Physics, Dover Publications, 1971 and 2019.

12. Howard Curtis, *Orbital Mechanics for Engineering Students*, 3rd edition, Elsevier Butterworth-Heinemann, Oxford, 2014.

13. NASA/JPL, *Voyager*, https://voyager.jpl.nasa.gov/

14. NASA/JPL, *Voyager*, Hyperbolic Orbital Elements, http://voyager.jpl.nasa.gov/science/hyperbolic.html

15. NASDA/JPL, Voyager Mission Status Bulletin, NASA and JPL, http://www.planetary.org/explore/resource-library/voyager-mission-status.html

16. Ashley Strickland, When Neptune got its stunning close-up: The Voyager 2 flyby, 30 years later, CNN, Space Science, https://www.cnn.com/2019/08/25/world/voyager-neptune-flyby-30-years-scn/index.html

17. R. J. Cesarone, A Gravity Assist Primer, *AIAA Student Journal,* Volume 27, Number 1, Spring 1989, 16–22. http://www.gravityassist.com/IAF3-2/Ref.%203-128.pdf

18. The *Cassini* Story 36 hours before Saturn Orbit Injection, https://www.c-span.org/video/?182494-1/saturn-mission-pre-arrival-overview

19. NASA Science, Solar System Exploration, *Cassini*,
http://saturn.jpl.nasa.gov or https://solarsystem.nasa.gov/missions/cassini/overview/

20. NASA/JPL/Space Science Institute, What did Cassini See During its Historic Mission to Saturn,
https://www.youtube.com/watch?v=E_Ono0-nNbI

21. Animation of Cassini Trajectory around Saturn,
https://commons.wikimedia.org/wiki/File:Animation_of_Cassini_trajectory_around_Saturn.gif

22. Jon Giorgini, Solar System Dynamics Group at JPL, *About Modern Orbit Determination*,
https://www.youtube.com/watch?v=tnNnGbZjqkg.

23. *MESSENGER* Mission to Mercury, http://messenger.jhuapl.edu/
http://www.nasa.gov/mission_pages/messenger/main/index.html#.VPHAhEsmk8N
https://www.jhuapl.edu/TechDigest/Detail?Journal=J&VolumeID=34&IssueID=1

24. James McAdams, Robert Farquhar, Anthony Taylor, and Bobby Williams, MESSENGER Mission Design and Navigation,
http://messenger.jhuapl.edu/Resources/Publications/McAdams.et.al.2007.pdf

25. *MESSENGER* Orbital Elements,
http://messenger.jhuapl.edu/About/mission-design/details-propulsion-activity/MOE.html

26. The home page of the Jet Propulsion Laboratory, http://www.jpl.nasa.gov/index.cfm

27. The home page of NASA, https://www.nasa.gov/

28. Dave Doody, *Deep Space Craft: An Overview of Interplanetary Flight*, Springer-Praxis Books in Astronautical Engineering, Praxis Publishing Ltd, Chichetser, UK, 2009.

29. Astronomy Picture of the Day Archive, https://apod.nasa.gov/apod/archivepix.html

Index

© Alexander J. Hahn 2020
A. J. Hahn (ed.), *Basic Calculus of Planetary Orbits and Interplanetary Flight,*
https://doi.org/10.1007/978-3-030-24868-0

Printed in the United States
by Baker & Taylor Publisher Services